Achieving Infinite Resolution:
A Gentle Look at the Role of Infinity in Calculus

Hassan Sedaghat

Copyright © 2020 by Hassan Sedaghat
All rights reserved.

Contents

Preface	**iii**
1 The Ubiquitous Infinity	**1**
1.1 Infinitely many numbers and infinity in numbers	1
1.2 What might we find at the end of an infinite process?	11
1.3 Infinity concealed: from areas inside circles to logarithms	17
1.4 ∞, 0, and nothingness in analysis	22
2 Sets, Functions and Logic: *the building blocks*	**31**
2.1 Sets and relations	31
2.2 Functions	42
2.3 Basic logical concepts and operations	55
3 Infinities: *infinitely many of them!*	**63**
3.1 Bijections, cardinality and Hilbert's hotel	64
3.2 An infinity of infinities: Cantor's theorem	67
3.3 Countable or uncountable?	69
4 Sequences: *taking infinitely many baby steps*	**77**
4.1 Infinite lists of numbers	77
4.2 Sequence types and plots	79
4.3 Convergent sequences: infinite resolution lenses	83
4.4 Divergent sequences: wandering about, near or far	92
5 The Real Numbers: *mostly irrational, but not lawless*	**93**
5.1 Rational numbers: a tiny but pervasive minority	94
5.2 From the rationals to the reals, in baby steps	98
5.3 The real numbers, at last!	104
5.4 Characteristics of the set of real numbers: completeness and more	106
5.5 The uncountability of real numbers: the irrational majority	113
6 Infinite Series: *adding how many numbers?*	**119**
6.1 A tale of two series and other oddities	119
6.2 Infinite series as sequence limits	123
6.3 The geometric series: beauty in simplicity	130

- 6.4 Testing for convergence (without calculating the sum) 134
- 6.5 The effects of sign changes and Riemann's rearrangement theorem 149
- 6.6 The real numbers revisited ... rational, irrational and transcendental 160

7 Derivatives: *changing by infinitely little* — 169
- 7.1 Measuring and calculating the velocity - without speedometers 169
- 7.2 Derivative and the tangent line: here comes infinity 174
- 7.3 Derivative formulas and higher derivatives 178
- 7.4 When derivatives fail to exist: more often than may seem 188
- 7.5 Continuity and singularities 191
- 7.6 Newton's method: fast convergence with a risk of singularities 203
- 7.7 The Mean Value Theorem and the shapes of functions 211
- 7.8 What about the ε and δ? 216

8 Integrals: *not just areas* — 221
- 8.1 From acceleration to velocity to position to ... *area?* 221
- 8.2 The integral: partition, add, take limit! 228
- 8.3 The Fundamental Theorem of Calculus: opposites annihilate! 243
- 8.4 Logarithmic and exponential functions: from areas to exponents 258
- 8.5 Numerical approximations of integrals: all you need is a computer 270
- 8.6 The improper Riemann integral: infinity made explicit 273

9 Infinite Series of Functions: *the wonders never cease!* — 285
- 9.1 The geometric series as a function series: amazing power at a low cost 286
- 9.2 Unexpected encounters with infinity 291
- 9.3 Exploring the hidden infinity: a thought experiment 297
- 9.4 Infinite sequences of functions: explore infinity's realm 299
- 9.5 Infinite series of functions: the magical infinity show! 318
- 9.6 Power series and Taylor expansions: transcendental functions explained 324
- 9.7 Fine tuning the infinite: l'Hôpital's rule 352
- 9.8 Trigonometric series and Fourier expansions 360
- 9.9 Continuous yet nowhere differentiable: Koch's snowflake and Weierstrass's function . 379

10 Infinity as the Link between Human Intuition and Reality — 385
- 10.1 Missing links in the human understanding of nature 385
- 10.2 The second missing link and numerical algorithms 386
- 10.3 The first missing link: human intuition 386
- 10.4 Connecting the components 388

11 Appendices — 391
- 11.1 Appendix: Archimedes's area argument and a modern derivation 391
- 11.2 Appendix: A Trigonometry refresher 396
- 11.3 Appendix: The proof of Cantor's power-set theorem 402
- 11.4 Appendix: Cantor's construction of the real numbers 403
- 11.5 Appendix: Discontinuity in a space of functions 415
- 11.6 Appendix: The derivative formula for $\sin x$ 421

11.7 Appendix: Proofs of some limit and derivative theorems in the text 425

References and Further Reading   **431**

Index   **432**

Preface

What we see on a television, computer or phone screen looks smooth and continuous. But digitial images on these screens are composed of *pixels*: tiny squares that are arranged in a grid to collectively form an image. The number of pixels per inch (or centimeter) is the *screen resolution*. More pixels per inch means greater resolution and a sharper and more vivid image. But the screen resolution is always a *finite number*: no matter how sharp and richly textured, every digital image turns into a jumble of colored or grey and white squares when we zoom in on or magnify a part of the image by a factor that is comparable to the resolution number.

What about non-digital images? How does a drawing on a piece of paper stand up to magnification? Or for that matter, smooth-looking surfaces like a bowling ball or polished marble? What happens when we zoom in on a part of some such non-digital image or physical object? Do we continue to see smooth textures no matter how much we zoom in with a lens like that of a microscope?

A surface, like that of paper, marble or a bowling ball looks continuous to the eyes but, again, it is not actually continuous. These material objects are made up of atoms, nature's three dimensional pixels. All forms of matter have finite resolution in this broader sense.

If matter is grainy at its core, *space and time* are not[1] To this day, even in quantum theory where matter and energy are discrete, space and time are considered continuous. If we zoom in on a patch of space over and over again, we never reach an underlying granular structure. Likewise, time is infinitely divisible; if we pick an interval of time, no matter how short, then half of it is still an interval of time that is shorter. Space and time both seem to have *infinite resolution*: there are no pixels of space or grains of time (sand-filled hour-glasses not withstanding to the contrary).

To deal with space and time it is necessary to use appropriate tools that deal with infinity. To this end, Isaac Newton and Gottfried Wilhelm Leibniz introduced calculus, an infinite-resolution tool that could resolve individual "points" of space and "instants" of time. After them, generations of mathematicians and other theoreticians developed the ideas further into fruitful new branches of mathematics like real analysis, complex analysis, topology, differential equations, differential geometry and so on. *We focus on real analysis in this book because that is where the infinite appears most explicitly.*

Using calculus (and analysis broadly) we resolve points and instants by essentially zooming in *an infinite number of times*. The tool for accomplishing this feat is the *limit*. The connection between limits and infinity is most readily brought out using *sequences*, basically strings of numbers. As we

---

[1] At least, not the familiar *intuitive* concept. The latest discoveries in physics suggest that our intuitive notions of space and time, like those of matter and radiation may need to be revised. The issue of intuition versus reality is discussed in the final chapter of this book.

see in this book, a *converging sequence* is analogous to a zoom feature on a camera or a microscope; but with a sequence we can zoom infinitely often!

In order to explain the role of infinity in calculus and beyond to a meaningful extent it is necessary to discuss some of basic ideas in the subject that motivated its use. This requires the inclusion of some of the standard material that is often covered in textbooks, like derivatives, integrals and infinite series. Familiarity with these topics is not a hard prerequisite to reading this book because I explain almost every topic from the start. Where standard topics are discussed, *the focus is on infinity* and how it is used to define the main ideas. So *the style and focus of the coverage here is different from what is found in typical textbooks on the subject*.

I also talk about interesting non-technical topics like paradoxes, thought experiments, historical anecdotes, and a few odd-ball notions and curiosities to illustrate (and in some cases showcase) how calculus manages to use infinity to pull off the infinite resolution feat mentioned above.

The topics that I have selected for this book illustrate the many *different* occurrences of infinity in calculus and in the broader field of *analysis* that grew out of calculus and now includes topics of great importance that go beyond calculus into such areas as differential equations and topology.

Chapter 1 is your starting point. It offers a broad perspective on the role of infinity in analysis via examples, stories, historical facts and non-technical prose. If you have not had prior exposure to calculus or analysis then this may be the main chapter for you; and if you do not want to get into analysis proper then a careful reading of chapters 1 and 3 along with a quick reading of 2, 4 and 5 should give you a basic but good idea of how infinity is woven inextricably into the fabric of our present day mathematics and mathematical theories in science.

If you wish to complete the whole journey then you can skip what may be familiar as you go, depending on your prior exposure to analysis. But unless you are an expert on the subject, I urge you to go through the chapters one by one; even if you have worked previously with such things as derivatives, series and integrals, it is likely that you have never seen them covered the way I present them here.

The exposition is intuitive rather than axiomatic. In particular, the set of real numbers that underlie the concept of space and intervals of time is carefully built from sequences of rational numbers. The properties of this set, such as its continuous nature are then derived from our construction. This approach is very different in content and style than in typical in textbooks on real analysis; these often define the real numbers via a series of field axioms plus the "axiom of completeness" which is typically expressed as the existence of a "least upper bound" for every bounded set. While the axiomatic approach is precise and expeditious for proving theorems and is more readily amenable to generalization, it is not an inspiring way to present the real numbers and their infinity-based nature.

I have taught mathematics in the USA for over 35 years at the college level. I earned my PhD in 1990 and was a full professor of mathematics by 2006. I have written two research monographs and over 60 research articles in peer-reviewed journals. My research and teaching have been based largely on real analysis and nonlinear difference equations; these are analogous to differential equations but with time and often also space being discrete sets. Difference equations also include recursions. More about me and my work can be found on the Internet. Infinity has been one of the best tools in my work for as long as I have been a mathematician.

With this I leave you to explore the infinite as you have likely not done before. I hope that you feel some of the joy that I felt in writing it!

H. Sedaghat

Chapter 1

The Ubiquitous Infinity

Infinity typically evokes a sense of mystery and awe. It is often considered a distant idea, apart from everyday thinking and experience, unreachable. But far from that, the mathematical tools like calculus that are routinely used by scientists and engineers in their work are based on infinity at a fundamental level. Its occurrence is so pervasive that if we try to take infinity out of the picture then the whole edifice crumbles to dust!

What exactly is *infinity*? Why is it so important in calculus and many other areas of mathematics, science and engineering? How do we go about exploring its uses in these areas without drowning in a sea of abstraction and technical detail?

In this first chapter I introduce the main topics that we consider in this book in summary fashion, using examples, simple numerical calculations and even paradoxes, stories, anecdotes and historical references. We discuss the different ways in which infinity appears in a field of mathematics called *analysis*, which includes calculus and forms the basis, and the modern context, for studying measure theory, probability and the foundations of statistics, the theories of differential and integral equations, operators on infinite dimensional spaces and much more. In physics, analysis is concerned with continuous matter, space, time and motion. This concept of continuity is clarified later in this book.

1.1 Infinitely many numbers and infinity in numbers

A quick look at very large numbers.

Our very first experience with infinity is that there are infinitely many *counting numbers, or whole numbers* $1, 2, 3, \ldots$. We also call them *the natural numbers* if 0 is added in. Before discussing other types of numbers like negative numbers, fractions, etc it is worth contemplating just how large a collection this one of counting numbers is. We can go down this list of numbers indefinitely by adding 1 to any number in the list; so if we keep counting up to a number, say, $1,000$ then $1,001$ is also in the list, $1,002$ is there and so on. But to get a better feel for the immensity of this list, let's proceed more rapidly using exponentiation:[1]

$$10, \; 10^{10}, \; 10^{10^{10}}, \; 10^{10^{10^{10}}}, \; \text{and so on} \tag{1.1}$$

The second number above is a 1 followed by 10 zeros or 10 multiplied by itself ten times, namely, ten billion; so far so good. The third number is 1 followed by a billion zeros (10 multiplied by itself a billion times); we do not have an official name for this so let's call it a *zillion* and keep going.[2] It is clearly pointless to keep inventing words because the list above keeps going on and on. Also consider that we haven't even come up with names for the numbers *in between* the four that are listed in (1.1); just these four entries have already taken us well past any number that we might encounter in practice in nature or in society.[3]

The fourth number is a 1 followed by a zillion zeros; what might we call *that*? How about a *gazillion* (maybe short for gargantuan zillion)? But what name might we give to the fifth number in this set? The tenth? How do we even write down the *zillion-th* number or imagine how large *that* is? After all, we only listed *four* numbers explicitly in (1.1)!

It is worth mentioning the famous *googol* here, which is 10^{100} (a 1 followed by 100 zeros). This funny name was made up by the young nephew of the American mathematician Edward Kasner in 1920. Regarding its similarity to "Google", the familiar search engine's name, it is said that when the early Google team was searching for available domain names they picked googol to highlight their product's ability to provide large quantities of information. But a Stanford graduate student misspelled googol and the name stuck once the domain name was registered. The headquarters of Google and its parent company Alphabet in Mountain View, California is named "Googleplex", in reference to goo*gol*plex, which is a 1 followed by a googol zeros, or $10^{10^{100}}$. You may, of course, *google* these terms for more historical and anecdotal information!

We can get a better sense of how large these numbers are by trying to write them down. 10^{10} is not hard at all: a 1 followed by 10 zeros. Writing googol down is more of a challenge (mainly to our patience) but still easy enough to write down a hundred zeros. So how about the googolplex? This requires writing down googol zeros after the 1. How hard can that be?

Suppose that we could write down ten zeros every second and keep going without slowing down. The time needed to complete the task of writing googol zeros is $10^{100}/10 = 10^{99}$ seconds. Here it is in years:

$$\frac{10^{99}}{3600 \times 24 \times 365} = 3.169863 \times 10^{91}$$

which certainly looks like a long time. One human lifetime doesn't even put a dent in this work!

How about the universe's lifetime? By some estimates, the universe will exist for another hundred trillion (10^{14}) years. Still, it takes

$$\frac{3.169863 \times 10^{91}}{10^{14}} = 3.169863 \times 10^{77}$$

---

[1] A well-known early use of exponentiation to count large quantities appeared in a treatise named "The Sand Reckoner" by the ancient Greek polymath Archimedes. Not having had access to Hindu-Arabic numerals, he extended the painfully inefficient and limited Roman numerals to estimate that it would take 10^{63} (in modern notation) grains of sand to fill a sphere with radius equal to the distance between the Earth and the Sun. For a more detailed description of what he did in this calculation, see "Infinity and the Mind" by Rudy Rucker.

[2] The second largest number with an official name is 10^{303} and is called a *centillion*. I will discuss the largest number with an official name shortly.

[3] Cryptography, the study of coding and decoding information, relates to social activities where very large numbers that are hundreds or even thousands of digits long are commonly used to encode or decipher sensitive information.

lifetimes of the universe to complete the task (never mind who or what is left to write all the remaining zeros after all the stars and black holes have disappeared).

Obviously, we have a major problem with finding the time needed to complete the task of writing all the zeros. But time isn't the only issue; we must also consider the enormous amount of matter (ink or graphite or just atoms) and the space in which to write or otherwise record all of these zeros. A calculation similar to the one above for time shows that there is not enough matter, or room, for writing all the zeros in our finite universe, as we currently understand it.

You may have noticed that the googolplex is tiny compared with our gazillion; but the gazillion itself is nothing compared with infinity. Let's call the numbers that are stuffed inside the "and so on" in (1.1) collectively the "gazillions". This is similar in essence to what the Hottentot (or Khoe-Khoe) tribe in southern Africa does for their own practical reasons; when explorers first came across this tribe over a century ago, they discovered that these people would count up to three but refer to all larger numbers simply as "many" or "a lot". For us perhaps a zillion (maybe even much less) qualifies as "a lot" (for practical purposes) and this is certainly an enormous number; but from it we are really no closer to "the end of the line" than the Hottentot; you can convince yourself by continuing the list in (1.1) that infinity is as far away from our gazillion as it is from the number 3.

Infinities – there's more than one?

The idea that infinity is not something unique may seem audacious. In spite of popular phrases referring to "infinity and beyond" we usually think of infinity as what lies beyond everything else and often call it "the infinite," signifying uniqueness.

The problem with this idea is that it does not specify what infinity is but rather what it is not. It isn't all of the billions of people on Earth, or all of the trillions of stars in the cosmos or any other large number or vast collection of things. We would like to think that the universe is infinite because if it weren't then we would have to face the uncomfortable and arguably unanswerable question: what lies beyond?

Our conceptualization of space and of existence, in material form or non-corporeal, is based on our sense experiences and the psychological and philosophical extrapolations of those experiences. But infinity has no known physical basis and cannot be registered by our sensory organs. So it has to be extrapolated from our experiences. Mathematicians are reluctant to rely on imprecise extrapolations of any sort, no matter how profound or delightful they may seem. This reluctance is based on millennia of hard experience that shows imprecise thoughts and arguments are too malleable: they do not withstand serious scrutiny and are often misused for personal gain or abused to mislead the gullible.

Infinity is precisely the sort of thing whose proper handling requires mathematical precision if we want to keep out of the realm of fantasy. In particular, we see that in mathematics and any other scientific subject that uses mathematics in a substantial way the idea of "many infinities" is as precise a concept as it is mind-boggling.

Let's explore the numbers between 0 and 1; these include fractions like 1/2, 4/7 etc where an integer is divided by a positive integer. These numbers are also called *rational numbers*.[4] Think about the number line, or the x-axis in algebra: the points on a stretch of this line from 0 to 1 represent all the numbers between 0 and 1. There are of course, larger parts of the number line, like

---

[4]The word rational in this context is rooted in the Latin word *ratio* meaning to reckon or to reason. Of course, the computational meaning that signifies fractions is the one intended here.

all numbers greater than 0 (all positive numbers). But the story of the little stretch from 0 to 1 is all that we need here.

To show that there are infinitely many numbers between 0 and 1 consider a thought experiment. Suppose we pick points (numbers) out of this segment one at a time like drawing little glass marbles out of a bag. If this goes on indefinitely then we expect that ultimately we will have withdrawn infinitely many numbers.

But which numbers do we begin with? An easy choice is to take out the reciprocals of the natural numbers:
$$\frac{1}{1}, \frac{1}{2}, \frac{1}{3}, \frac{1}{4}, \frac{1}{5}, \frac{1}{6}, \dots$$

Certainly all of these numbers are between 0 and 1 and there are infinitely many of them: looking at the denominators you see that there are no fewer numbers in this list than the totality of all natural numbers. But still infinitely many more points or numbers remain to be taken out like 0, $2/3$, $\pi/4$ and so on.

There are more clever ways of removing numbers: suppose that we knock them out of the line segment as follows:
$$\frac{0}{1}, \frac{1}{1}, \frac{1}{2}, \frac{1}{3}, \frac{2}{3}, \frac{1}{4}, \frac{3}{4}, \frac{1}{5}, \frac{2}{5}, \frac{3}{5}, \frac{4}{5}, \dots \tag{1.2}$$

Here we take out all the fractions with a fixed denominator before moving on to the next set of fractions (discarding all repetitions along the way, like $2/4$ which equals $1/2$). This method takes out *all* rational numbers between 0 and 1 (including 0 and 1). For example, the number $87/112$ is reached when removing all the fractions with denominator 112.

The infinite list in (1.2) is large enough to include every rational number between 0 and 1 but it leaves out all of the *irrational* numbers. These are numbers that cannot be written as a fraction, or a ratio of two integers. We can prove easily that there are infinitely many numbers of this type: choose any known irrational number between 0 and 1 like $1/\sqrt{2}$ which is approximately equal to 0.707. Now, the list
$$\frac{1}{\sqrt{2}}, \frac{1}{2\sqrt{2}}, \frac{1}{3\sqrt{2}}, \frac{2}{3\sqrt{2}}, \frac{1}{4\sqrt{2}}, \frac{3}{4\sqrt{2}}, \frac{1}{5\sqrt{2}}, \frac{2}{5\sqrt{2}}, \frac{3}{5\sqrt{2}}, \frac{4}{5\sqrt{2}}, \dots \tag{1.3}$$

that we get by multiplying every number in (1.2) by $1/\sqrt{2}$ is an infinite list of distinct irrational numbers between 0 and 1.

So there are no fewer irrational numbers between 0 and 1 than there are rational numbers. But notice that we can get more irrational numbers if we replace $\sqrt{2}$ in (1.3) by other irrational numbers that are larger than 1, say $\sqrt{3}$, $\sqrt[3]{2}$, π, etc. For each of these we obtain equally large lists of distinct irrational numbers!

You may have noticed phrases like "as many numbers as" or "no fewer numbers than" that I have used above. These are the kinds of words we naturally think of where we need to count but don't have numerals for it. When we apply these to our infinite lists of rational and irrational numbers above, notice that we count *by pairing the entries in each list together*. This idea is important when dealing with infinite sets because we can't use statements like "both lists have exactly a gazillion numbers in them".

Counting by pairing items in one list with items in another list predated the creation of counting numbers 1,2,3, etc. In very early times in human history our ancestors would count things using standard references like the fingers of their hands. If they could identify as many heads of cattle as the fingers of one hand then that meant 5 heads of cattle. When the numbers used were small there was no need to invent symbols or names for them.

But as human civilization grew so did the sizes of numbers that they used. The need for numbers with names and symbols probably became most urgent when merchants and governments wanted to count large numbers of items, be they coins, bags of wheat or heads of cattle. It would be rather hard for merchants to deal with hundreds or thousands of such items using fingers or other primitive counting methods.

As long as we are dealing with a finite (and not very large) number of items the existing numbers are just fine. In fact, the Romans found their cumbersome number system adequate for all their dealings. But Europe's needs eventually grew larger, along with its population, to the point that the Roman numerals were no longer adequate. So when the superior Hindu numerals arrived by the way of the Islamic civilization they were quickly adopted for usage in commerce and later in mathematics, too.

The Hindu-Arabic numerals are so versatile that they have proven sufficient even in our modern era of international, large-scale commerce. But as we saw earlier, numerals of any kind are useless when dealing with infinity. So, in a remarkable turnaround, we need to go back to the primitive ancestors' way of counting by pairing things up!

The German mathematician Georg Cantor (1845-1918) used that method creatively in the 1870's to show that the infinity of all irrational numbers between 0 and 1 is a *larger order of infinity* than the infinity of all the rational numbers. In other words, we can never pair up *all* the irrational numbers with the rational numbers: no matter how cleverly we pair things up, we always run out of rational numbers long before all the irrational ones are accounted for. This is certainly not a transparent point; we see exactly what it means later when discussing Cantor's original arguments in Chapter 3.

So, thanks to the irrational numbers there are different orders of infinity hidden in just the set of all numbers on the x-axis between 0 and 1. The irrational numbers turn out to be far more enigmatic than the rational ones and have an important role in making the x-axis the "continuous line" that it is. In analysis, this continuity property of the number line is called *completeness* and we discuss it in Chapter 5. Completeness is what we get when we put the rational and irrational numbers together; it is essential to successfully working with infinity in analysis: *it is what makes the magic of calculus possible!*

Infinity in each (irrational) number.

Calculus, and analysis more broadly, were born out of a marriage of algebra and geometry; it turned out to be a "marriage made in heaven" for mathematics. This union was not a singular event created by one person but an evolutionary process that was initiated by Descartes[5] and others once algebra and the Hindu-Arabic numerals were adopted and mastered in Europe.

An influential work that paved the way for the entrance of algebra in Europe was al-Khwarizmi's book *al-jabr*. Muhammad ibn Musa Al-Khwarizmi, or Khahrazmi (780-850) was an Iranian mathematician and a scholar in the *House of Wisdom* in Baghdad. The full title of his algebra book is: "al-kitab al-mukhtasar fi hisab al-jabr wal-muqabala" which translates into "The Concise Book on Calculation by Restoring and Balancing". This work appeared around 830 and was translated into

---

[5] Rene' Descartes (1596-1650) was a French philosopher and mathematician.

Latin in the twelfth century by Robert of Chester as *Liber Algebrae et Almucabola*.[6]

In contrast to geometry, algebra is not focused on continuous things; it studies ordinary numbers and number-like objects (like vectors in higher dimensions) in terms of their properties under operations like addition and multiplication. Descartes, Fermat[7] and others realized that numbers and vectors may be also regarded as *points* in a continuous space.

The question as to *how these points come together to form continuous objects, or the space itself is a profound issue* (the aforementioned completeness) and one that is of central concern to us here since it brings infinity into focus, front and center. It bothered mathematicians, philosophers, scientists and others for centuries; as we see later in this chapter, the philosopher Zeno's famous paradoxes are squarely focused on this issue.

Without the irrational numbers the number line (and more generally, curves, surfaces, etc) would be incomplete, that is, not continuous. The irrational numbers fill in the infinitesimal gaps left in the number line by the tidy but grainy rationals, not unlike a glue or filler that smoothes out holes and cracks. Together, rational and irrational numbers constitute the complete set of all *real numbers* that we also call the number line or the x-axis. We study this set in Chapter 5 where we see how to construct irrational numbers from (sequences) of rationals.

Irrational numbers don't have algebraically interesting properties by themselves but analysis would be literally incomplete without them. In large measure, the reason for this is that *each irrational number is an infinity onto itself*.

To explain this, first note that *every rational number has an infinite but repeating decimal expansion*. For example, 11/27 has a decimal form:

$$\frac{11}{27} = 0.407407407\cdots = 0.\overline{407} \tag{1.4}$$

We say that 11/27 has an infinite decimal expansion with *period* 3 because the 3-digit pattern 407 repeats. These few digits and the order in which they appear fully characterize the rational number 11/27. So rational numbers do not have a truly infinite character.

Calculators, of course, do not give an infinite expansion since they don't have infinite storage capacity; even if they did, it would be pointless to fill it all up with digits that we do not need in practice. They typically give an answer that is rounded off after a number of digits, like:

$$0.40740740740741$$

This number (with a finite number of digits) is very close to 11/27 but not quite equal to it. So how do we *know* that there is an infinite, repeating decimal form?

The answer is by *long division*.[8] While we rarely calculate with long division nowadays, it is

---

[6] Later in this chapter we come across another work of al-Khwarizmi regarding the Hindu-Arabic numerals that proved to be more influential outside mathematics.

[7] Pierre de Fermat (1607-1665), a French mathematician is best known for his "Last Theorem". Though not a mathematician or scientist by profession (he was a lawyer) he is known for many original contirbutions to science and mathematics; these include contributions to number theory and to the early development of calculus. Also his principle of "least time" was used later to solve problems in mechanics and optics and inspired the principle of least action.

[8] An alternative after-the-fact proof is as follows: let x be the infinite decimal expansion on the right hand side of (1.4) and observe that $1000x - x = 407$. Solving this equation for x gives the rational number $x = 407/999$ which reduces to 11/27 after canceling the common factor 37 from the numerator and the denominator. Notice that this argument extends to all repeating decimal expansions and shows them to be rational numbers. Some rational numbers have a finite number of digits, like $2/5 = 0.4$. It is also valid to write $2/5 = 0.4000\cdots = 0.4\overline{0}$ which can be said to have period 1. An alternative form is $2/5 = 0.3999\cdots = 0.3\overline{9}$, also with period 1.

available to us as an option when needed. Unfortunately, when it comes to irrational numbers this option is no longer available.

Calculators do have answers in (rounded) decimal forms for them though. For instance, a calculator's answer for $\sqrt{2}$ is

$$\sqrt{2} = 1.4142135623730950488016887242\cdots \qquad (1.5)$$

which in this case is accurate to 28 decimal places (the three dots do not usually appear on the calculator screen; I have added them here as a reminder that there are infinitely many digits).

But how do calculators find the digits of an irrational number?

How can we find, say, the 29th and 30th digits of $\sqrt{2}$? For this particular irrational number an iteration method called the *divide and average rule* that was known to the ancient Babylonians 4,500 years ago works very effectively.[9] We start with a convenient rational number r_1 as a first guess at $\sqrt{2}$, divide 2 by this number and average the result to get a new approximation

$$r_2 = \frac{r_1 + (2/r_1)}{2}$$

If we start with $r_1 = 1$ then we get $r_2 = (1+2)/2 = 3/2 = 1.5$. This process may be repeated using r_2 in place of r_1 to obtain a new approximation r_3.

With $r_2 = 1.5$ we get $r_3 = [1.5 + (2/1.5)]/2 = 1.416667$, rounding the answer to 6 decimal places to save space.[10] Next, using r_3 we get $r_4 = 1.414216$ which matches the answer in (1.5) to 5 decimal places. The next approximation is $r_5 = 1.414213562$ which is correct to at least 9 decimal places, and so on.

Why does this rule work? Are there similar rules for other irrational numbers?

We find the answers to these questions in Chapter 7 where we introduce the remarkable Newton-Raphson method, a widely used algorithm for solving nonlinear equations (the divide-and-average rule is a special case). As we just observed when finding an approximation to $\sqrt{2}$, the Newton-Raphson method generates good approximations to irrational numbers very fast. But to get the *actual or exact value* of the irrational number $\sqrt{2}$ *we must apply the method an infinite number of times.*

Iterative procedures provide one way of both verifying the presence of infinity in irrational numbers and of estimating them as accurately as we wish by calculating large numbers of digits of their decimal presentations. We discuss another important way in the next subsection that involves adding infinitely many numbers.

There are geometric ways of defining a relatively few irrational numbers like $\sqrt{2}$ and π that don't involve the infinite explicitly. For instance, we may visualize $\sqrt{2}$ geometrically as the diagonal of a square whose sides are one unit (meter, foot, etc); similarly, π is the length of the circumference of a circle of diameter one unit, or equivalently, the area of a disc of radius one. But such convenient alternative expressions don't exist for most irrational numbers and nowadays, even for the above few irrational numbers we use decimal approximations in practice.

An intriguing feature of the answer in (1.5) is the apparent lack of any patterns in the decimal expansion of $\sqrt{2}$. This makes it quite difficult to tell what the far-away digits of $\sqrt{2}$ are; for example,

---

[9]The Babylonians knew about the divide and average rule, presumably without any knowledge of calculus, and used it to approximate $\sqrt{2}$ to a high degree of accuracy.

[10]Actually, no rounding is necessary since at each step we could just keep using fractions, which are precise. I round them here to simplify the reading (and the writing). Why 6 decimal places and not, say, 3 or 4? Because the accuracy improves so quickly that we need more digits to show something is still happening after the 4th step!

it takes a lot of calculations to find the zillionth digit of $\sqrt{2}$, should this ever be needed. Of course, periodic patterns are not possible but that doesn't mean that an irrational number *must* have a totally unpredictable decimal expansion. Here is an easy-to-remember irrational number that is entirely orderly and predictable:

$$0.12345678910111213141516171819202122\cdots$$

because it is simply a concatenation of all non-negative integers. It was published in 1933 by D.G. Champernowne in his thesis and has become known as the "Champernowne number". Unlike $\sqrt{2}$ we know every digit of the Champernowne number; for instance, in the above listing we see that the 30th digit of this number is 2 and the 33rd is 1. By following the simple concatenation pattern we can tell that the 50th digit is 3; for the far-away digits, a simple counting rule takes care of the growing number of digits as we move up to 3-digit numbers, then to the 4-digit ones and so on.

The presence of infinity in each irrational number can make it hard to prove that a given number is irrational. While it is clear by its construction that the Champernowne number is irrational, and relatively easy to prove that $\sqrt{2}$ and a few other types of numbers are irrational, proving that π is irrational is more difficult, as it is to prove that $\sin 1$, $\log 2$, $2^{\sqrt{2}}$, $\pi^{\sqrt{2}}$ and similar numbers are irrational.[11]

To properly understand the role of infinity in irrational numbers we must be able to probe the farthest depths of each irrational number; the standard tools for doing this include infinite sequences and series, discussed in Chapters 4 and 6, respectively.

Adding infinitely many numbers.

Over 2400 years ago, the Greek philosopher Zeno of Elea proposed a series of provocative and enduring thought experiments to highlight problems with ancient Greeks' conceptions of time, space and motion. Zeno's ideas became popular as "paradoxes" and they appear in popular writings on infinity to this day.[12]

One of Zeno's most famous paradoxes involves an imaginary race between a tortoise and Achilles the sprinter (in some versions of this story a hare is mentioned instead of Achilles). He proposed that if the tortoise has a head start even by a tiny amount then Achilles will not be able to overtake it because each time he gets to a place the tortoise has already been, the latter will have moved forward by a little bit and therefore, it always stays ahead of Achilles.

Let's consider a slightly different version in more modern terms. A man walks from his house to a nearby store that is one kilometer (or mile) away. Zeno's argument stipulates that before he reaches the store the man needs to walk half the distance, or $1/2$; continuing his trek, he covers half of what remains or $(1/2)(1/2) = 1/4$, so he is $1/2 + 1/4 = 3/4$ of the way to the store. But he is still $1/4$ of the distance away so he has to walk half of that or $(1/2)^3 = 1/8$ (for a total of $7/8$ from the start) and so on. There is always some distance left, so he can never really reach the store.[13]

Of course, the man does reach the store eventually as we know from personal experience, just as Achilles would overtake the tortoise quickly, as we might imagine. It is said that the cynic Diogenes refuted Zeno by getting up and walking. But Zeno didn't need to be reminded of reality;

---

[11] You may wonder what $2^{\sqrt{2}}$ or similar numbers are and how they are calculated; we will return to this issue in Chapter 8 where we discuss the concept of exponential functions.

[12] Zeno's ideas come down to us through the writings of others like Aristotle.

[13] The walking man and the store are like the Achilles and the tortoise as viewed from the tortoise's frame of reference where Achilles approaches from behind but never actually reaches the tortoise, according to Zeno.

Ubiquitous Infinity

his proposals were aimed at the Greek way of thinking which treated infinity intuitively just as most people do today.

Let's take a closer look at what is taking place in the thought experiment. If we add up all the bits of distance that the man walking to the store travels then Zeno's scenario is summarized by the following equality:

$$\frac{1}{2} + \left(\frac{1}{2}\right)^2 + \left(\frac{1}{2}\right)^3 + \left(\frac{1}{2}\right)^4 + \cdots = 1 \qquad (1.6)$$

The right hand side is 1 because that is the total distance to the store (one kilometer or mile). The left hand side consists of the sum of all the bits of distance that the man covers on his way to the store; this sum is what we call an *infinite series*. The equality says that *adding infinitely many numbers can result in a finite number*.

This is by no means obvious, and it was even less clear back in Zeno's time. Certainly the Greeks didn't believe, any more than we do today, that adding infinitely many numbers *always* gives a finite number. For instance, if we add $1/2$ to itself (without including the powers) infinitely many times:

$$\frac{1}{2} + \frac{1}{2} + \frac{1}{2} + \cdots$$

then we get an infinitely large result (we will be more precise about this in Chapter 6); for example, the sum of the first 10 million halves is $10,000,000(1/2) = 5,000,000$. On the other hand, if we include the powers then no finite sum ever exceeds 1. The following table shows that the sums get closer and closer to 1 as we add more terms but they don't quite reach it:

Number of terms added	1	2	3	4	5	6	7
Sum of this many terms	0.5	0.75	0.875	0.9375	0.96875	0.984375	0.992188

The sums quickly approach 1; for instance, adding 30 terms takes us to 0.9999999990686774. But no matter how many terms we add, we never get past 1. This leads to the question:

What is it about (1.6) that makes it possible to add infinitely many numbers and get a finite number?

Evidently the powers play an important role: we are adding rapidly shrinking numbers in succession so the value of the sum grows by a smaller amount each time we add a new number. Without the powers, each $1/2$ that is added is no smaller than the previous one so the value of the sum grows by a fixed amount each time. This observation is important but not quite enough; we discover in Chapter 6 that the added numbers at each stage must shrink "fast enough" in a sense that I explain later.

Adding infinitely many numbers becomes more complex but also more fascinating when positive *and* negative numbers are involved. We must now consider the effects of cancellations between the positive and negative terms. Let's consider the following list of fractions

$$1, -\frac{1}{2}, \frac{1}{3}, -\frac{1}{4}, \frac{1}{5}, -\frac{1}{6}, \frac{1}{7}, \ldots$$

whose signs alternate between positive and negative. Now, suppose that you add these numbers in the order listed above, one after the other in the order given above and call the sum s:

$$s = 1 - \frac{1}{2} + \frac{1}{3} - \frac{1}{4} + \frac{1}{5} - \frac{1}{6} + \cdots \qquad (1.7)$$

We won't worry about what the total is here.[14] Now suppose that I add the numbers in a different order:

$$1 + \frac{1}{3} - \frac{1}{2} + \frac{1}{5} + \frac{1}{7} - \frac{1}{4} + \cdots \qquad (1.8)$$

where I add two odd fractions and then subtract an even one. We are both adding exactly the same numbers so it makes sense to expect that the results will come out the same. But do they?

They don't!

Suppose we divide all of the numbers in (1.7) by 2 to get

$$\frac{s}{2} = \frac{1}{2} - \frac{1}{4} + \frac{1}{6} - \frac{1}{8} + \frac{1}{10} - \frac{1}{12} + \cdots$$

When we add the two sides of this equality to those of above (1.7) and cancel out the equal fractions with opposite signs, we get

$$\frac{3s}{2} = 1 + \frac{1}{3} - \frac{2}{4} + \frac{1}{5} + \frac{1}{7} - \frac{2}{8} + \cdots$$
$$= 1 + \frac{1}{3} - \frac{1}{2} + \frac{1}{5} + \frac{1}{7} - \frac{1}{4} + \cdots$$

Notice that this is exactly the way that I added the numbers in (1.8)! But I get a different answer ($3s/2$) that is 50% larger than yours. We are forced to conclude that *the infinite series in (1.7) is different from that in (1.8)* just because their numbers (the *same* numbers) are added in different ways. Infinity has consequences!

It so happens that this is just the tip of an iceberg: choose any number that you like, 0, 1/2, $-4/3$, $\sqrt{2}$, π or whatever, and call your number r. A fascinating result that is known as the *Riemann Rearrangement Theorem* shows that by simply rearranging the fractions in (1.7) it is possible to make them add up to number r that you picked, not just to s or $3s/2$! We will discuss this amazing theorem in Chapter 6.

Before moving to the next topic, I mention one more important use of infinite series. Going back to our earlier observation that irrational numbers conceal the infinite in them, we showed how to find the digits of $\sqrt{2}$ using a recursion (Newton's method). The same approach does not work as well for π because the approximating numbers also turn out to be irrational like π itself. Fortunately, there are a variety of infinite series of *rational numbers* known that give the value of π to an arbitrary degree of accuracy.

The most well-known infinite series of rational numbers for π is the following:

$$\frac{\pi}{4} = 1 - \frac{1}{3} + \frac{1}{5} - \frac{1}{7} + \frac{1}{9} - \frac{1}{11} + \cdots \qquad (1.9)$$

By adding enough fractions on the right hand side we can obtain as may digits of $\pi/4$ as we would like. The approximating numbers are obviously rational since each is the sum of a finite number of fractions. This series was known to Leibniz and was popularized by him, but it has a much longer history; we will discuss this and similar series in Chapter 9 where we find that the series in (1.9) is rather inefficient (to get a reasonable approximation of π a large number of fractions is needed). We discuss more efficient series, one of which was known to Indian mathematicians in the 14th century, long before the discovery of calculus in Europe.

---

[14] We will see in Chapter 9 that $s = \ln 2 \simeq 0.693$, the natural logarithm of 2.

1.2 What might we find at the end of an infinite process?

As I mentioned earlier, the Greeks believed that the *continuum*, namely, any continuous object, including space and time was infinitely divisible. We can divide an idealized continuous object infinitely many times; the same is true of the distance between two points or an interval of time. This was not the difficulty; the problem was with what followed naturally: since after dividing something up it does not just vanish, *we are left with infinitely many things and none of them are divisible further*. These outcomes of the infinite process of dividing the continuum were called *indivisibles*, or "atomos" in Greek.

Exactly what were these "indivisibles"?

The answer is far from obvious and the idea was a source of controversy in the Greek culture. Greeks' way of dealing with this problem was to avoid the explicit use of infinity. They would allow for *potential infinity* which is never reached or completed. In particular, it was meaningless to speak of the continuum as an infinite collection of points having no sizes because that required an infinite division process. Euclid developed his famous geometry using axioms that made no mention of points as constituents of lines, planes, etc. And Aristotle declared that: "time is not composed of indivisible nows any more than magnitude is composed of indivisibles." And as we discover later in this chapter, Archimedes calculated the area inside a circle by carefully sidestepping infinity in his argument even though there is an infinite process involved in getting from polygons to a circle.

Infinity's curse: the strange tale of Generous and Greedy.

The idea of an infinite process ending in some fashion (in the sense of the *convergence of a sequence to a limit* that we clarify later) is of fundamental importance in analysis. I now describe an infinite process that leads to an unexpected and somewhat paradoxical result.

Consider a thought experiment where two elves, named Generous and Greedy, are told by the agent of a mysterious, powerful being that they will be given pouches of shiny gold nuggets, which they love, if they agree to a contract: every day, each of them gets two pouches of nuggets that they may deposit until the contract is fulfilled. But also every day they must give away one pouch to the destitute in their mythical land. If either one of the elves breaks the contract by not donating one pouch on any day then he or she will forfeit all of the nuggets they may have accumulated and will receive no additional gold.

Generous and Greedy agree and the pouches of gold are placed in deposit boxes in the Communal Elf Bank as they come in. The deposit boxes are endlessly long with infinitely many compartments as shown in Figure 1.1.

Generous and Greedy may each empty exactly one compartment of his or her own deposit box on any given day for donation purposes.

On Day 1, two pouches are deposited in each elf's box, a pouch containing a single nugget and another with two nuggets. The banker places the first pouch in compartment 1 and the second pouch in compartment 2 for each elf.

As a matter of principle, Generous empties compartment 2 and donates the two nuggets, putting the needs of the poor above her own interests. But Greedy empties compartment 1 of his deposit box and donates its content because to him the obvious thing to do is to keep the larger amount for himself while donating something to the poor at the same time. At the end of the first day, Generous has one gold nugget in the pouch in compartment 1 while Greedy has a pouch containing two nuggets in compartment 2.

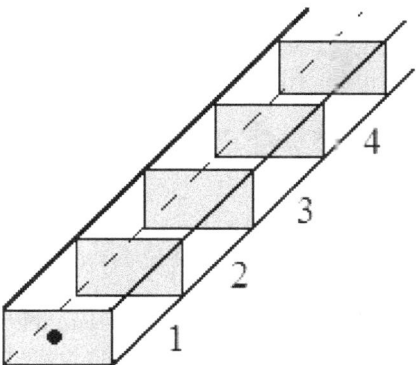

Figure 1.1: An open-ended Communal Elf Bank deposit box

The next day, the two elves find pouches of gold nuggets in compartments 3 and 4, containing 3 and 4 gold nuggets, respectively. Generous donates the larger amount (4 nuggets), the largest that she has, by emptying the 4th compartment in her deposit box; Greedy gives up the pouch with 2 nuggets, the smallest that he has, still in compartment 2 from the day before. At the end of this day, there are pouches in compartments 1 and 3 of Generous's deposit box but compartments 2 and 4 are empty. In Greedy's deposit box there are pouches in compartments 3 and 4 but compartments 1 and 2 are empty.

The following table shows the number of nuggets that Generous and Greedy end up with in the first two days after making their donations, as distributed in the compartments of their deposit boxes:

	Day 1	Day 2
Generous	1,0	1,0,3,0
Greedy	0,2	0,0,3,4

The zeros indicate empty compartments. The nuggets come in every day and the amounts deposited in each compartment increase by one nugget each day. Here is the situation in 3 days:

	Day 1	Day 2	Day 3
Generous	1,0	1,0,3,0	1,0,3,0,5,0
Greedy	0,2	0,0,3,4	0,0,0,4,5,6

At the end of the third day, compartments 1, 3, 5 of Generous's deposit box contains pouches of gold nuggets while compartments 2, 4, 6 are empty. She has kept a total of $1 + 3 + 5 = 9$ gold nuggets so far. In Greedy's deposit box compartments 4, 5, 6 contain pouches of gold nuggets while compartments 1, 2, 3 are empty. He has collected a total of $4 + 5 + 6 = 15$ gold nuggets.

This process continues. After a week, the elves have accumulated the following collections of gold nuggets, as distributed in the compartments of each deposit box:

	Day 7
Generous	1,0,3,0,5,0,7,0,9,0,11,0,13,0
Greedy	0,0,0,0,0,0,0,8,9,10,11,12,13,14

If they took out their gold now then Generous would have 49 gold nuggets and Greedy 77 gold nuggets. Greedy feels that his strategy (so obvious to him) is working and building much greater wealth than Generous's good-hearted but naive strategy.

After infinitely many days pass (elves are immortal) they are at last informed that the contract is fulfilled and there would be no further deposits of gold nuggets. They could at last claim whatever is left in their possession.

As Generous draws out her deposit box she finds infinitely many pouches in compartments $1, 3, 5, 7, \ldots$ (odd numbered compartments). She is thrilled with her boundless wealth and comforted by the thought that she donated even more to the needy.

Greedy also draws out his box but to his astonishment he finds one empty compartment after another. No matter how far down the deposit box he looks he is dismayed to find nothing but empty compartments! He screams a profanity and asks: what happened to my gold nuggets? He wonders: Did a misbegotten elf steal it all?

Greedy registers a complaint against the Bank with the Elf Authority and the matter is investigated. But the investigation comes to a speedy closure after checking with the Elf Charity Organization. In a meeting attended by Greedy, members of the Bank and the Elf Authority, a ECO representative named Grateful explains how *all of Greedy's gold* ended up in the ECO vaults–by Greedy's own action!

Elf Grateful noted that on the first day that Generous and Greedy came by, each one brought a pouch. Generous's pouch contained 2 gold nuggets and Greedy's one nugget. Both pouches were placed in the vault which was then locked. The next day Generous brought a pouch containing 4 nuggets and Greedy brought a pouch with 2 nuggets in it. These pouches were also placed in the ECO vault.

Grateful lists the number of gold nuggets brought to the ECO each day by our protagonists:

Generous: 2, 4, 6, 8, 10, 12, 14, 16, 18, 20, etc

Greedy: 1, 2, 3, 4, 5, 6, 7, 8, 9, 10, etc

By the end of day 10 Generous had donated twice as much as Greedy to charity; no surprises here. But notice that the above list doesn't end: in 20 days Greedy had managed to contribute not only all the gold that Generous had donated in the first 10 days but also what she had kept for herself, namely, the pouches containing odd numbers of nuggets. Grateful stressed that *since the donation process had never stopped, Greedy had managed (without realizing it) to not only catch up with Generous but in fact donate everything that he had been given!*

As he was leaving the meeting shaking his head, Greedy was overheard mumbling to himself: "*'twas infinity's curse!*"

The morale of this story (mathematically speaking) is that infinity has different rules that require careful mathematical scrutiny. What happened to Greedy is neither a paradox nor a problem in analysis; it is a strange side-effect of working with infinity that has no analog in the finite realm.

In analysis jargon, Greedy's fortune as a sequence of the daily states of his deposit box *converged to zero pointwise* despite the fact that the number of gold nuggets were growing infinitely large. We discuss this concept of convergence in Chapter 9 (see the discussion pertaining to Figure 9.8) where we discover that pointwise convergence is notorious for not preserving useful properties of functions, like being continuous or integrable. Now, to answer our earlier question: what we might expect to find at the end of an infinite process (a sequence of functions) is the "limit" to which the sequence converges, *if the limit exists.*

The ancient Greeks were so mistrustful of infinite processes that they never developed the concept of limit. It had to wait until the 19th century to clarify and fully explain what this notion was because only a proper understanding of this concept would explain the successes and failures of the important series that was introduced by Jean-Baptiste Joseph Fourier (1768-1830) in the year 1807. It would take another 70 years before mathematicians could resolve the major issues with convergence generally and with Fourier series in particular. We discuss the Fourier series in Chapter 9.

> **What is the length of a square's diagonal? Did Pythagoras miss something?**

The story of Generous and Greedy illustrates an odd side-effect of pointwise convergence. There are types of convergence that do not allow such outcomes. In Chapter 9 we discuss the most important of these that is known as *uniform convergence.*[15]

We discover in Chapter 9 that Greedy's deposit-box-state sequences, as well as most pointwise converging sequences, do not converge uniformly. In Greedy's case, even though his deposit box is increasingly depleted at one end, it isn't depleted everywhere; in fact, it gets increasingly loaded deeper into the deposit box. So while his wealth is zero in the end, it vanishes in a "non-uniform" fashion. We see in Chapter 9 that this non-uniformity can be attributed to a hidden occurrence of infinity in pointwise converging sequences that does not occur in uniformly converging sequences. But even uniform convergence cannot squeeze out all of infinity's strangeness. Here's a simple, yet disturbing example.

Consider the squares in Figure 1.2, each of which has sides 1 unit long.

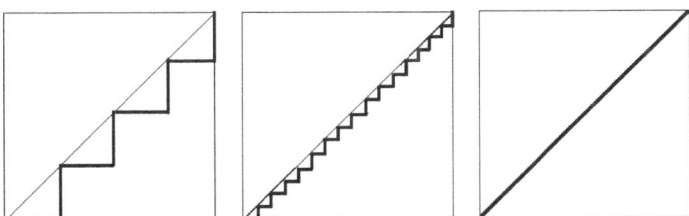

Figure 1.2: stairs-shaped curves converge uniformly to the diagonal

In the left and center squares we see stairs-shaped curves (thick lines) right under a diagonal. The lengths of both of these curves is the same: 2 units, because the treads add up to 1 going horizontally and the risers add up to 1 vertically. Adding more steps to existing ones doesn't change their total length although the size of each step shrinks.

As the number of steps grows infinitely large the stairs smoothen out into a straight line, namely, the diagonal of the square on the right in Figure 1.2. Since all the while the length of the staircase remains the same, it seems reasonable to conclude that the diagonal also has length 2.

But as we know that isn't the case; according to the Pythagorean theorem the length of the diagonal is $\sqrt{1^2+1^2}$ or $\sqrt{2}$ which is definitely less than 2. This makes sense thinking of each

---

[15] Uniform convergence is essential for developing a coherent theory of functions of a real variable, including the so-called transcendental functions of calculus, like logarithms and trigonometric functions.

individual staircase since the shortest distance between two points is a straight line in Euclidean geometry. But as the number of steps grows infinitely large, the stairs flatten out to a straight line segment while their lengths stay fixed at 2 ... *except at the very end!*

Something strange happens to the length of the staircases in the transition to infinity... there appears to be a sudden jump from 2 down to $\sqrt{2}$ in the lengths of staircase curves. In particular, we see that the staircase curves converge to the diagonal, so they become good approximations to the diagonal of the square; this is illustrated in the right hand panel in Figure 1.2. But *the lengths of the staircase curves do not approximate the length of the diagonal no matter how well the staircase curves resemble the diagonal!*

If we draw a staircase with billions of (tiny) steps then our drawing looks like a straight line, namely, the diagonal. But if we stop at, say, a zillion steps and magnify a small stretch of the diagram sufficiently then we eventually see an image that is like the center image in Figure 1.2. This happens whenever finite resolution is involved. But at the end of the infinite process of adding steps the result is indeed a straight line. If magnify this line gazillions of times the result is always the same: a straight line!

We discuss this problem in Chapter 9 where we see that the staircase shape converges to the diagonal line uniformly. So what is causing the anomalous change in length from 2 to $\sqrt{2}$?

This unsettling observation does have an explanation in analysis. The answer has to do with the concept of *variation*. A straight line is a curve of zero variation but a wiggly curve with many folds can have a large variation. Uniform convergence can't always tell the difference between the former and the latter but the length of a curve having many folds can be significantly larger than that of a straight line between the same pair of points. By adding more folds we can increase the length of the curve without changing the length of the straight line.

So how does this idea explain the staircase anomaly?

As we add more steps, we increase the variation of the staircase curve. But we are also making each step size smaller so as to fit them all between the corners of the square. In the above example, this is being done in such a way that the total length remains constant at 2. So there are two sets of numbers that balance each other: one is a sequence that approaches zero, namely, the sizes of individual steps and the other is a sequence that goes to infinity, namely, the total number of steps.

We discuss the details of this calculation in Chapter 9 after introducing the concept of *convergence for sequences of functions*. That calculation also reveals an interesting fact: we can modify the sizes of individual steps so that their total length *does what is expected* by converging to the length $\sqrt{2}$ of the diagonal line! This happens when the sequence that approaches zero (sizes of individual steps) overcomes the effect of the sequence that approaches infinity (the number of steps).

The opposite situation where the growth/decline rates of the competing sequences are *reversed* leads to a rather unsettling result. In this case, the lengths of the staircase curves can grow infinitely large (rather than stay fixed at 2) even as they ultimately fade into the diagonal of the square.

This leads to a paradoxical situation involving areas: suppose that we stop after enough steps have been added to make the total length of the staircase curves reach some large value, say, googol or 10^{100}. The resulting curve now looks like the diagonal line with an enormous number of tiny wiggles. Together with two sides of the square, the long wiggly curve bounds a roughly triangular region whose area is essentially equal to the area of the lower triangle in the square, namely, 1/2. However the *perimeter* of this region is googol+2 which is an enormous number. Can you imagine a plot of land roughly in the shape of a triangle whose area is about 1/2 square kilometer or mile but you will need a googol kilometers (miles) of fencing to surround it? But why stop at googol?

We can go on to zillions, gazillions and larger values. Ultimately, the perimeter gets infinitely large while the area converges to 1/2, the area of the triangle under the diagonal of the square!

This type of oddity is reminiscent of fractals, which are complex shapes generated by iterations. In Chapter 9 we discuss one such fractal called "Koch's snow-flake" that in fact has infinite length but bounds a finite area. But the staircase curves here don't end in a fractal! They flatten out to a simple straight line. In analysis jargon, this anomaly occurs because *length is a property of curves that is not preserved by uniformly converging sequences of functions*.[16]

This length anomaly also occurs with areas, though not areas of regions in a flat plane. But *areas of surfaces that bound finite volumes can be infinite*. We discuss a simple example, a *Tower of Boxes*, in Chapter 6 that has a finite volume but infinite surface area. Rather counter-intuitively, this set of boxes fits inside a single box of both finite volume and finite surface area.

I also discuss a less paradoxical version of this Tower of Boxes that has infinite height (but still finite volume) in Chapter 6 that is in the same category as the famous *Gabriel's Horn* (or *Torricelli's trumpet*), a horn-shaped surface of revolution with finite volume but infinite surface area.[17] See Figure 1.3.

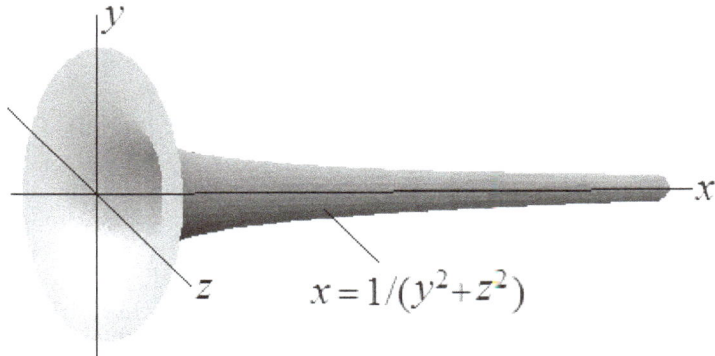

Figure 1.3: Gabrien's Horn (a surface of revolution)

Evalgelista Torricelli (1608-1647) was an Italian physicist and mathematician and a student of Galileo who is best known for inventing the barometer. He also worked on concepts of calculus that had started to attract the attention of mathematicians and scientists of the time, including Galileo, Kepler, Cavalieri, Fermat, Gregory, Wallis and Barrow, all before Newton and Leibniz[18] published their seminal works. Even earlier, Indian mathematicians had been working on infinite series in the 14th century; we touch upon some of the accomplishments of Indian mathematicians later in this book.

Infinite processes need not converge.

---

[16] In Appendix 11.5 the nature of the baffling sudden drop in lengths from 2 to $\sqrt{2}$ is explained in terms of the discontinuity of the length function relative to what is called the uniform metric on a space of functions.

[17] Also see the discussion of improper Riemann integration in Chapter 8 below. My Tower of Boxes with infinite height is a discrete analog that does not require any integration formulas or methods, just infinite series.

[18] Isaac Newton (1643-1727) is the renowned English mathematician and physicist. Gottfried Wilhelm Leibniz (1646-1716) is the equally renowned German mathematician and philosopher.

We have now seen that infinite processes may converge to unexpected limits. But must all infinite processes converge?

In fact, most infinite processes do not converge to a specific outcome. These types of processes can be interesting too. For example, an infinite process may simply cycle through a finite sequence of states over and over again, generating a periodic outcome. Still other processes may lead to nothing particular; they may move away out of bounds like a missile fired into outer space or wander about throughout some region of space like global weather patterns.

Such processes occur frequently in the study of *nonlinear dynamical systems*. Some such systems are known to exhibit chaotic behavior (like the unpredictable climate). Nonlinear dynamics is a fascinating subject that contains frequent encounters with infinity and substantially overlaps analysis and a few other areas of mathematics such as differential equations, difference equations and cellular automata.

For a simple example of a non-converging process, let's go back to the man walking to a store and assume that he uses his mobile phone as follows: He turns on his phone's screen when he is half way to the store to check messages and notifications; then he clicks the screen off when he is 3/4 (or $1 - 1/4$) of the way to the store. He clicks the phone back on when he reaches 7/8 (or $1 - 1/8$) of the way, then off again when he has gone $1 - 1/16$ of the way. He repeats this on-off cycle at points $1 - 1/2^n$ for $n = 1, 2, 3, \cdots$.[19]

Will his phone screen be on or off when he reaches the store?

This scenario is a restatement of the so-called "Thompson's lamp test" proposed in 1954 by the British philosopher J.F. Thompson in his study of "supertasks", i.e. completing an infinite number of tasks. What we have here is an on-off sequence, say, 0 for off and 1 for on:

$$0, 1, 0, 1, 0, 1, 0, 1, \ldots$$

The length of time during which the screen is on for each use shrinks with repeated use but these time intervals do not affect the answer to the above question. This sequence regularly cycles through the two numbers, or states, 0 and 1. It does not converge to any numerical value so the infinite process that it models has no specific or single outcome; at best, we can talk about the *probability* that the phone screen is on or off at the end of the journey (or very near the end). By contrast, Greedy's disappearing fortune was an infinite process that led to a single outcome (zero); likewise, the staircase curves flatten out to a line, again a specific outcome. We discuss non-converging (or divergent) sequences in Chapter 4 and divergent infinite series in Chapter 6.

1.3 Infinity concealed: from areas inside circles to logarithms

Circles need no introduction; but how do we find the area of the region inside one? Logarithms often frighten students when they first encounter them, but since the first appearance of logarithms in 1614 in the work of John Napier (1550-1617), they have been great tools for simplifying scientific and engineering calculations. We now take a look at how infinity figures in these concepts.

---

[19]Trying to imagine this process in real-world terms is futile; when n gets large enough the man will be turning his phone's screen on and off at a rate that is faster than the speed of light, assuming that his fingers and the phone's functions can accommodate nearly that rapid a pace in the first place. But in a mathematical thought experiment all these objections may be cast aside!

What is a regular polygon with infinitely many sides?

Archimedes who is arguably the brightest and most influential mathematician of antiquity, did some calculus (without thinking of it as such) over 2200 years ago to find the area inside a circle (or more precisely, the area of a disk). He used *regular polygons* (equal sides and equal angles) with many sides to approximate the circle from inside (inscribed) and outside (circumscribed). Figure 1.4 illustrates the point that the more sides an inscribed polygon has, the more closely its area (shaded) approximates the area inside the circle. The same is true for circumscribed polygons; see Appendix 11.1.

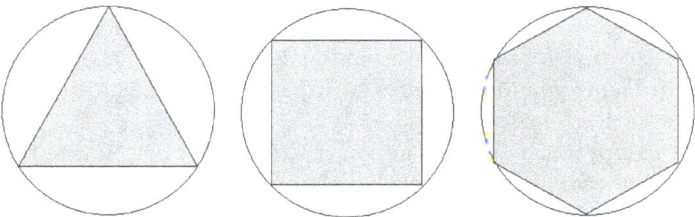

Figure 1.4: Inscribed regular polygons

Now, consider a circle of radius R with an inscribed regular N-gon, namely, a regular polygon with N sides each of length L_N. We see in Figure 1.4 that by choosing a larger N the length L_N of each side gets smaller since the perimeter of the N-gon cannot exceed the circumference of the circle. Ultimately, L_N reduces to zero as N becomes infinitely large, the polygon fades away into the circle. Indeed, it is impossible to imagine a polygon with infinitely many sides without visualizing a circle.

The perimeter of the N-gon is the product NL_N and its area is sum of the areas of all of the triangles shown in Figure 1.5. Notice that the number of triangles is N, same as the number of sides, so calculating the area of the N-gon does not involve infinity.

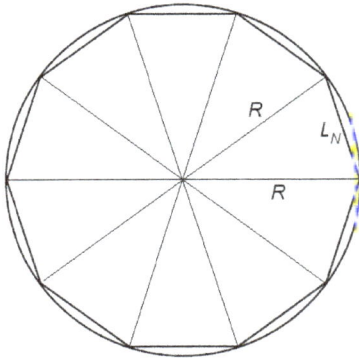

Figure 1.5: The partition of an inscribed polygon into triangles

Since each triangle in Figure 1.5 has the same area, say A_N, the area of the N-gon is NA_N. As the value of N becomes infinitely large, the area of the N-gon approaches the number πR^2, namely, the area of the circle, and NL_N approaches the circle's perimeter $2\pi R$. To see how subtle this is, set $R=1$ and consider how a quantity like NL_N might approach the irrational number 2π (of all numbers) as N becomes infinitely large while L_N approaches zero! Would it not be more plausible for NL_N to approach zero or maybe get infinitely large? These issues become clear after we discuss the concept of *limit* in Chapter 4.

The calculations associated with the Archimedes method are presented in Appendix 11.1 together with an account of Archimedes's original (infinity-subdued) reasoning (called the "method of exhaustion"). It is remarkable that Archimedes worked all this out without having access to trigonometry or such conveniences as the Hindu-Arabic numerals.

An expression like: "L_N reduces to zero as N becomes infinitely large" has a clear enough intuitive content since we can look at the figures and mentally extrapolate or imagine what happens. In more complicated problems where no figures or other mental images are available the meanings of such statements are far less clear.

Archimedes's approach (if not the exact argument that he used) is one of the earliest known examples of *integration*, an idea developed some 1900 years later independently by Newton and Leibniz. As we will see in Chapter 8 the basic idea in integration is using the infinity to bridge the gap between *polygonal paths* and *smooth curves*.

Why do trigonometric functions and logarithms seem mysterious?

When we are presented with a power function like x^3 with an integer power or a sum like x^3+2x, we have no difficulty making sense of these: the first function multiplies a number x by itself three times; the second adds twice x to the first.

These definitions and calculations involve a *finite number of algebraic operations*, like addition and multiplication. But when we come across trigonometric functions like $\sin x$ and $\cos x$ or the logarithm $\log x$ and the like, they seem enigmatic by comparison. How do we calculate the values of these quantities? Of course, a calculator with designated buttons spits out some numbers but where do those numbers come from?

Strange at it may seem to modern students and most other people, trigonometric functions, logarithms and other such quantities were created to help us understand and solve important scientific, engineering and even bureaucratic problems in government and business. Even through the 1970's extensive tables of values of these transcendental functions were supplied with most introductory mathematics textbooks as well as books in science and engineering. Historical accounts of many of these functions are freely available on the Internet nowadays.

Trigonometric functions are defined with the aid of a right triangle in which x is one of the two acute angles (not 90 degrees).[20] If we divide the length of the side in front of the angle x by the length of the hypotenuse (the side in front of the 90 degree angle) then we get the value of $\sin x$.

Notice that this definition is *geometric* in nature, not algebraic. But when we need to calculate the sine of a specific angle, say, $\sin 20°$ we don't grab some manufactured right triangle and measure its sides and angles; instead, we typically pick up a calculating device and punch in a few designated keys. A few digits pop up on the screen that we take to be $\sin 20°$. We rarely think about where these digits come from, let alone how they relate to the geometric definition of $\sin x$. Although we need not be overly concerned with technical details, having some idea about where geometry and algebra meet calculators can help make us feel more comfortable with trigonometric functions.

---

[20] See Appendix 11.2 for reminders on trigonometry, if needed.

There is a similar disconnect with the logarithm. We define $\log x$ as the number y with the property that $10^y = x$. An equality like this routinely occurs in scientific, engineering and other calculations. For example, consider when a rumor "goes viral". Suppose that, of the people who hear about or view the rumor on their devices in the first hour, at least 10 broadcast it. Every hour, each broadcast is further broadcasted by at least 10 more people, so in two hours no fewer than 100 or 10^2 people know about it, in three hours $10^3 = 1,000$ people and so on. How long does it take for at least three million people to hear this rumor?

To answer this question, we find y such that $10^y = 3,000,000$. The value of y that solves this equation is[21]

$$y = \log 3,000,000 = 6 + \log 3 \quad \text{hours}$$

(less than six and a half hours). As with the trigonometric functions, to obtain the value of something like $\log 3$ (approximately 0.477) nowadays we just use a calculating device.

I should emphasize that the answers given by calculators or computers are *rarely* exact when it comes to functions like $\sin x$ or $\log x$. They are imprecise in almost exactly the same sense that the machine results for $\sqrt{2}$ or π are: the exact answers, typically irrational, involve an infinite number of digits. If all that we know of $\sin x$ or $\log x$ is their original definitions and the digital approximations of their values that are generated by machines then it should not come as a surprise that these functions seem so ... *strange*.

Functions like $\sin x$, $\log x$, etc are actually related to the nice and simple power functions but only through infinity. Think of them as irrational functions; a more common name for them is *transcendental functions*. We will see in Chapter 9 that *each of these functions can be expressed as an infinite series of power functions*, or a *power series*. For instance, we see in Chapter 9 that

$$\sin x = x - \frac{x^3}{6} + \frac{x^5}{120} - \frac{x^7}{5040} + \cdots$$

From this infinite series we may calculate the value of $\sin x$ for any value of x. For example, to calculate $\sin 20°$ we first change 20 degrees to the radian measure (because the infinite series above is given in the standard form where x is in radians):

$$20° = 20 \times \frac{\pi}{180} = \frac{\pi}{9}$$

and then insert $x = \pi/9$ radian in the infinite series for $\sin x$ to get the numerical estimate for $\sin 20°$.

As you might imagine, the result has an infinite number of digits since it is an irrational number. But a nice feature of the infinite series is that we may use only a *finite number of fractions* of the infinite series to get a *numerical approximation*. For instance, if we use just the above 4 explicitly listed fractions then we get the approximation

$$\frac{\pi}{9} - \frac{(\pi/9)^3}{6} + \frac{(\pi/9)^5}{120} - \frac{(\pi/9)^7}{5040}$$

With $\pi/9$ being approximately 0.349, the above sum is approximately equal to 0.342 which agrees rather well with what you might get for $\sin 20°$ on a calculator with the "sine" button (in the degree mode, or $\sin(\pi/9)$ in the radian mode).

---

[21] I discuss logrithms and their inverse, the exponential functions in Chapter 8.

Ubiquitous Infinity

The important point about these calculations is that, in principle (given enough time and some patience) we can estimate the value of $\sin x$ without using any calculating devices or other resources. We deal similarly with other transcendental functions such as $\log x$. Other than the hidden infinity, there is really nothing mysterious about these quantities!

We devote Chapter 9 to the topic of infinite sequences and series of functions not only because they lead to amazing results as we have already noted in this chapter but also because their convergence involves an additional hidden occurrence of infinity that goes beyond what we allow for adding infinitely many numbers. This further occurrence of the infinity distinguishes between uniform convergence (usually nice, property preserving) and pointwise convergence (typically weird).

It all comes down to converging infinite sequences...

Let's recall what we have discussed so far in this chapter. We encountered occurrences of infinity in each irrational number and found two ways of dealing with it: we could iterate a recursion such as the divide and average rule for $\sqrt{2}$, or use an infinite series such as the one for π (in Chapter 6 we show that when an infinite series adds to a number s then this number is the limit of a converging sequence of finite sums). Incidentally, the familiar expansion of an irrational number is just one way of writing it as an infinite series; for instance,

$$\pi = 3.14159\ldots = 3 + \frac{1}{10} + \frac{4}{10^2} + \frac{1}{10^3} + \frac{5}{10^4} + \frac{9}{10^5} + \cdots$$

Next, in the story of elves we discussed an infinite process that ends in a specific outcome that is considered the limit of the sequence of deposit box states (even though that outcome was unexpected). After that in the case of staircase paths we discovered that their ultimate length is the limit of a sequence of numbers (the lengths of the paths) even though lengths did not match in the limit.

We then encountered the problem of finding the area of a disk using the areas of a sequence of polygons. And after that we explained transcendental functions like $\sin x$ and $\log x$ using infinite series of power functions, which as we discover in Chapter 9 are limits of sequences of functions.

These examples show that various important concepts in analysis can be stated in terms of converging sequences. In the chapters ahead we discover that we can actually define all of the basic concepts like continuity, derivatives, integrals and infinite series using converging sequences. I use sequences throughout this book to define and study concepts that involve infinity.

The most compelling argument for using sequences is the fact that they are *natural tools of approximation*. In practice we rarely work with actual irrational numbers in scientific measurements or calculations, nor do we usually obtain exact solutions to differential equations modeling natural phenomena because familiar formulas or finite descriptions do not exist for them. We use approximations obtained on digital computers; by using enough terms or entries of the sequences that converge to the anticipated numbers or functions we can get an approximation that is as accurate as is *practically* necessary for the problem at hand.

The idea that irrational numbers can be represented as (the limits of) sequences of rational numbers also forms the basis of Cantor's construction of real numbers. We discuss this construction in Chapter 5 where we identify irrational numbers with sequences of rational numbers that converge to them. Much of the technical detail has to do with the fact that each irrational number is the limit of infinitely many sequences of rational numbers but the basic idea is quite simple: *identify an irrational number with an infinite sequence of rational numbers.* This is consistent with viewing irrational numbers in terms of infinite non-repeating decimal expansions and also the clearest indication of how the infinity is hidden in each irrational number!

Occurrences of infinity are not all handled using sequences. A more fundamental framework for dealing with the infinite is set theory; we briefly discuss infinity in this context in Chapter 3.

1.4 ∞, 0, and nothingness in analysis

What is ∞?

Early in the history of calculus, infinity was often treated intuitively in calculations and derivations. The English mathematician and clergyman John Wallis (1616-1703) introduced the enduring symbol ∞ and its reciprocal $1/\infty$ in a 1656 paper where he refers to the latter as "infinitely little".

$1/\infty$ is often used as a symbol for *infinitesimal* quantities, which are not zero but are too small to measure in any practical sense. Leibniz occasionally treated the infinitesimal as a quantity that could be omitted from calculations if it was multiplied by another infinitesimal as if the product were 0 because he thought of these combinations as negligible quantities.[22] In particular, the square of an infinitesimal quantity is zero even though the infinitesimal itself is not.

For a modern analogy consider a simple calculator that displays only 8 digits and cuts all digits beyond 8. Now suppose that we multiply the small number 0.00001 by itself (i.e. square it). The machine returns 0 rather than 0.0000000001 since the digit 1 gets cut off. In this example, the number 0.00001 is not so small to be discarded by our calculator but its square is.

Calculations where infinitesimals are discarded basically work in the same way. Consider the fraction
$$\frac{(x+dx)^2 - x^2}{dx}$$
where x is a variable and dx is a number that can be arbitrarily small but not zero.[23] This type of fraction, called a "difference quotient" is encountered routinely in our discussion of derivatives in Chapter 7.

Let's simplify this difference quotient:
$$\frac{(x+dx)^2 - x^2}{dx} = \frac{x^2 + 2xdx + (dx)^2 - x^2}{dx} = \frac{2xdx + (dx)^2}{dx} \qquad (1.10)$$

If we choose a small value, say, $dx = 0.00001$ and use the primitive calculator then $(dx)^2 = 0$ is returned so that
$$\frac{(x+dx)^2 - x^2}{dx} = \frac{2xdx}{dx} = 2x \qquad (1.11)$$

Notice that once $(dx)^2$ is dropped the dx cancels out and its actual value does not matter; in particular, it can be arbitrarily small and the result $2x$ remains valid no matter how small the value of dx is. This idea extends far beyond this particular fraction and makes the concept of derivative possible. But there is another way of approaching the issue of derivative that I discuss below.

---

[22] Newton also referred to infinitesimal quantities, which he termed "evanescent". But his notation was more obscure than Leibniz's and the idea is less apparent in his work.

[23] Here I use dx, Leibniz's notation for an *infinitesimal change* in x.

Ubiquitous Infinity

The idea that infinity could be an actual quantity was hard enough to swallow but when infinitesimals appeared they seemed impossible to digest. For instance, if $1/\infty$ were a positive real number then it would have to be smaller than every positive number. It would be smaller than 0.1 because ∞ is larger than 10; it would be smaller than 0.001 because ∞ is larger than 1,000. And so on. The only meaningful value that $1/\infty$ could sensibly have would be 0. But this meant that $1/\infty$ couldn't symbolize the *nonzero* infinitesimals!

These issues understandably led to confusion and skepticism. The Irish philosopher and theologian George Berkeley (1685-1753) criticized this state of affairs in his 1734 book "The Analyst", referring to infinitesimals as "the ghosts of departed quantities" because of the apparently arbitrary omission of quantities like $(dx)^2$ above by the mathematicians of his time from their calculations. Berkeley writes: "If we lift the veil and look underneath ... we shall discover much emptiness, darkness and confusion ... direct impossibilities and contradictions." He concluded that if mathematicians could accept such things as the omission of $(dx)^2$ above then they should not be hesitant about accepting divinity and religion because what they did was based more on faith than on logical arguments, Euclid-style.

Undaunted by the criticism and the confusion, prominent mathematicians including Euler, d'Alembert and the Bernoullis, followed Newton and Leibniz to use the infinitesimal all the way through the 18th century. Calculus was new and exciting then and it worked just too well in solving science and engineering problems for the early practitioners to worry about its foundations. But after the rush passed and mathematicians started to look inward they managed to clear things up by the end of the 19th century. By then infinitesimals had been abandoned and calculus was finally given a logically consistent foundation using the concept of limit.

The new approach involved finding the limit as dx *approaches* 0 in the following sense: rather than dropping $(dx)^2$ in (1.10) notice that there is a dx that is common to both terms in the numerator of the fraction. Let's factor, or pull this out to get:

$$\frac{(x+dx)^2 - x^2}{dx} = \frac{dx(2x+dx)}{dx}$$

It is important that $dx \neq 0$ in this discussion so we may cancel the left-hand dx in the numerator with the one in the denominator to get

$$\frac{(x+dx)^2 - x^2}{dx} = 2x + dx$$

Noteworthy here is that we managed to remove the dx in the denominator through routine algebraic manipulations (no infinity or infinitesimals involved). The last quantity $2x + dx$ is not 0 but it can be as close to 0 as we want by choosing the value of dx small enough.

Is this a good answer? Though not as clean looking as the earlier calculation, saying that we get $2x$ because dx is essentially 0 is no worse in practice, and less problematic conceptually than setting its square equal to 0 without justification.

The early practitioners of calculus were routinely omitting quantities from their calculations or derivations of formulas by appealing to intuition only, much like scientists (still) derive formulas or terminate calculations by omitting what they consider *negligible* within their respective contexts. However, there are no semantic contexts in mathematics so every step must be *logically* justified.

Defining the concepts of limit and convergence so as to apply without contradictions to all fractions and functions requires careful development and we devote a generous amount of space to developing the most important ideas. In analysis as it is currently practiced (and which we follow in this book) *magnitude* means a positive real number (like 1, 3/2, $\sqrt{2}$, π, googol, etc) that denotes a quantity like distance, volume, wave amplitude, etc or the size of something like a number or a vector. Other types of numbers (negative, imaginary, etc) are not valid indicators of magnitude. The number 0 is also not a valid indicator of magnitude but it plays a special role that we discuss below.

The symbol ∞ does not indicate a mathematical object in analysis but when paired with an *arrow* as in $n \to \infty$ or $f(x) \to \infty$, these symbols together have a precise meaning as "the quantity (or magnitude) n or $f(x)$ grows arbitrarily large, or is unbounded". We use ∞ quite often in this capacity in Chapter 4 and beyond.

Is 0 the opposite of ∞? Does it indicate nothingness?

When we go to buy groceries and the store is out of peaches, the owner doesn't say "I have zero peaches left" when we ask for them; instead he says "I have no peaches left". To say that there are 0 things is to say that there are no things, or nothing. Indeed, for most daily activities zero need not be invoked and for the longest time in human history, zero was ignored completely.

The mathematics of ancient Egypt, acclaimed in the Mediterranean region and adopted by the Greeks was devoid of zero. Mathematics of the ancient Greece expanded greatly on that of the Egyptians but it too avoided zero. From both the Egyptian and Greek points of view, zero meant nothing and was therefore, not worthy of consideration. The Greeks were also uncomfortable with the connection between zero and the infinite; we discuss this connection shortly.

One of the consequences of this avoidance of zero was that the Egyptian calendar, and later, the Greek and Roman calendars did not contain a zeroth year. Not long ago in the modern era, this was the cause of some confusion about whether the start of the 21th century should be the year 2000 or 2001. It is also the reason why the years in the 21st century start with the digits 20 instead of 21, the years in the 20th century start with 19, etc. All this because the first one hundred years after the birth of Christ were called the first century rather than the zeroth.[24]

The avoidance of zero and the associated nothingness by the Greek civilization reached its zenith with Aristotle. The adoption of his doctrine by the Christian church was partly responsible for the stagnation of western science and mathematics until the 17th century. Zero had already been introduced as a number to the West some 400 years earlier with the work of the Italian mathematician Leonardo of Pisa, better known as Fibonacci.

Fibonacci traveled to North Africa where he learned mathematics from the Muslims. His book *Liber Abaci* ("Book of Calculation") was published in 1202 and was among the earliest books to introduce the Hindu-Arabic numerals (including zero) to the Western world. The concept of zero as a number, as well as negative numbers had been known centuries earlier to Hindu, Chinese and other mathematicians in Asia but they were novelties in the West in Fibonacci's time.

---

[24] It is worth mentioning here that the Mayan calendar did not suffer from this mismatch. They always started counting with zero: the zeroth day of a month, the zeroth month of the year and so on. They even had a symbol for zero that they also used as a place holder. In this respect they were like the ancient Babylonian, Sumerian and other civilizations of ancient Mesopotamia who also devised a symbol for zero to serve as a place holder. These civilizations did not consider 0 to be a number in the same sense as 1, 2, 3, etc because 0 does not represent a magnitude for physical objects or phenomena. That conceptual leap was taken by the Indians centuries later, as discovered in the work of Brahmagupta, a seventh century mathematician and astronomer who also stated some rules for calculating with 0.

The most influential mathematical work at the time was the book *On Calculation with the Hindu Numerals* by al-Khwarizmi the author of the aforementioned book *al-jabr* (algebra). The book on Hindu numerals made him famous in the West where he became known by the Latin version of his name, *Algorithmi*, later the origin of the word "algorithm" (his other work on algebra was more significant within mathematics proper). It is also in part because of this "best seller" that the Hindu numerals spread in the West and became known as Arabic numerals (with some modifications).

Until the work of Fibonacci, as well as other sources spreading the new ideas from Islamic scholars[25] counting in the West was done mostly using the cumbersome Roman numerals and before that, the Greek and Egyptian ones. Calculations were performed using devices like the abacus. Merchants and bankers quickly adopted the superior Hindu-Arabic numerals, but the rest of the West, including mathematicians, reacted more slowly. When the latter finally came around and adopted the new numbers (along with the mastering of algebra and its union with geometry in Descartes's coordinate system) the road to calculus was finally discovered.

Far from nothing, 0 has a precise meaning in modern mathematics as a *number*. Of course, it is a special number, being the only one with the properties $0 + a = a$ and $0 \times a = 0$ for all numbers a. Defining 0 in this way lets us calculate with numbers as we normally do, makes it possible to distinguish between numbers like 12 and 102 by serving as a placeholder, and lets us define things like the origins of coordinate systems and other important concepts like the "null subspace" of a linear operator in a vector space.

But how is 0 related to infinity?

The distinction between the *number* 0 (a precise algebraic concept) and the *magnitude* zero (a loose representation of nothingness) is a subtle one.

The special algebraic properties of 0 make division by 0 undefined. The reason for this is simple: every number r except zero has a reciprocal $1/r$ which is the only number with the property that $r \times (1/r) = 1$.[26] If 0 had a reciprocal $1/0$ in the same vein then we would have $0 \times (1/0) = 1$. But remembering the property that multiplication by 0 kills every number, it follows that $0 \times (1/0) = 0$. This leads to the contradiction that $0 \times (1/0)$ is both 1 and 0; that is, $1 = 0$. To avoid this contradiction we must leave $1/0$ undefined and let 0 be the only number that has no reciprocal.

But there is more to 0 than its algebraic characteristics. If x is a number that is very close to 0 then $1/x$ has a very large magnitude: for instance, if $x = 0.000001 = 10^{-6}$ then $1/x = 10^6 = 1,000,000$. Since we can get as close to 0 as we wish, the reciprocals can be arbitrarily large in magnitude. We conclude that in terms of magnitude *approaching 0 is tantamount to approaching ∞ via reciprocals*.

This observation may suggest that $1/0 = \infty$. This can be a helpful abbreviation for the observation that I highlighted in the last paragraph but it does not make mathematical sense; as a definition of either ∞ or $1/0$ it is circular since neither side of the equality is a well-defined concept.

---

[25]In its golden age, Islamic civilization brought together the mathematical achievements of the ancient Greece with those of Eastern mathematicians, including those from India and China. Muslim mathematicians studied and translated texts in mathematics and astronomy from both the East and the West and added significant ideas of their own in the process, such as what we see in al-Khwarizmi's work as well as in works by other famous Muslim mathematicians and scientists, like Ibn al-Haytham or Alhazen (965-1040), Abu Reyhan Biruni (973-1050), Ibn Sina or Avicenna (980-1037), Omar Khayyam (1048-1131) and Nasir al-Din Tusi (1201-1274), among others. Khayyam, who is more widely known for his Rubaiyat poetry in modern times, was actually better known as a mathematician in his own time. He made significant contributions in geometry, analytic geometry and the theory of cubic equations.

[26]This also makes it possible to define division more generally using multiplication: to divide a number s by r we multiply $s \times (1/r)$ and abbreviate this as s/r.

In science, engineering and life in general, when a magnitude is too close to 0 to be measured or to make a practical difference then it is often considered to *be* 0. This practical way of inflating 0 to include all numbers that are very close to it has to do with *the magnitude* being "practically 0" rather than *the number* 0.

The idea of a "negligible magnitude" is remarkably vague and strongly context dependent. What may seem to the naked eye an insignificant point of light in the night sky turns into a galaxy containing billions of sun-like stars when viewed through a telescope. A flea or a bed bug, barely visible to the naked eye, turn into scary monsters when viewed under a microscope. A sharp, beautiful digital photo on a monitor depicting a rich tapestry of colors and smooth shapes turns into a collection of monochrome square pixels when we zoom sufficiently on some spot.

In mathematics, iterations serve as telescopes or microscopes. In an often modest series of steps they can magnify the tiniest quantities into gargantuan sizes or reduce an immensity to an insignificant spec. To illustrate the way that iterations work consider a simple experiment. Suppose that we want to filter out all contaminants from a jar of drinking water. If we do a perfect job, which means that 0 contaminant is left in the jar, then the water will be pure and remains that way as long as it is sealed from the environment. In this case the amount, or magnitude of the contaminant is the actual number 0.

But in real life doing a perfect job is rare and usually impossible. So suppose that we miss a clump of fungus or bacteria that is too tiny to be caught by our filter, say, a clump with a diameter of about 0.000000001 (a billionth) of a centimeter (less than half a billionth of an inch). If this contaminant doubles its size every minute then in 1 minute the clump's diameter is 0.000000002; in 2 minutes the diameter grows to $0.000000002 \times 2 = 0.000000004$ and so on–still small but increasing in size. In a quarter of an hour, the size is

$$0.000000001 \times 2^{15} = 0.000000001 \times 32768 = 0.00032768$$

which may have grown large enough to be filtered but still too small to be seen. In just half an hour the diameter grows to

$$0.000000001 \times 2^{30} = 0.000000001 \times 1073741824 = 1.073741824$$

which, at about one centimeter (just under half an inch) is certainly noticeable as a greenish blob. *A mere 30 iterations of the doubling process has blown up the diameter 10 billion fold*, from the level of *practically 0* to *definitely not 0*. We see that in this sense *nearly zero* is as far from the actual 0 as gazillions are from infinity.

This aspect of iteration also occurs routinely in the form of error propagation when doing scientific and engineering calculations using computers which as I mentioned earlier, round answers off due to finite storage and memory limitations. A small error introduced in some process that is repeated will grow large enough to impede or even stop the process.

The mathematical concepts 0 and ∞ are not quite the same as the philosophical opposites *nothing* and *everything*. The number 0 is as precisely defined a concept in mathematics as any other number but in analysis ∞ is neither a number nor a magnitude. As we saw earlier in this chapter, infinity has been the subject of much debate and controversy in mathematics and only recently (since the late 19th century) it has gained a precise meaning via set theory, where incidentally a concept of nothingness was also developed that was different from zero.

The concept that is associated with nothingness within mathematics is the *empty set*; that is, a set containing no elements. The empty set itself is not nothing. It serves the purpose of dealing with nonexistence in a logical way; for instance, to say that the equation $x = x + 1$ has no solutions means that "the set of all solutions of this equation is the empty set".

As we see in Chapter 2 below, sets turn out to be rather complicated things to define in a consistent way, perhaps because at some level even mathematicians have to deal with basic philosophical issues. The burden of doing so has fallen largely on the shoulders of set theorists and logicians. Throughout the 20th century, they managed to come up with a few basic axioms to develop, in a logically consistent way, the present day mathematics, a superstructure that works in a seemingly flawless and amazingly effective way in science and engineering.

Mathematicians discovered that logic alone is not enough[27] and to get things off the ground they must use axioms, which are essentially mathematical hypotheses (or agreements or articles of faith) that lead via the usual rules of deduction to consistent results. Consistency (that is, non-occurrence of contradictions) is paramount and a non-negotiable requirement for any axiom scheme for mathematics. If our legal code could be so designed then only a single grand session of Congress would be enough to write and pass such laws without the need to ever meet again!

Infinitesimals resurrected! But not as Leibniz imagined them.

The infinitesimal had been largely abandoned as a mathematical concept by the end of the 19th century, but it was not forgotten. Some mathematicians continued to believe that it was possible to make both the infinite and the infinitesimal magnitudes mathematically meaningful concepts. This turned out to be a lot more challenging than the early pioneers of calculus had thought. It was not until the mid-20th century, after set theory and mathematical logic had sufficiently evolved, that these concepts could be resurrected in logically consistent ways.

Two major ideas took hold. The first of these defines both ∞ and $1/\infty$ as *new kinds of numbers*. Mathematicians had defined various different types of new numbers in earlier times that turned out to be very useful, such as the negative numbers and 0. They also invented numbers beyond the real numbers. For example, there are no real numbers that solve the equation $x^2 + 1 = 0$ and for centuries mathematicians had been perfectly content with that. But they later decided that this and similar equations can have solutions that are new kinds of numbers after the work of Cardano (1501-1576) and his contemporaries on the roots of cubic polynomials presented a greater urgency for them. Their work showed that certain combinations involving square roots of negative numbers came out to be ordinary real numbers. So a new class of numbers, namely, *the complex numbers* were defined to accommodate these types of algebraic combinations and equations. The new numbers included all of the ordinary real numbers.

With this in mind, what if we could enlarge the set of real numbers to include a class of "infinitely little" numbers, or infinitesimals that are not zero but are smaller than all positive numbers in some sense?

We need to make sense of what is meant by the statement "smaller than all positive numbers without being 0". Here we are speaking of an ordering of numbers. Finding a set of numbers that

---

[27] One of the major discoveries of the 20th century mathematical logic was Godel's incompleteness theorem, a result that convinced mathematicians that their craft could not prove everything. The incompleteness theorem states that mathematics cannot be both consistent and complete at the same time. Mathematicians decided that admitting that they cannot prove everything (incompleteness), though a bitter pill to swallow, was a better alternative to madness (inconsistency).

enlarged the set of real numbers and preserved their ordering was not easy.

There is a feature of the usual way in which the real numbers are ordered that prevents the occurrence of infinity and infinitesimals: the so-called *Archimedean property* that configures the integers and their reciprocals as the skeleton on which the real numbers are built; we discuss these matters in Chapter 5. If we do not require that this property hold then it is indeed possible to define both infinite and infinitesimal *numbers*. These numbers whose entire collection can be ordered in a non-Archimedean sense and contains all the real numbers are called the *hyperreal numbers*. Hyperreal numbers were introduced by Abraham Robinson in the 1960's who worked out the details of their mathematical and logical structure in the area of mathematics that has come to be known as *non-standard analysis*. The methods that Robinson used to create the hyperreal numbers belong to the branch of mathematical logic known as *model theory*. A substantial amount of work has since been done in non-standard analysis and Robinson's own 1966 book "non-standard Analysis" is a good exposition to this subject. Although not needed for the development of calculus, non-standard analysis has led to a large body of research and the discovery of many interesting mathematical results.

Another approach to infinitesimals that was also developed in the latter part of the 20th century did not involve infinite numbers and came to be known as *smooth infinitesimal analysis*. Like non-standard analysis, this approach required the creation of a new logical structure through model theory and two highly active areas of mathematics known as *category theory* and *differential geometry*. The theory grew out of the work of mathematicians F.W. Lawvere and Anders Kock in the 1970's.

Smooth infinitesimal analysis is, in a sense, more drastic than non-standard analysis. The latter basically enlarges the set of real numbers by adding both infinite and infinitesimal numbers (as reciprocals or algebraic inverses) but it uses mostly standard mathematical logic. Smooth infinitesimal analysis on the other hand requires a rewrite of basic logic since we must give up the "law of excluded middle" which says that *every statement (or logical proposition) is either true or false* (there is no middle ground). While this may seem like a natural step towards general thinking (there are always shades of grey) it has serious consequences for mathematical thinking.

The most significant consequence of denying the law of excluded middle is that *proof by contradiction* is no longer valid.[28] But it so happens that many important results of analysis (and of mathematics beyond analysis) are proved by contradiction! Furthermore, the analytical results that require proof by contradiction are not esoteric abstractions but pillars on which a substantial amount of applied mathematics is based; a familiar example that we discuss later is proving that $\sqrt{2}$ is irrational. Another interesting example is Euclid's proof that there are infinitely many prime numbers that we discuss in Chapter 3.

The modification of logic in smooth infinitesimal analysis is essential for its definition of infinitesimals. We normally think that if a number is neither negative nor positive then it is 0. But since we cannot prove this conclusion without the law of excluded middle, there is room for the infinitesimals to exist as *nonzero numbers that are indistinguishable from 0*. This statement is the best that can be proved with the logic of smooth infinitesimal analysis because there is no law of excluded middle to draw sharp distinctions between different types of numbers.

---

[28] As we see in Chapter 2, to prove a statement or hypothesis by contradiction we assume that it is false. Then using standard logical deduction we show that this assumption of falsehood leads to a contradiction like $1 = 0$ or more generally, A and not A (recall our earlier argument in this section that 0 cannot have a reciprocal). With the law of excluded middle in place, we may avoid the contradiction by accepting the truth of the statement or hypothesis. We use proof by contradiction quite often in this book.

The principles of smooth infinitesimal analysis make the number line, or more generally, the continuum *indecomposable*; in particular, it is impossible to divide the continuum into a collection of (infinitely many) points. And I should also mention the fact that the theory allows for a nonzero infinitesimal c to have a zero square: $c^2 = 0$. This might have interested Leibniz and his contemporaries although they would have likely had problems working without the law of excluded middle.

Non-standard analysis and smooth infinitesimal analysis have successfully resurrected the concept of infinitesimal in different ways and created infinite and infinitesimal numbers while avoiding the logical pitfalls that the pioneers of calculus ran into. A great deal of work has already been done in these fields and research continues today, focused mostly on mathematical issues. However, from both methodological and practical points of view these new approaches introduce substantial logical and mathematical extras (denying the law of excluded middle, having to deal with non-Archimedean fields, etc) that make them rather abstract and technical to work with for most people, including most mathematicians. They belong primarily to the domain of pure mathematics.

The new infinitesimals are not the vague, intuitive concepts that Leibniz and other pioneers of calculus had thought of; their infinitesimals were vague, intuitive notions similar in nature to the negligible but finite quantities that scientists routinely dispose of in their derivations and calculations. And like the scientists, the early practitioners relied on the practical success of their results in solving scientific and engineering problems to validate their methods. By the end of the 19th century the standard development of analysis had resolved the logical difficulties and done so without introducing infinite or infinitesimal magnitudes. That is the road that we travel on in this book.

Chapter 2

Sets, Functions and Logic: *the building blocks*

This chapter and Chapters 3 and 4 supply a limited amount of background or reference material that may be helpful in understanding the rest of the book. You can read this material carefully in a linear fashion, either because you are rusty or unfamiliar with it or for maximum pickup, or alternatively, decide to forge ahead to later chapters. In the latter case, it is enough to scan through the material in Chapters 2-4 just so you know what to return to for reference, if needed.

Doing the exercises is optional here and elsewhere in this book; they do help to grasp ideas more firmly and some of them contain related ideas that are not discussed in the main text, but you can learn a lot without them too!

2.1 Sets and relations

The most basic definition of a *set* is just what comes to mind: a collection of objects. The set of all people working in an office, the set of all office furniture, the set of all water bottles in the office are all valid examples of sets. The objects that constitute a set are called its *elements* (or *members*). Sets that are too large to list all their elements explicitly are described by a shared property; for example, the set of all New Yorkers or the set of all stars in the Milky Way galaxy. The *set notation* often uses braces; for example, in the aforementioned office we may have the following sets of workers and water bottles:

$$W = \{\text{Amy, Bob, Cathy, David}\},$$
$$B = \{\text{Amy's bottle, David's bottle, spare bottle}\}$$

assuming that Bob drinks soda and Cathy drinks tea. For large sets, or sets whose elements are not fixed, the notation reflects the shared property; for example,

$$N = \{P : P \text{ is a New Yorker}\}, \qquad M = \{S : S \text{ is a star in the Milky Way}\}$$

In the above set notation, read N as: *the set of all P such that P is a New Yorker*; M is read similarly.

Membership and equality

Being an element of a set is usually indicated by the symbol $\in$. So if A is a set and a is an element of A then we write $a \in A$ and say that *a is a member of A* or that *a belongs to A*. For instance,

$$\text{Bob} \in W, \quad \text{Amy's bottle} \in B, \quad \text{sun} \in M$$

which we read as: Bob is an element of W and so on. We may list the elements of a set in any order; rearranging the elements of a set does not change it. Thus

$$W = \{\text{Amy, Cathy, Bob, David}\}$$

is exactly the same set of four people that I listed previously. Generally, we do not repeat the elements in set notation, so

$$W = \{\text{Bob, Bob, Amy, Cathy, David}\}$$

is just an awkward way of writing W. What if David leaves and is replaced by another Bob? In this case we must distinguish between the two Bobs in W. For example, we could use their nicknames, or middle names or just write OBob for the original Bob and NBob for the new Bob. The new W is then

$$W = \{\text{OBob, NBob, Amy, Cathy}\}$$

We can now say what we mean by two sets being equal.

> **Equal sets.** Two sets A and B are equal if every member of A is a member of B and vice versa. More precisely,
>
> $$A = B \quad \text{means:} \quad \text{for every } x, \text{ if } x \in A \text{ then } x \in B \text{ and conversely.}$$

Cardinality: how many elements are there?

The number of elements in a (finite) set is the *cardinality* of the set; we denote it by the symbol $\#$.[1] Thus, $\#W = 4$ and $\#B = 3$. Notice that there are no precise values for the numbers of elements of N or M at any given time. If time *is* considered in the definitions of N or M, say,

$$N(t) = \{P : P \text{ is a New York resident on January 1 of year } t\}$$

then from year to year $\#N(t)$ is a specific number at the start of each year but it changes from year to year.

In working with sets, it is also necessary to consider a set that contains no elements: { }. This set being unique and special, has been given a name and a symbol:

> The unique set containing no elements is the *empty set*. It is denoted by $\varnothing$. Note that $\#\varnothing = 0$.

In mathematics the empty set is also used to indicate non-existence. For example, the equation $x^2 + 1 = 0$ doesn't have any solutions that are real numbers so we say that *the set of all real solutions*

---

[1] Cardinality or cardinal number of a set is an abstract concept whose general and precise definition can be found in texts on axiomatic set theory. Other symbols that are also used for cardinality include the bar notation, as in $|W| = 4$ and the Hebrew aleph $\aleph$ as in $\aleph(W) = 4$.

Sets, Functions and Logic

of this equation is empty. On the other hand, the same equation has two imaginary (or complex) solutions $\pm i$ where $i = \sqrt{-1}$ so we say that the set of all solutions of this equation is $\{-i, i\}$.

Subsets.

Any collection of elements of a given set S is a *subset* of S. In particular, S is considered a subset of itself and the empty set is trivially contained in every set. These two exceptional subsets are called *improper*. Every other subset is *proper*.

We write $A \subset S$ if A is a subset of S whether proper or improper. For example, $\mathbb{N} \subset \mathbb{Z}$ and $\mathbb{N}$ is obviously a proper subset of $\mathbb{Z}$. The subset relation $\subset$ is also called the *containment relation* or alternatively, *set inclusion*.

Set inclusion lets us express the equality of sets as *double containment*:

$$A = B \text{ means } A \subset B \text{ and } B \subset A$$

Double containment turns out to be a convenient way of proving statements about sets; we discuss a couple of examples later on.

Note that if a set is finite then the total number of its subsets is also finite. For example, the set W of workers above has the following subsets (proper and improper):

$\varnothing$, {Bob}, {Amy}, {Cathy}, {David}, {Bob, Amy}, {Bob, Cathy}, {Bob, David},
{Amy, Cathy}, {Amy, David}, {Cathy, David}, {Bob, Amy, Cathy},
{Bob, Amy, David}, {Bob, Cathy, David}, {Amy, Cathy, David}, W

a total of 16, or 2^4.

We discuss the subsets of infinite sets in the next chapter.

Exercise 1 *List all possible subsets of the set of water bottles B. How many subsets does B have?*

Exercise 2 *Let's generalize: suppose that a set S has n elements. How many subsets (proper and improper) does S have? How do you explain your answer?*

Infinite sets.

Sets do not have to be finite; for example, the set of all whole, or *natural numbers*

$$\mathbb{N} = \{1, 2, 3, \ldots\}$$

is an infinite set (it is not uncommon to include 0 in $\mathbb{N}$). The open-ended *triple-dots notation* in a mathematical context such as the definition of a set, or in an equation usually means "so on, ad infinitum".

Here is one way to define infinite sets:[2]

---
[2] Another definition is given in the next chapter.

> An *infinite set* is *a set that is not contained in any finite set.*

We can be sure that such sets exist because we have already identified one: $\mathbb{N}$.[3] Similarly, the set of *all integers* $\mathbb{Z}$ (in German *zahlen* means numbers) is infinite and consists of all natural numbers together with their negatives and zero:

$$\mathbb{Z} = \{\ldots, -2, -1, 0, 1, 2, \ldots\}$$

In set theory it is not so clear that infinite sets must exist because sets like $\mathbb{N}$ and $\mathbb{Z}$ are not primitive concepts and must be constructed; so the existence of infinite sets is guaranteed axiomatically.

The cardinality of $\mathbb{N}$ is usually denoted by the first letter of the Hebrew alphabet:

$$\#\mathbb{N} = \aleph_0$$

which is *aleph null*.[4] Unlike the symbol ∞ the number $\aleph_0$ has a precise meaning; using a simple argument of Cantor's, we will see later that there is a vast hierarchy of infinities in mathematics in which $\aleph_0$ happens to be the smallest possible. We will also see later that $\#\mathbb{Z} = \aleph_0$ which may seem surprising since $\mathbb{Z}$ seems to have twice as many elements as $\mathbb{N}$ (but then, what is "twice infinity"?)

Set operations.

Four *set operations* are of fundamental importance as they apply to all sets. To avoid confusion, we apply these operations to subsets of a fixed *universal set* U.[5] These are (a) *intersection* (or *meet*) of sets, (b) *union* (or *join*) of sets, (c) the *complement* of a subset and (d) *product* of sets. Let's see what these concepts mean.

> For each pair of sets A, B in U (subsets of U) their *intersection* $A \cap B$ is the set of all elements (of U) that are in both A and B.

For example, If A is the set of all school boys in New York City and B is the set of all fourth graders in NYC then $A \cap B$ is the set of all NYC school boys in fourth grade. Here, depending on the context of interest, U may be the set of all elementary school students in NYC, or the set of all students in NYC, or the set of all elementary school students in North America, or even the set of all human beings.

The intersection operation is visually illustrated in the right hand panel of Figure 2.1 using a *Venn diagram*.

> For each pair of sets A, B in U (subsets of U) their *union* $A \cup B$ is the set of all elements (of U) that are in A or in B.

Here the "or" is inclusive, meaning "or both". For example, if A is the set of all male human beings and B is the set of all females then $A \cup B$ is the set of all human beings (of specified gender).

---

[3] We have not defined what $\mathbb{N}$ is technically, a nontrivial matter in set theory where it is defined in different but logically equivalent ways. One definition is based on Peano's axioms, named after the Italian mathematician Giuseppe Peano (1858-1932).

[4] The reason for using aleph is likely because it was introduced by Cantor, who is widely acknowledged to be the founder of modern set theory.

[5] This is no loss of generality. If we have any collection of sets of interest then the set of all elements in all the sets of the collection is the universal set U and each set in the collection is a subset of U.

Sets, Functions and Logic

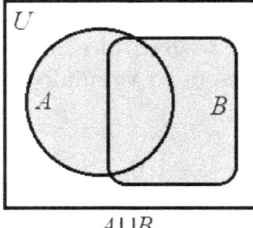
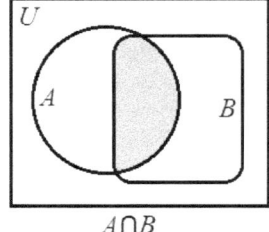

Figure 2.1: Venn diagrams for union (left) and intersection

Here, depending on the context of interest, U may be the set of all human beings, or the set of all primates, or the set of all mammals or even the set of all living things on Earth. See the left panel of Figure 2.1 for a visual illustration of the union operation using a Venn diagram.

> The *complement of A in U* is the set A^c of all elements (of U) that are *not* in A. In symbols:
> $$A^c = \{x : x \notin A\}$$

For instance, the complement of the set of all even numbers in $\mathbb{N}$ is the set of all odd numbers

$$\{1, 3, 5, \cdots\} = \{2n - 1 : n \in \mathbb{N}\}$$

Note that $(A^c)^c = A$ since an element of U that is not in A^c has to be in A.

It is sometimes necessary to consider what is in a set A but not in another set B, that is, $A \cap B^c$; this is the *relative complement* of B in A and it is often written as $A \backslash B$. In words, this set is what is left of A when we punch B out; it is the ordinary complement B^c if $A = U$ so that $B^c \subset A$.

The relative complement is useful when the underlying set is unclear. For instance, the complement of all even numbers in $\mathbb{Z}$ is different from that in $\mathbb{N}$.

Exercise 3 *Consider the following subsets of $\mathbb{Z}$:*

$$A = \{2k : k \in \mathbb{Z}\} = \{\ldots, -6, -4, -2, 0, 2, 4, 6, \ldots\}$$
$$B = \{k \in \mathbb{Z} : -9 \leq k < 9\} = \{-9, -8, \ldots, 7, 8\}$$

Find each of the following sets:

$$A \cap B, \quad A^c = \mathbb{Z} \backslash A, \quad \mathbb{N} \backslash A, \quad B \cap \mathbb{N},$$
$$B \cup \mathbb{N}, \quad (B \cup \mathbb{N})^c, \quad B^c \cap \mathbb{N}^c$$

Exercise 4 *Write the set B in Exercise 3 as the intersection of two infinite subsets of $\mathbb{Z}$.*

Exercise 5 *What is the relative complement $A \backslash B$ if $A \subset B$? What is the relative complement if A and B do not intersect?*

Basic logical relations hold between the intersection and the union of two sets through their complements. The following result connects intersections, unions and complements and is named after the English mathematician Augustus De Morgan (1806-1871).

> **De Morgan laws:** *The following equalities are true for all sets A, B in U (subsets of U)*
> $$(A \cup B)^c = A^c \cap B^c \quad \text{and} \quad (A \cap B)^c = A^c \cup B^c \qquad (2.1)$$

These equalities are visually illustrated by Venn diagrams in Figure 2.2; the shaded regions in each panel show the equal sets in (2.1).

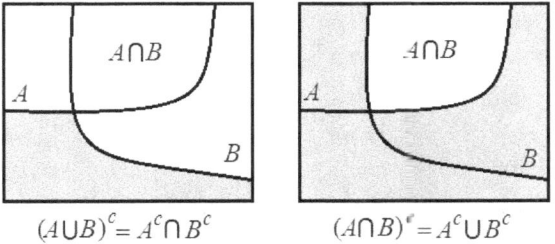

Figure 2.2: Venn diagram illustrating De Morgan's laws

It is easy to see why De Morgan laws are true. In the first equality, if x is a member of $(A \cup B)^c$ then x is not in $A \cup B$. Because the union contains both A and B we conclude that x is not in A and x is not in B, and this is the same as saying that x is in both A^c and B^c, so $x \in A^c \cap B^c$. We didn't pick x as any special member of $(A \cup B)^c$ so this argument applies to all members of $(A \cup B)^c$ and shows that $(A \cup B)^c \subset A^c \cap B^c$.

Next, if we show that the reverse containment is also true then we have proven the first equality in (2.1) by double containment (2.1). So start with any $x \in A^c \cap B^c$ and retrace the steps in the above argument but going in the opposite direction: x is in both A^c and B^c so that x is not in A and x is not in B. So x is not in the union $A \cup B$ which means that $x \in (A \cup B)^c$. Because x could be any member of $A^c \cap B^c$ we have shown that the reverse containment $A^c \cap B^c \subset (A \cup B)^c$ is also true and our proof is complete!

The second equality in (2.1) is proved similarly.

Exercise 6 *Prove the second De Morgan law.*

The equalities in (2.1) represent logical statements in the following sense: replace the sets by propositions p and q, their union by "*or*" and their intersection by "*and*". The complementation is the negation, or "*not*"; we discuss these more fully in the section on logic below. Then the first equality in (2.1) reads as:

"*not[p or q]*" is equivalent to "*not p and not q*"

Sets, Functions and Logic

This statement is rather self-evident in the usual context of daily life. For instance, if we know that a ball is not green *or* red then the ball is not green *and* it is not red. Similarly, the second equality reads as:

"*not[p and q]*" is equivalent to "*not p or not q*"

For instance, if someone is not a girlscout then that person is not a scout or not a girl.

The above ideas of taking the union or intersection extend to any collection of sets in U. If $A_1, A_2, \ldots, A_n \subset U$ then

$$A_1 \cap A_2 \cap \ldots \cap A_n = \{x : x \in A_1 \text{ and } x \in A_2 \text{ and } x \in A_3 \text{ and } \cdots x \in A_n\}$$

A more concise and maybe easier to read way of writing the above intersection is

$$\bigcap_{i=1}^{n} A_i = \{x : x \in A_i \text{ for all values of } i = 1, 2, \ldots, n\} \tag{2.2}$$

For instance, if $A_i = \{1, \ldots, i\}$ for $i = 1, 2, \ldots, 10$ then $\bigcap_{i=1}^{10} A_i = \{1\}$ since 1 is the only number that is in all ten of the A sets.

The intersection of a number of nonempty sets may be empty. For example, if D is the set of all Americans who vote for democrats, R the set of all Americans who vote for republicans and V is the set of all undecided voters (may lean one way or the other in an election) then $D \cap R \cap V = \varnothing$ since $D \cap R = \varnothing$ (nobody can vote for both a democrat and a republican) even if $D \cap V \neq \varnothing$ and $R \cap V \neq \varnothing$.

> If the intersection of a pair of sets is empty then the sets are *disjoint*.

The sets D and R mentioned above are disjoint. Notice that

$$A \cap A^c = \varnothing$$

for every set A which says that every set is disjoint from its complement.

Similarly, if $A_1, A_2, \ldots, A_n \subset U$ then

$$A_1 \cup A_2 \cup \ldots \cup A_n = \{x : x \in A_1 \text{ or } x \in A_2 \text{ or } x \in A_3 \text{ or } \cdots x \in A_n\}$$

A more concise way of writing the above union is

$$\bigcup_{i=1}^{n} A_i = \{x : x \in A_i \text{ for some value of } i = 1, 2, \ldots, n\} \tag{2.3}$$

For instance, for the ten sets A_i above, $\bigcup_{i=1}^{10} A_i = \{1, \ldots, 10\}$ because every integer from 1 to 10 is in some (at least one) of the A sets.

The above concepts of intersection and union of a collection of sets are valid for infinite collections, too. We discuss those in the next chapter.

The fourth basic set operation is the following.

> The *product (direct, or Cartesian)* of two sets A and B in U is the set $A \times B$ of all ordered pairs (a, b) with a from A and b from B.

The aforementioned pair (a,b) is *ordered* because it is not the same as (b,a), except trivially when $a=b$.

Note that if A (or B) is empty then so is $A \times B$ because there are no ordered pairs with the first element in A (or second element in B).

Exercise 7 *(a) Let $A = \{0,1,2\}$ and $B = \{x,y\}$. Find $A \times B$ and $B \times A$. How many elements does each product have?*

(b) Suppose that A is a set with m elements and B is a set with n elements. How do you show that $\#A \times B = mn$?

When we talk about the Cartesian or coordinate xy-plane in algebra with the usual x-axis and y-axis we are talking about a collection of ordered pairs (x,y) where both x and y are real numbers. The set of all real numbers $\mathbb{R}$ is an infinite set that we study in detail later. For now, the intuitive view of $\mathbb{R}$ as the "real number line" (x-axis or y-axis) of algebra works.

Like intersection and union, the product operation extends to any number of sets. If $A_1, A_2, \ldots, A_n$ are given sets then

$$\prod_{i=1}^{n} A_i = A_1 \times A_2 \times \ldots \times A_n = \{(x_1, x_2, \ldots, x_n) : x_i \in A_i \text{ for } i = 1, 2, \ldots, n\}$$

The objects $(x_1, x_2, \ldots, x_n)$ are called *n-tuples* or *n-vectors* depending on the context. For example, in calculus when $n=3$ and $A_i = \mathbb{R}$ for all three values of i then

$$\prod_{i=1}^{3} A_i = \mathbb{R} \times \mathbb{R} \times \mathbb{R} = \{(x_1, x_2, x_3) : x_1, x_2, x_3 \in \mathbb{R}\}$$

is the familiar three dimensional space. The product of a set with itself n times is usually abbreviated as a power A^n (this is the case $A_i = A$ for $= 1, 2, \ldots, n$ in the definition of the product of n sets). Thus, $\mathbb{R} \times \mathbb{R} \times \mathbb{R} = \mathbb{R}^3$.

Intervals.

Connected or single-piece subsets of $\mathbb{R}$ are especially important sets in analysis; they are called *intervals*. For instance, the set of all real numbers, or points on the x-axis that are between -2 and 3.5, including -2 but excluding 3.5 is an interval that is written as $[-2, 3.5)$. This set contains not only the integers $-2, -1, 0, 1, 2, 3$ but also other real numbers like $-3/2$ or $\sqrt{2}$, π or 3.49. There is a variety of intervals depending on whether or which end-point is included.

If a and b are real numbers and $a < b$ then we have the following types of intervals:

$$(a,b) = \{x \in \mathbb{R} : a < x < b\} \quad \text{(open interval)}$$
$$[a,b] = \{x \in \mathbb{R} : a \leq x \leq b\} \quad \text{(closed interval)}$$

In the case of an open interval the two end-points a and b are excluded but in a closed interval both end-points are included. When only one of the end-points is included, we have the *half-open* (or *half-closed*) intervals:

$$[a,b) = \{x \in \mathbb{R} : a \leq x < b\}, \quad (a,b] = \{x \in \mathbb{R} : a < x \leq b\}$$

Sets, Functions and Logic

The above intervals are all *bounded intervals* because they begin and end with a real number.

Disconnected or multiple-piece sets of numbers do arise in practice too and sometimes it is possible to write them as unions of intervals. For example, consider *all numbers whose squares are larger than 1 but smaller than 4*. The number 3/2 is such a number, as is $-3/2$ since $(-3/2)^2 = 9/4 = 2.25$. But 0 is not such a number because $0^2 = 0$ is not larger than 1; so the numbers that satisfy the italicized condition are not a connected set. You can readily check by squaring that the numbers that work are those in the following union of intervals:

$$(-2, -1) \cup (1, 2)$$

Notice that the wording of the italicized statement excludes the four integers ± 1, ± 2. This union of intervals is not itself an interval.

Intervals that are not bounded, or *unbounded intervals* are also routinely used in analysis and they may be open or closed; for instance:

$$[a, \infty) = \{x \in \mathbb{R} : x \geq a\}, \quad (-\infty, a] = \{x \in \mathbb{R} : x \leq a\}$$

are closed unbounded intervals.

We use the symbol ∞ here to highlight the fact that $[a, \infty)$ contains all real numbers starting from a and including arbitrarily large numbers. We can't close the interval at ∞ because ∞ is not a number; it is just shorthand for *there is no endpoint on the right* which we recognize as the potential infinity of arbitrarily large numbers. Similarly, $-\infty$ is shorthand for *there is no endpoint on the left*, that is, the potential infinity of negative numbers with arbitrarily large magnitudes.

With this in mind, here are the open unbounded intervals:

$$(a, \infty) = \{x \in \mathbb{R} : x > a\}, \quad (-\infty, a) = \{x \in \mathbb{R} : x < a\}$$

The set of all real numbers can also be written as an interval: $\mathbb{R} = (-\infty, \infty)$.

It is worth noticing that every bounded interval is the intersection of two unbounded ones; for instance,

$$[a, b) = [a, \infty) \cap (-\infty, b)$$

Exercise 8 *Write each of the bounded intervals below as the intersection of two unbounded intervals:*

$$[3, 8], \quad (-\pi, \pi), \quad \left(-\frac{1}{2}, 0\right]$$

Relations.

We define relations as sets. Specifically,

> If S is a nonempty set then *a relation R in S is any nonempty subset of $S \times S = S^2$.*

For example, let H be the set of all human beings at a given time and consider the *motherhood relation* in H: a is related to b if a is b's mother. This relation, call it R_m, consists of all pairs (a, b) where a is b's mother. Then $R_m \subset H \times H$. In particular, (Bob's mom, Bob) $\in R_m$ if Bob's mother is alive. However, (Bob, b) is not an element of R_m for any human being b since Bob, a man, cannot be a mother. So R_m is a proper subset of $H \times H$. Further, if Joan is a mother of Alex, Molly and Betty then all three pairs (Joan, Alex), (Joan, Molly) and (Joan, Betty) are in R_m.

An important mathematical relation in $\mathbb{N}$ is "divisibility" $R_d \subset \mathbb{N} \times \mathbb{N}$ which is defined as $(m, n) \in R_d$ if m divides n (the remainder of the division is zero). Thus $(2, 2)$ and $(2, 6)$ are both in R_d since $2/2 = 1$ and $6/2 = 3$ with zero remainders in both cases. On the other hand, $(2, 5)$ is not in R_d because when we divide 5 by 2 we get a remainder of 1.

It is sometimes convenient to write aRb which we read as "a is R-related to b" if $(a, b) \in R$. For divisibility the common symbol used is $m|n$ for "m divides n" or $(m, n) \in R_d$.

Equivalence relations and classes.

Among the most frequently encountered relations in mathematics are the *equivalence relations*.

Any relation $\sim$ having the following three properties is an equivalence relation:
(i) $x \sim x$ for every $x \in S$ (*reflexive property*)
(ii) If $x \sim y$ then $y \sim x$ for all $x, y \in S$ (*symmetric property*)
(iii) If $x \sim y$ and $y \sim z$ then $x \sim z$ for all $x, y, z \in S$ (*transitive property*)

Equivalence relations often occur in everyday life. For example, when you go to the store and see two shirts that cost the same then the shirts are "cost equivalent": $x \sim y$ if shirt x has the same price as shirt y (even if x and y have different colors, are not the same size, etc). You can easily check that all three properties (a)-(c) above hold. Similarly, having the same color or the same size also define equivalence relations among shirts, regardless of their prices. On the other hand, the motherhood relation above is decidedly *not* an equivalence relation; it does not satisfy any one of (a)-(c).

Exercise 9 *Explore (a)-(c) above for the "best friends" relation: $a \sim b$ if a and b are best friends (assume that each person is his or her own best friend).*

Exercise 10 *Is the divisibility relation in $\mathbb{N}$ an equivalence relation? If not then which properties hold and which do not?*

Recall the cost equivalence relation above. For each shirt s consider all the shirts with the same price tag (say, p) as s. This group of shirts is a set E_s. Notice that if a shirt t is priced at q and $q \neq p$ then no shirt in E_t (all of which are priced q) can be in E_s, and conversely. Therefore, the two sets E_s and E_t are disjoint: $E_s \cap E_t = \emptyset$ if s and t have different prices. Further, the union of all of these disjoint sets gives all the shirts in the store because every shirt has a price tag. This means that the sets E_s partition all the shirts into disjoint sets classified according to cost. The sets E_s above are called *equivalence classes* or *cells*.

Every equivalence relation in a set S has its own collection of equivalence classes that partition S into disjoint subsets. If $\sim$ is an equivalence relation in S then for each element of s we define the equivalence class of s as the following subset of S

$$[s] = \{x \in S : x \sim s\}$$

Sets, Functions and Logic

The brackets notation is common and we stick to it from now on. As the above example of shirts illustrates, these equivalence classes have special properties that I list here:

> *If $\sim$ is an equivalence relation in a nonempty set S then:*
> *(a) $[s] = [t]$ if $s \sim t$ and $[s] \cap [t] = \varnothing$ if $s \not\sim t$ (s and t are not equivalent).*
> *(b) The union of $[s]$ for all $s \in S$ is the whole set S.*

Order relations.

When we look at positive integers, we think of 1 as being less than 2, 2 less than 3 and so on. This arrangement of numbers is a way of *ordering* the set $\mathbb{N}$. The essential property of an ordering relation is that it be *transitive*; if 1 is less than 6 and 6 is less than 13 then 1 must be less than 13. So order relations share the transitive property with equivalence relations discussed above. But unlike equivalence relations, an order relation can never be symmetric since that would remove the essence of ordering, namely, one element coming before or after another.

To identify the correct property, recall the set inclusion relation $\subset$ that we discussed earlier ($A \subset B$ if every element of the set A was also an element of the set B). If $A \subset B$ and $B \subset A$ then A and B are the same sets by double-containment, that is, $A = B$. Let's use this as a template for our next definition.

> A relation $\preceq$ is *antisymmetric* if "$x \preceq y$ and $y \preceq x$ imply $x = y$".

This leads to the following definition of order relation:

> A relation $\preceq$ is an *ordering* or an *order relation* on a set S if it is both transitive and antisymmetric. If $x \preceq y$ and $x \neq y$ then write $x \prec y$ and call $\prec$ a *strict ordering*.

We use the symbol $\succeq$ to mean the opposite or *reverse ordering* (but not the negation) of $\preceq$ in the sense that $x \succeq y$ whenever $y \preceq x$. Clearly, $\succeq$ is an order relation.

The usual ordering $\leq$ for integers is, as you might think, an antisymmetric order relation. An interesting observation about the strict relation $<$ is that it can be defined through algebra: $m < n$ if $n - m$ is positive, i.e. $n - m \in \mathbb{N}$. For instance, $2 < 7$ because $7 - 2 = 5$ is positive. This observation will be important in our construction of the set of real numbers.[6]

It is important to realize that the above definition of ordering is "partial" in the sense that it does not preclude the possibility that for *some* pairs of elements x, y we have *neither $x \preceq y$ nor $y \preceq x$*. This may occur for set inclusion $\subset$ which orders the subsets of a set. For instance, consider the subsets $\{1,2\}$ and $\{2,3\}$ of $\mathbb{N}$. Note that neither $\{1,2\} \subset \{2,3\}$ holds nor $\{2,3\} \subset \{1,2\}$. Therefore, the above definition is said to define a *partial ordering*. Set inclusion is then a partial ordering. But the usual ordering of integers is not partial. If we take any pair m, n of integers then we can say that either $m \leq n$ or $n \leq m$ must be true. Because in this book we are interested in this type of ordering, we define:

> A relation $\preceq$ is a *total ordering* on a set S if it is transitive, antisymmetric and *total*; i.e. for all $x, y \in S$, either $x \preceq y$ or $y \preceq x$.

Notice that a total ordering $\preceq$ is a reflexive relation. The totality property can also be expressed in terms of the strict relation $\prec$ slightly differently as:

---

[6] If 0 is included in $\mathbb{N}$ then the non-strict ordering can also be defined in this way: $m \leq n$ if $n - m$ is positive or zero.

> The relation $\prec$ is a *strict total ordering* on a set S if it is transitive and for every pair $x, y \in S$ *exactly one* of the following is true: $x \prec y$ or $y \prec x$ or $x = y$.

Because one of three possibilities must be the case, this observation means that $\prec$ has the *trichotomy property*.

It is worth mentioning here that for a *total* ordering the symbol $\succeq$ does in fact mean the negation of $\preceq$; however, for the strict case the negation of $\prec$ is not $\succ$ even when total: the negation of $x \prec y$ is $x \succ y$ or $x = y$ due to trichotomy, if $\prec$ is a total ordering.

Ordered sets.

The set $\mathbb{N}$ of all natural numbers has its usual order relation $\leq$ (or $<$). We call $\mathbb{N}$ an ordered set. More generally, we define:

> A nonempty set S together with an order relation $\leq$ forms an *ordered set*. S is a *totally ordered set* if $\leq$ (or $<$) is a total ordering.

We will find later that the set of all real numbers, or the number line, is totally ordered. This total ordering property of the set of real numbers is one of their identifying characteristics and fundamentally important in analysis.

2.2 Functions

When making a list of weekly activities on a calendar, we may list a number of things to do for each day of the week. Suppose that for a particular week, Sunday to Saturday, Emily has the following schedule where only the number of things she does each day is shown, not her actual activities:

Sunday	Monday	Tuesday	Wednesday	Thursday	Friday	Saturday
2	4	5	4	3	4	3

The above list is an example of a *function* that assigns a single number to each day of the week. Let us label this calendar function F_c and take note of the fact that F_c generates a natural number in return for a day of the week. We write this as $F_c(\text{Sunday}) = 2$, $F_c(\text{Monday}) = 4$, $F_c(\text{Tuesday}) = 5$ and so on. The seven weekdays constitute a set with seven elements:

$$\text{Week} = \{\text{Sunday, Monday, Tuesday, Wednesday, Thursday, Friday, Saturday}\}$$

This set is the *domain* of F_c. Since all the values of F_c are natural numbers it is a numerical function. We think of F_c as a correspondence between its domain and the set $\mathbb{N}$ of natural numbers. We write this as

$$F_c : \text{Week} \longrightarrow \mathbb{N}$$

Other examples of functions from daily life are easy to find: the correspondence between a set of objects and their weights is a numerical function; likewise, objects' volumes are also examples of numerical functions whose domains are physical objects. A moving car's position on the map is a function of time; here the domain is the number line since time is measured as a number but the function's values are vectors with two components: longitude and latitude.

Here is a simple definition of function that will be enough for most of what we do here.

Sets, Functions and Logic 43

> A function $f : D \to S$ is an *assignment* of a *unique* element $f(x)$ in a set S to each element x of D. The set D on which f is defined is its *domain* and the set S that contains the function values $f(x)$ is the *range* of f.

For the above calendar function F_c the domain is the set Week of all weekdays and the range is any set of positive integers that contains the numbers 2, 3, 4, 5, say, $\mathbb{N}$.

The word "unique" above works only one way: f can only pick one element in the range for each given element in its domain but many (or even all) elements in the domain can be assigned a single element in the range.

For instance, $F_c(\text{Monday}) = 4$ and $F_c(\text{Wednesday}) = 4$ too. For the weight function, an object cannot have two different weights but certainly many objects can have the same weight. Similarly, the moving car can be in only one location at each given instant of time but it can be in the same location many times; in fact, if it stops moving then it will be in the same location for all subsequent times!

A correspondence that assigns more than one element in the range to a single element in the domain is not a function, but it is a relation. Every function $y = f(x)$ can be written as a set of ordered pairs (x, y). As such, all functions are special types of relations where if (x, y) and (x, z) are both related via f then $y = z$.

For instance, in an auditorium that is not fully booked people may use an extra chair for their coats or other belongings so the assignment of chairs to people is not a function. In this case, if we reverse the arrow, so to speak, and consider the assignment of people to chairs then we have a function because two people do not use the same chair.

Visualization: graphs, plots and odd-even symmetry.

The arrow symbol is widely used probably because functions in calculus, where the domains are typically infinite sets, are often thought of as *mappings*. consider the square function $f(x) = x^2$ from algebra where x is an ordinary real number. Here the domain of f is the set of all real numbers $\mathbb{R}$ (the x-axis). Using arrows:

$$f : \mathbb{R} \longrightarrow \mathbb{R}$$

The graph of this function is a bowl-shaped curve called a *parabola*; see Figure 2.3.

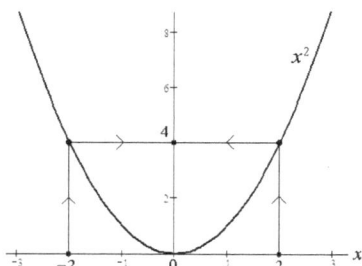

Figure 2.3: The graph of the square function is a parabola

The square function maps $\mathbb{R}$ into itself; notice the arrows: -2 is mapped to $(-2)^2 = 4$ just as 2 is. The fact that the square x^2 is never negative implies that the square function can only cover a proper subset of $\mathbb{R}$, namely the set of all non-negative numbers or the interval $[0, \infty)$.

Another feature of the graph of the square function is its *symmetry relative to the y-axis*: the half on the left of the y-axis is the mirror image (or reflection across the y-axis) of the half to the right; see Figure 2.3. This is easy to explain: the reflection of x across the y-axis is $-x$, so if we check the function values:

$$f(-x) = (-x)^2 = (-x)(-x) = x^2 = f(x)$$

we see that the action of f on $-x$ is exactly the same as its action on x. Functions having this property are called *even functions*. In symbols:

$$f(-x) = f(x)$$

An important even function from trigonometry that we discuss later is the cosine function $\cos x$; we encounter this fact again when discussing the infinite series for $\cos x$.

There also *odd functions*; an odd function $f(x)$ satisfies the identity

$$f(-x) = -f(x)$$

The cube function x^3 is odd because $(-x)^3 = (-x)(-x)(-x) = -x^3$; see Figure 2.4.

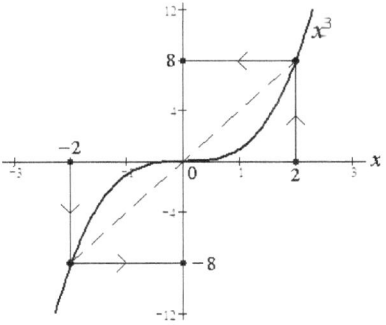

Figure 2.4: The cube function is an odd function

In this case, the graph of the function is *symmetric relative to the origin*: every point on the graph to the left or below the origin is the reflection across the origin of a point on the graph to the right or above the origin; this is shown in Figure 2.4.

Exercise 11 *Specify whether the following functions are even, odd or neither:*

$$\text{(a) } x^9 \qquad \text{(b) } x^9 + 3 \qquad \text{(c) } x^4 + 3 \qquad \text{(d) } x^4 + 3x \qquad (2.4)$$

Having odd or even powers is not a requirement for a function to be odd or even; for instance, the sine function $\sin x$ is an important odd function that we encounter later. The reason for using the terms odd and even in this context is that multiplying such functions works analogously to multiplying odd and even integers, in the sense that multiplying an even number by an odd one

Sets, Functions and Logic

produces an odd number while multiplying two odd or two even numbers yields an even number. For functions, if $f(x)$ is odd and $g(x)$ is even then:

$$fg(-x) = f(-x)g(-x) = -f(x)g(x) = -fg(x)$$

so the product $fg(x)$ is an odd function. Similarly, if $f_1(x)$ is odd and $g_1(x)$ is even then:

$$ff_1(-x) = f(-x)f_1(-x) = -f(x)[-f_1(x)] = (-1)^2 f(x)f_1(x) = ff_1(x)$$
$$gg_1(-x) = g(-x)g_1(-x) = g(x)g_1(x) = gg_1(x)$$

so the products ff_1 and gg_1 are both even functions.

Functions as mappings.

To better appreciate the mapping aspect of functions, we examine the square function's effect on intervals of numbers. Consider first the interval [0,1]. As Figure 2.3 shows, the square function is increasing to the right of the origin 0; also $0^2 = 0$ and $1^2 = 1$. So the square function maps [0,1] onto itself. But the sagging shape of the parabola indicates that this interval is not left intact. Compare the effect of the square function on each of the two half-intervals [0,1/2] and [1/2,1], each having length 1/2. Since

$$f(0) = 0, \quad f\left(\frac{1}{2}\right) = \left(\frac{1}{2}\right)^2 = \frac{1}{4}, \quad f(1) = 1$$

you can see (Figure 2.5) that [0,1/2] is *compressed* by $f(x) = x^2$ to [0,1/4] that has length 1/4 while [1/2,1] is *stretched* to [1/4,1] with length 3/4.

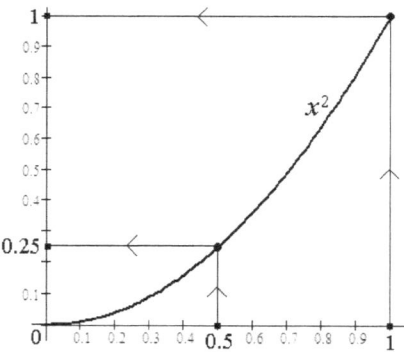

Figure 2.5: Highlighting the mapping aspect of the square function

Further, Figure 2.3 shows that if an interval of unit length is chosen further away from the origin then its squared image is stretched more: [1,2] is stretched to [1,4], [2,3] to [4,9] and so on. The square function stretches (or compresses) an interval of fixed length by different amounts.

Exercise 12 *Consider the function $f(x) = 2x - 1$ which is a linear-affine function.*

(a) Show that the interval $[0,1]$ is mapped to the interval $[-1,1]$ which is twice as long as $[0,1]$. Extend this observation to an arbitrary interval $[a,b]$ where $a<b$ is mapped to an interval twice as large. Note that the length of $[a,b]$ is $b-a$; what is $f(b)-f(a)$?

(b) Repeat the above calculations for $f(x) = -x/3 + 1$. What can you conclude about the general $f(x) = mx + c$ where m and c are arbitrary numbers?

Exercise 13 Consider the exponential function $f(x) = 2^x$. Graph this function if you have a graphing device or utility to see that it is an increasing curve.

(a) Calculate the values of $f(x)$ for $x = -1, 0, 1, 2, 3$. Compare the effect of f on the intervals $[-1,0]$, $[0,1]$, $[1,2]$ and $[2,3]$; which ones are compressed and which ones are stretched?

(b) You may notice in (a) that f stretches each interval by double the amount that it stretches the previous interval. Explain this using the fact that $2^{x+1} = 2^x 2^1 = 2(2^x)$ for every value of x.

Composition of functions.

Functions can be combined in a variety of ways. Numerical functions can be added or multiplied because we can add and multiply numbers in the usual way: if $f(x) = x^2$ and $g(x) = 3x - 1$ then

$$f(x) + g(x) = x^2 + 3x - 1, \qquad f(x)g(x) = x^2(3x-1) = 3x^3 - x^2$$

But the most fundamental and universal way of combining functions is by combining them as follows.

> Let $f : D \to S$ and $g : S \to T$ be two given functions (notice the shared set S). Then for each $x \in D$ the action of f gives $f(x) \in S$. Since g acts on the set S the action of g following f gives $g(f(x)) \in T$. This combined action is called the *composition of f and g* and commonly written as $g \circ f$. It is a single function $g \circ f : D \to T$ from D into T.

The above definition is illustrated schematically in Figure 2.6.

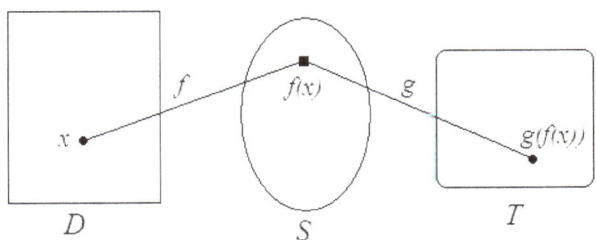

Figure 2.6: Composition of functions

For example, consider again the functions $f(x)$ and $g(x)$ above. Then

$$g \circ f(x) = g(f(x)) = 3f(x) - 1 = 3x^2 - 1$$

Sets, Functions and Logic

A natural question regarding function composition is whether it is *commutative*; that is, if done in reverse order we get the same result. Addition and multiplication are after all commutative operations. But function composition is *not* commutative.

To begin with, $f \circ g$ need not even be defined just because $g \circ f$ is; see Exercise 16. But even when $g \circ f$ and $f \circ g$ are both defined they are usually not equal. Going back to the functions $f(x)$ and $g(x)$ above, notice that since
$$D = S = T = \mathbb{R}$$
both $g \circ f$ and $f \circ g$ are defined. But the latter is not the same as the former:
$$f \circ g(x) = f(g(x)) = g(x)^2 = (3x-1)^2 = 9x^2 - 6x + 1$$

To illustrate composition of functions in a practical situation, consider a stream of water that is filling a ditch at a known rate. We know the volume of water $V(t)$ in the ditch at any time t by using a meter to measure the amount going into the ditch. If we also know enough about the shape of the ditch then we can determine the height h of water inside in terms of its volume, say $h(V)$. The composition of $h(V)$ with $V(t)$ gives the height of water as a function of time $h(V(t))$ or $(h \circ V)(t)$.

For instance, if the ditch is roughly cone-shaped with apex angle $90°$ then $h(V) = c\sqrt[3]{V}$ where $c = \sqrt[3]{3/\pi} = 0.985;$ [7] see Figure 2.7.

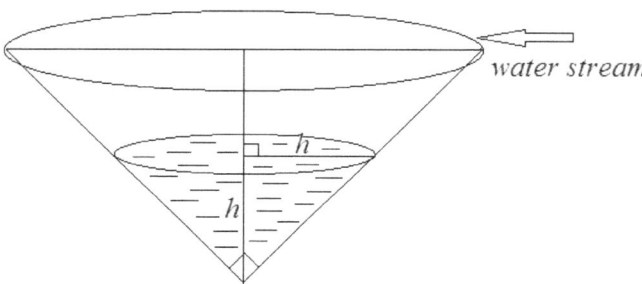

Figure 2.7: Cone-shaped ditch being filled with water

If the water enters the ditch at a steady rate of 2 cubic meters per minute then the volume of water in the ditch t minutes after water first enters it is $V(t) = 2t$ cubic meters. So
$$h(V) = 0.985\sqrt[3]{V} \quad \text{and} \quad V(t) = 2t$$
and these compose to form a single function of time as
$$(h \circ V)(t) = h(V(t)) = 0.985\sqrt[3]{V(t)} = 0.985\sqrt[3]{2t} = 1.26\sqrt[3]{t} \quad \text{meters}$$

Now we can easily calculate the height at different times. One minute after water enters the ditch the height of the water has risen to $1.26\sqrt[3]{1} = 1.26$ meter; 8 minutes later the height is $1.26\sqrt[3]{8} = 2.52$ meters (twice as high). It will take 27 minutes for the height to be three times as high because $\sqrt[3]{27} = 3$.

---

[7] This comes from the formula for the volume of a cone $V = \pi r^2 h/3$. If the cone's apex angle is $90°$ as in Figure 2.7 then $r = h$ and $V = \pi h^3/3$ which we can solve for h.

Exercise 14 *Generalize the example above: If water enters the cone-shaped ditch at a steady rate of ρ cubic meters per minute where $\rho > 0$ then the height of water in the ditch at any time t is given by*

$$(h \circ V)(t) = \sqrt[3]{\frac{3\rho t}{\pi}}$$

Function iteration. A function $f : D \to S$ may be composed with itself any number of times to give $f \circ f$, $f \circ f \circ f$ and so on provided that $S \subset D$ (do you see why this is necessary? See Exercise 16 below). It is common to use the notation f^n for $f \circ f \circ \cdots \circ f$ (n times) to denote this *iteration* of the function. This iteration describes the repetition of a process. For instance, suppose that a form of bacteria doubles its numbers (by splitting in half) every so often. Thus $f(P) = 2P$ for any given population size P. If the original population size is P_0 then the population at first splitting of each bacterium is $f(P_0) = 2P_0$. Later, when splitting occurs again,

$$f^2(P_0) = f(f(P_0)) = f(2P_0) = 2^2 P_0$$

and so on:

$$f^3(P_0) = f(f(f((P_0)))) = f(2^2 P_0) = 2^3 P_0$$
$$\vdots$$
$$f^n(P_0) = 2^n P_0$$

Exercise 15 *Consider the square function $f(x) = x^2$. Find $f \circ f$ and $f \circ f \circ f$. Can you guess a formula for f^n?*

Exercise 16 *Let $f(\theta) = \cot \theta = (\cos \theta)/(\sin \theta)$ be the cotangent function which is defined on the set $D = \{\theta : 0 < \theta < \pi\}$. Then $f(\pi/2) = \cot(\pi/2) = 0$, which is not in D. What is $f \circ f(\pi/2)$? Note that with D as its domain the range of $\cot \theta$ is all of the real numbers.*

Image and Inverse image sets.

Consider a function $f : D \longrightarrow S$. This means that f maps each point x of the domain D into an element $y = f(x)$ of the range S. We call y the *image* of x (under f). For instance, if f is the square function $f(x) = x^2$ then the image of $x = 2$ is $y = f(2) = 4$.

The entire range S is not covered in this way even if all elements of D are exhausted. For instance, under the square function $f(x) = x^2$ only the non-negative numbers are covered by mapping all of the real numbers. These non-negative numbers are the images of all real numbers (the domain).

> For each nonempty subset A of the domain D the *image set* of A is the subset of S (range) consisting of those elements that are images of some element in A. In symbols,
>
> $$f(A) = \{y \in S : y = f(x) \text{ for some } x \in A\}$$

Sets, Functions and Logic

For instance, the image of the set $A = \{-3, 0, 2, 3\}$ under the square function $f(x) = x^2$ is

$$f(A) = \{f(-3), f(0), f(2), f(3)\} = \{0, 4, 9\}$$

The image set $f(D)$ is a subset of the range S but it need not be equal to all of S. In the case of the square function the image set of its entire domain is

$$f(D) = f(\mathbb{R}) = \{y \in \mathbb{R} : y = x^2 \text{ for some } x \in \mathbb{R}\} = \{y \in \mathbb{R} : y \geq 0\}$$

where the last inequality holds because every non-negative real number is the square of some real number, possibly negative, like $3 = (-\sqrt{3})^2 = (\sqrt{3})^2$. The interval notation gives a more succinct expression:

$$f((-\infty, \infty)) = [0, \infty)$$

If $f : D \longrightarrow S$ is a given function then for some y in S there may not be any x in D such that $f(x) = y$; and if such an x exists then it need not be unique. This necessitates defining a backward mapping.[8]

> For a function $f : D \longrightarrow S$ and each element y in S, the *inverse image set* of y is the set $f^{-1}(y)$ of *all* $x \in D$ for which $f(x) = y$. In symbols:
>
> $$f^{-1}(y) = \{x \in D : f(x) = y\}$$

If we think of an element y as a person then $f^{-1}(y)$ may indicate y's parents or ancestors, depending on how f is defined.

Here's a numerical example: if $f(x) = x^2$ is the square function then the inverse image of 4 is $f^{-1}(4) = \{-2, 2\}$ because $f(2) = f(-2) = 4$.

Note that $f^{-1}(y)$ is a *subset* of the domain D for every y in S. This idea is illustrated in Figure 2.8.

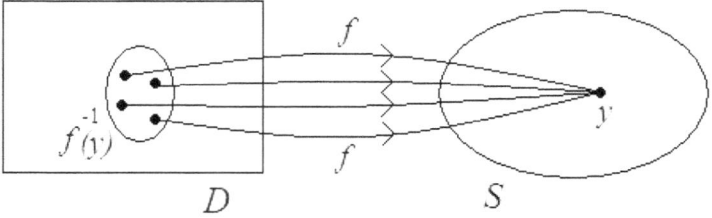

Figure 2.8: Generic illustration of an inverse image

If f is the square function $f(x) = x^2$ and $D = S = (-\infty, \infty)$ then for each $y < 0$ there is no x such that $f(x) = y$ because this would mean that $x^2 < 0$ which is impossible for a real number x. On the other hand if $y > 0$ then $f(x) = y$ means $x^2 = y$. Taking square roots we find two possible

---

[8]The notation $f^{-1}(y)$ should not be read as the value of the inverse *function* at the point y. In fact, the inverse image $f^{-1}(y)$ is generally a set not a number. We discuss inverse functions later.

values $x = \pm\sqrt{y}$ that work. Finally, if $y = 0$ then the only x which gives $f(x) = 0$ is $x = 0$. These observations are summarized below:

$$f^{-1}(y) = \{-\sqrt{y}, \sqrt{y}\} \text{ if } y > 0$$
$$f^{-1}(0) = \{0\} \text{ and } f^{-1}(y) = \varnothing \text{ if } y < 0$$

Exercise 17 *Let $f : D \to S$ be a function. Define a relation in D as follows: $x \sim y$ if $f(x) = f(y)$. For example, if $f(x) = x^2$ then $-2 \sim 2$ because $f(-2) = 4 = f(2)$.*
(a) Verify that $\sim$ is an equivalence relation.
(b) For each $x \in D$ explain why the equivalence class of x is $f^{-1}(y)$ where $y = f(x)$.
(c) Consider the function $f(x) = x^3 - x$. What is the equivalence class of 0? The equivalence class of 6? Sketching a graph of $f(x)$ can be helpful.

One to one functions.

As we saw in the example of the square function above, the inverse image of a single element may contain more than one element or no elements at all. In fact, for the square function the only number whose inverse image is just one number is 0. But some functions have the special property that $f^{-1}(y)$ is a single element (or possibly empty) for every y in the range of f. This means that such f can only map a *single* element in their domain to an element in their range. Such functions are reasonably called *one to one*.

An example of a one to one function is the assignment of one social security number to each (living, legal) resident in America. Other daily life examples include assigning unique jersey numbers to the members of a sports team or matching one man with one woman in a dating service.

> A function $f : D \longrightarrow S$ with domain D is *one to one* or *injective* on D if for each pair of elements u, v in D the following is true:
>
> $$f(u) = f(v) \implies u = v \qquad (2.5)$$

The symbol $\implies$ means "implies" and so it is called the *implication* symbol. For a simple illustration of how to use (2.5) consider $f(x) = 2x - 1$. Here D and S are both the entire number line $(-\infty, \infty)$. If u, v are any pair of numbers that satisfy $f(u) = f(v)$ then we quickly derive

$$2u - 1 = 2v - 1$$
$$2u = 2v$$
$$u = v$$

This shows that (2.5) is true for arbitrary pairs of numbers in D so $2x - 1$ is a one to one function.

An equivalent way of stating the above definition is sometimes useful:

$$u \neq v \implies f(u) \neq f(v) \qquad (2.6)$$

Sets, Functions and Logic

This is the *contrapositive* of the original definition above, which is *logically equivalent*. (see the next section below). On the other hand, the *converse* statement

$$u = v \implies f(u) = f(v) \tag{2.7}$$

that we get by simply reversing the implication in (2.5) is *not equivalent* to the original. In fact (2.7) is true for *every function* f (one to one or not) since a function can only assign one value to each element of its domain.

Exercise 18 *Explain why if f is a one to one function then for every y in the range the inverse image $f^{-1}(y)$ is either a single number or the empty set.*

When the graph of a function is available there is a graphical way of telling whether it is one to one. For instance, if you look at the graph the function $f(x) = x^3$ in Figure 2.4 then you can tell that it is a one to one function because every horizontal line that crosses the graph, does so at just one point.

> **The horizontal line test:** *Starting from any y on the y-axis, draw a horizontal line. If this line crosses the graph of the function $y = f(x)$ at some point then drop a perpendicular to the x-axis from this point on the curve to get the value of x that corresponds to the y you started with. If there is a horizontal line that crosses the graph of $f(x)$ at more than one point then the function is not one to one because there are more than one value on the x-axis that correspond to just one value on the y-axis. But if all horizontal lines that cross the graph do so at just one point then the function $f(x)$ is one to one.*

The horizontal line test shows that the square function x^2 is *not* one to one.

Monotone functions.

The horizontal line test shows, intuitively, that every function that is either increasing or decreasing is automatically one to one (injective).

The idea of increasing or decreasing functions is not limited to numerical functions and applies to functions on any ordered set. For example, suppose that a bakery store is having a promotion in the form of handing out free cookies to people who arrive in a queue one day. The first 10 persons get 3 cookies each, the second ten get 2 cookies each and everybody else gets one cookie while supplies last. This promotion defines a decreasing function with domain consisting of the queue of people. The arrival order of the patrons is how the domain is ordered as a set; can you describe the range of this function?

Here is a precise definition of monotone functions.

> **Increasing, decreasing and monotone functions.** Let $f : D \to S$ be a function where D is totally ordered by a relation $<$ and S is totally ordered by a relation $\prec$. Then f is an *increasing* function on the set D if for every pair of elements x, y in D:
>
> $$x < y \Rightarrow f(x) \prec f(y) \qquad (2.8)$$
>
> Similarly, f is a *decreasing* function on D if
>
> $$x < y \Rightarrow f(y) \prec f(x) \qquad (2.9)$$
>
> If the relation $\prec$ is replaced by its *non-strict* form $\preceq$ in the above implications then we say that f is *non-decreasing* (or *non-increasing*). We call a function *monotone* on D if it is either non-decreasing or non-increasing on D.

The function $f(x) = x^3$ is increasing on the entire number line $(-\infty, \infty)$; see Figure 2.4. Its negative $-x^3$ is decreasing on $(-\infty, \infty)$. Both are monotone functions.

The square function is neither increasing nor decreasing on $(-\infty, \infty)$ as we see in Figure 2.3; therefore, x^2 is not monotone on $(-\infty, \infty)$. However, x^2 is increasing on the interval $[0, \infty)$ and decreasing on $(-\infty, 0]$.

A function may be monotone but not strictly; for example,

$$f(x) = \begin{cases} x^2, & \text{if } x < 0 \\ 0, & \text{if } x \geq 0 \end{cases}$$

is decreasing on $(-\infty, \infty)$, but not strictly since $f(x) = 0$ is constant for all positive numbers x. An extreme example of a monotone function is a *constant function*

$$f(x) = c$$

where c is any fixed number. This function is trivially non-increasing *and* non-decreasing everywhere.

The following result is general and quite useful.

> *A function $f : D \to S$ that is either increasing on D or decreasing on D is one to one.*

This is easy to see if we visualize the graph of an increasing or decreasing function in the xy-plane and use the horizontal line test. But notice that the theorem does not require either D or S to be sets of numbers. So how do we prove it?

> Suppose that f is increasing on D and let u, v be an arbitrary pair of elements in D. If $u \neq v$ then either $u < v$ or $v < u$ since $<$ is a total ordering on D. If $u < v$ then by (2.8) $f(u) < f(v)$ so in particular, $f(u) \neq f(v)$. This shows that (2.6) is true; the same argument works if $v < u$ since we can just switch u and v. So f is one to one. A similar argument works if f is decreasing.

The *converse* of the above theorem is not true! It is possible for a one to one function to not be monotone and a simple example of this type of function is $f(x) = 1/x$. This is one to one (consider its graph and the horizontal line test) but not monotone; for instance, $-1 < 1$ and $1 < 2$ but $f(-1) = -1 < 1 = f(1)$ while $f(1) = 1 > 1/2 = f(1/2)$.

Sets, Functions and Logic

Exercise 19 *Suppose that a function f is decreasing everywhere on its domain D and prove that the function must be one to one.*

It may seem that for every function on the number line $(-\infty, \infty)$ it must be possible to split the line into small enough intervals such that on each interval the function is monotone. This is possible to do for the functions that typically appear in a standard calculus textbook, including trigonometric functions of arbitrarily short periods. But we see later that there are infinite sets of functions that are not monotone on any interval, no matter how short. We cannot visualize such functions but they appear quite naturally in analysis, usually as a result of a limiting process. So the credit, or blame, goes to infinity, of course!

Bijections.

Suppose that a function covers its entire range, which means that *every* element of the range is assigned to some element of the domain. Such a function is *onto* its range, or more technically, the function is *surjective*.

> A function $f : D \longrightarrow S$ is onto S, or surjective, if the image set of D is all of S, that is, $f(D) = S$. In words, for every y in S there is (at least) one x in D such that $y = f(x)$. A function that is *both* one to one and onto is called a *bijection*.

As a simple example of a bijection, consider $f(x) = 2x - 1$. We previously saw that this function is one to one. Let's show that it is also onto. The domain and range are the same set, namely, the entire number line so $D = S = (-\infty, \infty)$.

Pick an arbitrary y in this set; we need to find a value of x such that $y = 2x - 1$. The obvious thing to do is to solve the last equation for x to get $x = (y+1)/2$. This value satisfies the definition of onto functions for our chosen y, which was picked arbitrarily. So we may justifiably claim that for every y in $S = (-\infty, \infty)$ there is $x = (y+1)/2$ in $D = (-\infty, \infty)$ such that $y = f(x)$. Therefore, our function is onto as well as one to one, so it is a bijection.

This argument can be extended to all affine functions

$$f(x) = ax + b \tag{2.10}$$

to prove that these functions are bijections of the whole number line $(-\infty, \infty)$ for each given pair of real numbers a, b provided that $a \neq 0$. Can you see what happens if $a = 0$?

Exercise 20 *Generalize the argument given above to show that the function $f(x)$ in 2.10 is a bijection if $a \neq 0$.*

We previously saw examples of functions that are one to one but not onto, such as the exponential function 2^x of Exercise 13 which is not onto $(-\infty, \infty)$ since 2^x is always positive.

There are also plenty of functions that are onto but not one to one. If you have a graphing device consider graphing $f(x) = x^3 - x$. You will notice that the graph covers all of the y-axis (the range) but this function is not one to one as its graph fails the horizontal line test for some numbers on

the y-axis. In particular, if $y = 0$ then $x^3 - x = 0$ which is a simple algebraic equation that can be solved as follows:
$$x(x^2 - 1) = 0 \Rightarrow x = 0 \text{ or } x^2 = 1 \Rightarrow x = 0 \text{ or } \pm 1$$

Therefore, three numbers $0, 1, -1$ on the x-axis are all assigned the same number 0 on the y-axis by the function f. If you think of the x-axis itself as a horizontal line then this line crosses the graph of $f(x)$ at three distinct points.

Bijections are important in mathematics as they represent a common feature of maps that express the equivalence of mathematical structures, like the so-called isometries of metric spaces; these are transformations of metric spaces that preserve distances between points, such as rotations or translations of the xy-plane.

Exercise 21 *Consider the finite sets $A = \{1, 2\}$ and $B = \{a, b, c\}$.*

(a) Specify a one to one function that maps A into B; can you specify any function, one to one or not, from A onto B? Why not?

(b) Specify a function that maps B onto A; can you specify any one to one function from B into A? Why not?

(c) Can there be a bijection from A to B? Why not?

The above exercise points to another essential feature of bijections: their role as counting tools.

> **Counting with bijections.** *One of the most important uses of bijections is to count elements of sets by setting up a unique matching between elements of a set whose cardinality is known and another less familiar set.*
> *If A and B are finite sets with $\#A = m$ and $\#B = n$ then a bijection from A onto B exists only when $m = n$. This is easy to see: if $f : A \longrightarrow B$ is a bijection then every element a in A is assigned to exactly one b in B, namely, $b = f(a)$. So B has at least as many elements as A which means that $m \leq n$. On the other hand, f is also onto B so that for each element b in B there is some a in A that is assigned to b. This means that A has at least as many elements as B so $m \geq n$. Therefore, it must be true that $m = n$. This type of reasoning does not extend immediately to infinite sets since we do not have the sizes m and n that can be readily compared to each other. In particular, if a finite set is a proper subset of another then these sets cannot have the same number of elements. However, this is not the case for infinite sets: it often easy to find bijections between an infinite set and proper infinite subsets of it. I will discuss this and related issues in Chapter 3.*

Exercise 22 *Consider a relation $\sim$ (among subsets of a given set U) defined as $A \sim B$ if there is a bijection $f : A \longrightarrow B$ where A and B are subsets of U. Verify that $\sim$ is an equivalence relation (this relation is called "equipollence" and as we see later, it is important when dealing with infinite sets in Chapter 3).*

Sets, Functions and Logic 55

Inverse functions.

If f is a bijection then as we saw above the inverse image $f^{-1}(y)$ of every element in S is a single element $\{x\}$ in D. This means that f^{-1} can be viewed as a function $f^{-1} : S \longrightarrow D$. This is the *inverse function* of f. Rather than writing $f^{-1}(y) = \{x\}$ we drop the braces and write $f^{-1}(y) = x$.

Calculating inverse functions of bijections is not an easy task generally and often requires advanced methods. However, simple examples do exist. We say earlier that $f(x) = 2x-1$ is a bijection. For each y there is a number $x = (y+1)/2$ that is mapped to y by the function $f(x)$. So we conclude that $f^{-1}(y) = (y+1)/2$.

Exercise 23 *(a) Find the inverse function of the affine function $f(x) = 2x - 6$.*
(b) Sketch the graphs of both f and f^{-1} in the same set of coordinate axes. Also sketch the graph of the line $y = x$. What do these graphs tell you?

The following is an important result about the inverse functions that we use later on.

> If $f : D \longrightarrow S$ is any bijection of a pair of sets D and S then the following are true:
> (i) $f^{-1} : S \longrightarrow D$ is a bijection;
> (ii) $f^{-1}(f(x)) = x$ and $f(f^{-1}(y)) = y$ for every $x \in D$ and $y \in S$.

Exercise 24 *Explain why (a) and (b) above are true.*

Important examples of inverse functions in analysis are the logarithmic and exponential functions that we discuss in Chapter 8.

2.3   Basic logical concepts and operations

At various points earlier in this chapter we came across logical terms like implication, equivalence, converse, contradiction, etc. We even proved a couple of simple statements. In this section, I present a few terms and concepts from mathematical logic that are essential to a meaningful, even if minimal, discussion of some mathematical topics in later chapters. It is true that many ideas in calculus, including some that involve infinity, can be illustrated by diagrams and concrete examples and we put this thought to good use. But some ideas, especially ideas that involve the infinity in some of its incarnations, can be properly understood only through the use of logical deduction; a simple example of this is the proof (by contradiction) that $\sqrt{2}$ is irrational.

Analysis, like the other major branches of mathematics, uses *binary logic* (every statement is either *true* or *false*) to deduce its results and communicate them. The advantage of this approach is two-fold: (a) it is *precise*; there is no middle ground and once a statement is proven true it is never false, and (b) it makes *proof by contradiction* possible.

A substantial number of results in analysis that involve infinity would not be provable (that is, it is not possible to prove them true or false) without using proof by contradiction.

The discussion in this section is informal, relying on ordinary language rather than the usual precise and economical symbolism of mathematical logic. If you are curious about the details and value purity and precision then you should definitely check a textbook on mathematical logic and preferably also one on axiomatic set theory. The presentation here is focused on logic's role as a language and communication tool for analysis; it bears some similarity to the standard legal codes and language that lawyers use in presenting cases in a court of law (but without the stress–no fines or jail terms to worry about here!).

The fundamental connectives.

Mathematical statements can be combined, as logical propositions, to create more complex statements using the *conjunction* "and", the *disjunction* "or" (inclusive), and the *implication*. "$\Rightarrow$"

> If p and q are two statements or propositions then:
> (a) the conjunction $[p \text{ and } q]$ is true only when both p is true and q is true, otherwise, it is false;
> (b) the disjunction $[p \text{ or } q]$ is false only when both p is false and q is false, otherwise, it is true;
> (c) the implication $[p \Rightarrow q]$ is false only when p is true and q is false, otherwise, it is true;
> (d) p is (logically) *equivalent* to q, denoted $[p \Leftrightarrow q]$ if $[p \Rightarrow q \text{ and } q \Rightarrow p]$.
> Where confusion is unlikely, we drop the brackets from the implication and just write $p \Rightarrow q$.

You may have noticed from the above description that a false statement implies anything. In mathematics this makes a false statement worthless. But in real life there may be significant consequences; for instance, in a court of law a criminal may be set free or an innocent defendant may be convicted by a "false premise" if care is not taken. In politics, false statements (whether deliberate lies or not) are commonly used to direct (or deceive?) uncritical voters to support (or oppose) a particular candidate or idea that they might oppose (respectively, support) if provided with the truth.

Bertrand Russell wins an argument! *People who have not dealt significantly with math or with logic may find it hard to accept that a false statement can, with sufficient cleverness, imply anything. Consider this story about the mathematician, logician and philosopher Bertrand Russell (1872-1970): At a party he tried to explain this exact point about to someone who kept rejecting it, but eventually agreed to accept it if Russell could prove that "$1 = 0$" implies "Russell was the Pope". After a moment's reflection, Russell said: if $1 = 0$ then $2 = 1$". Pope and I are two, so Pope and I are one.*

For amusement, consider using Russell's idea to show that "$1 = 0$" implies that all human beings are 5 feet tall; or that all Americans are mathematicians. Just show that $1 = 0$ implies $N = 1$ for any positive integer N and the rest should be easy!

Sets, Functions and Logic 57

> **Proving necessity and sufficiency:** *In mathematical statements of theorems and their proofs, we read $p \Rightarrow q$ as "p implies q" by which we mean "if p is true then q is true". Also, $p \Leftrightarrow q$ is written as "p (is true or false) if and only if q (is true or false)". The combination if and only if is often abbreviated as iff. To prove a statement containing "if and only if" we must prove the truth of both the "if" part $p \Rightarrow q$ and the "only if" part $q \Rightarrow p$. The order in which this is done is not important and usually dictated by pedagogy. In mathematics an iff statement is also called a necessary and sufficient condition.*

Truth tables.

A simple way of checking the validity of abstract logical statements is provided by means of their *truth tables*. Each statement p is assigned one of two possible "truth values": T for true and F for false. The definitions of fundamental connectives are more comprehensively illustrated using their truth tables.

For conjunction (and) we have the following table:

p	q	p and q
T	T	T
T	F	F
F	T	F
F	F	F

This shows what (a) above stated as well as other possible combinations that lead to the false outcome. For disjunction (inclusive or) we have the table:

p	q	p or q
T	T	T
T	F	T
F	T	T
F	F	F

The only combination that yields a false outcome is when both p and q are false. For implication the truth table is as follows:

p	q	$p \Rightarrow q$
T	T	T
T	F	F
F	T	T
F	F	T

Using combinations of just these 3 truth tables we can create many more new truth tables. For example, consider the compound statement

$$[p \Rightarrow q \text{ and } p] \Rightarrow q \qquad (2.11)$$

In addition to the two columns for p and q we need to have columns for the conjunction and the implications:

p	q	$p \Rightarrow q$	$p \Rightarrow q$ and p	$[p \Rightarrow q$ and $p] \Rightarrow q$
T	T	T	T	T
T	F	F	F	T
F	T	T	F	T
F	F	T	F	T

To see why the entry in first row of truth values and fourth column is a T we check the entries in the first row for $p \Rightarrow q$ and p: they both have the value T so their conjunction also comes up T. In the second, third and foruth rows of the same column one of $p \Rightarrow q$ or p has the value F so their conjunction also has value F. In the final column for the outermost implication we consider the truth values for $[p \Rightarrow q$ and $p]$ and those for q, in that order, and find that the combination T⇒F never occurs. So according to the truth table for implication the final column can only have T as valid entries.

A truth table such as the last one that we considered with all entries of its final column being T shows that the statement in the final column is always true. In this case we have a *tautology*.

Exercise 25 *Use a truth table to how that the following statement is a tautology:*

$$p \text{ or } [\text{not } p] \tag{2.12}$$

This statement is often called the "law of excluded middle" because it says that either a statement or its negation must be always true; there is no other alternative or "middle ground". In a multi-valued logic where there are other alternatives to true or false, (2.12) is not a tautology and (2.13) is not a contradiction. In such logics proofs by contradiction may not work which is a major issue that brings into question what is considered knowable and the extent to which a concept is thought to be understood.

Quantifiers.

Also essential are the two *quantifiers* for statements that contain *variables* like x (x may be a number, vector or any other valid concept):

> **Quantifiers.** The *universal quantifier*: $[\text{for all } x]p(x)$ means that the statement or proposition $p(x)$ is *true for all values* of the variable x;
> The *existential quantifier*: $[\text{there exists } x]p(x)$, or $[\text{for some } x]p(x)$, means that the statement or proposition $p(x)$ is *true for some value* of the variable x;
> The notation $[\text{there exists } x]$ is usually read as: "there exists (or there is) an x such that".
> The symbolic abbreviations $\forall$ for "for all" and $\exists$ for "there exists" are often used in more advanced or more formal contexts but we will not use them.

Propositions or statements containing more that one variable often occur in math. If $p(x, y)$ is a statement with two variables then we may use more than one quantifier to "bound" the variables, as in $[\text{for all } x][\text{there exists } y]p(x, y)$. A concrete example is:

"for all x, there exists a y such that $x + y > 0$"

Sets, Functions and Logic 59

Note that any $y > -x$ will do in this case, so in general no assumption of uniqueness is involved for the "there exists" quantifier. Of course, if I replaced the inequality $>$ with equality $=$ then only a unique $y = -x$ would work.

Negation.

The definition is straightforward:

The *negation* of a statement p is $[not\ p]$.

It is evident that the negation of p is true when p is false and vice versa.

The negations of conjunctions, disjunctions, implications and the quantifiers are obtained as follows:

$not[p\ and\ q] \Leftrightarrow [not\ p\ or\ not\ q]$;
$not[p\ or\ q] \Leftrightarrow [not\ p\ and\ not\ q]$;
$not[p \Rightarrow q] \Leftrightarrow [not\ p\ or\ q]$;
$not[for\ all\ x]p(x) \Leftrightarrow [there\ exists\ x]not\ p(x)$, or perhaps easier to read: $[for\ some\ x]not\ p(x)$
$not[there\ exists\ x]p(x) \Leftrightarrow [for\ all\ x]not\ p(x)$, or perhaps easier to read: $not[for\ some\ x]p(x) \Leftrightarrow [for\ all\ x]not\ p(x)$

The negation of implication can be stated more informally as: $p \not\Rightarrow q$ or "p does not imply q" if "p is true and q is false".

We sometimes need to consider the negation of composite statements where the above list comes in very handy. For instance, consider the mathematical statement:

For all x, there is y such that $p(x, y)$ is true.

The negation of this may be worked out as follows:

$$not[for\ all\ x[[there\ exists\ y]p(x,y)]] \Leftrightarrow there\ exists\ x[not[for\ some\ y]p(x,y)]$$
$$\Leftrightarrow there\ exists\ x[for\ all\ y]p(x,y)$$

The last statement can be written less formally as:

There is x such that $p(x, y)$ is true for all y.

Notice the absence of the word "not" in the above negation. Switching between quantifiers is commonplace in analysis arguments.

Converse, contrapositive and contradiction.

Related to the implication are the following two concepts that arise often in mathematical writing:

The *converse* of $p \Rightarrow q$ is $q \Rightarrow p$;
The *contrapositive* of $p \Rightarrow q$ is $[not\ q \Rightarrow not\ p]$

The truth of $p \Rightarrow q$ has no bearing on its converse $q \Rightarrow p$ which may be true or false; for instance, assuming that "if we eat tainted food then we get sick" is a true statement, its converse "if are sick then we ate tainted food" is certainly open to question. However, the contrapositive of $p \Rightarrow q$ is logically equivalent to it:

$$[\text{not } q \Rightarrow \text{not } p] \Leftrightarrow [p \Rightarrow q]$$

This is easy to show using the truth table method:

p	q	not q	not p	not $q \Rightarrow$ not p
T	T	F	F	T
T	F	T	F	F
F	T	F	T	T
F	F	T	T	T

Notice that the right-most column has the same truth values as the the truth table for $p \Rightarrow q$ regardless of the truth values of p or q. Therefore, we have equivalence.

Exercise 26 *Prove each of the following equivalences:*
$[p \Rightarrow q] \Leftrightarrow \text{not}[p \text{ and } [\text{not } q]]$; $[p \text{ or } q] \Leftrightarrow \text{not}[\text{not } p \text{ and not } q]$
These statements show that implication and disjunction may be defined in terms of negation and conjunction.

A *contradiction* is a statement that is always false. The most basic contradiction is the statement:
$$p \text{ and } [\text{not } p] \qquad (2.13)$$

Let's check the truth table for this statement:

p	not p	p and [not p]
T	F	F
F	T	F

Notice that the only truth value in the right-most column is F. This is generally true of contradictory statements which turn out to be the opposites of tautologies.

Proof by contradiction, or indirect proof. This usually involves showing that negating the conclusion to be proved leads to (2.13) with respect to some hypothesis p. We will do several proofs by contradiction later on. A proof done by contradiction is also called an *indirect proof*. This type of argument is often unavoidable when working with infinity.

Rules of inference.

Proofs in mathematics proceed via *rules of inference*. The most fundamental is the following:

Modus ponens: $[p \Rightarrow q \text{ and } p] \Rightarrow q$, or in words:
"if p implies q and p is true then q is true".

For instance, suppose that "if it is raining then we need an umbrella". This implication can be written as $p \Rightarrow q$ if p is the statement "it is raining" and q is "we need an umbrella". Now the

Sets, Functions and Logic

modus ponens says something rather obvious, that is, if we grant the implication and also grant that it is raining then it follows that we need an umbrella. Indeed, modus ponens is always true (or a *tautology*) as we saw in the truth table for (2.11) above.

Many mathematical theorems are proved by applying the rules of inference multiple times if necessary, along with the other facts that are mentioned above. A specific example from calculus is the following: we show later that "if $f(x)$ is a continuous function then it is integrable". We also see that the function $f(x) = \sin(x^2)$ is continuous, since it is a composition of two continuous functions $\sin x$ and x^2. Therefore, by the modus ponens, $\sin(x^2)$ is integrable. Inferences of this sort are mathematically useful information; for instance, there is no known elementary formula for the integral of $\sin(x^2)$.

A similar statement that is always true (can be verified using a truth table) is the *transitivity of implication*, another rule of inference which is also called the *(hypothetical) syllogism*:[9]

$$\boxed{\textbf{Syllogism (hypothetical):}\ [p \Rightarrow q \text{ and } q \Rightarrow r] \Rightarrow [p \Rightarrow r]}$$

Here is an example from calculus: we prove later that "if a function $f(x)$ has a derivative then it is continuous" and "if $f(x)$ is continuous then it is integrable". From these statements we infer that "if $f(x)$ has a derivative then it is integrable". Again it is easier to find the derivative of a function than its integral so the information here is useful. For instance, using basic derivative rules, not only can we show $\sin(x^2)$ has a derivative but in fact calculate its derivative as $2x\cos(x^2)$. So again we can tell that $\sin(x^2)$ is integrable without knowing what its integral is. This knowledge comes to us indirectly through logical relations from a relatively simple derivative formula.

---

[9]The term "hypothetical" is used because the syllogism involves a hypothetical statement, if p then q. This is a little different from the more familiar "categorical syllogism" consisting of a trio of statements like: "All mammals are apes" and "Uncle Joe is a mammal" from which it follows that "Uncle Joe is an ape" (this conclusion is true if the statement: "all mammals are apes" is true).

Chapter 3

Infinities: *infinitely many of them!*

The broadest definition of infinity is a negation: *not finite*. But this doesn't define anything specific, so to gain some understanding of the concept we may either appeal to intuition or pursue other possibilities.

Intuition *does* help us distinguish between *infinite things that look different* to the mind, like $\mathbb{N}$ and $\mathbb{R}$, both of which are actually *idealizations* of what our intuition tells us. We visualize $\mathbb{R}$ as a straight line (the x-axis), a sort of idealized long, thin rod. But we think of $\mathbb{N}$ as a discrete collection of points, or numbers 1, 2, 3, etc on the x-axis. Both $\mathbb{R}$ and $\mathbb{N}$ are infinite but in what precise sense are they different?

The above question strikes at the heart of the matter as far as analysis is concerned. The description that I gave used imprecise language (idealized rod, discrete points) that does not stand to scrutiny. For example, the numbers in $\mathbb{N}$ are also points in $\mathbb{R}$ so we conclude that the latter is a larger collection of objects than the former. But if they are both infinite then what does it mean to call one larger than the other? In fact, the same question arises regarding other special numbers. For instance, there are more positive integers than even positive integers 2, 4, 6, etc but again, there are infinitely many even positive integers.

It seems then that to understand the difference between $\mathbb{N}$ and $\mathbb{R}$ we need to go beyond intuition and develop sufficiently precise concepts and language that stands up to scrutiny.

In the first chapter of this book, we drew a distinction between number and magnitude. We noted that while magnitude can be described as a positve number, there are numbers that represent no magnitudes, such as 0 or the negative numbers. The distinction between set and magnitude is also important in the sense that unlike infinite magnitude which is not intuitively comprehensible, we can readily hold the abstract concept of an infinite set in mind as a *single* object that by assumption contains an infinite number of other objects. For example, imagine a bucket of sand which is thought of as a single object (one bucket) that contains a very large number of sand grains; to imagine, or handle the bucket and its content we don't need to visualize or manipulate every single grain of sand. In a similar vein, we can think of sets as being "actually infinite" without having to imagine, or work with infinite-sized objects. Physical reality is inconceivable without physical measurements, something that we cannot apply to infinite (or infinitesimal) magnitudes.

The above idea suggests that we may define infinity and work with it in a consistent way using sets rather than magnitudes. The core of what we discuss in this chapter comes from Georg Cantor's ideas and publications in the 1870's. The discussion falls far short of what we see nowadays in a book on axiomatic set theory but it is precise enough to clarify the issues that I mentioned above.

3.1 Bijections, cardinality and Hilbert's hotel

You may have had the unpleasant experience of wanting to book a room at a hotel or a seat on an airliner during the peak travel season but finding that no rooms, or seats, are available. These mundane problems belong to our finite world; in a realm where the infinity is a thing, these issues aren't!

There is always a vacancy at Mr. Hilbert's *Hotel Infinity*!

Let's start with a thought experiment: consider a hotel with an infinite number of rooms and call it *Hotel Infinity*.[1] Even with a firm policy of allowing only one guest per room, Mr. Hilbert discovered that he could always find vacancies in his unusual place without throwing anyone out; here is how: If all rooms are full when a new guest arrives then he moves the guest in Room 1 to Room 2, moves the guest in Room 2 to Room 3 and so on. In this way Room 1 is made available and all existing guests still have rooms of their own.

What about the guest in the last room? Well, there is no "last room" so no problem! That's infinity for you!

I should also emphasize that this shifting of guests can be done repeatedly, so even more guests can be accommodated should more of them arrive. Throughout this process, the hotel remains the same (no new construction necessary) the number of rooms does not change and the policy of one guest per room is strictly enforced. Like a magician, the proprietor always finds empty rooms by moving the guests around even when the place is verified as fully booked! As we soon discover, Hilbert's hotel can be more paradoxical yet: it can accommodate infinitely many new guests not just any (finite) number of them.

This story shows that *subsets of infinite sets can have as many elements as their parent sets*. Let's see what this means next.

Bijections and cardinality: counting things by pairing them up.

Important in studying infinite sets is to find a consistent way to compare the cardinalities (or sizes) of infinite sets. This is done using bijections which we discussed in the previous chapter.

> A pair of sets A and B have the same cardinal number, or $\#A = \#B$, if there is a bijection from A onto B.

For finite sets this amounts to saying that A and B have the same number of elements. But for infinite sets the situation is more complex. In particular, an infinite set can have the same cardinal number as some of its proper subsets. In fact, following Cantor, we may define infinite sets as just such sets![2]

---

[1] This hotel was introduced as a thought experiment by the German mathematician David Hilbert (1862-1943) who referred to it as the "Grand Hotel" in a 1924 lecture on infinity. Hilbert's lecture was not published but it was popularized by the famous physicist George Gamow in his classic 1947 book "One Two Three... Infinity" and since then by many other authors.

[2] Defining infinite sets is no simple task in set theory and Cantor's definition is not fault-proof. If you are curious about a more precise definition and further discussions then consider reviewing a text on axiomatic set theory.

Infinities

> **Infinite set:** (Cantor) *A set is infinite if it can be put in a one to one correspondence (bijection) with a proper subset of itself.*

Consider Hilbert's hotel again. The set of guests in the original fully booked hotel is a *proper* subset of the set of guests in the hotel after new guests arrive. This is not paradoxical because if we label the first guest who arrives after the hotel is "fully booked" as g_0 and let $g_1, g_2, g_3, \ldots$ be the guests already in rooms 1,2,3, etc then we can set up a matching $f : \{0, 1, 2, 3, \ldots\} \to \mathbb{N}$ between the new number of guests and the set of rooms:

$$f(g_n) \to n+1, \quad n = 0, 1, 2, 3, \ldots$$

This f is a bijection with $f(g_0) = 1$ with $n = 0$ (the new guest is placed in room 1), $f(g_1) = 2$ with $n = 1$ (the guest already in room 1 is moved to room 2) and so on. We see that the set $\{0, 1, 2, 3, \ldots\}$ has the same cardinality as its proper subset $\mathbb{N}$; that is, the now greater number of guests can be matched evenly with the same number of rooms.

Exercise 27 *(a) Suppose that in response to endless complaints by inconvenienced guests, Hotel Infinity proprietor D. Hilbert decides on a less invasive plan to accommodate new guests. He moves the guest in room 1 to room 10, the guest in room 10 to room 100, the guest in room 100 to room 1000 and so on and offers a discount on rooms that are powers of 10. What function f describes this reassignment of guests? Notice that positive integer powers of 10 can be matched in a one to one way with the positive integers!*

(b) Later on, a huge bus with infinitely many passengers arrives at the hotel. Unfazed, Hilbert moves the guest in room 1 to room 2, moves the guest in room 2 to room 4, the guest in room 3 is moved to room 6, the guest in room 4 to room 8 and so on. In this way, he relocates all the guests to even-numbered rooms and makes all the odd-numbered rooms vacant and ready to receive the infinitely many new guests. What function f describes this reassignment of guests?

Next, let's show that the sets $\mathbb{N}$ and $\mathbb{Z}$ (all integers) have the same cardinality $\aleph_0$. Note that $\mathbb{N}$ is a disjoint union of the two sets of even numbers and odd numbers. These two subsets of $\mathbb{N}$ can be evenly matched with the positive and non-positive integers as shown below:

$$\underset{\text{odd numbers}}{\underset{\longleftarrow}{\begin{array}{ccccccc} \cdots & -2 & -1 & 0 & 1 & 2 & \cdots \\ & 5 & 3 & 1 & 2 & 4 & \end{array}}} \underset{\text{even numbers}}{\underset{\longrightarrow}{}}$$

This assignment can be coded as a function $f : \mathbb{Z} \to \mathbb{N}$ in the following way:

$$f(n) = \begin{cases} 2n, & \text{if } n > 0 \\ -2n+1, & \text{if } n \leq 0 \end{cases}$$

For instance, since $0 \leq 0$ we have $f(0) = -2(0) + 1 = 1$ as shown in the table; similarly, $f(1) = 2(1) = 2$ since $1 > 0$ and $f(-1) = -2(-1) + 1 = 3$ since $-1 \leq 0$. The existence of a bijection between $\mathbb{N}$ and $\mathbb{Z}$ proves that

$$\#\mathbb{Z} = \#\mathbb{N} = \aleph_0$$

I emphasize that the above bijection is not unique; see Exercise 28. However, the existence of *any one bijection* is enough to show that two sets have the same cardinality.

Exercise 28 *Show that the following function is also a bijection form $\mathbb{Z}$ to $\mathbb{N}$:*

$$f(n) = \begin{cases} 2n+1, & \text{if } n \geq 0 \\ -2n, & \text{if } n < 0 \end{cases}$$

Which integers does this function map onto the odd numbers?

Equipollent sets.

From Exercise 22 recall that the existence of a bijection between two sets A and B gives an equivalence relation $A \sim B$. This is just the *equipollence relation* so A and B are *equipollent sets*.

In particular, the symmetry condition $A \sim B$ implies $B \sim A$ because the inverse function $f^{-1}: B \to A$ is also a bijection. As an example, the function $f: \mathbb{N} \to \mathcal{E}$ defined as $f(n) = 2n$ is a bijection onto the set $\mathcal{E}$ of all even numbers. Its inverse is obtained by solving the equation $2n = k$ for n to get $n = k/2$; this gives the inverse $f^{-1}(k) = k/2$ which is a function from $\mathcal{E}$ onto $\mathbb{N}$.

Exercise 29 *Prove that the set of all odd positive integers, call it $\mathcal{O}$, is equipollent to $\mathbb{N}$ by finding a bijection $f : \mathbb{N} \to \mathcal{O}$. Also calculate f^{-1}.*

Counting the prime numbers.

I end this section on the note that sometimes it is not easy to find a bijection between sets that we know by other means to be equipollent. As an example, consider the set of all prime positive integers: $P = \{2, 3, 5, 7, 11, \ldots\}$. Recall that a positive integer p is a *prime* if $p \neq 1$ and the only integers that divide p are 1 and p itself. An explicit bijection $f : \mathbb{N} \to P$ as in Exercises 29 and 28 is not known in this case since we know of no formula that gives every element of P.[3] But we do know that P *is an infinite proper subset of* $\mathbb{N}$. This has been known since Euclid, whose idea was essentially the following:

Euclid's Theorem. *There are infinitely many prime numbers.*

By way of contradiction, (proof by contradiction!) suppose that there are finitely many prime numbers, which we may list in the order of increasing magnitude as follows:

$$2 = p_1 < 3 = p_2 < 5 = p_3 < \cdots < p_k \text{ (the largest prime)}$$

Note that the product $p_1 p_2 \cdots p_k$ is divisible by every one of these prime numbers. Further, the number $n = p_1 p_2 \cdots p_k + 1$ is not prime by our assumption since it is larger than p_k. So there is a prime p (at least one) that divides n. This p must be one of the prime numbers $p_1, \ldots, p_k$ (assumed to be all of the primes) so it also divides the product $p_1 p_2 \cdots p_k$. Therefore, p divides the difference $n - p_1 p_2 \cdots p_k = 1$. But this contradicts the fact that no integer divides 1 (other than 1). This contradiction occurred because we assumed that there are finitely many primes; therefore, this assumption must be false. We conclude that $\#P$ is larger than every finite number but, of course, no larger than $\aleph_0$; so $\#P = \aleph_0$; that is, P is equipollent to $\mathbb{N}$.[4]

---

[3] If such a bijection $f(n)$ were known then it would determine every prime number–a number theorist's dream come true!

[4] Technically, this conclusion needs further justification. We discuss this issue later when discussing countable sets.

Infinities

3.2 An infinity of infinities: Cantor's theorem

It is time to answer the question: is there any infinite cardinal number other than $\aleph_0$? Cantor showed that there was, using the set of all subsets, or the *power set* of $\mathbb{N}$. Let us use the notation $\mathcal{P}(S)$ for the power set of S.

In Exercise 2 we see that if S has n elements then it has 2^n subsets, including S itself and the empty set. Therefore, $\#\mathcal{P}(S) = 2^n$. In particular, $\#\mathcal{P}(S) > \#S$ for every finite set S. It is worth examining how much larger the exponential function 2^n is compared to n; the table below lists a few values (the last three are approximate):

n	10	20	40	80	120
2^n	1,024	1,048,576	1.099×10^{12}	1.209×10^{24}	1.461×10^{48}

A set with only 20 elements has over a million subsets; a 40-element set has over a trillion subsets; and we run out of familiar words for the number of all subsets of 80-element sets! The number of subsets becomes ridiculously large even when the number of elements is quite modest.

So, how about $\#\mathcal{P}(\mathbb{N})$?

Based on the above discussion, it's reasonable to expect that $\#\mathcal{P}(\mathbb{N})$ is not only greater than $\#\mathbb{N}$ but vastly greater. But in what sense? We are dealing with infinite sets so cannot even assume that $\#\mathcal{P}(\mathbb{N}) > \#\mathbb{N}$ just because this inequality is true for finite sets; after all, if S is a finite set then $\#\mathcal{P}(S)$ may be unimaginably large in numerical terms, but it is still a finite integer. Furthermore, while we have an order relation $>$ for finite cardinal numbers, it is not obvious that a similar relation exists for infinite cardinals.

These issues are complicated matters, requiring axiomatic set theory for their complete resolution, including the definition of an order relation for infinite cardinals. However, our goal here is more modest so we proceed by reasoning as follows: for every set S there is an obvious one to one function $S \to \mathcal{P}(S)$, namely, the function that assigns to each element $a \in S$ the single-element set $\{a\}$ in $\mathcal{P}(S)$. Of course, this function is not onto since it misses many other elements of $\mathcal{P}(S)$, including $\varnothing$ or S. So $\#S$ must be less than or equal to $\#\mathcal{P}(S)$; the next result shows that it is in fact less than, not equal to.

> **Cantor's power-set theorem**: For every set S, whether finite or infinite, $\#S$ is larger than $\#\mathcal{P}(S)$.

A more precise statement for Cantor's Theorem and its proof (by contradiction!) are discussed in Appendix 11.3.

In particular, Cantor's theorem implies that:

$\#\mathcal{P}(\mathbb{N})$ *is an infinity that is greater than* $\aleph_0$.

Borrowing the notation for finite sets, it is common to write

$$\#\mathcal{P}(\mathbb{N}) = 2^{\aleph_0}$$

I emphasize that here $2^{\aleph_0}$ is just notation and *not* "2 raised to the power $\aleph_0$". To get a better feeling for how much bigger $\mathcal{P}(\mathbb{N})$ is than $\mathbb{N}$ (and gain a better sense of how $2^{\aleph_0}$ compares to $\aleph_0$) consider the following sampling of sets that are all in $\mathcal{P}(\mathbb{N})$.

The power-set $\mathcal{P}(\mathbb{N})$ contains all finite subsets of $\mathbb{N}$ (of which there are infinitely many) like

$$\{1, 2, 3, \ldots, 10^{10^{10}}\}, \quad \{10, 200, 3000, 4 \times 10^4, \ldots 9 \times 10^9\}, \quad \text{etc}$$

$\mathcal{P}(\mathbb{N})$ also contains the infinity of all *co-finite* subsets of $\mathbb{N}$, namely, the complements of finite sets, like

$$\{10^{10^{10}} + 1, \ 10^{10^{10}} + 2, \ 10^{10^{10}} + 3, \ldots\},$$
$$\{1, 2, \ldots, 9, 11, 12, \ldots, 199, 201, \ldots, 9 \times 10^9 + 1, \ 9 \times 10^9 + 2, \ldots\}, \quad \text{etc}$$

$\mathcal{P}(\mathbb{N})$ also contains infinite sets that are not co-finite, like the sets of all odd numbers and all prime numbers:

$$\{1, 3, 5, 7, 9, \ldots\}, \quad \{2, 3, 5, 7, 11, \ldots\}$$

$\mathcal{P}(\mathbb{N})$ also contains multiples of fixed integers, like

$$\{2, 4, 6, 8, \ldots, 2n, \ldots\}, \quad \{3, 6, 9, 12, \ldots, 3n, \ldots\}, \quad \{4, 8, 12, 16, \ldots, 4n, \ldots\}, \quad \text{etc}$$

$\mathcal{P}(\mathbb{N})$ also contains sets of powers of integers, like:

$$\{1, 4, 9, 16, \ldots, n^2, \ldots\}, \quad \{1, 8, 27, 64, \ldots, n^3, \ldots\}, \quad \text{etc}$$

Exercise 30 *Can you think of some other infinite sets in $\mathcal{P}(\mathbb{N})$? How about if you add 1 (or 2 or any other fixed positive integer) to every number in each of the sets in multiples or powers? And how about their complements? You can indeed go far creating distinct sets this way (distinct doesn't mean disjoint, so overlapping of sets is not a problem). Doing so will enhance your sense of just how much larger the set $\mathcal{P}(\mathbb{N})$ and its cardinal number $2^{\aleph_0}$ are compared to $\mathbb{N}$ and $\aleph_0$.*

The above discussion shows that the infinite cardinal $\aleph_0$ is less than the infinite cardinal $2^{\aleph_0}$. If you are wondering whether there are any infinite cardinals *between* these two cardinals then you are in good company!

For a finite integer n there are many integers between n and 2^n so it seems plausible that there may be other (infinite) cardinals between $\aleph_0$ and $2^{\aleph_0}$. On the other hand, a moment's reflection should convince you that it is not easy to come up with an infinite set whose cardinal number is larger than $\aleph_0$ and smaller than $2^{\aleph_0}$.

It turns out, to claim that there are *distinct* cardinals between $\aleph_0$ and $2^{\aleph_0}$ is tantamount to refuting a basic hypothesis of modern mathematics, namely, the *continuum hypothesis*, a so-called *undecidable* proposition in set theory.[5]

As far as our work in this book is concerned, whether we accept the continuum hypothesis or ignore it makes no difference because the sets that we are interested in either have cardinal number $\aleph_0$ or $2^{\aleph_0}$.

---

[5] A statement is called "undecidable" if it can be neither proved nor disproved using the axioms of set theory (commonly the Zermelo-Fraenkel axioms).

Infinities 69

We see later that $2^{\aleph_0}$ is in fact the cardinality of the set of all *real* numbers. But for now notice that the power-set theorem yields an *infinite chain of infinities* by repeatedly taking power sets:

$$\#\mathbb{N} < \#\mathcal{P}(\mathbb{N}) < \#\mathcal{P}(\mathcal{P}(\mathbb{N})) < \#\mathcal{P}(\mathcal{P}(\mathcal{P}(\mathbb{N}))) < \cdots$$

or equivalently,

$$\aleph_0 < 2^{\aleph_0} < 2^{2^{\aleph_0}} < 2^{2^{2^{\aleph_0}}} < \ldots \tag{3.1}$$

Sets having cardinal number $2^{2^{\aleph_0}}$ do arise often enough in analysis; an example is the set of all functions defined on an interval like [0,1]. However, our work here does not require us to work with these types of sets.

3.3 Countable or uncountable?

The very first two cardinals in (3.1), namely, $\aleph_0$ and $2^{\aleph_0}$ are the most common infinite cardinals because many familiar sets including $\mathbb{N}$ and $\mathbb{R}$ have these cardinalities. We now discuss the main difference between $\aleph_0$ and $2^{\aleph_0}$.

Every finite set is clearly *countable*, since in principle we can start with any element and keep counting till we run out of elements.

We also think of the set $\mathbb{N}$ as countable; not literally of course, but only in the sense that its elements can be listed one after the other in the form of a sequence. The same listing of consecutive elements is possible for any set that is equipollent to $\mathbb{N}$ for if $f : \mathbb{N} \to S$ is a bijection then $f(1), f(2), f(3), \ldots$ is a listing of elements of S. So, since $\mathbb{N}$ has cardinality $\aleph_0$ we define:

> A set S is *countably infinite* or *denumerable* if S has cardinality $\aleph_0$. If S is either finite or countably infinite then S is *countable*. If S is infinite but *not* equipollent to $\mathbb{N}$ then such a set is not countable, or *uncountable* for short.

In this book, and more generally, in analysis we are primarily interested in two types of infinity: countable and uncountable; by the latter we mean sets with cardinality $2^{\aleph_0}$ or greater.

Saying that $2^{\aleph_0}$ is larger than $\aleph_0$ is not exactly the same as saying 2^n is larger than n. We can get from n to 2^n in a finite number of steps through a string of integers. But how do we get from $\aleph_0$ to $2^{\aleph_0}$? There is a way, but it is not obvious why only that works until we examine some related issues along the way.

Intersections and unions of infinitely many sets.

The concepts of union and intersection of sets that we defined earlier for finite sets in (2.3) and (2.2) respectively, extend to infinite collections of sets in a straightforward way.

Consider a nonempty set $\mathcal{I}$ which may be countable or uncountable and a collection of sets A_i indexed by $\mathcal{I}$. We define the union of this indexed collection of sets as

$$\bigcup_{i \in \mathcal{I}} A_i = \{x : x \in A_{i_0} \text{ for some } i_0 \in \mathcal{I}\} \tag{3.2}$$

The "for some" here means for at least one A_i (say A_{i_0}) but possibly more, maybe all sets A_i. The intersection is defined similarly:

$$\bigcap_{i \in \mathcal{I}} A_i = \{x : x \in A_i \text{ for all } i \in \mathcal{I}\} \tag{3.3}$$

De Morgan laws hold for arbitrary unions and intersections too and as such, serve as basic results in analysis.

> **De Morgan laws (generalized):** *The following equalities are true:*
> $$\left(\bigcup_{i \in \mathcal{I}} A_i\right)^c = \bigcap_{i \in \mathcal{I}} A_i^c \quad \text{and} \quad \left(\bigcap_{i \in \mathcal{I}} A_i\right)^c = \bigcup_{i \in \mathcal{I}} A_i^c \tag{3.4}$$

An easy way to remember these equalities is to notice that distributing the complementation symbol over a union switches it to intersection and vice versa.

Often in calculus-related discussion, $\mathcal{I}$ turns out to be $\mathbb{N}$ (or some set equipollent to $\mathbb{N}$). In this case, since $\mathbb{N}$ is countable and its elements can be listed sequentially, the notation is less abstract:

$$\bigcup_{i \in \mathbb{N}} A_i = A_1 \cup A_2 \cup A_3 \cup \cdots \tag{3.5}$$

$$\bigcap_{i \in \mathbb{N}} A_i = A_1 \cap A_2 \cap A_3 \cap \cdots \tag{3.6}$$

It is worth mentioning that in practice we often run into situations when $\mathcal{I}$ is not $\mathbb{N}$ but some set equipollent to it. For example, the starting index may be 0 or some other integer less than 1 as in $\mathcal{I} = \{-2, -1, 0, 1, 2, \ldots\}$ or the starting index may be greater than 1 as in $\mathcal{I} = \{3, 4, 5, \ldots\}$. As we saw earlier, these index sets are equipollent to $\mathbb{N}$ so there are no essential differences between them and $\mathbb{N}$ as far as indexing goes.

Infinite unions and intersections often give surprising answers. For example, consider the subsets $A_i = \{i, i+1, i+2, \ldots\}$ of $\mathbb{N}$, each an infinite set. What is the intersection of all of these sets?

$$\bigcap_{i \in \mathbb{N}} A_i = \{1, 2, 3, \ldots\} \cap \{2, 3, 4, \ldots\} \cap \{3, 4, 5, \ldots\} \cap \cdots$$

If this infinite intersection has a number n in it then n is in every one of the constituent sets which means that $n \geq i$ for every i in $\mathbb{N}$. But since there is no such n in $\mathbb{N}$ it follows that:

$$\bigcap_{i \in \mathbb{N}} A_i = \varnothing$$

even though each of the sets A_i is an infinite set.

Let $A_1, A_2, A_3, \ldots$ be given sets. If $\#A_i$ is a finite cardinal number for every index $i \in \mathbb{N}$ then we expect that the cardinality of $\bigcup_{i \in \mathbb{N}} A_i$ is $\aleph_0$ since we can just list all elements of all of the sets one after another and get a countable list. Now, what if each A_i is equipollent to $\mathbb{N}$ (or copies of $\mathbb{N}$)?

It turns out that the cardinality of $\bigcup_{i \in \mathbb{N}} A_i$ is still $\aleph_0$; in fact, the following stronger statement is true since it implies that possible overlaps between the sets A_i are not to blame:

> **(CU)** *A countably infinite union of mutually disjoint, countably infinite sets is countably infinite.*

Evidently then, we cannot get from $\aleph_0$ to $2^{\aleph_0}$ by taking unions!

Infinities

Proving (CU) is easy but first let's prove another interesting statement that may seem unrelated:

> The direct product $\mathbb{N} \times \mathbb{N}$ has cardinality $\aleph_0$.

Notice that with this we are showing in effect that $\aleph_0^2 = \aleph_0$. To see that $\mathbb{N} \times \mathbb{N}$ is equipollent to $\mathbb{N}$ let's use an idea of Cantor's. We simply list the elements of $\mathbb{N} \times \mathbb{N}$, namely, ordered pairs of integers in a layered pattern like a matrix; then we count them in a zigzag fashion:

$$\begin{array}{ccccccc}
(1,1) & \longrightarrow & (1,2) & & (1,3) & \longrightarrow & (1,4) & \cdots \\
& \swarrow & & \nearrow & & \swarrow & & \nearrow \\
(2,1) & & (2,2) & & (2,3) & & (2,4) & \cdots \\
\downarrow & \nearrow & & \swarrow & & \nearrow & & \swarrow \\
(3,1) & & (3,2) & & (3,3) & & (3,4) & \cdots \\
& \swarrow & & \nearrow & & \swarrow & & \nearrow \\
(4,1) & & (4,2) & & (4,3) & & (4,4) & \cdots \\
\downarrow & \nearrow & \vdots & \swarrow & \vdots & \nearrow & \vdots
\end{array}$$

Notice that each row is just a copy of $\mathbb{N}$; the arrows define a path starting from (1,1) and proceed along the diagonals in such a way that the sum of the two coordinates along each diagonal equals a fixed integer. The first diagonal is just (1,1) and its coordinates add to 2, then the second diagonal has two pairs (1,2) and (2,1) and the coordinates of each pair add to 3, the third diagonal has three pairs with coordinates of each adding to 4 and so on. If we list the consecutive pairs along the zigzag path we get

$$(1,1), (1,2), (2,1), (3,1), (2,2), (1,3), (1,4), \ldots$$

This sequence of ordered pairs contains all of $\mathbb{N} \times \mathbb{N}$ and, as listed, is in a one to one correspondence with $\mathbb{N}$. This visually defined bijection shows that $\mathbb{N} \times \mathbb{N}$ is equipollent to $\mathbb{N}$.

Exercise 31 *If you are analytically inclined and looking for a more convincing bijection than the visual one above then consider proving that $\mathbb{N} \times \mathbb{N}$ is equipollent to $\mathbb{N}$ by verifying that the following function $f : \mathbb{N} \times \mathbb{N} \to \mathbb{N}$ is a bijection:*

$$f((m,n)) = 2^{m-1}(2n-1) \quad m,n \in \mathbb{N}$$

Exercise 32 *(a) Let $f : \mathbb{Z} \to \mathbb{N}$ be a bijection, say as in Exercise 28. Define the function $g : \mathbb{Z} \times \mathbb{N} \to \mathbb{N} \times \mathbb{N}$ as*

$$g((m,n)) = (f(m), n) \quad m \in \mathbb{Z}, \ n \in \mathbb{N}$$

Show that g is a bijection and conclude from it that $\mathbb{Z} \times \mathbb{N}$ has cardinality $\aleph_0$.

(b) You may alternatively verify that $\mathbb{Z} \times \mathbb{N}$ is equipollent to $\mathbb{Z}$ by showing that the following function is a bijection:

$$f((m,n)) = 2^{m-1}(2n-1), \quad m \in \mathbb{N}, \ n \in \mathbb{Z}$$

Now that we know $\mathbb{N} \times \mathbb{N}$ is equipollent to $\mathbb{N}$ the proof of (CU) is straightforward.

Since each A_i is countably infinite its elements can be put into one to one correspondence with $\mathbb{N}$ and we can write out its elements as a sequence:

$$A_i = \{a_{i,1}, a_{i,2}, a_{i,3}, a_{i,4}, \ldots\}$$

We do this for every i and stack up the sets as follows:

$$A_1 = \{a_{1,1}, a_{1,2}, a_{1,3}, a_{1,4}, \ldots\}$$
$$A_2 = \{a_{2,1}, a_{2,2}, a_{2,3}, a_{2,4}, \ldots\}$$
$$A_3 = \{a_{3,1}, a_{3,2}, a_{3,3}, a_{3,4}, \ldots\}$$
$$A_4 = \{a_{4,1}, a_{4,2}, a_{4,3}, a_{4,4}, \ldots\}$$
$$\vdots$$

Now compare the indices of the elements $a_{m,n}$ with the entries of $\mathbb{N} \times \mathbb{N}$ listed above to notice that the assignment $(m,n) \to a_{m,n}$ defines a bijection (recall also that the sets A_i in (CU) are mutually disjoint so no two $a_{m,n}$ are identical). It follows that (CU) is true.

Exercise 33 *Explain how (CU) also implies the more general statement "a countable union of countable sets is countable." In this case, allow the countable sets to overlap and some or all of them to be finite; recall that a countable union may also be a finite union.*

Since a countable unions of sets with cardinality $\aleph_0$ again has cardinality $\aleph_0$, we may well wonder if there is any *countable* operation that gets us from $\aleph_0$ to $2^{\aleph_0}$; it turns out that there is!

The Cartesian product of infinitely many sets.

Let us consider $\mathbb{N}$ again, the prototype set with cardinality $\aleph_0$. We have already seen that $\mathbb{N} \times \mathbb{N}$ has cardinality $\aleph_0$ and by writing $\mathbb{N} \times \mathbb{N} \times \mathbb{N}$ as $(\mathbb{N} \times \mathbb{N}) \times \mathbb{N}$ we see that $\mathbb{N} \times \mathbb{N} \times \mathbb{N}$ still has cardinality $\aleph_0$. Clearly doing this a *finite* number of times does not change the cardinality so

$$\#(\underbrace{\mathbb{N} \times \mathbb{N} \times \cdots \times \mathbb{N}}_{n \text{ times}}) = \aleph_0$$

We can also rephrase this symbolically as

$$\aleph_0^n = \aleph_0, \quad n \in \mathbb{N}$$

for every positive integer n. But what if we take the direct product of $\mathbb{N}$ with itself $\aleph_0$ times?

To answer this question, first it is necessary to explain what a direct product of a countable infinity of sets means. Let us look at the definition of a finite direct product again, where I should emphasize that the sets A_i need not be countable:

$$\prod_{i=1}^{n} A_i = \{(x_1, x_2, \ldots, x_n) : x_i \in A_i \text{ for } i = 1, 2, \ldots, n\}$$

Infinities

The objects that make up this set are called n-tuples, or n-dimensional vectors. Every vector is essentially a truncated or finite sequence.

What if we do not truncate?

Think of each sequence $x_1, x_2, x_3, \ldots$ as a "vector with $\aleph_0$ components"; then the set of all sequences is a natural candidate for the direct product of a countable infinity of sets:

$$\prod_{i \in \mathbb{N}} A_i = A_1 \times A_2 \times A_3 \times \cdots = \{(x_1, x_2, x_3, \ldots) : x_i \in A_i \text{ for each } i \in \mathbb{N}\}$$

In the special case that $A_i = A$ are all the same set the common notation is the power notation $A^{\mathbb{N}}$; this is shorthand for the set of all sequences in the single set A, or:

$$A^{\mathbb{N}} = A \times A \times A \times \cdots = \{(x_1, x_2, x_3, \ldots) : x_i \in A \text{ for each } i \in \mathbb{N}\}$$

Now, if we pick $A = \mathbb{N}$ then we have:

> *The direct product of $\mathbb{N}$ with itself a countable infinity of times is the set $\mathbb{N}^{\mathbb{N}}$ of all sequences of positive integers.*

So now it remains to find the cardinality of $\mathbb{N}^{\mathbb{N}}$. If we write the cardinal number of $\mathbb{N}^{\mathbb{N}}$ symbolically as $\aleph_0^{\aleph_0}$ then this does seem to be larger than $2^{\aleph_0}$. Let's see if our hunch is valid!

It is easier to work with a $\subset$ *subset* of $\mathbb{N}^{\mathbb{N}}$, specifically, the set of all binary sequences of 0's and 1's. This set is the product of the set $\{0, 1\}$ with itself a countable infinity of times; symbolically, $\{0, 1\}^{\mathbb{N}}$. If you suspect that this set has cardinality $2^{\aleph_0}$ because $\{0, 1\}$ has cardinality 2 then you're right! We show the truth of this by proving the following equivalent statement:

> **(S)** *The set $\{0, 1\}^{\mathbb{N}}$ is equipollent to the power set $\mathcal{P}(\mathbb{N})$.*

The idea is simple. We need to find a bijection $f : \{0, 1\}^{\mathbb{N}} \to \mathcal{P}(\mathbb{N})$ by assigning a unique subset of $\mathbb{N}$ to each and every binary sequence. The natural thing to consider is using the indices of the terms in each sequence. For instance, the sequence $0, 1, 0, 1, \ldots$ has a 1 for every even index 2,4,6,... and a 0 elsewhere. So let's define the set that corresponds to the sequence $0, 1, 0, 1, \ldots$ to be the set of all even numbers $\{2, 4, 6, \ldots\}$. Using the symbol f for this bijection we can write

$$f(0, 1, 0, 1, 0, 1, \ldots) = \{2, 4, 6, \ldots\}$$

Similarly, the finite set $\{1, 5, 8\}$ is assigned to the sequence 1,0,0,0,1,0,0,1,0,... which has all zeros after the eighth place. In symbols,

$$f(1, 0, 0, 0, 1, 0, 0, 1, 0, \ldots) = \{1, 5, 8\}$$

More generally,

$$f(x_1, x_2, x_3, \ldots) = \{n \in \mathbb{N} : x_n = 1\} \tag{3.7}$$

It is not hard to see that the function f in 3.7 is a bijection.

What we have established so far is that $\{0, 1\}^{\mathbb{N}}$ has cardinality $2^{\aleph_0}$. The same holds for any set with two distinct elements $\{a, b\}$ as you can easily show.

Exercise 34 *Assume that $a \neq b$ and specify a bijection between $\{a, b\}^{\mathbb{N}}$ and $\{0, 1\}^{\mathbb{N}}$.*

Subsets and countability.

If a set S is countable and $A \subset S$ then must A be also countable? This makes sense and it is true. By assumption there is a bijection $f : S \to \mathbb{N}$. Define the function $g : A \to \mathbb{N}$ by $g(n) = f(n)$ for every $n \in A$. Then g is one to one and the image set $g(A)$ is a countable set because it is a subset of $\mathbb{N}$. So $g : A \to g(A)$ is a bijection and we conclude that A is a countable set. It is useful to summarize this fact:

> **(CS)** *Every subset of a countable set is countable.*
> As an obvious corollary, we also have the following equivalent contrapositive version:
> **(US)** *If a set S has an uncountable subset then S is uncountable.*

Next, note that every countable (finite or countably infinite) set with at least two elements contains a set like $\{a, b\}$. It follows that if A_n is a countable set for every positive integer n then the product $\prod_{i \in \mathbb{N}} A_i$ must have no fewer elements than the set $\{a, b\}^{\mathbb{N}}$. Therefore, (US) above implies that $\prod_{i \in \mathbb{N}} A_i$ is uncountable.

> **(PS)** *The direct product of a countable infinity of countable sets, each containing at least two elements is uncountable with cardinality $2^{\aleph_0}$.*

Functions as points in an infinite product of sets.

I close this section by extending the concept of direct product further to *all* possible index sets. Recall that every sequence in A is actually a *function* $f : \mathbb{N} \to A$. Now think of an arbitrary set $\mathcal{I}$ as our index set and a collection of sets A_i for $i \in \mathcal{I}$. Define the set of all functions $f : \mathcal{I} \to \bigcup_{i \in \mathcal{I}} A_i = A$ to be the direct product of the A_i, or in symbols:

$$\prod_{i \in \mathcal{I}} A_i = \{f : \mathcal{I} \to A : f(i) \in A_i \text{ for every } i \in \mathcal{I}\}$$

In particular, if all $A_i = A$ are the same set then we obtain:

> *The set of all functions* from $\mathcal{I}$ into A is
> $$A^{\mathcal{I}} = \{f : \mathcal{I} \to A : f(i) \in A \text{ for every } i \in \mathcal{I}\}.$$

For example, $A = \mathbb{N}^{\mathbb{R}}$ is the set of all functions with domain $\mathbb{R}$ taking values in $\mathbb{N}$, like the *step function* $f(x) = [x]$, the greatest integer that is less than or equal to x ($[2.69] = 2$, $[-0.5] = -1$ and so on). If $\mathcal{I} = \mathbb{N}$ then we refer to $A^{\mathbb{N}}$ as the set of all sequences (with range) in A; for example, $\mathbb{Z}^{\mathbb{N}}$ is the set of all sequences of integers. On the other hand, $A^{\mathbb{Z}}$ is the set of all doubly infinite sequences in A; for instance, $\mathbb{R}^{\mathbb{Z}}$ is the set of all doubly-infinite sequences of real numbers, like

$$\ldots, -2\pi, -\pi, 0, \pi, 2\pi, \ldots$$

I emphasize that this *entire* sequence is just *one element* of $\mathbb{R}^{\mathbb{Z}}$.

Infinities

Exercise 35 *Explain in words what the following sets are and give examples of some elements in each:*
$$\mathbb{R}^{\mathbb{N}}, \quad \mathbb{R}^{\mathbb{R}}, \quad \mathbb{N}^{\mathbb{Z}}$$

In a field of analysis called *functional analysis* the A in $A^{\mathcal{I}}$ is often a subset of the real or complex numbers and $\mathcal{I}$ is an infinite set, like $\mathbb{N}$, an interval $[a,b]$, etc. In these cases $A^{\mathcal{I}}$ often has additional structure like a metric (distance function) that turns it into a *space* of functions, or a *function space*. Important examples of function spaces include Hilbert spaces that are basic to quantum physics.

We do not explore functional analysis here because it requires a substantial amount of technical background. To get a flavor of functional analytic topics, check Appendix 11.5 where I discuss the discontinuity of the length function that is defined on the space of continuous functions from $\mathcal{I} = [0,1]$ into $A = [0,\infty)$. The discussion in this appendix is tied to the problem of staircase curves that we discussed in Chapter 1.

Chapter 4

Sequences: *taking infinitely many baby steps*

In this chapter we discuss the basics of sequences of numbers. In particular, we focus on convergent sequences, our primary tools for exploring the concept of limit. In Chapter 6 we use sequences to define infinite series of numbers and later in Chapter 9 we move up to sequences and series of functions.

Intuitively, a convergent sequence is like a camera with an infinite resolution lens. I will discuss this analogy in Section 4.3 where we discuss the concepts of convergence and limits. In Chapter 5 we discover that irrational numbers can be identified with convergent sequences of rational numbers. This is the main idea behind Cantor's construction of real numbers!

4.1 Infinite lists of numbers

As we saw earlier, a sequence is a function whose domain is either the set $\mathbb{N}$ or a set that is equipollent to $\mathbb{N}$. If S is a nonempty set, a function

$$f : \mathbb{N} \longrightarrow S$$

has values $f(n)$ in S for every $n \in \mathbb{N}$ and S is therefore the *range* of f; we often say that f is a *sequence in S* somewhat inaccurately for reasons that become clear later on. If, for instance, S is the set $\mathbb{Z}$ of all integers then a sequence $f : \mathbb{N} \to \mathbb{Z}$ is a sequence of integers, such as $f(n) = n^2 - 5n$. The first few values are listed explicitly below:

n	1	2	3	4	5	6	7	$\cdots$
$f(n)$	-4	-6	-6	-4	0	6	14	$\cdots$

In this book, the range S of a sequence is usually the set $\mathbb{R}$ of all real numbers but S can be any nonempty set, finite or infinite.

Finite lists.

If I arrange a large quantity of shirts of three different colors, say, red (R), green (G) and blue (B) in this exact order I get the list

$$R, G, B, R, G, B, R, G, B, \ldots, R, G, B$$

I might say that I have a "sequence of shirts". This finite list is not strictly a sequence because the domain of a sequence must be equipollent to $\mathbb{N}$.

There are actually no proper (infinite) sequences of objects in the universe as we know it since the number of all known particles (of matter or radiation) is finite. If you started with any subatomic particle in the universe and listed all of them in a sequence like the above shirts, the list would stop when you ran out of particles. It would be a long list to be sure, but still far from infinite. I will use the word "list" for finite sequences so as to reserve the term sequence for the main, infinite case where the domain is all of $\mathbb{N}$.

Sequences as infinite strings of numbers.

Even though sequences are special types of functions it is generally more useful to think of them as *strings of numbers or points* rather than as mappings. To develop the proper notation, suppose that s_n is the object in S that corresponds to $f(n)$, so that $f(n) = s_n$. Then it is often helpful to list the first few values of the sequence f explicitly

$$s_1, s_2, s_3, \ldots \tag{4.1}$$

Each of these s values is called a *term* of the sequence. Thus, s_1 is the first term, s_2 is the second term, and so on. For an unspecified n the term s_n is just called the *n-th term* of f. This term is also usually the formula that defines the sequence itself.

For an example of how the representation (4.1) can be useful consider the sequence f defined by

$$f(n) = s_n = \frac{1 + (-1)^n}{2} \tag{4.2}$$

The n-th term of this complicated-looking f is the ratio or fraction on the right hand side. The first four terms are calculated easily:

$$s_1 = \frac{1+(-1)^1}{2} = \frac{1-1}{2} = 0, \quad s_2 = \frac{1+(-1)^2}{2} = \frac{1+1}{2} = 1$$
$$s_3 = \frac{1+(-1)^3}{2} = \frac{1-1}{2} = 0, \quad s_4 = \frac{1+(-1)^4}{2} = \frac{1+1}{2} = 1$$

so that the sequence starts off as $0, 1, 0, 1, \ldots$ Because $(-1)^n = -1$ for odd n (multiplying -1 by itself an odd number of times) all odd terms of the sequence work out to zero. Similarly, $(-1)^n = +1$ for even n so all even terms work out to 1. We see that this is a very simple sequence by listing just a few terms of it. Its range is just the two-element set $\{0, 1\}$ while, of course, the sequence itself is a function on $\mathbb{N}$, or an infinite string of 0's and 1's:

$$0, 1, 0, 1, 0, 1, \ldots$$

A sequence whose range consists of a single element like $\{c\}$ is called a *constant sequence*. It is boringly simple

$$c, c, c, \ldots$$

Sequences

This is usually (and aptly) called *trivial*. Note that for each given c there is precisely one constant sequence, namely, the one shown above. However, with just two distinct terms, like 0 and 1, there are $2^{\aleph_0}$ 0-1 sequences, because as we saw in Chapter 3 the cardinal number of the set $\{0,1\}^{\mathbb{N}}$ of all 0-1 sequences is $2^{\aleph_0}$.

Exercise 36 *Let x be any fixed real number and define the set $\mathbb{N}x = \{nx : n \in \mathbb{N}\} = \{x, 2x, 3x, \ldots\}$ as the range of a sequence. Write the first 6 terms of the sequence for each of the values: $x = -2, 1/2, \sqrt{2}$ (you should get three different sequences). What sequence do you get if $x = 0$?*

Sets versus sequences.

We write the set $\mathbb{N}$ as $\{1, 2, 3, \ldots\}$ using braces. This looks just like the *sequence* $f(n) = n$ which we write as $1, 2, 3, \ldots$ without the braces. The distinction is important in that the elements of a set can be written in any order: for instance, $\{2, 1, 3, 4, \ldots\}$ is identical to $\mathbb{N}$. However, the sequence $2, 1, 3, 4, \ldots$ (call it f_1) is not the same as $1, 2, 3, 4, \ldots$ namely, the sequence f because $f_1(1) = 2$ while $f(1) = 1$ and $f_1(2) = 1$ while $f(2) = 2$. So $f_1 \neq f$. Of course, f and f_1 have the same range $\mathbb{N}$.

This is also a good place to recall that a sequence (always infinite) may well have a finite range; for instance, $0, 1, 0, 1, \ldots$ is an infinite repeating sequence while its range is a set consisting of just two numbers: 0 and 1. The different sequence $0, 0, 1, 0, 0, 1, \ldots$ with the pattern 0,0,1 repeating has the exact same range.

4.2 Sequence types and plots

Sequences of real numbers may be plotted on a real number line, like the x-axis. This is sometimes helpful but visually difficult to assess because we lose track of which point comes before another; for instance, if we plot the sequence in (4.2) on the x-axis we only see two points, one at 0 and the other at 1.

Periodic sequences.

We often plot a sequence in a space-time diagram as a time series. The indices, which are positive integers in $\mathbb{N}$ are listed on the horizontal or x-axis, which is the analog of the time axis and above each integer n, the value of s_n is marked vertically as a point. To make the plot easier to understand, we also connect consecutive points of the sequence plot using line segments (these connecting lines are visual aids and not parts of the sequence). For example, Figure 4.1 shows the plot of the following sequence up to $n = 12$ (I used the special-angle values in Appendix 11.2 to get the fractions shown)

$$s_n = \cos\frac{\pi n}{3}: \quad \frac{1}{2}, -\frac{1}{2}, -1, -\frac{1}{2}, \frac{1}{2}, 1, \frac{1}{2}, -\frac{1}{2}, -1, -\frac{1}{2}, \frac{1}{2}, 1, \ldots \qquad (4.3)$$

Notice from the figure, as well the list of the first 12 elements above that there is a pattern: the first 6 terms repeat. We say that this *sequence is periodic with period six*. The sequence in (4.2) is periodic with period two. You can easily plot this sequence by hand. In general:

> **Periodic sequence:** A sequence s_n is *periodic with period k* if $s_{n+k} = s_n$ for every n. If $k = 1$ then we have a *constant sequence*. A sequence that does not start off periodic but becomes periodic beyond a certain index is *eventually periodic*.

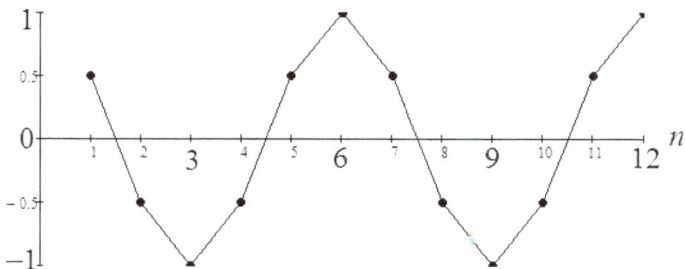

Figure 4.1: A periodic sequence with period 6

There is no actual or physical time involved in the definition of a sequence and "eventually" means "after a certain index". In some applications, the index does track time (like seconds, hours, days, etc) so in these contexts the term "eventually" has its usual temporal meaning.

Exercise 37 *Consider the sequence $s_n = \sin(\pi n/4)$.*

(a) Calculate and list the first 16 terms of this sequence (use the special-angle values in Appendix 11.2). Do you notice any pattern to the behavior of the terms? Is this a periodic sequence? Of what period? Using a plotting device, or by hand, plot the first 16 terms of the sequence and connect the dots as in Figure 4.1.

(b) Verify that $s_{n+8} = s_n$ for all n using the fact that $\sin x$ is a periodic function with period 2π.

(c) Can you figure out the period of $\sin(\pi n/8053)$? You will find the approach in (b) easier than that in (a)–a lot, if not using math software–and as a bonus, in the process of calculation you also prove that this sequence is indeed periodic! Start by letting k be a potential period and figure out what k has to be so that $s_{n+k} = s_n$ for all n.

The following sequence is eventually constant (period 1)

$$\frac{3}{2}, -\frac{1}{2}, \frac{1}{2}, 0, \frac{2}{3}, \frac{1}{3}, \frac{3}{4}, 1, \frac{3}{2}, 2, 2, 2, \ldots$$

We sometimes call the initial segment that is not periodic the *transient part*; again, physical time is not implied in this definition of "transient" outside of applications.

Exercise 38 *Suppose that a sequence is defined as follows:*

$$s_n = 10\cos\frac{\pi n}{3} + \max\left(\frac{50}{n^2}, 1\right)$$

where $\max(x, y)$ means the larger of the two numbers x and y. Explain why this sequence is eventually periodic, what the eventual period is and list the entire transient part.

Sequences

Is $\sin n$ periodic?

Let's close this section by taking a closer look at the sequence $s_n = \sin n$. Is this a periodic sequence? This question is sensible given that for a real variable x the *function* $\sin x$ is periodic and has period 2π. However, the same is not true if the variable is an integer!

To see why the *sequence* $\sin n$ is not periodic let's use a simple proof by contradiction. Let's suppose that $\sin n$ is periodic. Then, in particular, the value $\sin 0 = 0$ must repeat regularly. Recall that $\sin x = 0$ only if $x = k\pi$ for some integer k. But if n is an integer then $n = k\pi$ only when $k = 0$ because π is irrational. So $\sin n = 0$ only when $n = 0$, contradicting our earlier conclusion that the value 0 repeats. It follows that $\sin n$ is not a periodic sequence.

Exercise 39 *Prove that $\cos n$ is not a periodic sequence.*

If the range S of a sequence is an ordered set with an ordering $\preceq$ then some sequences never reverse their direction; if a term s_n is larger (or smaller) than the preceding term s_{n-1} then the next term s_{n+1} is larger (respectively, smaller) than s_n. These sequences are simple but important in calculus.

I start with the basic definitions. To keep technical issues to a minimum while maintaining our focus on what matters, I assume that S is the set $\mathbb{R}$ of real numbers or some subset of it.

Monotone sequences.

Sequences are functions so, recalling the definition of monotone functions earlier, we know what a monotone sequence already is. But it helps to see it in the sequence-specific notation too.

> A sequence s_n of real numbers is *non-decreasing* if $s_{n+1} \geq s_n$ for every n, or *non-increasing* if $s_{n+1} \leq s_n$ for every n. With strict inequalities, s_n is *increasing* if $s_{n+1} > s_n$ for every n or *decreasing* if $s_{n+1} < s_n$ for every n. If a sequence in $\mathbb{R}$ is either non-decreasing or non-increasing then it is a *monotone sequence*.

For example, the sequence $s_n = 2n$ is increasing because

$$s_{n+1} = 2(n+1) = 2n + 2 > 2n = s_n$$

for every n. The first few terms of this sequence are $2, 4, 6, 8, \ldots$, in agreement with our conclusion. On the other hand, $s_n = 1/n$ is decreasing because

$$s_{n+1} = \frac{1}{n+1} < \frac{1}{n} = s_n$$

for every n. The first few terms of this sequence are $1, 1/2, 1/3, 1/4, \ldots$ or (rounded to two decimals): $1, 0.5, 0.33, 0.25, \ldots$

Note that every constant sequence is simultaneously non-decreasing and non-increasing, hence (trivially) monotone. A sequence that is not monotone from the start may become monotone after a finite number of terms.

> If there is a positive integer N such that a sequence s_n is non-decreasing (non-decreasing, monotone) for $n > N$ then s_n is *eventually non-decreasing (respectively, eventually non-increasing, eventually monotone)*.

Exercise 40 *For the following sequence, show that $s_{n+1} - s_n > 0$ for every n, thereby establishing that this sequence is increasing:*
$$s_n = \frac{n-1}{n}$$

Exercise 41 *Calculate and plot s_1 through s_8 for each of the sequences below:*

(a) $\quad s_n = \frac{(-1)^n}{n}$ \quad\quad (b) $\quad s_n = n + \frac{(-1)^n}{n}$ \quad\quad (c) $\quad s_n = \frac{n}{2} - \frac{(-1)^n}{n}$

Use your plot to tell which of these sequences is monotone or eventually monotone.

Oscillating sequences.

As you might imagine there are sequences that are not monotone and they occur often enough to also have been given a name.

> A sequence that is *not* eventually monotone is an *oscillating sequence.*

We have already seen a type of oscillating sequence. All periodic sequences with period 2 or greater are certainly oscillating but oscillating sequences are generally not periodic. For example, the sequence
$$s_n = (-1)^n n^2 \tag{4.4}$$
is oscillating due to repeated sign changes but not periodic because the magnitudes of its terms get larger as n does, so there is no repetition:

n	1	2	3	4	5	6	$\cdots$
s_n	-1	4	-9	16	-25	36	$\cdots$

A more subtle example of an oscillating sequence that is not periodic is $\sin n$ that we discussed above.

Note that eventually monotone sequences and eventually periodic sequences are special types of sequences. So it must be that most sequences of real numbers are oscillating but not eventually periodic!

The so-called *chaotic sequences* that are generated by nonlinear recursions are among oscillating sequences that are not eventually periodic. A familiar example is the discrete "logistic equation":
$$x_{n+1} = ax_n(1 - x_n), \quad n = 1, 2, 3, \ldots$$
with the value of the fixed parameter a between 3.7 and 4. If the initial value x_1 is in the interval $[0, 1]$ then all generated numbers x_n are also in the same interval and they oscillate unpredictably in the sense that slightest changes in the initial value quickly spread to the entire interval.

Sequences

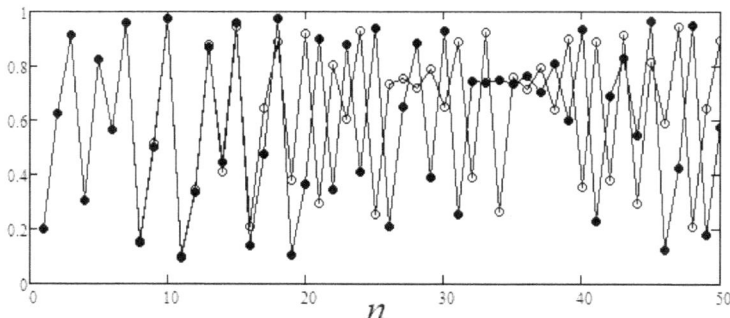

Figure 4.2: A sequence that exhibits chaotic oscillations

Figure 4.2 shows the time series plots of the first 50 terms of two sequences generated by the iteration of the logistic equation above with parameter value $a = 3.9$. The starting values of these sequences are close to each other: $x_1 = 0.2$ (hollow circles) and $x_1 = 0.2001$ (filled circles).

We see in Figure 4.2 that for about a dozen terms the two sequences are in close agreement, but then they begin to diverge from each other and by the 20th term, the two sequences oscillate independently and apart from each other; they have completely forgotten how close they were in the beginning! If we had chosen x_1 a lot closer to 0.2, say, $x_1 = 0.20000001$ then it would take a few more steps before the two sequences diverge but this separation happens eventually for most initial values of x_1 in [0,1]. This tendency of nearby points to move away from each other is a characteristic feature of chaos that has its own name: *sensitivity to initial values*.

This *sensitivity* is a core feature of chaotic systems that in popular literature is known as the "butterfly effect". The similarity of the famous Lorentz attractor to the wings of a butterfly may have helped make the term enduring.

4.3 Convergent sequences: infinite resolution lenses

Consider the sequence
$$s_n = 1 + \frac{1}{n}$$

Notice that as the value of n gets larger the fraction $1/n$ approaches 0; therefore, s_n approaches 1. For example, if $n = 10$ then we find that $s_{10} = 1 + 1/10 = 1.1$ but for $n = 10,000$ we get $s_{10000} = 1.0001$. We say that s_n converges to 1 as n goes to infinity.

The word "converge" in this context is synonymous with "approaches" or "gets arbitrarily close to". It doesn't mean that s_n eventually equals 1; in fact, s_n never equals 1 exactly.

If we stop at a certain value of n, say, $n = 10,000$ then s_n is an approximation of the limiting value (1 here). But if we keep increasing n without ever stopping then we are going beyond approximation. Our study of sequences in this chapter sheds some early light on this important conceptual distinction. You likely sense that if we don't stop at any value of n then we are flirting with infinity at some level. But since we never have $n = \infty$ the point about convergence of sequences is a subtle one.

The infinite resolution lens.

In the physical world we cannot "let n go to infinity" but by stopping at a suitably large value of n we can approximate physical quantities "well enough" or "with sufficient accuracy" for whatever purpose we may have in mind. The idealization of this concept of approximation is *getting arbitrarily close* to some number by never stopping the march of the value of n towards infinity.

Sequences are the natural tools for accomplishing this objective. To illustrate the way that sequences work in an intuitive way, consider a thought experiment.

Suppose that we have a camera, say, on a cell phone.[1] We use it to take a photo of a friend and find that the digital image is sharp and clear. Let's call the current camera setting zoom factor $n = 1$ and the image of your friend as the outcome z_1.

Next, we notice a small bird with beautiful colors sitting on the upper branches of a nearby tree and decide to photograph it. With the $n = 1$ setting the image comes out unclear and hard to make out. So by using our fingers or clicking a button we zoom in to factor $n = 2$ and now obtain a sharp and clear image z_2 of the bird.

When pointing the camera at the bird, we happen to notice a ladybug on the wall of a nearby building. A zoom factor $n = 3$ then gives us a clear image z_3 of this ladybug. Further zooming resolves the ladybug's details. A zoom factor $n = 4$ shows minute details of the wing structure and higher values of n will even show the molecular and atomic structure. At this point, laws of nature stop us from further zooming because according to quantum theory, subatomic particles are not tiny granular objects that can be pinpointed in space. Nature has finite resolution!

But at this time we notice a shiny dot appearing near the side of the building, about half way up from the ground to the roof. Is it a small, remote-controlled drone hovering by the side of the building? We use our miraculous lens to zoom in on the dot and see ... a dot again! We increase the zoom factor over and over again and take new images but each time we see just a shiny dot.

So is the infinite zooming capability of no use in this case?

Not quite! We did a number of things worth highlighting. First, we answered our question above: the object was not a drone. Secondly, *because we used an infinite resolution lens* we also established that the shiny dot was not a distant star. A far away star might not be resolvable using conventional lenses but it would eventually show up as a large ball of fire if we use high enough resolution.[2]

We conclude that the shiny dot is a geometric point of space with no length, area or volume.

A mathematically important outcome of our infinite zooming process is the fact that *it pinpointed the location of the shining point in space* alongside the building.

Let's align the x-axis with the vertical edge of the building, with 0 at where the edge meets the ground and 1 where the edge ends in the roof. See Figure 4.3.

We now quantify the zooming process; each time that we zoom let's say the the image is magnified by a factor of 10. This is equivalent to magnifying the edge of the building 10 fold so each tenth of a unit (0.1 in length) is stretched to 1 unit. Then we identify the interval that is 0.1 unit long and contains the location of the shining point and discard all other intervals; see Figure 4.3. If a part of the edge appears in the image then this part is magnified 10 times in the new image so we can see details like narrow cracks in the concrete.

---

[1] Or a telephoto lens, though for our purposes the most sophisticated telephoto zoom lens is no more effective than a cell phone camera.

[2] Ignoring, as we may in this thought experiment, the optically detrimental effects of Earth's atmosphere, possible interstellar gas and dust in between us and the star and other physically pertinent matters.

Sequences

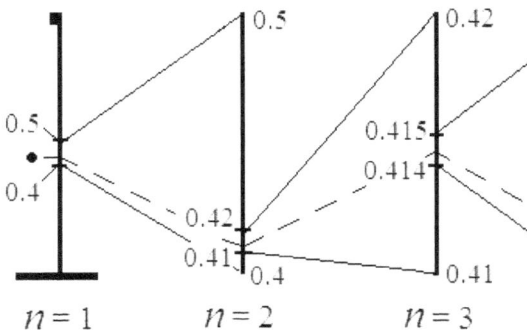

Figure 4.3: Illustrating the zooming process

To make our discussion more concrete, suppose that the dot is between 0.4 and 0.5 on the vertical scale along the side of the building. Let z_1 represent the stretch from 0 to 1, corresponding to the zoom factor $n = 1$. When we zoom once the magnified image (10 times larger) of the stretch from 0.4 to 0.5 is then z_2 (zoom factor $n = 2$). Suppose that we see the spot in the second tenth from the bottom, between 0.41 and 0.42. Now we have narrowed the location of the spot to within 2 decimal places! Zooming in again, we discover that the object falls between 0.414 and 0.415 in the new image z_3; see Figure 4.3. Now the location of the spot is narrowed further to within 3 decimal places.

As we continue to zoom in, the images that we create by going through the above process become the terms of our "zooming sequence" z_1, z_2, z_3,... This is a sequence of ever-shrinking intervals of numbers. The length of z_2 is $1/10$ the length of z_1, z_3 is $1/10$ the length of z_2 or $(1/10)(1/10) = (1/10)^2$ the length of z_1, and so on.

After zooming in n times, the length of z_n is $(1/10)^n$ that of z_1, which has length 1. Since $(1/10)^n$ approaches 0 as n goes to infinity, we conclude that only one number can be common to all the infinitely many zooming intervals. This number gives the *exact location* of the shiny dot in space. This intuitive conclusion is logically supported by a property of the set of real numbers called *the Nested Intervals Property* (NIP).[3]

It is worth emphasizing that the exact location could not be pinpointed after zooming only a finite number of times because at each stage we got an interval of numbers. In digital terms, these are pixels not geometric points.

We can extract a numerical sequence s_n out of the zooming sequence by simply taking the lower (or upper) end point of each interval z_n. For example, $z_1 = [0, 1]$ so $s_1 = 0$; $z_2 = [0.4, 0.5]$ so $s_2 = 0.4$; $z_3 = [0.41, 0.42]$ so $s_3 = 0.41$ and so on. This sequence converges to the location of the dot (say, $x = \sqrt{2} - 1$).

Convergence and limits.

Let's start with some basic terminology. The location of the shiny dot in the thought experiment above, or the number 1 that the sequence $s_n = 1 + 1/n$ converges to, are called *limits*. The notation

---

[3] We discuss this and other important properties of real numbers in Chapter 5. The NIP is equivalent to the most characteristic of all properties of the real numbers, namely, the *completeness property* that gives a precise meaning to their continuous appearance.

"$s_n \to 1$ as $n \to \infty$" is self descriptive with the arrow $\to$ having its usual "goes to" meaning. Think of the notation $n \to \infty$ as indicating that the value of the index n exceeds every given positive integer: a billion, a googol, a zillion, whatever. But crucially, *n does not actually reach an infinite magnitude* ∞ which does not exist; remember that in analysis only *sets* can be infinite!

We commonly use the *limit notation*

$$\lim_{n \to \infty} s_n = 1 \quad \text{or} \quad \lim_{n \to \infty} \left(1 + \frac{1}{n}\right) = 1$$

as shorthand for the phrase "*the limit of* $1 + 1/n$ *as n goes to infinity is* 1". As I mentioned above, intuitively this means that no matter how small a distance anyone picks at any time, say, ε (the Greek letter *epsilon*, a context-standard in analysis) we can find an index N large enough (since $n \to \infty$) so that

$$1 - \varepsilon < 1 + \frac{1}{n} < 1 + \varepsilon \quad \text{if} \quad n > N$$

or equivalently,

$$1 - \varepsilon < s_n < 1 + \varepsilon \quad \text{if} \quad n > N \tag{4.5}$$

The integer N is a "threshold" that once the index n crosses it, the inequalities (4.5) are guaranteed thereafter. They ensure that $1 + 1/n$ is at most ε units away from 1 from either side of 1. This may seem overstated: since $1 + 1/n$ is always greater than 1 why bother with $1 - \varepsilon$? The generality is needed to take care of similar cases like $1 - 1/n$ which is always less than 1 or $1 + (-1)^n/n$ which can be on either side of 1 depending on whether n is odd or even.

Our task is to find a threshold N for indices n, given any $\varepsilon > 0$.[4] Adding a -1 to the inequalities in (4.5) to cancel out the 1's we get

$$-\varepsilon < s_n - 1 < \varepsilon \quad \text{if} \quad n > N \tag{4.6}$$

or equivalently,

$$-\varepsilon < \frac{1}{n} < \varepsilon \quad \text{if} \quad n > N$$

This gives a link between the index n and the arbitrarily small number ε; for example, if a friend picks $\varepsilon = 0.003$ then all you would need to do is choose n large enough that

$$\frac{1}{n} < 0.003$$

is always true. Flipping the fractions (and the direction of inequality) we obtain

$$n > \frac{1}{0.003} \simeq 333.33$$

So you might pick $N = 334$, or any *integer* that is larger than 333.33 to satisfy your friend's criterion for being "really close". The same argument works if ε is given any (positive) value as a measure of a tiny distance. So we just consider the inequality

$$\frac{1}{n} < \varepsilon$$

---

[4] We solve the inequalities (4.5) in terms of n in order to determine N; here, n is a variable like the proverbial x is in an algebraic inequality, say, $3x + 1 > 7$. When you solve this you get a numerical (or sometimes symbolic) answer, like $x > 2$ in this case. Here x is like n and 2 is like N (of course, N must be an integer).

Sequences 87

and flip everything to get
$$n > \frac{1}{\varepsilon}$$

Then any *integer* N that is larger than $1/\varepsilon$ works for any choice of ε (it is not necessary that N be uniquely defined). When this is possible we say that the limit exists. Before stating the formal definition of limit for a converging sequence, I point out that the two inequalities in (4.6) can be written succinctly as one inequality using the absolute value:[5]

$$|s_n - 1| < \varepsilon$$

> **Convergence:** A sequence s_n of real numbers *converges* to a real number s if for every $\varepsilon > 0$ we can find a positive integer N such that for all indices $n > N$
>
> $$|s_n - s| < \varepsilon \qquad (4.7)$$
>
> We call s the *limit* of s_n and write
>
> $$\lim_{n \to \infty} s_n = s$$

The above definition of limit conveys the intuition of numbers s_n approaching a particular number s as the index n goes to infinity. Basically, it shows that the distance between s_n and s, namely, the number $|s_n - s|$ gets smaller than any arbitrarily small positive number ε that is chosen if n larger than some threshold N, a positive integer that depends on ε.

The latter number can be used to define convergence. First, let's observe that if we determine some N that works then all larger N, like $N+1$ or $2N$ also work because once s_n reaches within an ε of s it stays within the ε distance of s for all larger indices n. Since the set $\mathbb{N}$ of all positive integers is well-ordered, *a least or smallest value* of N that works exists too. If ε is fixed then such a least index is uniquely defined. For instance, for the sequence $s_n = 1/n$ above, $N = 334$ is the one and only *least index* that works for $\varepsilon = 0.003$. If $\varepsilon = 0.0002$ then the least index that works is

$$N = \frac{1}{0.0002} = \frac{10000}{2} = 5000$$

For each positive real number x let $[x]$ denote *the least integer that is greater than or equal to* x. For example, $[1/0.003] = 334$, and $[1/0.0002] = 5000$; in short, the least index that works for an arbitrary ε in this example is $[1/\varepsilon]$.[6] We are now ready to define a concept that becomes especially important when studying sequences of functions in Chapter 9.

> **The ε-index of a convergent sequence.** Assume that a sequence s_n converges to a real number. For each $\varepsilon > 0$, *the least index N_ε for which (4.7) holds is the ε-index of the sequence s_n*.

Note that $N_\varepsilon < \infty$ for each $\varepsilon > 0$ when the sequence converges, although it is possible (and generally true) that $N_\varepsilon \to \infty$ as $\varepsilon \to 0$. For example, $N_\varepsilon = [1/\varepsilon]$ for the sequence $s_n = 1/n$ and $1/\varepsilon$

---

[5] Recall that the absolute value of a number x is just the magnitude of x stripped of its sign. Thus $|-2| = 2$ and $|+2| = 2$. More generally, $|x| = x$ if $x \geq 0$ and $|x| = -x$ if $x < 0$.

[6] The notation [ ] defines a function whose domain is the set of all positive real numbers; its range is the set of all non-negative integers $0, 1, 2, \ldots$ This function is often called the "least integer function". The domain of [ ] may be extended to all of the real numbers (and it range to all integers) but we do not need this extension here.

goes to infinity as ε goes to 0. Further, N_ε is a decreasing function of ε since if $\varepsilon_1 < \varepsilon_2$ and (4.7) holds for ε_1 then it certainly holds for ε_2 as well, which means that $N_{\varepsilon_1} \geq N_{\varepsilon_2}$.

The concept of ε-index is useful for relating the idea of convergence of a sequence to the finiteness of the ε-index, especially for sequences of functions. We return to this idea later on.

Exercise 42 *Use the above definition of convergence to explain why the constant sequence $c, c, c, \ldots$ converges to the number c. Can you see why $N_\varepsilon = 1$ for every $\varepsilon > 0$?*

Exercise 43 *Use an argument involving ε and N to prove that for any (fixed) real number c*

$$\lim_{n \to \infty} \frac{c}{n} = 0$$

Exercise 44 *Consider the two sequences*

$$s_n = 1 - \frac{1}{n^2} \qquad s'_n = 1 + \frac{1}{\sqrt{n}}$$

(a) List the first six terms $s_1, s_2, \ldots, s_6$ of s_n explicitly; round your answers to 3 decimal places
(b) For $\varepsilon = 0.003$ find a positive integer N such that $|s_n - 1| < \varepsilon$ for all $n > N$. What is N_ε?
(c) Find N_ε for any (unspecified) $\varepsilon > 0$ to prove that $\lim_{n \to \infty} s_n = 1$
(d) Repeat (a)-(c) for s'_n.

The above definition does not require introducing infinitely large or infinitely small magnitudes or quantities. The "arbitrarily small" variable number ε is always a *finite* real number. The notation $n \to \infty$ means that n gets "arbitrarily large", but all n past the *always finite* threshold N are bundled as a *set* of numbers and not treated individually.

Further, as we see in the limit notation itself, two types of infinities are related to each other via convergence. There is $n \to \infty$ which involves the countable infinity of $\mathbb{N}$, and the more subtle approaching of the terms s_n to s in the set of real numbers (or even larger sets). In particular, we define irrational numbers later using convergent sequences of rational numbers and manage the infinity in each irrational number with the help of convergent sequences.

The algebra of limits.

Sequences are functions and as such, they can be added, multiplied, etc; in short, the algebra of functions applies to sequences as well. If we have two sequences $s_1, s_2, s_3, \ldots$ and $s'_1, s'_2, s'_3, \ldots$ then we can perform the following algebraic operations:

$$s_1 + s'_1, s_2 + s'_2, s_3 + s'_3, \ldots \quad \textit{addition term by term, abbreviated as} \quad s_n + s'_n$$

$$s_1 s'_1, s_2 s'_2, s_3 s'_3, \ldots \quad \textit{multiplication term by term, abbreviated as} \quad s_n s'_n$$

Subtraction and division are defined using addition and multiplication, respectively as: $s_n + (-s'_n)$ and: $s_n(1/s'_n)$ if $s'_n \neq 0$ for all n.

Sequences

Now, suppose that both s_n and s'_n converge to real numbers. What can we say about the convergence of their sum, product, etc? The following gives some expected answers.

Convergence and algebraic operations. *Assume that $s_n \to s$ and $s'_n \to s'$ as $n \to \infty$. Then:*

(a) The sum of the two sequences $s_n + s'_n$ is a convergent sequence and
$$s_n + s'_n \to s + s'$$

(b) The product of the two sequences $s_n s'_n$ is a convergent sequence and
$$s_n s'_n \to s s'$$

(c) The quotient of the two sequences s_n/s'_n is a convergent sequence and
$$\frac{s_n}{s'_n} \to \frac{s}{s'}$$

if $s' \neq 0$ and $s'_n \neq 0$ for all n.

The proofs of these statements are given in Appendix 11.7. If we set $s = \lim_{n\to\infty} s_n$ and $s' = \lim_{n\to\infty} s'_n$, then the above list can be written more compactly as follows.

The algebra of limits. *If $\lim_{n\to\infty} s_n$ and $\lim_{n\to\infty} s'_n$ both exist and are real numbers then:*

(a) $\lim_{n\to\infty} (s_n + s'_n) = \lim_{n\to\infty} s_n + \lim_{n\to\infty} s'_n$

(b) $\lim_{n\to\infty} (s_n s'_n) = \left(\lim_{n\to\infty} s_n\right) \left(\lim_{n\to\infty} s'_n\right)$

(c) $\lim_{n\to\infty} \frac{s_n}{s'_n} = \frac{\lim_{n\to\infty} s_n}{\lim_{n\to\infty} s'_n}$ if $\lim_{n\to\infty} s'_n \neq 0$ and $s'_n \neq 0$ for all n

---

These properties help us calculate limits. For example,

$$\lim_{n\to\infty} \left(2 - \frac{7}{n}\right)^4 = \lim_{n\to\infty} \left[\left(2 - \frac{7}{n}\right)\left(2 - \frac{7}{n}\right)\left(2 - \frac{7}{n}\right)\left(2 - \frac{7}{n}\right)\right]$$

$$= \lim_{n\to\infty} \left(2 - \frac{7}{n}\right) \lim_{n\to\infty} \left(2 - \frac{7}{n}\right) \lim_{n\to\infty} \left(2 - \frac{7}{n}\right) \lim_{n\to\infty} \left(2 - \frac{7}{n}\right)$$

$$= \left[\lim_{n\to\infty} \left(2 - \frac{7}{n}\right)\right]^4$$

$$= \left[\lim_{n\to\infty} 2 - \lim_{n\to\infty} \frac{7}{n}\right]^4$$

$$= (2 - 0)^4 = 16$$

Can you tell which properties I used to go from the first line to the second and from the third to the fourth?

We can use the idea in this example to show that for any real number c and any positive integer

$$\lim_{n\to\infty} \frac{c}{n^m} = 0 \qquad (4.8)$$

since we can write

$$\lim_{n\to\infty} \frac{c}{n^m} = \lim_{n\to\infty} \left[c\left(\frac{1}{n}\right)^m\right] = \lim_{n\to\infty} c \lim_{n\to\infty} \left(\frac{1}{n}\right)^m = c\left(\lim_{n\to\infty} \frac{1}{n}\right)^m = c(0)^m = 0$$

You may wonder, what if m is not a positive integer, but say, a positive rational or even irrational number? In such a case the above reasoning does not work because we cannot write the power as repeated multiplication. However, *(4.8) is true if m is any positive real number* and to show it we may use the *definition* of convergence similarly to Exercises 43 and 44 earlier.

The next example introduces another useful idea for calculating limits of certain fractions.

$$\lim_{n\to\infty} \frac{1+9n-2n^5}{7n^2-n^4+6n^5} = \lim_{n\to\infty} \frac{\frac{1+9n-2n^5}{n^5}}{\frac{7n^2-n^4+6n^5}{n^5}}$$

$$= \lim_{n\to\infty} \frac{\frac{1}{n^5}+\frac{9n}{n^5}-\frac{2n^5}{n^5}}{\frac{7n^2}{n^5}-\frac{n^4}{n^5}+\frac{6n^5}{n^5}}$$

$$= \lim_{n\to\infty} \frac{1/n^5 + 9/n^4 - 2}{7/n^3 - 1/n + 6}$$

The key idea was dividing the numerator and denominator by *the highest power in the denominator*, here n^5. Next, apply Property (c) of limits to distribute the limit over the fraction:

$$\lim_{n\to\infty} \frac{1+9n-2n^5}{7n^2-n^4+6n^5} = \frac{\lim_{n\to\infty}(1/n^5) + \lim_{n\to\infty}(9/n^4) - \lim_{n\to\infty} 2}{\lim_{n\to\infty}(7/n^3) - \lim_{n\to\infty}(1/n) + \lim_{n\to\infty} 6}$$

$$= \frac{0+0-2}{0-0+6}$$

$$= -\frac{1}{3}$$

You might be wondering if there are other ways of calculating the above limit; there are, the most well-known of which is called *L'Hôpital's rule* that we discuss in Chapter 9.

Did you know that calculus is possible because a limit theorem fails?

The nonzero conditions in (c) of the above theorems are necessary since we don't want to divide by 0 on either side of the equation. But the case where $s'_n \ne 0$ for all n and

$$\lim_{n\to\infty} s_n = \lim_{n\to\infty} s'_n = 0 \qquad (4.9)$$

is especially important in calculus. In a sense, it includes the concept of *derivative* that we study later. For now, consider an example where

$$s_n = \left(2+\frac{1}{n}\right)^2 - 4 \quad \text{and} \quad s'_n = \frac{1}{n}$$

Sequences 91

This sequence s_n appears in the definition of derivative of $f(x) = x^2$ as the difference $f(2 + 1/n) - f(2)$. Using the algebra of limits we calculate

$$\lim_{n\to\infty} s_n = \lim_{n\to\infty} \left[\left(2 + \frac{1}{n}\right)^2 - 4\right] = (2+0)^2 - 4 = 0 + 0 = 0$$

so evidently (4.9) holds in this case. The right hand side in (c) yields $0/0$; it may be tempting to say this must be 1 since both the numerator and the denominator are the same number. But $0/0$ cannot be assigned a fixed meaning for a good reason: in this particular case, the left hand side of (c) works out to

$$\lim_{n\to\infty} \frac{s_n}{s'_n} = \lim_{n\to\infty} \frac{4/n + 1/n^2}{1/n} = \lim_{n\to\infty} \left[\left(\frac{4}{n} + \frac{1}{n^2}\right)\frac{n}{1}\right] = \lim_{n\to\infty}\left(4 + \frac{1}{n}\right) = 4$$

This may seem weird but it is actually a huge blessing:

> *In the algebra of limits theorem, because (c) fails when (4.9) holds, we can define the concept of derivative and start building calculus up!*

The Squeeze Theorem.

I end this section with a basic and often used theorem. Its odd name is quite apt; to illustrate, consider a thought experiment. Imagine three cars that are moving in an alley. To avoid collisions, the car in the middle must keep ahead of the car that is following it but cannot move past the car that is in front. Now, suppose that the car ahead and the car behind gradually slow down so as to come to a stop in front of a building. Then the car in the middle has to slow down too and eventually stop in front of the same building.

The cars ahead and behind essentially "squeeze" the one in the middle, forcing it to follow. If we take snapshots of the three cars in, say, one-second intervals and line up the photos side by side then we see three interlaced sequences (images of the three cars). One of the sequences is sandwiched or squeezed in between the other two. The inevitable outcome may be stated as a theorem.[7]

> **The Squeeze Theorem.** *Suppose that a_n, b_n and c_n are sequences of real numbers such that $a_n \leq b_n \leq c_n$ for every index n. If the flanking sequences a_n and c_n both converge to the same limit L then b_n also converges to L.*

This theorem is proved in Appendix 11.7. To see how useful this simple result is, consider the sequence

$$s_n = \frac{\sin n}{n}$$

What is the limit of this sequence (as $n \to \infty$)? The Squeeze Theorem provides a quick answer once we find suitable flanking sequences. This is done easily enough here; the sine function oscillates between 1 and -1, that is,

$$-1 \leq \sin n \leq 1$$

for every value of n; when we divide these inequalities by n we obtain

$$-\frac{1}{n} \leq \frac{\sin n}{n} \leq \frac{1}{n}$$

---
[7]The Squeeze Theorem is also called the *Sandwich Theorem* which is also an apt name.

The flanking sequences are $-1/n$ and $1/n$; both of these converge to 0 as $n \to \infty$ so we conclude that s_n converges to 0 as well.

4.4 Divergent sequences: wandering about, near or far

We close this chapter by introducing some common classifications of sequences. A sequence that can be contained in a bounded interval is a bounded sequence. More precisely:

> **Bounded sequence:** A sequence s_n is *bounded* if there are real numbers a and b such that $a \leq s_n \leq b$ for all indices n. An *unbounded sequence* is naturally a sequence that is not bounded.

For instance, the sequence in (4.4) is unbounded.

Every convergent sequence of real numbers is bounded; this is easy to see: if $\lim_{n \to \infty} s_n = s$ then the terms of the sequence cluster around s so the sequence fits in some interval centered at s. Also every periodic sequence is bounded since such a sequence has only a finite number of distinct terms (no more than its period).

For the purposes of classification, we also define:

If a sequence s_n does not converge then it *diverges* or is a *divergent sequence*.

A periodic sequence with period at least 2 is divergent. For example, consider the sequence $(-1)^n$ which has period 2. If n is even then $s_n = 1$ while if n is odd then $s_n = -1$. So no matter how large an N we pick, for $n \geq N$ half the time s_n is 1 and half the time it is -1 so s_n cannot approach any one number as limit. You may have also noticed that this example shows that not every bounded sequence is convergent.

A more dramatic type of divergence is the following:

> **Divergence to infinity:** A sequence s_n of real numbers *diverges to* ∞ (or to $-\infty$) if for every positive real number M there is a positive integer N such that $s_n \geq M$ (respectively, $s_n \leq -M$) for all $n \geq N$.

In other words, s_n diverges to ∞ if its terms eventually exceed every positive real number; we can also see that s_n diverges to $-\infty$ if $-s_n$ diverges to ∞. It is sometime said that s_n "approaches" ∞ (or $-\infty$) as if it were converging to some well-defined magnitude. This technically incorrect usage is intuitively appealing as long as it doesn't mislead.

Every sequence that diverges to infinity is clearly not bounded, or is unbounded. The converse of this statement is false because there are unbounded sequences that do not diverge to infinity; consider the sequence in (4.4) for instance. Half of the terms diverge to ∞ while the other half diverge to $-\infty$. The sequence as a whole diverges but neither ∞ nor to $-\infty$.

Chapter 5

The Real Numbers: *mostly irrational, but not lawless*

After reading and working through a calculus book, we develop some familiarity with the real numbers and Real numbers consist of both rational numbers (ratios of integers) and irrational numbers. We start with a discussion of the rational numbers, which can be considered a skeletal structure for the real number line. Intuitively, if we think of the rational numbers as sand grains lined up in a row then the irrational numbers are like a very thin stream of epoxy glue spread over the sand grains that binds them and forms a hardened, continuous line when dry. Like the glue, the irrational numbers do not have definitive, clear-cut values while, like sand grains, the rationals are far more definitive in value. But as we will see, the image is still not right; there is no finite or discrete scale that captures an essential property of rational numbers called *density* (between every pair of rational numbers there is another rational number) and yet, rational numbers fall short of making a continuous line because a little like the sand grains without the glue, they are not sufficiently bonded together.

In order to properly capture the essence of the number line, mathematicians replace numbers with other objects of which there are enough to account for all real numbers. These objects are usually either sets or sequences. There are three well-known approaches to defining the real numbers.

One approach due to Dedekind[1] uses *sets* rather than sequences. Corresponding to each irrational number are all the rational numbers that are smaller and all the rational numbers that are larger. Such collections of rational numbers are called *Dedekind cuts*. There are enough of these special cuts to account for all the irrational numbers. For example, the number $\sqrt{2}$ uniquely corresponds to the totality of all the rational numbers that are smaller than it. If q is such a rational number then it may be defined by the inequality $q^2 < 2$ for $q > 0$ plus all $q \leq 0$; for instance, 1.3 is in the cut because $1.3^2 = 1.69 < 2$. Infinity comes in by considering not just one such q but all of them simultaneously as a set; for instance, all negative rationals and 0 are in the cut for $\sqrt{2}$, as are the numbers 1.4, 1.41, 1.414, etc.

The second approach, due to Cantor, uses *sequences* of rational numbers. In this approach we identify each irrational numbers with all sequences of rational numbers that converge to it. This

---

[1] Richard Dedekind (1831-1916) was a German mathematician and a doctoral student of Gauss.

correspondence is not unique; an infinite number of sequences of rationals typically converge to every irrational number. But all of these sequences are *equivalent* in a precise sense and every irrational number can be uniquely associated with the collection of all such convergent sequences of rationals.

Each of the above approaches has its technical advantages; Cantor's approach is more closely aligned with our goals here given the importance of sequences both in theory and in practice, and also for greater uniformity in the presentation of later material in this book. In particular, Cantor's approach provides a natural context for introducing the important concept of *Cauchy sequence*, a cornerstone of modern analysis.

Most real analysis and advanced calculus textbooks use neither of the above approaches. They use a more expedient axiomatic approach where the set of all real numbers is defined as a *totally ordered algebraic field that also satisfies a completeness axiom* (the "least upper bound" axiom). We discuss these concepts below after finishing Cantor's construction.

The axiomatic approach is a better fit to the classroom's time constraints where a substantial amount of technical material must be covered in a limited amount of time. The axiomatic approach is quite efficient for proving theorems. However, this method is highly abstract attributes no intrinsic character to individual real numbers. In particular, it glosses over the infinity's presence in each irrational number and the huge size disparity between the sets of rational and irrational numbers. These are topics that we won't want to miss in this book!

5.1 Rational numbers: a tiny but pervasive minority

The set of all ratios of pairs of integers:

$$\mathbb{Q} = \left\{ \frac{m}{n} : m \in \mathbb{Z},\ n \in \mathbb{N} \right\}$$

is the set of all *rational numbers*.

Note that when we fix $n = 1$ we obtain the set $\mathbb{Z}$ of all integers as a proper subset of $\mathbb{Q}$. We take a careful look at the rational numbers and some of their properties in this section.

In discussing the rational numbers, I assume the usual properties of natural numbers and integers at an intuitive level rather than by axioms. No form of infinity is required for defining *individual* integers, and since every rational number is a quotient of two integers, no infinities are used in defining *individual* rational numbers either.

The set of rational numbers is countable.

Recall that the set $\mathbb{Q}$ of rational numbers is countable as a subset of $\mathbb{Z} \times \mathbb{N}$ which we earlier saw was countable in Chapter 3.

Rational numbers form a totally ordered field.

Ordinary addition and multiplication of rational numbers are defined the way we remember them from past algebra experience:

$$\frac{j}{k} + \frac{m}{n} = \frac{km + jn}{kn}, \quad \frac{j}{k} \frac{m}{n} = \frac{jm}{kn}$$

Here $j, m \in \mathbb{Z}$ and $k, n \in \mathbb{N}$. Since sums and products of integers are again integers, the above definitions indicate that sums and products of rationals are again rationals.

The Real Numbers 95

This discussion may seem to suggest that $\mathbb{Q}$ not particularly more structured than $\mathbb{N}$ or $\mathbb{Z}$; but $\mathbb{Q}$ is in fact a more structured set. From an algebraic point of view, $\mathbb{Q}$ is a *field* but neither $\mathbb{N}$ nor $\mathbb{Z}$ are fields. In particular, when we divide two numbers in $\mathbb{Q}$ the result is in $\mathbb{Q}$ (as long as we do not divide by 0) because if $p = m/n$ and $q = j/k$ are rational numbers then

$$\frac{p}{q} = p\frac{1}{q} = \frac{m}{n}\frac{k}{j} = \frac{mk}{nj}$$

is again rational; so operations as simple as taking the average of two numbers

$$\frac{q+p}{2}$$

are possible in $\mathbb{Q}$ but not in $\mathbb{Z}$ or $\mathbb{N}$; for instance, the average of 2 and 7 is 4.5 which is not an integer. Also, we can subtract two numbers in $\mathbb{Q}$ (or $\mathbb{Z}$) as $q - p = q + (-p)$ but not in $\mathbb{N}$ which does not contain the negatives of its elements.

Recall that the set $\mathbb{N}$ has a built-in ordering $<$. This can be extended to a total ordering of all the integers in $\mathbb{Z}$ in a straightforward way by requiring that

$$m < n \quad \text{if } n - m \in \mathbb{N}$$

For instance, $-3 < -1$ because $-1 - (-3) = 2$ is a natural number. Similarly, $-1 < 0$ and $2 < 5$. A similar thing happens in $\mathbb{Q}$ but less directly; we say that if m/n and j/k are rationals then

$$\frac{m}{n} < \frac{j}{k} \quad \text{if } km < jn \text{ in } \mathbb{Z}$$

For instance, $1/2 < 2/3$ because $3 < 4$ and $-1/2 < -1/3$ because $-3 < -2$. Notice that if we fix $n = k = 1$ then we recover the ordering of integers.

By adding the equality option to $<$ we obtain its reflexive extension $\leq$.

Exercise 45 *Recalling the definition of a total ordering in Chapter 2, verify that the relation $<$ defined above does, in fact, totally order $\mathbb{Q}$.*

Finally, $\mathbb{Q}$ is an *ordered field* not just an ordered set. This means that *addition and multiplication preserve the ordering* in the sense that for p, q in $\mathbb{Q}$ the following hold:
 (a) if $p < q$ then $p + r < q + r$ for every rational number r
 (b) if $p < q$ and r is positive ($r > 0$) then $pr < qr$

These properties can be verified using the (assumed) properties of the integers. For example, to see why (a) is true note that the hypothesis $p < q$ is equivalent to $q - p > 0$. Now add and subtract r (adding a net value of 0) to the left hand side to obtain $q + r - p - r > 0$ which is equivalent to $q + r > p + r$. We proved what was claimed. (b) is proved similarly.

There are no consecutive rational numbers...

We speak of consecutive integers because for each n in $\mathbb{N}$ or in $\mathbb{Z}$ there are no integers between n and $n + 1$. But we don't speak of "consecutive rationals" because between every pair of rational numbers a and b, no matter how close they are to each other, there is another rational number, say, their average $(a+b)/2$. So *rational numbers are densely packed*; if you picked any rational number q there is no other rational that sits immediately next to q with no other rationals in between. Later we see that rationals have a much more important density property, in that they are dense in the vastly larger set of all *real* numbers.

If we tried to physically plot all rational numbers as tiny dots on a line, say, between 1 and 2 then we would reach a point before long when the dots would start touching each other with infinitely many rationals still left to plot. Each dot is actually a tiny disk of finite radius; even with super high resolution finite is still finite. If we could plot disks of radius 10^{-9} (a billionth of a unit, say a millimeter) the interval of unit length from 1 to 2 would completely fill up with just a billion dots when there are in fact infinitely many rational numbers still left to insert.

And yet, the set of rational numbers has gaps or holes!

The number $\sqrt{2}$ with the approximate value of 1.4142 is between 1 and 2. I have often said that this number is irrational; bear with me one more time and I will prove this fact before ending this subsection.

Now if $\sqrt{2}$ is not rational, and yet it is between 1 and 2 then taking it out of the number line leaves a hole without affecting any rational number. We know that there are rational numbers to the left and also to the right of $\sqrt{2}$ say, 1.4141 and 1.4143. We also saw above that the rational numbers are densely packed: there are not two consecutive rational numbers. If we insisted that the number line contain only the rational numbers then our line would have holes corresponding to irrational numbers like $\sqrt{2}$.

More precisely, $\sqrt{2}$ is a number r with the property that $r^2 = 2$. Recall that the square function x^2 is an increasing function when $x > 0$. We can use this fact to get rational approximations for $\sqrt{2}$. We start with the observation that $1^2 = 1$ and $2^2 = 4$. Since $r^2 = 2$ and 2 is between 1 and 4, we conclude that $\sqrt{2}$ is between 1 and 2. It is not quite as large as 1.5 since $1.5^2 = 2.25$, but it is larger than 1.4 since $1.4^2 = 1.96$. Since 1.4 is rational (we can write it as 14/10 or 7/5 in reduced form) and closer to $\sqrt{2}$ than 1.5 is, let's set $q_1 = 1.4$ as a first rational approximation to $\sqrt{2}$. Next, compute the average of 1.4 and 1.5, which is 1.45 and notice that $1.45^2 = 2.1025$ so $\sqrt{2}$ is between 1.4 and 1.45. Now, 1.96 is still closer to 2 than 2.1025 is, so we calculate the average of 1.4 and 1.45. This gives 1.425 whose square is 2.030625, now closer to 2 than 1.96 is. Thus a second approximation in $\mathbb{Q}$ to $\sqrt{2}$ is $q_2 = 1.425$. Continuing in this way we generate a sequence of approximations in $\mathbb{Q}$

$$q_1, q_2, q_3, \ldots = 1.4, 1.425, 1.4125, \ldots \tag{5.1}$$

that gets closer and closer to $\sqrt{2}$, which as we have seen is *not* in $\mathbb{Q}$.

The above discussion shows that although there is an infinity of rational numbers between 1 and 2 in $\mathbb{Q}$ that accumulate around $\sqrt{2}$ and get very close to it, they do not actually reach it. This leaves a gap in $\mathbb{Q}$ on the number line where $\sqrt{2}$ is. This gap, or hole has no finite width; if we thought of it as a tiny disk then its radius would be zero. Nevertheless, if $\sqrt{2}$ is removed from the number line then we are left with a gap that contains no number.

To fill this gap, we must add $\sqrt{2}$ to the set $\mathbb{Q}$ and create a slightly larger set than the rational numbers. But how do we fill in *all* the gaps in $\mathbb{Q}$? Don't we need to figure out all irrational numbers first? That sounds like "the chicken or the egg" conundrum (which came first?).

These questions are not easy to answer; in fact, there is nothing obvious or simple about the real numbers because infinity is so intricately woven into their fabric. But we do have a clue in the

The Real Numbers   97

approximation of $\sqrt{2}$ above; specifically, the terms q_n of the sequence in (5.1) are rational numbers but they are practically indistinguishable from $\sqrt{2}$ when the index n gets very large. It is hard to distinguish the rational numbers in the tail end of the sequence from $\sqrt{2}$.

So why not identify $\sqrt{2}$ with the tail end of the sequence?

This identification is essentially what Cantor's construction of real numbers is about. We study this construction in the next section; for now, let's approximate $\sqrt{2}$ by a sequence of rational numbers as in the following exercise.

Exercise 46 *Here is a recursion, known since the time of the Babylonians some 4500 years ago, that generates a sequence of rational numbers that converges to $\sqrt{2}$ rather quickly:*

$$q_n = \frac{1}{2}\left(q_{n-1} + \frac{2}{q_{n-1}}\right) \tag{5.2}$$

This is just the divide and average rule stated as a recursion. Starting from some positive rational number q_0 we calculate the next term q_1 by dividing 2 by q_0 and then finding the average of $2/q_0$ and q_0. We repeat this process of dividing and averaging to find q_2 using q_1 and continue to higher index terms to generate a sequence recursively. Starting with $q_0 = 1$, calculate down to q_4; you should find that q_4 agrees with a calculator value of $\sqrt{2} \simeq 1.41421356237$ to at least 10 decimal places! Repeat these calculations starting with $q_0 = 1/2$. How many times do you need to repeat now to get at least ten decimal places correct?

Before ending this section, let's prove the irrationality of $\sqrt{2}$ (as promised earlier). This is one of the best known examples of a proof by contradiction that also says something significant beyond pedagogical issues.

> The number $\sqrt{2}$ is not rational.

Suppose that there are (positive) integers m and n such that $m/n = \sqrt{2}$. We can assume that m and n share no common factors (m/n is in *reduced form*); if not then you can first cancel out all common factors till you get a reduced form, without changing the fraction's value. Now, squaring both sides gives

$$\frac{m^2}{n^2} = \left(\frac{m}{n}\right)^2 = \left(\sqrt{2}\right)^2 = 2 \quad \text{so:}$$
$$m^2 = 2n^2$$

The occurrence of 2 on the right hand side of the second equality above means that m^2 is an even number. Since the square of an odd number is again odd, it follows that m is even, say, $m = 2k$ where k is another positive integer. Then

$$2n^2 = (2k)^2 = 4k^2 \quad \text{so:}$$
$$n^2 = 2k^2$$

Again, this implies that n is even too. But now m and n do have a common factor of 2, contrary to our assumption that they had no common factors. The only way of avoiding this contradiction is if $\sqrt{2}$ is not rational.

5.2 From the rationals to the reals, in baby steps

Our goal in this section is to start from the set $\mathbb{Q}$ of all rational numbers and build the set of all real numbers $\mathbb{R}$, including all of the irrationals, from $\mathbb{Q}$ in a systematic way.

What we have to work with are the rational numbers. You may recall from the last section that we can approximate irrational numbers using sequences of rationals. But in this chapter we don't know the irrationals yet; so if a sequence of rationals does not converge to another rational number then what does it converge to?

For one thing, some sequences do not converge at all; for example, the sequences $q_n = n^2$ (or $1^2, 2^2, 3^2, \ldots$) and $q_n = (-1)^n$ (or $-1, 1, -1, 1, \ldots$) do not approach any number. So what would distinguish the sequence in Exercise 46 from any old sequence of rationals if we had no idea what $\sqrt{2}$ was? How can we tell if a sequence converges if we don't know a limit for it?

Our first step on the way to the real numbers is to identify these special sequences of rational numbers.

Cauchy sequences of rational numbers.

Earlier we saw how to estimate $\sqrt{2}$ using a sequence of rational numbers generated by the repeated applications of the divide and average rule. The more terms we used, the closer the approximation became. This leads to the *practical* question: *how many terms of the rational sequence should we use to get a desired accuracy, say, 10 decimal places?* The answer is not obvious because we do not know the exact digital value of $\sqrt{2}$.

Here is the key observation: *as the terms of the rational sequence approach $\sqrt{2}$ they also get closer to each other.*

If we use many terms of the sequence then the later terms form a dense, tiny cluster of rationals around $\sqrt{2}$. The distance between any pair of terms q_m and q_n with large enough indices m and n is as small as we need it to be: a target threshold for a desired level of accuracy. For 10 decimal place accuracy this threshold is 0.00000000005.

More precisely, we must reach an index N that is large enough that the difference between any pair of terms q_m and q_n is less than 0.00000000005 as long as m, n both exceed N. Let's illustrate this idea using the recursion (5.2). Iterating the recursion five times generates the first five terms of the approximating sequence as:

n	q_n
1	1.5
2	1.41666666667
3	1.41421568627
4	1.41421356238
5	1.41421356237

I have rounded the rational approximations to 11 decimal places for ease of reading.[2] The difference between q_4 and q_5 is 0.00000000001 which hits our target threshold since it is indeed less than 0.00000000005. If you continue repeating the divide and average process then you will find that after the 5th term listed above ($N = 5$) subtracting any pair of terms q_m and q_n from 6th on up indeed comes up less than 0.00000000005 (and getting smaller and smaller).

---

[2] The recursion actually generates exact rational terms (quotients of integers) which are: $q_1 = (1/2)/(1+2/1) = 3/2$, $q_2 = (1/2)/(3/2 + 2/[3/2]) = 17/12$, etc.

The Real Numbers 99

This idea is not limited to approximating $\sqrt{2}$. Every converging sequence of rational numbers has the same property: due to clustering around the limit, subtracting terms with large enough indices results in numbers that are as close to zero as desired. Like any other good idea, we formalize this property by giving it a name.[3]

> **Cauchy sequence:** A sequence of rational numbers $q_1, q_2, q_3, \ldots$ is called *Cauchy* (pronounced *coshi*) if the terms of the sequence approach each other with increasing index; more precisely,
> $$|q_m - q_n| \to 0 \text{ as } m, n \to \infty \qquad (5.3)$$

The bars as usual give the absolute value of the difference which tells us how far apart the terms q_m and q_n are, regardless of their signs. The symbol $m, n \to \infty$ may be read "m and n go to infinity"; they may do so together or in tandem, but in general they are independent of each other.

The quantity $|q_m - q_n|$ is not strictly zero for all large m and n, only *getting close* to zero. I give the precise definition of a Cauchy sequence in Appendix 11.4 where I give the technical details of Cantor's construction of $\mathbb{R}$ from $\mathbb{Q}$ for the analytically inclined. For a simple example that we can examine now, consider the sequence

$$q_n = \frac{n-1}{n}, \quad n = 1, 2, 3, \ldots \qquad (5.4)$$

The first few terms of this sequence (rounded to two decimal places) are

n	1	2	3	4	5	6	7	8	9
q_n	0	0.5	0.67	0.75	0.8	0.83	0.86	0.88	0.89

Let us compare the distances between various terms:

$$|q_2 - q_3| = |0.5 - 0.67| = |-0.17| = 0.17$$
$$|q_5 - q_9| = |0.8 - 0.89| = |-0.09| = 0.09$$
$$|q_7 - q_9| = |0.86 - 0.89| = |-0.03| = 0.03$$

Notice that the distances between terms get smaller when we pick *both* terms farther away, that is, terms with larger indices. For large enough indices the difference

$$|q_{100} - q_{1000}| = \left|\frac{99}{100} - \frac{999}{1000}\right| = \frac{9}{1000} = 0.009$$

is small even though $m = 100$ and $n = 1000$ are rather far apart.

> The important thing that makes a sequence "Cauchy" is that the distance between *any pair of terms* is nearly zero provided that *both indices* m and n are large enough.

The sequence in (5.4) seems to be Cauchy. It takes just a little bit of algebra to verify this; you can work it out in Exercise 47.

---

[3] Named after Augustin-Louis Cauchy (1789-1857) one of the founders of modern analysis. He may be better known to some people for his integral formula in complex analysis.

Exercise 47 *To show that the sequence in (5.4) is Cauchy, we can write it in the equivalent way*

$$q_n = 1 - \frac{1}{n}$$

and then show that

$$|q_m - q_n| = \left|\frac{1}{n} - \frac{1}{m}\right|$$

and then complete the proof.

Exercise 48 *(a) A sequence q_n of rational numbers is eventually constant if there is a rational number q and an index N such that $q_n = q$ for all $n \geq N$. Is this a Cauchy sequence?*

(b) Consider a periodic sequence q_n of rational numbers with period $k \geq 2$. Can this type of sequence be Cauchy?

The following fact is easy to prove; if a x_n is a sequence that converges to a limit x then for all large values of the index n the terms x_n are all ε units away from x so no two terms can be no more twice this amount away from each other. The sequence is therefore, Cauchy.

> Every sequence that converges is a Cauchy sequence.

The converse of the above statement, namely, "every Cauchy sequence converges" is not always true. For instance, we saw earlier that the divide and average rule generates a sequence of rational numbers that does not converge to a rational number. Of course, it does converge to an *irrational* number, so in a sense, a Cauchy sequence "would" converge if there were a limit in the set of interest. If this set is just the set of all rational numbers then sequences end up in a hole. But when the irrational numbers are inserted so as to fill in all the holes, the converse is true!

The critical issue is how do we get to *all* of the irrational numbers. We see how in the next section. At this point it seems relevant to address a more basic issue:

> **When is a sequence not Cauchy?** A sequence q_n is *not* Cauchy when there is some value of ε, say, ε_0 such that *no matter how large an integer N we pick, we can always find a pair of indices $n, m \geq N$ such that $|q_n - q_m| \geq \varepsilon_0$.*

For example, consider the periodic sequence $q_n = (-1)^n$. If we pick any positive integer N, no matter how large, then there is a pair of integers, say, $m = N$ and $n = N+1$ such that

$$|q_n - q_m| = |(-1)^{N+1} - (-1)^N| = 2$$

because one of N or $N+1$ is odd while the other is even so the difference $(-1)^{N+1} - (-1)^N$ is always ± 2 with absolute value 2. Now, if ε_0 is any number between 0 and 2 then (5.5) fails for every (finite) integer N.

Finally, let's consider the ε-index for Cauchy sequences. You may skip this discussion without a loss of continuity and proceed with our construction of the real numbers, which doesn't require this concept. I am adding these notes here to illustrate how the infinity lurks behind some of the most fundamental notions in mathematics, namely, Cauchy sequences and the completeness property of

The Real Numbers 101

the set of all real numbers. We will discuss the ε-index in a more profound way in Chapter 9 when studying sequences of functions.

> **The ε-index.** We can be more precise in our definition of a Cauchy sequence: A sequence q_n (of rational numbers for now) is Cauchy if for every $\varepsilon > 0$ there is a positive integer N such that
> $$|q_n - q_m| < \varepsilon \quad \text{for all } m, n \geq N \tag{5.5}$$
> The *smallest or least integer* N_ε that satisfies (5.5) is *the ε-index* of the sequence q_n.

Comparing this definition with the earlier definition of ε-index we see a major difference: this one does not require the value of a limit! This is significant because we can calculate the ε-index for sequences like the one generated by the recursion (5.2) in Exercise 46 without knowing what the (irrational) limit is. For instance, based on the table of values that we listed earlier, we see that $N_\varepsilon = 5$ for $\varepsilon = 0.0000000005$.

Calculating N_ε as a function of ε (i.e. for all values of ε) is not easy if a limit is not known or does not exist because we need to find what happens as the difference between m and n goes to infinity. Even for a simple sequence like the one in (5.4) the calculation requires dealing with indeterminate forms and finding complicated limits (see the following exercises).

Of course, we can determine some of the values of N_ε using a digital computing device. The fact that N_ε is an integer makes such a computer-aided calculation easier since its value jumps from one integer to the next as ε decreases continuously.

Exercise 49 *(a) Use a calculator or a computer to tabulate enough of the terms of the sequence $q_n = 9/n^2$ to be able to determine N_ε for $\varepsilon = 0.05$; see our discussion above. Use your table of values to also find N_ε for $\varepsilon = 0.1$. Repeat these calculations of N_ε for the sequence $r_n = (2n^2 - 9)/n^2$.*

(b) Use the limit theorems for sequences that we discussed earlier to show that $\lim_{n\to\infty} q_n = 0$ and $\lim_{n\to\infty} r_n = 2$. Use these facts to find N_ε as a function of ε for each sequence (see our earlier discussion of the ε-index for convergent sequences). Then easily calculate N_ε for $\varepsilon = 0.1$, $\varepsilon = 0.05$ and $\varepsilon = 0.0001$.

Exercise 50 *This optional exercise is meant to illustrate the difficulty of finding N_ε as a function of ε without the use of limits even for a simple sequence. Feel free to work out the details if you are in the mood for a little bit of hiking in rugged terrain (must feel comfortable with algebraic manipulations of some complexity and remember l'Hôpital's rule for the indeterminate form $0 \times \infty$ from calculus). This excursion is not relevant to our later work.*

(a) Recall the sequence $q_n = (q_n - 1)/q_n$ in (5.4) that we showed to be Cauchy. Suppose that $m < n$ and set $n = m + k$ where k is a positive integer. Show that
$$|q_n - q_m| = \frac{k}{m(m+k)}$$

(b) Next, show that if we solve
$$\frac{k}{m(m+k)} < \varepsilon$$

for m then we obtain

$$m > \frac{k}{2}\left(\sqrt{1+\frac{4}{\varepsilon k}} - 1\right)$$

(c) Show that the expression on the right hand side of the above inequality is an increasing function of k and use l'Hôpital's rule to find its maximum value, which is also its limit:

$$\lim_{k\to\infty} \frac{k}{2}\left(\sqrt{1+\frac{4}{\varepsilon k}} - 1\right) = \frac{1}{\varepsilon}$$

Conclude that $N_\varepsilon = \lceil 1/\varepsilon \rceil$ where $\lceil x \rceil$ is the least integer that is greater than or equal to x.
(d) Verify your result in (c) by calculating N_ε the easy way using the fact that $\lim_{n\to\infty} q_n = 1$.

If a sequence isn't Cauchy then no smallest N exists to serve as the ε-index. We may conclude that the ε-index is infinite. This observation suggests that we may define a Cauchy sequence as follows:

> **Cauchy sequence in terms of ε-index.** *A sequence q_n is Cauchy if and only if its ε-index N_ε is finite for every $\varepsilon > 0$.*

Real numbers as equivalence classes of rational Cauchy sequences.

The insight that irrational numbers are associated with sequences of rational numbers is the basis for our construction of real numbers from the rational ones. We have seen that every Cauchy sequences of rational numbers tends to congregate or accumulate around a single number. The limit may be rational again, as in the case of (5.4) where the sequence approaches 1 or irrational as we observed when approximating the value of $\sqrt{2}$ using the divide and average rule.

Now, more than one Cauchy sequence approaches the same number; for instance, when using the divide and average rule, different rational initial values for q_0 yield different sequences, all of which approach $\sqrt{2}$. Which of these sequences should represent $\sqrt{2}$?

These sequences must be equivalent in the sense that any one of them should be as good as any other. This naturally takes us to equivalence relations that we discussed earlier.

Let's collect all Cauchy sequences of rational numbers in a set $\mathcal{C}$ of sequences. Define a relation $\sim$ in $\mathcal{C}$ as follows:

> **Equivalent Cauchy sequences.** *Let $s = q_1, q_2, q_3, \ldots$ and $s' = q'_1, q'_2, q'_3, \ldots$ be sequences in $\mathcal{C}$. We say that $s \sim s'$ if $|q_n - q'_n| \to 0$ as $n \to \infty$.*

Think of $s \sim s'$ intuitively as stating that *the tail ends of s and s' meet*. For example, the sequence $q_n = 1 + 1/n$ is equivalent to the constant sequence $1, 1, 1, \ldots$ since $|(1 + 1/n) - (1)| = |1/n| \to 0$ as $n \to \infty$. The tails of these two sequences approach each other until they are indistinguishable!

Why is the relation $\sim$ is an equivalence relation in the set $\mathcal{C}$?

The reflexive and symmetric properties are obvious; the transitive case goes like this: if $s \sim s'$ and $s' \sim s''$ then both $|q_n - q'_n| \to 0$ and $|q'_n - q''_n| \to 0$ as $n \to \infty$. This means that the distance between q_n and q'_n and the distance between q'_n and q''_n are both approaching zero as $n \to \infty$; in order for q'_n to stay arbitrarily close to both q_n and q''_n in this process, the distance between q_n and q''_n has to go to zero too. So $|q_n - q''_n| \to 0$ as $n \to \infty$, which means that $s \sim s''$.

The Real Numbers 103

For each sequence s in $\mathcal{C}$ let's write $[s]$ for its equivalence class. For example, the sequence $q_n = 1/n$ is in $[0]$; in fact, $[1/n] = [0]$ since by symmetry, the constant sequence $0, 0, \ldots$ is a member of the equivalence class $[1/n]$. There are infinitely many sequences in $[0]$; in particular, sequences that are equal to 0 after some point, like $3, 2, 1, 0, 0, \ldots$ are also in $[0]$ as well as Cauchy sequences that are equal to 0 infinitely often, like the sequence $q_n = [1 + (-1)^n]/n$ (or $0, 1, 0, 1/2, 0, 1/3, \ldots$). On the other hand, if a Cauchy sequence is *not* in $[0]$ then its terms cannot accumulate around 0. For example, the constant sequence $1, 1, 1, \ldots$ is not in $[0]$ and neither is the sequence $s_n = 1 + 1/n$ that converges to 1.

Exercise 51 *Consider the sequence in (5.4); let's call this sequence s. Can you explain which of the following sequences are equivalent to s and which are not? Remember that all equivalent sequences must accumulate around the same number:*

$(a) \quad 1, 1, 1, \ldots \qquad (b) \quad 1 + \dfrac{1}{n} \qquad (c) \quad 2 - \dfrac{1}{n} \qquad (d) \quad \dfrac{1 + (-1)^n}{2}$

Finding the rational numbers in a larger set.

Each rational number q generates the constant sequence $q, q, q, \ldots$ which is trivially Cauchy, hence in $\mathcal{C}$. Every Cauchy sequence of rational numbers whose tail end meets this sequence is in the equivalence class $[q]$. The collection of all equivalence classes $[q]$ of rationals is in a one to one correspondence with the set $\mathbb{Q}$ of rationals, because if q and p are distinct rationals then $[q]$ and $[p]$ are disjoint sets (the tail end of no Cauchy sequence of rationals can meet *both* of the constant sequences $q, q, q, \ldots$ and $p, p, p, \ldots$ when p is not equal to q).

> We use the notation $[\mathcal{C}]$ for *the set of all equivalence classes* of the relation $\sim$. Then we may think of the collection all equivalence classes $[q]$ of the rational numbers as a copy of the set $\mathbb{Q}$ in $[\mathcal{C}]$.

This idea seems a bit abstract but it is not hard to absorb. Think of $[\mathcal{C}]$ as a vast collection of boxes filled with (very long) chains, each chain being a Cauchy sequence of rational numbers. Recall that we identify each rational number q with $[q]$, the class of all Cauchy sequences that are equivalent to the constant sequence $q, q, q, \ldots$. The special classes $[q]$, are the "rational-type boxes" and we may think of each ordinary rational *number q* as *a label* for a rational-type box in $[\mathcal{C}]$. This is basically how we identify the ordinary rational numbers in $[\mathcal{C}]$!

After using up all of our labels in $\mathbb{Q}$, there are still many boxes in $[\mathcal{C}]$ left without a label; for instance, all Cauchy sequences generated by the divide and average rule are in a single equivalence class, or box, that we may label $\sqrt{2}$. Similarly, there is the box labeled π and so on. These leftover boxes in $[\mathcal{C}]$ are classes that cannot be labeled by rational numbers, so we may think of them collectively as "irrational-type boxes" (equivalence classes).

I must emphasize that each irrational class or box is still filled with *rational* sequences, but there is a significant omission. *The irrational-type boxes do not contain any constant rational sequences $q, q, q, \ldots$.* We can use familiar symbols like $\sqrt{2}$ or π to label the irrational-type boxes whose contents have special meanings to us; but as we discover soon, there are far too many irrational-type boxes to bother labeling them.

5.3 The real numbers, at last!

With the preceding discussion in mind, we now define:

> A *real number* is the equivalence class of a Cauchy sequence of rational numbers. We use the symbol $\mathbb{R}$ for the set of real numbers rather than the notation $[\mathcal{C}]$.

It may seem strange to think of a number like $\sqrt{2}$ as a *collection* of sequences (not even an individual sequence); but all the properties of real numbers that we are familiar with hold for these equivalence classes! We see how in this section.

Real numbers form a totally ordered field.

Recall that the set of all rational numbers $\mathbb{Q}$ is a totally ordered field. If $\mathbb{R}$ is to be a field then it must have two operations, addition and multiplication that must be compatible with and extend the ordinary addition and multiplication in the set $\mathbb{Q}$ of rationals.

The operations of addition and multiplication are defined as follows: for each pair of real numbers $r = [q_n]$ and $r' = [q'_n]$

$$r + r' = [q_n + q'_n] \qquad rr' = [q_n q'_n]$$

The brackets above indicate equivalence classes of sequences of rational numbers that are obtained by adding or multiplying two sequences. If both r and r' are rationals then each can be the equivalence class of a constant sequence; say, if $r = [2, 2, 2, \ldots]$ and $r' = [-1/3, -1/3, -1/3, \ldots]$ then

$$r + r' = \left[2 - \frac{1}{3}, 2 - \frac{1}{3}, 2 - \frac{1}{3}, \ldots\right] = \left[\frac{5}{3}, \frac{5}{3}, \frac{5}{3}, \ldots\right]$$

$$rr' = \left[2\left(-\frac{1}{3}\right), 2\left(-\frac{1}{3}\right), 2\left(-\frac{1}{3}\right), \ldots\right] = \left[-\frac{2}{3}, -\frac{2}{3}, -\frac{2}{3}, \ldots\right]$$

The same rules apply to equivalence classes of irrational numbers although sequences that correspond to the irrationals are not eventually constant. For instance, in Chapter 9 we discover that the rational sequence

$$q_n = 1, 1 - \frac{1}{3}, 1 - \frac{1}{3} + \frac{1}{5}, 1 - \frac{1}{3} + \frac{1}{5} - \frac{1}{7}, \ldots = 1, \frac{2}{3}, \frac{13}{15}, \frac{76}{105}, \ldots$$

converges to $\pi/4$, an irrational number. This lets us define numbers like the sum $-1/3 + \pi/4$ and the product $(-1/3)(\pi/4) = -\pi/12$ as

$$-\frac{1}{3} + \pi = \left[-\frac{1}{3}, -\frac{1}{3}, -\frac{1}{3}, \ldots\right] + \left[1, \frac{2}{3}, \frac{13}{15}, \frac{76}{105}, \ldots\right]$$

Adding the corresponding numbers in the brackets gives

$$\left[-\frac{1}{3} + 1, -\frac{1}{3} + \frac{2}{3}, -\frac{1}{3} + \frac{13}{15}, -\frac{1}{3} + \frac{76}{105}, \ldots\right] = \left[\frac{2}{3}, \frac{1}{3}, \frac{8}{15}, \frac{123}{315}, \ldots\right]$$

Likewise,

$$-\frac{\pi}{12} = \left[-\frac{1}{3}, -\frac{1}{3}, -\frac{1}{3}, \ldots\right]\left[1, \frac{2}{3}, \frac{13}{15}, \frac{76}{105}, \ldots\right] = \left[-\frac{1}{3}, -\frac{2}{9}, -\frac{13}{45}, -\frac{76}{315}, \ldots\right]$$

The Real Numbers

Bear in mind that an equivalence class such as $[-1/3, -1/3, \ldots]$ consists of an *infinite number* of Cauchy sequences of rational numbers, each *equivalent* to, but not the same as the constant sequence $-1/3, -1/3, \ldots$. For example, the sequence $q_n = 1/n - 1/3$ whose first few terms are:

$$\frac{2}{3}, \frac{1}{6}, -\frac{1}{12}, -\frac{2}{15}, \ldots$$

is equivalent to the constant sequence $-1/3, -1/3, \ldots$ because

$$\left| q_n - \frac{-1}{3} \right| = \left| \frac{1}{n} - \frac{1}{3} + \frac{1}{3} \right| = \left| \frac{1}{n} \right| = \frac{1}{n} \to 0 \quad \text{as } n \to \infty$$

The same is true if $1/n$ is replaced with $2/n$ or $3/n$ and so on, showing that there are infinitely many sequences in $[-1/3, -1/3, \ldots]$.

The above two operations on the set $[\mathcal{C}]$ of all equivalence classes of Cauchy sequences of rational numbers do have the same properties as the familiar addition and multiplication of real numbers. In particular, the classes $[0]$ and $[1]$ are in fact like the ordinary 0 and 1 that we encounter in school; if $r = [q_n]$ is any real number then

$$r + [0] = [q_n] + [0] = [q_n + 0] = [q_n] = r$$
$$r[0] = [q_n][0] = [q_n 0] = [0] = 0$$
$$r[1] = [q_n][1] = [q_n 1] = [q_n] = r$$

The negative is also easy to define:

$$-r = -[q_n] = [-q_n]$$

We use the equivalence class $[-q_n]$ of the rational sequence of negatives to define $-r$. If $r \neq 0$ then the reciprocal is defined similarly as[4]

$$\frac{1}{r} = r^{-1} = [q_n]^{-1} = [q_n^{-1}] = \left[\frac{1}{q_n} \right]$$

For instance, if $r = \ln 2$ then

$$\frac{1}{r} = \frac{1}{\ln 2} = \left[1, 2, \frac{6}{5}, \frac{12}{7}, \ldots \right]$$

We now have a *field* of real numbers $\mathbb{R}$ that contains the rational numbers $\mathbb{Q}$ (or rather, a copy of $\mathbb{Q}$ in the form of classes $[q]$ equivalent to constant rational sequences).

Next, we introduce an ordering on $\mathbb{R}$ that is compatible with the usual ordering $<$ of $\mathbb{Q}$ as previously defined. This means that the intended ordering of $\mathbb{R}$ when restricted to the rational numbers must be the same as the integer-induced ordering of $\mathbb{Q}$ that we discussed earlier.

In order to motivate the definitions below, let us take another look at the ordering of $\mathbb{Q}$; since $p < q$ is equivalent to $0 < q - p$ we see that to say $p < q$ is to say that "$q - p$ is positive". It is now easy to define what it means to say something is "positive" in $\mathbb{R}$.

---
[4] We need to take care that $q_n \neq 0$ for any n. The condition $r \neq 0$ does not guarantee this, since every equivalence class $[q_n]$ contains sequences having some zeros. This issue is not hard to resolve; see Appendix 11.4.

> The real number $r = [q_n]$ is *positive* if the rational Cauchy sequence $q_1, q_2, q_3, \ldots$ is eventually positive *and* its terms do not approach 0.

For example, consider the sequence $q_n = 1/4 - 1/n$ of rational numbers whose first few terms are:

$$-\frac{3}{4}, -\frac{1}{4}, -\frac{1}{12}, 0, \frac{1}{20}, \frac{1}{12} \ldots$$

All terms after the fourth are positive and approach $1/4$ as $n \to \infty$ so the equivalence class of this sequence is a positive real number. Indeed, this sequence is equivalent to the rational number $1/4$ which we deem to be positive. On the other hand, even though *all* terms of the sequence $1/n$ or

$$1, \frac{1}{2}, \frac{1}{3}, \frac{1}{4}, \ldots$$

are positive, it does not represent a positive real number because it converges to 0 and therefore, it is in the equivalence class $[0]$.

Now, in analogy with the ordinary rational numbers, we define a relation $\prec$ in $\mathbb{R}$.

> Let $x = [p_n]$ and $y = [q_n]$. Then: $x \prec y$ if $[0] \prec [q_n] - [p_n]$, that is: $x \prec y$ if $[q_n] - [p_n] = [q_n - p_n]$ is positive.

For all x and y in $\mathbb{R}$ the following are true (see Appendix 11.4 for proofs): If $x \prec y$ then $x + r \prec y + r$ for every r in $\mathbb{R}$. If $x \prec y$ then $xr \prec yr$ for every positive r in $\mathbb{R}$. The relation $\prec$ is total: if $x \neq y$ then either $x \prec y$ or $y \prec x$ (not both). The relation $\prec$ is transitive: if $x \prec y$ and $y \prec z$ then $x \prec z$

In addition to the above ordering properties, it is an important fact that $\prec$ *is compatible with the usual ordering $<$ of the rational numbers* because:

> If p, q are rational numbers such that $p < q$ then $[p] \prec [q]$.

To see why, notice that $p < q$ implies that the tail ends of constant sequences $p, p, p, \ldots$ and $q, q, q, \ldots$ do not approach each other. Thus $[p] \neq [q]$. Further, $[q - p]$ is positive so $[0] \prec [q - p] = [q] - [p]$. It follows that $[p] \prec [q]$ as claimed.

I follow the customary (and convenient, even if not precise) practice of using the symbol $<$ for the ordering of real numbers instead of $\prec$ since confusion is unlikely.

5.4 Characteristics of the set of real numbers: completeness and more

Recall that being an ordered field was also a feature of the set of rational numbers. So what is different about the set $\mathbb{R}$? It is certainly larger than $\mathbb{Q}$ but exactly how much larger? What gives $\mathbb{R}$ its continuous appearance?

We answer these questions and discuss some related issues in this section. Our focus is on $\mathbb{R}$ *as a set* rather than on individual real numbers. Further insights about real numbers come later when discussing infinite series in Chapters 6 and 9.

The Real Numbers

Having constructed the set of real numbers, we no longer need to strain our minds by thinking of real numbers as equivalence classes of sequences. Although we have not formally proven many basic facts about these objects, it is harmless for our purposes in this book to go back to our intuitive view of the real numbers. In Appendix 11.4 we do prove some of the properties of real numbers (as equivalence classes of sequences) that justify this faith.

The density of the set of rational numbers.

As we discovered, every real number, rational or irrational, is associated with a set of equivalent, rational Cauchy sequences. Let us take a closer look at this aspect; if $x = [q_n]$ then all sequences that are equivalent to q_n accumulate around the same limit x which we intuitively identify as a point on the number line, (or the x-axis). If r is another point on the line and $r < x$ then there is a Cauchy sequence of rationals q'_n equivalent to q_n whose terms are larger than r but eventually closer to x than r is; that is, there is an index N such that $r < q'_N < x$.[5] This observation justifies the following conclusion:

> *The rational numbers are dense in $\mathbb{R}$; that is, between any two distinct real numbers there is a rational number.*

This statement actually implies something stronger that you can easily verify in the following exercise by applying the above theorem repeatedly.

Exercise 52 *Let x and r be distinct real numbers. Explain why there has to be infinitely many rational numbers between x and r.*

The fact that rationals are dense in the set of real numbers implies the same about irrationals.

> *The irrational numbers are dense in $\mathbb{R}$; that is, between any two distinct real numbers there is an irrational number.*

To see why the irrational numbers are dense, first note that if q is any positive rational number then $q/\sqrt{2}$ is irrational (see Exercise 53 below). Now if $0 < x < y$ then $x\sqrt{2} < y\sqrt{2}$ and we know that there is a rational number q such that $x\sqrt{2} < q < y\sqrt{2}$. Dividing by $\sqrt{2}$ thus gives $x < q/\sqrt{2} < y$ and we have found an irrational number between x and y. If $x \leq 0 < y$ then there is a rational q' such that $0 < q' < y$ and we have seen that we can find an irrational between q' and y. Such an irrational is obviously also between x and y. Finally, if $x < y \leq 0$ then $0 \leq -y < -x$ and we can find an irrational number, say, z where $-y < z < -x$. Now $-z$ is the number we are looking for since $x < -z < y$.

---

[5]To be precise, in the set $[\mathcal{C}]$ we have $r = [p_n]$ for some sequence $[p_n]$ of rational numbers so we identify q'_N with its constant sequence equivalence class $[q'_N, q'_N, \ldots]$ to get

$$[p_n] < [q'_N, q'_N, \ldots] < [q_n]$$

Exercise 53 *Suppose that x, y are irrational numbers and q is rational ($q \neq 0$). Explain why $x + q$ and xq are always irrational but $x + y$ and xy may be irrational or rational. In particular, this shows that irrational numbers do not form a field under ordinary addition and multiplication, and thus they are not as structured as the rational numbers.*

Density is a very useful property; in particular, the following result implies that every real number can be approximated by a sequence of rational numbers as well as a sequence of irrational numbers. We will discover later that something similar is true in a general metric space, a fact with far-reaching implications in both pure and applied mathematics.

> For every real number r there is a sequence q_n of rational numbers and a sequence x_n of irrational numbers such that
> $$r = \lim_{n \to \infty} q_n = \lim_{n \to \infty} x_n$$

Here, we only know the limit r; so how do we find the sequences q_n and x_n? Recall that r is the equivalence class of infinitely many sequences so putting together any one of these is all we require; *we are not looking for specific sequences* here. So here is how we go about it:

By density, there is a rational number q_1 between $r + 1/2$ and $r + 1$; and there is a rational q_2 between $r + 1/3$ and $r + 1/2$, and a rational q_3 between $r + 1/4$ and $r + 1/3$, and so on. This process generates a sequence of distinct numbers

$$r + 1 > q_1 > r + \frac{1}{2} > q_2 > r + \frac{1}{3} > q_3 > r + \frac{1}{4} > \cdots > r$$

Since $r < q_n < r + 1/n$ for every $n = 1, 2, 3, \ldots$ and $1/n \to 0$ as $n \to \infty$, the Squeeze Theorem implies that q_n converges to r. We do not need to know what each number q_n is specifically, but only have a recipe for defining it. Repeating this same argument with all q_n changed to x_n proves the statement about irrationals.

Exercise 54 *To illustrate the above construction, which is a very common type of construction in analysis, consider $r = 1$. Then choose q_1 between $1 + 1/2 = 3/2 = 1.5$ and $1 + 1 = 2$; pick any rational number in this range, for example, 1.6 and call it q_1. Then move closer to the limit 1 and pick q_2 between $1 + 1/3 = 1.33$ and 1.5; say, $q_2 = 1.4$. Can you come up with two more rationals q_3 and q_4 in the sequence using this idea.*

The Archimedean property.

Another simple but important consequence of the density of rational numbers in $\mathbb{R}$ is the following property of the *ordering* of real numbers:[6]

---

[6] This property of real numbers was named after Archimedes by the mathematician O. Stolz in the 1880s. Archimedes in turn credited it to Eudoxus. There are abstract number fields that do not have this property, such as the hyperreal numbers of nonstandard analysis. As we mentioned earlier, these are number fields that contain the real numbers as well as infinitesimals and their reciprocals, the infinite numbers.

The Real Numbers

> *For every positive real number x there is a positive integer n such that $n > x$.*

This seems to be saying something rather obvious; for instance, if we pick any real number, say, $x = 72015.3894\cdots$ then the integer $n = 72016$ or any larger integer works. But this observation does not extend to *all* real numbers because we have not proved yet that every real number has a decimal expansion.

Why is the Archimedean property true? For rational numbers this is easy to see: if $q = k/m$ then since $m \geq 1$ we see that $q \leq k$ so the integer $n = k+1$ works. If r is any positive real number then by the density property of rational numbers, there is a (positive) rational number q such that $r < q < r+1$. If n is a positive integer such that $n > q$ then by transitivity, $n > r$ and we are done!

The Archimedean property in particular implies that there is no real number that is larger than all the integers; the integers go as far away as the real numbers do! This insight may not rock you to your core but remember the not-so-obvious fact that this property of real numbers is closely tied to the density of rational numbers!

Also worth a mention here is that the Archimedean property is often used in the following way when working with limits:

> If r is any positive real number then for every real $\varepsilon > 0$, *no matter how small*, there is a positive integer n such that $r/n < \varepsilon$ (set $x = r/\varepsilon$ in the statement of the Archimedean property).

So if $r = 1$ and we pick any very small number, say,

$$\varepsilon = 0.00000000000000123 = 1.23 \times 10^{-15}$$

then dividing r by the integer $n = 10^{15}$ (a quadrillion) gives $r/n = 10^{-15}$ which is less than the above value of ε.

Exercise 55 *Suppose that x and y are distinct positive real numbers. Why must there be an integer n such that $ny > x$? Note that x/y is a real number.*

> **The real numbers are complete (success! *no more gaps*)**

The set of real numbers as constructed above has the following property that distinguishes it from $\mathbb{Q}$ and uniquely characterizes it among totally ordered fields:

> **Completeness Property**: *The set $\mathbb{R}$ is complete, that is, every Cauchy sequence of real numbers converges to a real number.*

This statement that we prove in Appendix 11.4 should not seem surprising since we built the real numbers to be essentially those points around which Cauchy sequences accumulate.

Among other things, completeness implies that $\mathbb{R}$ has no gaps or holes because by its very construction it contains the limits of all Cauchy sequences, the only sequences that can possibly converge.

We should also keep in mind that the set $\mathbb{Q}$ of all rational numbers is not complete since as we observed earlier, there are sequences of rational numbers that do not converge to rational numbers.

Least upper bounds and nested intervals.

The completeness property has deep consequences that are essential for proving other profound statements that form a logical foundations on which our mental intuition of continuum may be based. I discuss two of the most fundamental consequences of completeness in this section to illustrate how intricate this logical foundation is.

The first property has to do with bounded sets. Consider a nonempty set S of real numbers. If we can find a real number u such that every number x in S is less than or equal to u then we call u an *upper bound* for S. For example, take the set S of all rational numbers that are less than 4, or stated symbolically,

$$S = \{r : r \text{ is rational and } r < 4\}$$

has a natural upper bound, namely, 4. This upper bound is not unique; for instance, 5 is also an upper bound, as is 4.01, 14/3, 2π, etc. In fact, every real number greater than or equal to 4 is an upper bound. On the other hand, π is not an upper bound because there are numbers in S that are bigger than π, like 3.5. But if your instincts tell you that there is something special about 4 with regard to S then you've hit on something important! More on this soon.

A *lower bound* of a set is a real number that is less than or equal to every number in S. For example, a lower bound for the interval $(0, \infty)$ is the real number 0. Like upper bounds, a lower bound is not unique.

A set can have upper bounds but no lower bounds. For instance, the above set S of rational numbers has no lower bounds since it contains all negative rational numbers.[7] Similarly, the interval $(0, \infty)$ is a set having lower bounds but no upper bound.

A set that has both an upper bound and a lower bound is a *bounded set*. For instance, every bounded interval such as $[0, 500)$ is a bounded set. A more interesting example is the set of all rational numbers whose squares are less than 2:

$$T = \{q : q \text{ is rational and } q^2 \leq 2\}$$

This is a bounded set with a lower bound -3 and an upper bound 3 (among others). We can improve on this by realizing that every number in this set is less than $\sqrt{2}$ and greater than $-\sqrt{2}$.

On the other hand, no real number that is less than $\sqrt{2}$ can be an upper bound of T; if x is a real number and $-\sqrt{2} < x < \sqrt{2}$ then because the rational numbers are dense we can find a rational q such that $x < q < \sqrt{2}$. Notice that such a q is a number in T so x cannot be an upper bound. A similar reasoning shows that no real number that is greater than $-\sqrt{2}$ can be a lower bound of T.

The numbers $\pm\sqrt{2}$ are special upper and lower bounds of the set T. We call $\sqrt{2}$ the *least upper bound* of T simply because it is the smallest possible upper bound. Similarly, we call $-\sqrt{2}$ the *greatest lower bound* of T.

---

[7]If x is any negative real number then $-x$ is positive so by the Archimedean property there is a positive integer n such that $-x < n$. Multiplying by -1 gives $x > -n$ and $-n$ is a rational number less than 4. So $-n$ is in S and x cannot be a lower bound of S.

The Real Numbers 111

Let's summarize these definitions for easy reference:

> For a nonempty set S of real numbers:
> (a) an *upper bound* is a real number u such that $x \leq u$ for all x in S; the smallest possible upper bound is the *least upper bound*, also called the *supremum* of S and written $\sup S$.
> (b) a *lower bound* is a real number l such that $l \leq x$ for all x in S; the largest possible lower bound is the *greatest lower bound*, also called the *infimum* of S and written $\inf S$.
> (c) If S has an upper bound then we say that it is bounded from above; if it has a lower bound then we say that S is bounded from below. If S is bounded from above and from below then S is a *bounded set*.

For the set S mentioned earlier, the number 4 is the supremum of S. Because S has no lower bounds, it cannot have an infimum. Similarly, for the set T mentioned above, $\sup T = \sqrt{2}$ and $\inf T = -\sqrt{2}$. Obviously, T is a bounded set.

If S has a supremum or least upper bound r then this must be unique. If r' were also a least upper bound then we can argue that $r' \leq r$ because r is in particular an upper bound. Similarly, $r \leq r'$ and it follows that $r' = r$.

The supremum and the infimum of a set S, if they exist, are unique.

Exercise 56 *Find the supremum and the infimum of each of the following sets if they exist. If a set doesn't have a supremum or an infimum then explain why.*

$$S_1 = [0, 500) \qquad S_2 = [0, \infty) \qquad S_3 = \{q : q^3 \leq 2\}$$

Is $\sup S$ in S? The answer is sometimes, but not usually. The above discussion and examples make it quite clear that neither $\sup S$ nor $\inf S$ need be a number in S. For example, 500 is the supremum of *[0,500)* and 0 is its infimum but this interval does not include 500. More interesting is the set T above; it has both a supremum $\sqrt{2}$ and an infimum $-\sqrt{2}$ but neither of these is rational so cannot be a number in T.

If S is a finite set though then it always has both a largest and a smallest element and both of these are always in S. There is no drama or mystery for finite sets!

Sets include ranges of sequences. For example, if $x_n = 1, 1/2, 1/3, \ldots$ then its range is the set $\{1, 1/2, 1/3, \ldots\}$. This is a bounded set whose least upper bound is 1 and greatest lower bound is 0 (because 0 is a lower bound and no other number can distinctly fit in between 0 and some number in the set). This observation leads to the following idea.

> Let x_n be a sequence of real numbers. Then the upper bound, lower bound, supremum and infimum of x_n are the corresponding concepts for the range of the sequence as a set.

The distinction between a sequence (a function) and its range (a set) should not cause any confusion. Although a sequence always has an infinite number of terms, its range may be finite. An example is $x_n = (-1)^n$ which is the infinite string $-1, 1, -1, 1, \ldots$ whose range is the two-element set $\{-1, 1\}$. This sequence is therefore bounded with $\sup x_n = 1$ and $\inf x_n = -1$.

A useful corollary of the Archimedean property is the following theorem; see Appendix 11.4 for the proof.

> **Bounded implies Cauchy (BC).** Let x_n be a non-decreasing sequence of real numbers so that $x_1 \leq x_2 \leq x_3 \leq \cdots$ If x_n is bounded from above then x_n is Cauchy and converges to $\sup x_n$.

For an example, consider the increasing sequence $x_n = 1 - 1/n$ which converges to its supremum 1. The supremum is 1 because 1 is larger than every number in the sequence and no other real number fits distinctly between 1 and a term of the sequence.

Exercise 57 *Let x_n be a non-increasing sequence of real numbers that is bounded from below. Explain why such a sequence must converge to a real number. What is that number?*

Think about what kind of a sequence $-x_n$ is and consider applying Theorem (BC). If working with an example is helpful, then consider $x_n = 1/n$ first.

Another consequence of the completeness property involves *nested intervals*, or intervals within intervals. For example, the intervals $[1/3, 1/2]$, $[0.35, 1/2]$, $[0.35, 0.45]$, $[0.4, 0.41]$ are nested since each is contained in the one before it. The following is an important fact about infinite sequences of nested intervals and leads to the second of the aforementioned consequences of completeness (see Theorem (LUBS) below).

> **Nested Intervals Property (NIP):** *Consider a sequence of nested closed intervals $[a_1, b_1] \supset [a_2, b_2] \supset [a_3, b_3] \supset \cdots$ If $\lim_{n \to \infty}(b_n - a_n) = 0$ then the intersection of all the intervals is a single real number:*
> $$\bigcap_{n=1}^{\infty} [a_n, b_n] = \{r\} \qquad (5.6)$$

This is easy to prove: consider the non-decreasing sequence $a_1, a_2, a_3, \ldots$ This is bounded by any one of the b_n so it has a least upper bound, say α, by Theorem (BC). Similarly, the non-increasing sequence $b_1, b_2, b_3, \ldots$ has a greatest lower bound β by Exercise 57. So we can say that:
$$a_1 \leq a_2 \leq a_3 \leq \ldots \leq \alpha \leq \beta \leq \ldots \leq b_3 \leq b_2 \leq b_1$$

In particular, $0 \leq \beta - \alpha \leq b_n - a_n$ for every index n. Since $\lim_{n \to \infty}(b_n - a_n) = 0$ the Squeeze Theorem implies that $\beta - \alpha = 0$. So $\alpha = \beta$ and this common value is the number r in (5.6).

The nested intervals property holds because the completeness of real numbers ensures that there are no gaps or holes. Within the set of just the rational numbers (which is incomplete) the intersection in (5.6) can be empty. For example, if a_n is a sequence of rational numbers that is increasing and converges to $\sqrt{2}$ from below and b_n is a sequence of rational numbers that is decreasing and converges to $\sqrt{2}$ from above then:
$$a_1 < a_2 < a_3 < \cdots < \sqrt{2} < \cdots < b_3 < b_2 < b_1$$

The Real Numbers 113

and we have a sequence of nested intervals with rational endpoints:

$$[a_1, b_1] \supset [a_2, b_2] \supset [a_3, b_3] \supset \cdots$$

whose intersection is $\sqrt{2}$, an irrational number.[8] If we ignore irrational numbers like $\sqrt{2}$ then the intersection of these closed intervals will be empty!

The NIP is often useful in proving other properties of sets of real numbers. For example, we can use it to prove the generalization of Theorem (BC) above to all sets of real numbers.

This result is general enough that if it is taken as an axiom then we can reverse-prove completeness from it.

This reverse approach turns out to be more succinct and easier to use in proving theorem. Understandably, it is the preferred approach by most analysis textbook authors.

> **The Least Upper Bound Property for sets (LUBS).** *Every nonempty set S of real numbers that is bounded from above has a least upper bound or supremum (in $\mathbb{R}$). Likewise, if S is bounded from below then it has a greatest lower bound or infimum.*

The proof of the first claim uses the NIP and is rather technical but the second claim follows easily from the first using multiplication by -1 and the fact that inequalities are reversed by this operation. If S has a lower bound B then $x \geq B$ for every x in S. It follows that $-x \leq -B$ so the set $-S = \{-x : x \in S\}$ has an upper bound $-B$. Now Theorem (LUBS) guarantees a least upper bound U for $-S$ and you can easily show that $-U$ is a least upper bound for S.

Notice that completeness is essential in Theorem (LUBS). For example, the set $\{q : q \text{ is rational and } q^2 < 2\}$ has no least upper bound in the (incomplete) set of rational numbers.

5.5 The uncountability of real numbers: the irrational majority

Here is a summary of the construction of real numbers discussed above:

> We elevated each rational number q to an equivalence class $[q]$ of the constant sequence $q, q, q, \ldots$ We added the equivalence classes $[q_n]$ of all *other* Cauchy sequences (not eventually constant) that correspond to the irrational numbers and showed that the enlarged set could be made into a totally ordered field that contained the (equivalence classes of) rational numbers. We showed that the new, larger set which we called the set of real numbers was complete; that is, it had no gaps or holes.

Now is the time to answer the question: *how big is the set of all real numbers?* Since the set of rational numbers is countable, this question may be phrased more precisely as: *is the set of real numbers countable or uncountable?*

Let's find out next.

---

[8]Notice that $\sqrt{2}$ is both the least upper bound or supremum of the sequence a_n and the greatest lower bound or infimum of the sequence b_n.

Cantor's diagonal argument

One of the most famous proofs of the uncountability of real numbers was given by Cantor and has come to be known as the "diagonal argument". To use this simple argument we need the idea that every real number has a decimal expansion in terms of the integers 0 through 9. We prove this fact in Chapter 6 after some exposure to infinite series. At that time we will also see another proof of the uncountability of real numbers that achieves more than the diagonal argument since it actually gives the cardinal number of $\mathbb{R}$ as $2^{\aleph_0}$, not just uncountable.

> We limit our attention to all real numbers that lie between 0 and 1, namely, the open interval $(0,1)$. As a proper subset of $\mathbb{R}$, if $(0,1)$ is uncountable then so is $\mathbb{R}$ (see (US) above).
>
> The diagonal argument is a proof by contradiction. We assume that $(0,1)$ is countable and show that this assumption leads to a contradiction.
>
> Every real number r in $(0,1)$ has a decimal expansion $0.d_1 d_2 d_3 \ldots$ where each digit d_i is an integer between 0 and 9. If $(0,1)$ were countable then all of its elements could be listed using a one to one correspondence with positive integers:
>
> $$\begin{array}{ll} 1 & 0.d_{11} d_{12} d_{13} \ldots d_{1n} \ldots \\ 2 & 0.d_{21} d_{22} d_{23} \ldots d_{2n} \ldots \\ 3 & 0.d_{31} d_{32} d_{33} \ldots d_{3n} \ldots \\ \vdots & \quad \vdots \\ n & 0.d_{n1} d_{n2} d_{n3} \ldots d_{nr} \ldots \\ \vdots & \quad \vdots \end{array}$$
>
> Consider the real number $r = 0.d_1 d_2 d_3 \ldots$ with d_n defined as follows using the diagonal entries in the above list:
>
> $$d_n = \begin{cases} 1 & \text{if } d_{nn} = 0 \text{ or } 9 \\ 0 & \text{if } d_{nn} \neq 0 \text{ or } 9 \end{cases} \quad n = 1, 2, 3, \ldots$$

This number r cannot be in the above list because it is different from every number in the list in at least one digit! Since the existence of the above list was implied by the assumption that $(0,1)$ was countable, that assumption had to be false. Therefore, we conclude that:

The uncountability of real numbers: *The interval $(0,1)$ and thus, the set $\mathbb{R}$ of all real numbers is uncountable.*

Although $(0,1)$ and $\mathbb{R}$ are both uncountable, you may wonder whether $\mathbb{R}$ has a larger cardinal number because

$$\cdots \cup (-2,-1) \cup (-1,0) \cup (0,1) \cup (1,2) \cup \cdots \cup \mathbb{Z} = \mathbb{R}$$

and each of the infinity of intervals listed on the left hand side of the equation has the same size as $(0,1)$; see Exercise 58 below. Nevertheless, both $(0,1)$ and $\mathbb{R}$ have the same cardinal number. The equipollence is established by the standard bijection

$$\frac{1}{\pi} \tan^{-1} x + \frac{1}{2}$$

The Real Numbers

that maps ℝ one to one onto $(0,1)$; see Figure 5.1.

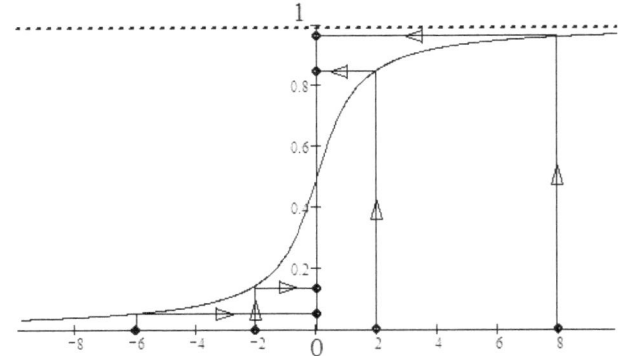

Figure 5.1: The interval $(0,1)$ is equipollent to $(-\infty, \infty)$

A simple modification of the above inverse tangent function in the form

$$\frac{2}{\pi} \tan^{-1} x$$

shows that the set of all *positive* real numbers is also equipollent to the interval $(0,1)$. If you have a graphing device consider sketching the graph for a stretch of positive values of x starting from 0 to see the shape of the function and how it works.

In the following exercise you verify the useful fact that all intervals of real numbers are equipollent to $(0,1)$ and thus also to ℝ.

Exercise 58 *(a) Consider an arbitrary open interval of real number (a,b) from a to b where $a < b$. Verify that the function $f : (a,b) \to (0,1)$ defined as*

$$f(x) = \frac{x - a}{b - a}$$

is a bijection, thus confirming that (a,b) is equipollent to $(0,1)$. Note that $f(x)$ here is a linear function.

(b) Determine the specific form of $f(x)$ for $(a,b) = (-1, 3)$ and $(a,b) = (0, 1/4)$. Consider sketching the simple graph of $f(x)$ to get a better feel for how this function works.

ℝ is continuous, thanks to the irrational numbers!

The uncountability of real numbers gives us a low-hanging fruit to pluck.

Recall that the set of all real numbers is the *disjoint* union of all irrational numbers and all rational numbers. We also know now that ℝ is uncountable and ℚ is countable. If the irrational numbers were countable too, then their union with the rationals, namely $\mathbb{R}$, would be countable by (CU) above. This is not the case so we conclude:[9]

> **Uncountability of the irrationals**: *The set of all irrational numbers is uncountable.*

This result, together with the countability of the rationals, point to a striking fact about real numbers: the overwhelming majority of real numbers are irrational! This is all the more significant because the rational numbers are the only ones that we can concretely deal with: all measurements that we take in nature by any type of instrument, any set of data that is ever saved in texts or in the "cloud" or any other type of computer memory, algorithms that our machines can ever process, all of these can only handle *rational* numbers. This revelation does not surprise us given the knowledge that the irrational numbers are infused with infinity.

A unique feature of the rational numbers that makes calculus and the continuum amenable to calculations is their *density* in $\mathbb{R}$.

As our construction above shows, each irrational number can be approximated to an arbitrary degree of accuracy by a sequence of rational numbers. We do not need to know all the digits in $\sqrt{2}$, π and the like in order to calculate with them or to record the results involving them, because we can always find rational numbers that are indistinguishable from them for all practical purposes. Rational numbers may be a tiny minority but fortunately, they are pervasive!

While the rational numbers give algebraic structure to the real numbers and, through density, provide approximation power, it is the irrational numbers that fill the gaps that the rationals leave open on the number line. Being limits of sequences of rational numbers, the irrationals are the fillers, the glue that bind the rationals to make a continuous (or complete) whole out of the real numbers.

Algebraic and transcendental numbers.

We close this chapter by discussing a classification of real numbers that enlarges the set of rational numbers a little bit by appending some irrational numbers to it. This brings algebraic structure to some irrational numbers although the newly structured irrationals remain a small minority among all the irrational numbers.

You may remember the polynomials from algebra; they are built from numbers and positive-integer powers of a variable x:

$$P(x) = c_0 + c_1 x + c_2 x^2 + \cdots + c_n x^n$$

We say that $P(x)$ is a *polynomial of degree* n, assuming that the number c_n (called the leading coefficient) is not zero. Examples are $2 + 3x$ (with degree 1), $x^2 - x - 1$ (degree 2), x^5 (degree 5) and $7/4$ (degree 0, like all numerical constants).

A *root* of $P(x)$ is a number r that makes $P(x)$ zero if we set $x = r$. For instance, if we set $2 + 3x = 0$ and solve for x we find the root $x = -2/3$. To check, insert this value of x in the polynomial and notice that $2 + 3(-2/3) = 2 - 2 = 0$. Though less obvious, you may recall that the roots of $x^2 - x - 1$ are found using the good old "quadratic formula" to be

$$x = \frac{1 \pm \sqrt{5}}{2} = \frac{1}{2} \pm \frac{\sqrt{5}}{2} \tag{5.7}$$

---

[9] A technical term for this type of proof by contradiction is a *cardinality (or countability) argument*.

The Real Numbers 117

The *Fundamental Theorem of Algebra*[10] states that a *polynomial of degree n has at most n distinct roots*. Some, or all of the roots may be repeated (or multiple) roots; for example, $x^5 = 0$ has a single solution 0 which is also a 5-times repeated root (a root of "multiplicity" 5). Complex (non-real) roots often occur too; for example, the quadratic formula gives the roots of $x^2 - x + 1$ as two "complex conjugate" numbers

$$\frac{1}{2} \pm \frac{\sqrt{3}}{2}i, \quad i = \sqrt{-1} \tag{5.8}$$

An *algebraic number* is a root of a polynomial with *rational coefficients* (the numbers c_0, c_1, etc are rationals). For example, the numbers in (5.7) are algebraic, or more precisely, *real* algebraic to distinguish them from the complex algebraic numbers like those in (5.8).

The set of real algebraic numbers is considered to be an extension of the set of rational numbers because because each rational number m/n is a root of the polynomial $nx - m$ with rational coefficients, so it is an algebraic number. The set of algebraic numbers contains a lot of irrational numbers but still the following is true:

> *The set of all algebraic numbers is countable.*

This is not surprising; the rationals are countable so finite sets of them (in the form of polynomial coefficients) are still countable as a countable union of countable sets by Exercise 33 above. Further, each polynomial comes with a finite number of roots so we are still able to use Exercise 33 to complete the proof. Although not all algebraic numbers are real numbers, it is clear that the set of all real algebraic numbers must be countable too; see (CS) above.

This means that most real numbers are actually *not algebraic*. Real numbers that are not algebraic are called *transcendental*.

By the previous discussion and (US) above, we have:

> *All transcendental numbers are irrational and the set of all transcendental numbers is uncountable.*

It is astonishing that even though the vast majority of irrational numbers are transcendental, conjuring up transcendental numbers or verifying that any given irrational number is transcendental is typically hard. For example, is π algebraic or transcendental?

It is trivially a root of the polynomial $x - \pi$ so is the coefficient $c_0 = \pi$ rational? Well, it is known that π is not rational[11] and we need a polynomial with rational coefficients to show that a number is algebraic. We did not know for sure if *any* number was really transcendental until 1844 when Liouville came up with such numbers (on demand!) for the first time; we discuss Liouville's numbers briefly in Chapter 6.

That π was transcendental was finally proved in 1882 by the German mathematician Ferdinand Lindemann . The most famous transcendental number (in calculus) that we will come across often is $e \simeq 2.71828$, named in honor of Leonhard Euler (1707-1783). It was proved to be transcendental in 1873 by the French mathematician Charles Hermite (1822-1901) of the "Hermitian operators" fame; these operators are important in quantum theory and elsewhere in science and mathematics.

---

[10]This theorem shows that the set of complex numbers has an algebraic property that the real numbers lack: all polynomials factor completely in the set of complex numbers. There is a simple proof of this important result by Liouville who used complex analysis. The algebraic proof applies in more general contexts than real or complex numbers. Both of these proofs require a level of conceptual and technical preparation that go beyond our focus here.

[11]The Swiss polymath J.H. Lambert proved the irrationality of π in 1768.

So we see that "transcendental" is an apt name for these abundant, yet highly illusive numbers. It is even hard to tell whether simple algebraic combinations of known transcendental numbers are again transcendental. For example, although we know that each of e and π is transcendental, we don't know yet if $e + \pi$ or $e\pi$ are also transcendental.

Exercise 59 *It is usually easier to prove that a number is NOT transcendental by showing that it is algebraic. For instance, if $r = \sqrt{3 + \sqrt{2}}$ then $r^2 = 3 + \sqrt{2}$ so*

$$(r^2 - 3)^2 = 2$$
$$r^4 - 6r^2 + 9 = 2$$
$$r^4 - 6r^2 + 7 = 0$$

We see that r is a root of the polynomial $x^4 - 6x^2 + 7$. Since this has rational coefficients, r is algebraic. Use a similar argument to show that the following numbers are algebraic too:

$$\sqrt[4]{2}, \qquad 1 + \sqrt{3}, \qquad \sqrt[3]{2 - \sqrt{3}}, \qquad \frac{\sqrt{2}}{1 + \sqrt{3}}$$

Chapter 6

Infinite Series: *adding how many numbers?*

Our aim in this chapter is to clarify what is meant by "adding infinitely many numbers". In particular, we will see that infinite series are rarely computed exactly, but when they are, it is never done by actually adding infinitely many numbers. The key words here are *limit* and (finite) *partial sums*. There are some results about series that I will not discuss in this book; you usually find those in a standard calculus textbook.

6.1 A tale of two series and other oddities

When describing Zeno's scenarios earlier we encountered an infinite sum, namely,

$$\frac{1}{2} + \left(\frac{1}{2}\right)^2 + \left(\frac{1}{2}\right)^3 + \left(\frac{1}{2}\right)^4 + \cdots = 1 \tag{6.1}$$

According to this equation an infinity of numbers add up to the number 1. The left hand side is a special case of the following more general expression

$$r + r^2 + r^3 + r^4 + \cdots \tag{6.2}$$

where we take *any* nonzero real number r and multiply it repeatedly by itself and add the results as we go. In the special case of (6.1), $r = 1/2$.

The expression (6.2) is an example of a *geometric series* that we discuss later in more detail.

If the expression (6.2) *does add up to some number* S (and it may not, depending on the value of r) then it is possible to figure out what the number S is using simple algebra: we multiply S by r to get

$$Sr = (r + r^2 + r^3 + r^4 + \cdots)r = r^2 + r^3 + r^4 + \cdots$$

If we subtract Sr from S then

$$S - Sr = r + r^2 + r^3 + r^4 + \cdots \\ - r^2 - r^3 - r^4 - \cdots$$

On the right hand side everything except the first r cancels out and we are left with:

$$S - Sr = r$$

Now, we first factor out the common S and then divide the result by $1 - r$ to get:

$$S = \frac{r}{1-r}$$

This gives the formula

$$r + r^2 + r^3 + r^4 + \cdots = \frac{r}{1-r} \tag{6.3}$$

In particular, if $r = 1/2$ the right hand side of (6.3) works out to 1 and verifies (6.1).

A tale of two series: a numerical comparison.

You may have noticed above that as we keep adding fractions, we are adding smaller ones as we proceed; for example, when $r = 1/2$ the first fraction is $1/2$, the second is $1/4$ which is half of the first, the third is $1/8$ or half of the second and so on. So as we keep adding new fractions, we accumulate less at each step as we go on. Further, the individual fractions $1/2, 1/4, 1/8$ etc go to 0, rapidly becoming negligible in size. For example, the tenth term is $(1/2)^{10}$ which is less than 0.00098, and the 20th term is about 0.00000095.

In summary, the observations we made in the last paragraph suggest that to be able to get a number from the sum of infinitely many numbers, it is necessary that the numbers that we add should get smaller and not only that, they must approach 0.

Necessary, yes; *but not sufficient*. To illustrate this essential point, let's compare (6.2) numerically with the following series:

$$1 + \frac{1}{2} + \frac{1}{3} + \frac{1}{4} + \frac{1}{5} + \frac{1}{6} + \cdots \tag{6.4}$$

This is an important enough series that like (6.2) it has been given a name: the *harmonic series*. You can see that in (6.4) we are also adding numbers that get smaller and approach 0. But there is a subtle difference between the two series that makes them entirely different; it is possible to get a sense of this difference by doing a numerical experiment.

Consider $r = 0.9$ in (6.3). Then

$$0.9 + 0.9^2 + 0.9^3 + \cdots = \frac{0.9}{1 - 0.9} = 9 \tag{6.5}$$

This shows that adding more and more of the powers of 0.9 gets us closer and closer to the number 9; if we add a lot of them then we get quite close. Adding 100 numbers and rounding the answer to 6 decimal places gives

$$0.9 + 0.9^2 + 0.9^3 + \cdots + 0.9^{100} = 8.999761$$

Infinite Series

This is different from 9 by less than 0.00024. Notice also that the left hand side of (6.5) will always be less than 9 no matter how many numbers we add because the right hand side of (6.5) is always greater than the left side if we retain only a finite number on the left.

By contrast, adding a finite number of fractions of (6.4) can eventually get past 9 (or any other number); for instance, adding 4550 fractions gives

$$1 + \frac{1}{2} + \frac{1}{3} + \cdots + \frac{1}{4549} + \frac{1}{4550} \simeq 9.000208$$

I have rounded the answer to 6 decimal places but it is clear that the sum exceeds 9. For comparison,

$$0.9 + 0.9^2 + 0.9^3 + \cdots + 0.9^{4550} = 8.999999999999996$$

Very close to but still less than 9, as we noted above. Using a computer software you can explore things a little further; adding a finite number of fractions in (6.4) gets you past larger numbers. For example,

$$1 + \frac{1}{2} + \frac{1}{3} + \cdots + \frac{1}{12399} + \frac{1}{12400} \simeq 10.0027$$

which is obviously larger then 10. If we keep adding more of the fractions in this series we continue to move forward; passing 11, then 12 and so on.

You may well ask how do we know that the harmonic series can reach and pass even larger numbers. We saw that we needed to add 4550 fractions to pass 9 but then we needed 12400 fractions (7850 additional terms) to get past 10. Adding a million fractions gives a total of about 14.39 and a billion fractions add to about 21.3; getting larger of course, but at a frustratingly slow pace.[1] Numerical work alone comes up short in convincing a skeptic that the sum of the harmonic series does in fact get infinitely large.

Fortunately, there is a clever argument that proves adding enough fractions of the harmonic series indeed does get us past any number whatsoever.[2] Arrange the numbers in (6.4) in blocks of size a power of 2 as follows:

$1 + \frac{1}{2} +$	$> \frac{1}{2}$
$\frac{1}{3} + \frac{1}{4} +$	$> \frac{1}{4} + \frac{1}{4} = \frac{1}{2}$
$\frac{1}{5} + \frac{1}{6} + \frac{1}{7} + \frac{1}{8} +$	$> \frac{1}{8} + \frac{1}{8} + \frac{1}{8} + \frac{1}{8} = \frac{1}{2}$
$\vdots$	$\vdots$

The above table has an infinite number of rows and the sum of the numbers in each row is at least $1/2$, so the sum in (6.4) is greater than

$$\frac{1}{2} + \frac{1}{2} + \frac{1}{2} + \cdots$$

which is $1/2$ added to itself repeatedly without end. It is clear now that the sum in (6.4) can get past any number if enough fractions are added in. We may need to add billions of trillions of terms to pass 100 say, but since $1/2$ added 200 times reaches 100 it follows that by adding enough fractions,

---

[1] The growth of sum of the numbers is logarithmic as we add one number at a time. Equivalently, the number of fractions we need to add in order to pass a given number grows exponentially.

[2] This argument predates calculus, dating back to around 1350 and the French bishop Nicholas (or Nicole) Oresme.

the sum does eventually pass 100. This settles one question but leads to another, even more urgent question:

Why do we never pass a certain number in the case of (6.3) but not in the case of (6.4), even though in both cases the numbers that we are adding approach 0?

This is one of several questions that require greater clarity in what we mean when we talk about adding infinitely many numbers. I will soon lay a bit of ground work on which to stand, but first, a few more oddities!

Formulas have restrictions!

Let us go back to (6.3). We derived this formula *assuming* that the series on its left does add up to a number; making this assumption was important because (6.3) is not valid for *all* values of r. For example, if we insert $r = 1$ into (6.3) we get:

$$1 + 1^2 + 1^3 + 1^4 + \cdots = \frac{1}{0} \tag{6.6}$$

Since $1 = 1^2 = 1^3$ etc, the left hand side is 1 added to itself infinitely often, so intuitively this side of (6.6) is infinitely large. But the right hand side is problematic since division by 0 is algebraically meaningless.

It can get even worse: if we set $r = 2$ in (6.3) then we obtain

$$2 + 2^2 + 2^3 + 2^4 + \cdots = \frac{2}{1-2}$$
$$2 + 4 + 8 + 16 + \cdots = -2 \tag{6.7}$$

The last "equality" is sufficiently absurd to convince us that (6.3) is not valid when $r = 2$ (or any number larger than 1).

Exercise 60 *(a) Let $r = -2$ in (6.3) and calculate the first 6 terms of the sum. Do you think that the answer from (6.3) is plausible?*

(b) Repeat (a) but with $r = -1/2$.

(c) Based on your conclusions in (a) and (b) and the earlier observations above, for what values of r do you think (6.3) is valid? We will find the range of acceptable values for r later in the section on geometric series.

Warning! Signs keep changing.

In addition to the cautionary notes about adding infinitely many numbers that I mentioned above, another set of issues comes up when dealing with series that contain infinitely many numbers with both positive and negative signs. For instance, if we set $r = -1$ in (6.3) then we obtain

$$(-1) + 1 + (-1) + 1 + \cdots = -\frac{1}{2} \tag{6.8}$$

The alterations of signs occurs because -1 when raised to an even power gives $+1$ and when raised to an odd power it gives -1. Notice that there are no infinite magnitudes here and the answer

Infinite Series 123

$-1/2$ does not seem as implausible as those in (6.6) or (6.7). But $-1/2$ isn't an intuitively clear result either; how can the left hand side of (6.8) add up to $-1/2$ when it looks more like it should be zero with all the possible cancellations? Here is how we might add up the left hand side of (6.8) to get zero:

$$[(-1)+1] + [(-1)+1] + \cdots = 0 + 0 + \cdots = 0 \qquad (6.9)$$

Since the calculation method that gave us (6.3) did not seem to give valid results for all values of r like those in (6.6) or (6.7), so then maybe we ignore the result $-1/2$ and choose 0 as the "valid answer" the way it was obtained in (6.9)? If only it were that simple!

Here is yet another possible answer: suppose that we keep the first -1 aside in (6.8) but pair up the rest of the numbers to get:

$$(-1) + [1+(-1)] + [1+(-1)] + \cdots = -1 + 0 + 0 + \cdots = -1$$

It is hard to argue that this answer is less valid than 0 because it was derived in the same way, only with a slightly different grouping. As long as there are no restrictions on how we add the numbers on the left hand side of (6.8) we can add them in any manner.

Why not get more creative? For example, starting after the first number -1, let's add an even number of $+1$ terms followed by adding an odd number of -1 terms as in

$$\underbrace{-1}_{1} + \underbrace{(1+1)}_{2} + \underbrace{(-1-1-1)}_{3} + \underbrace{(1+1+1+1)}_{4} + \cdots = -1 + 2 - 3 + 4 - 5 + \cdots$$

Since we have an *infinite supply* of $+1$ and -1 terms, *if there are no restrictions* on how we add them up then the series on the right hand side is a possible outcome. The last series may also be added in a variety of ways. For instance, if we add the numbers one pair at a time then

$$-1 + 2 - 3 + 4 - 5 + \cdots = (-1+2) + (-3+4) + (-5+6) + \cdots$$
$$= 1 + 1 + 1 + 1 + \cdots$$

Adding 1 to itself an infinite number of times give infinity! Certainly this does not seem like a better answer in (6.8) than any of the others that we discussed, but how do we rule it out?

In the next section we begin a clean discussion of infinite series from scratch that resolves the above issues.

6.2 Infinite series as sequence limits

The above discussion suggests that if the signs of terms alternate then the result of infinite summation depends on the way the numbers are added. Even when the signs do not alternate the sum may fail to be a number, and when a number is found using some derivation, it may not be correct. These problems were exactly what led mathematicians in the 19th century to clarify the concept using something that we met earlier, namely, sequences.

The partial sums: sequences to the rescue!

Suppose that we have an infinite sequence of real numbers

$$a_1, a_2, a_3, \ldots, a_n, \ldots \qquad (6.10)$$

and we want to add all of these numbers. We do it in a step-by-step manner by creating *another* sequence of real numbers as follows:

$$s_1 = a_1$$
$$s_2 = a_1 + a_2$$
$$s_3 = a_1 + a_2 + a_3$$
$$\vdots$$
$$s_n = a_1 + a_2 + \cdots + a_n$$

Each of the finite sums above is a part of the total, or the sum of all the infinite number of terms. As the value of the index n increases the finite sums will get closer to the value of the infinite series, if such a value exists. We formalize this promising idea.

> The finite sum in each of the above rows is called a *partial sum* of the terms in (6.10). The n-th term s_n is called the *n-th partial sum*; The sequence
>
> $$s_1, s_2, s_3, \ldots, s_n, \ldots$$
>
> is *the sequence of partial sums* of the original sequence in (6.10). If as $n \to \infty$ the sequence s_n converges to a real number s then we say that the *infinite series*
>
> $$a_1 + a_2 + \cdots + a_n + \cdots$$
>
> *converges* and call the limit s *the sum of the infinite series:*
>
> $$a_1 + a_2 + \cdots + a_n + \cdots = s = \lim_{n \to \infty} s_n$$
>
> The numbers $a_1, a_2, \ldots$ are called the *terms of the series* and the generic term a_n that typically gives the formula for the sequence is called the *n-th term*.

This is a good place to introduce the *sigma notation* $\sum$ which is the standard shorthand for summation, finite or infinite. We write the finite sum as

$$a_1 + a_2 + \cdots + a_n = \sum_{k=1}^{n} a_k$$

The infinite series is written as

$$a_1 + a_2 + \cdots + a_n + \cdots = \lim_{n \to \infty} \sum_{k=1}^{n} a_k = \sum_{k=1}^{\infty} a_k$$

Because sequences of partial sums may converge or diverge we use the same terminology for series.

> **Convergence:** *The infinite series* $a_1 + a_2 + \cdots + a_n + \cdots$ **converges** *to a real number s if the infinite sequence of partial sums* $s_1, s_2, \ldots, s_n, \ldots$ *converges to s.*
>
> **Divergence:** *The infinite series* $a_1 + a_2 + \cdots + a_n + \cdots$ **diverges** *if the partial sums sequence* $s_1, s_2, \ldots, s_n, \ldots$ *does not converge to a real number.*

Let's use the above definitions to re-examine the series on the left hand side of (6.8). The partial sums sequence of this series is

$$s_1 = -1,$$
$$s_2 = -1 + 1 = 0$$
$$s_3 = -1 + 1 - 1 = -1$$
$$s_4 = -1 + 1 - 1 + 1 = 0$$

and so on. The numbers $s_1, s_2, s_3, s_4, \ldots$ oscillate with period 2 (repeating every other term). This pattern of numbers does not approach any real number s so the sequence, and thus, the series diverges.

Here is a special type of divergence that involves infinity directly.

Infinite values: *The infinite series $a_1 + a_2 + \cdots + a_n + \cdots$ **diverges to ∞** (or to $-\infty$) or has infinite value if its sequence of partial sums $s_1, s_2, \ldots, s_n, \ldots$ diverges to ∞ (respectively, $-\infty$).*

Recall that a sequence $r_1, r_2, \ldots, r_n, \ldots$ of real numbers diverges to ∞ if for every positive real number M the terms of the sequence eventually exceed M; that is, there is an index N such that $r_n > M$ for all $n \geq N$. Also $r_1, r_2, \ldots, r_n, \ldots$ approaches $-\infty$ if $-r_1, -r_2, \ldots, -r_n, \ldots$ approaches ∞.

Also, where it is not misleading we write $\sum_{k=1}^{\infty} a_k = \infty$ or $\sum_{k=1}^{\infty} a_k = -\infty$ as appropriate. For example, we may write

$$1 + 1 + 1 + \cdots = \infty$$

because the sequence $s_n = n$ diverges to infinity.[3]

Keep in mind that the symbols $\pm\infty$ are abbreviations; to avoid confusion, we must not mistake them for numerical magnitudes!

Exercise 61 *Consider the infinite series*

$$1 + 1 + 1 - 1 - 1 + 1 + 1 + 1 - 1 - 1 + \cdots$$

in which the pattern $1 + 1 + 1 - 1 - 1$ keeps repeating.

(a) Calculate and list the values of the 15 partial sums $s_1, \ldots, s_{15}$. Can you guess what s_{20} is without any further additions? How about s_{100}? s_{1000}?

(b) What is special about every 5th partial sum $s_5, s_{10}, s_{15}, \ldots$? How do s_n compare with these if $n \neq 5, 10, 15, \ldots$?

(c) Use your observations in (a) and (b) to explain why the above series diverges to ∞. This is not surprising because the repeating pattern $1 + 1 + 1 - 1 - 1$ has a net positive value of 1; but we now have a proof rather than a mere hunch.

---

[3] Here's how the proof goes: if M is any given real number then (by the Archimedean property!) there is an integer $N > M$. So $n > M$ for every $n \geq N$, that is, $s_n > M$ for all $n > N$.

Defining infinite sums using *the sequence* of partial sums avoids the difficulties mentioned earlier without compromising the usefulness of the concept of infinite series or their potential applicability to various scientific or engineering models. *We are not free to add the terms of a series any way we like*; that freedom exists only for finite sums.

There is now just one way of adding infinitely many numbers: we start from the first term and add the other terms successively in the order given by the sequence in (6.10).

The telescoping series.

To illustrate how the partial sum sequence is used to find the *actual* sum (not just an estimate) of an infinite series, consider:

$$\sum_{n=1}^{\infty} \frac{1}{n(n+1)} = \frac{1}{(1)(2)} + \frac{1}{(2)(3)} + \frac{1}{(3)(4)} + \frac{1}{(4)(5)} + \cdots \qquad (6.11)$$

We interpret this series as the sum of the terms of the sequence

$$\frac{1}{(1)(2)}, \frac{1}{(2)(3)}, \frac{1}{(3)(4)}, \frac{1}{(4)(5)}, \ldots$$

The partial sums of this sequence are

$$s_1 = \frac{1}{(1)(2)}$$
$$s_2 = \frac{1}{(1)(2)} + \frac{1}{(2)(3)}$$
$$s_3 = \frac{1}{(1)(2)} + \frac{1}{(2)(3)} + \frac{1}{(3)(4)}$$

and so on. The n-th partial sum is the sum of the first n numbers:

$$s_n = \frac{1}{(1)(2)} + \frac{1}{(2)(3)} + \frac{1}{(3)(4)} + \cdots + \frac{1}{(n)(n+1)} \qquad (6.12)$$

It is unclear from the right hand side above whether s_n converges to any real number. But notice that each of the fractions breaks down as

$$\frac{1}{(n)(n+1)} = \frac{1}{n} - \frac{1}{n+1}$$

If we use this split form in (6.12) then we discover a remarkable set of cancellations:

$$s_n = \left(\frac{1}{1} - \frac{1}{2}\right) + \left(\frac{1}{2} - \frac{1}{3}\right) + \left(\frac{1}{3} - \frac{1}{4}\right) + \cdots + \left(\frac{1}{n} - \frac{1}{n+1}\right) = 1 - \frac{1}{n+1} \qquad (6.13)$$

Notice how all the in-between terms cancel out, leaving the first and last terms only. Now it is easy to see that $s_n \to 1$ as $n \to \infty$ because $1/(n+1) \to 0$. By the definition of convergent series we can now declare that

$$\sum_{n=1}^{\infty} \frac{1}{n(n+1)} = 1$$

Infinite Series 127

If you have the time then it is worthwhile to check this result by adding the series up to some large number (that is, calculate a big partial sum) using a calculator or a computer.

Because of the dramatic cancellations which caused the expression (6.13) for s_n to fold in, the series in (6.12) is called a *telescoping series*.[4]

Exercise 62 *Here is another telescoping series:*

$$\sum_{n=1}^{\infty} \frac{2}{n(n+2)}$$

Use an argument similar to the one above to show that this series adds up to 3/2. To see the cancellations in this case, consider listing a partial sum that has enough numbers in it, say s_{10}, in a form that is similar to (6.13).

If you feel like taking a more substantial step further with the telescoping series, then consider extending this line of reasoning to the series

$$\sum_{n=1}^{\infty} \frac{k}{n(n+k)}$$

where k is any fixed (but unspecified) natural number. What do you think the above series adds up to? Examining $k = 3$ can be helpful if you get stuck.

Another paradox in the style of Zeno. There is a simple interpretation of the telescoping series that may help us remember its simple and natural character. Imagine a man painting a column of length 1 unit by first painting the right half from 1 down to 1/2, then he paints the part from 1/2 to 1/3 which amounts to $1/2 - 1/3 = 1/(2)(3)$, then on to the segment 1/3 to 1/4 which amounts to $1/(3)(4)$ and so on. He will have painted the entire column when he is finished painting each of the infinitely many bits of it. Zeno might say that the man should never finish painting because he will always have a little bit more left!

Finitely equivalent infinite series.

If we rearrange *finitely* many numbers of the sequence in (6.10) and then sum the new sequence we obtain the same answer for the series because a finite number of rearrangements in the sequence does not affect the partial sums beyond a certain index. For example, suppose that the rearrangements are confined to at most N numbers in (6.10). Then a_{N+1}, a_{N+2} etc are the same and not changed so the partial sums s_{N+1}, s_{N+2} etc are the same for both the old sequence and the rearranged sequence. Whatever one partial sums sequence converges to, the other must converge to the same number.

---

[4]The numbers or terms of the telescoping series are related to the so-called *triangular numbers*: 1, 3, 6, 10, 15, etc. An easy way to visualize the triangular numbers is to think of the way bowling pins are set up for a game: we see that they form a triangle, hence the name. The n-th triangular number is $n(n+1)/2$. The reciprocal of this is $2/n(n+1)$ which is twice a term in the telescoping series. From the formula for the telescoping series we infer that *the sum of the reciprocals of all triangular numbers is just 2.*

For example, suppose that we rearrange the first three terms of the telescoping series in (6.11) to get the series

$$\frac{1}{(3)(4)} + \frac{1}{(1)(2)} + \frac{1}{(2)(3)} + \frac{1}{(4)(5)} + \cdots$$

Then for this series the partial sums s_1 and s_2 are different from the first and second partial sums of (6.11) but

$$s_3 = \frac{1}{(3)(4)} + \frac{1}{(1)(2)} + \frac{1}{(2)(3)} = \frac{1}{(1)(2)} + \frac{1}{(2)(3)} + \frac{1}{(3)(4)}$$

So s_3 is the same as the 3rd partial sum of (6.11). In fact, we can see that s_n is the same as the n-th partial sum of (6.11) for all $n \geq 3$.

By a similar reasoning, if we *change* finitely many numbers in (6.10) then the numerical value of the series may change but not whether it converges or not. If the old series converges (or not) then the new series does the same thing because the partial sums for the two sequences beyond some index N are different by a finite amount, namely, the difference between the old s_N and the new one. I summarize this useful fact for future reference:

> **Finite equivalence (F)** *If two infinite series are different in finitely many terms and one series converges (or diverges) then the other series converges (respectively, diverges).*

Exercise 63 *Suppose we change the first three terms of the telescoping series (6.11) by setting all of them equal to 0 to get the series*

$$\frac{1}{(4)(5)} + \frac{1}{(5)(6)} + \frac{1}{(6)(7)} + \cdots$$

Explain why this series converges and that its value is:

$$1 - \left[\frac{1}{(1)(2)} + \frac{1}{(2)(3)} + \frac{1}{(3)(4)}\right] = \frac{1}{4}$$

Formally, the observation (F) defines an equivalence relation on the set of all infinite series of real numbers. Define a relation $\sim$ as:

$$\sum_{n=1}^{\infty} a_n \sim \sum_{n=1}^{\infty} b_n$$

whenever $a_n = b_n$ for all but finitely many indices n. Then $\sim$ is a reflexive relation because every series is different from itself in zero (which is finite) number of indices. That $\sim$ is symmetric is obvious; as for transitivity, if $\sum_{n=1}^{\infty} a_n \sim \sum_{n=1}^{\infty} b_n$ then there is a finite index N_1 such that $a_n = b_n$ for $n \geq N_1$ and if $\sum_{n=1}^{\infty} b_n \sim \sum_{n=1}^{\infty} c_n$ then there is an index N_2 such that $b_n = c_n$ for $n \geq N_2$. So if N is the larger of N_1 and N_2 then $a_n = c_n$ for $n \geq N$ which means that $\sum_{n=1}^{\infty} a_n \sim \sum_{n=1}^{\infty} c_n$ and shows that $\sim$ is transitive.

Two series are finitely equivalent if the relation $\sim$ holds between them. Now (F) says that *if two infinite series of real numbers are finitely equivalent then they both converge or they both diverge.*

Infinite Series 129

The shrinking tail detail.

Suppose that an infinite series $\sum_{k=1}^{\infty} a_k$ converges to a real number s. Then the partial sums $s_n \to s$ as $n \to \infty$. Since every convergent sequence is Cauchy, this means that the sequence of partial sums is Cauchy; so $|s_n - s_m| \to 0$ as $m, n \to \infty$. Without loss of generality, take $n > m$ and notice that

$$s_n = \sum_{k=1}^{n} a_k = \underbrace{a_1 + a_2 + \cdots + a_m}_{s_m} + a_{m+1} + \cdots + a_n = s_m + a_{m+1} + \cdots + a_n$$

Therefore,

$$|s_n - s_m| = |a_{m+1} + a_{m+2} + \cdots + a_n| = \left| \sum_{k=m+1}^{n} a_k \right|$$

If we let $n \to \infty$ while keeping m fixed then $s_n \to s$ so we get

$$|s - s_m| = \left| \sum_{k=m+1}^{\infty} a_k \right|$$

The sum inside the absolute value is *the tail* of the original infinite series $\sum_{k=1}^{\infty} a_k$. If we now let $m \to \infty$ then $|s - s_m| \to 0$ and we see that the shrinking tail also converges to zero. To summarize:

The shrinking tail of a convergent series: *If an infinite series $\sum_{k=1}^{\infty} a_k$ converges then its tail $\sum_{k=m+1}^{\infty} a_k$ converges to zero as $m \to \infty$.*

Intuitively, this statement says that if the whole series approaches a real number then its tail shrinks to zero as we take out more and more of the numbers in the series. The reason this idea makes sense is because there is a finite number, namely s out of which we are taking out the numbers a_k one at a time until nothing is left of s.

The contrapositive version of the shrinking tail statement above is the following which is sometimes useful to know:

If the tail of an infinite series does not converge to zero then the series diverges.

Exercise 64 *Let's take a look at the tail of the telescoping series in (6.11):*

$$\sum_{k=m+1}^{\infty} a_k = \sum_{k=m+1}^{\infty} \frac{1}{k(k+1)}$$
$$= \frac{1}{(m+1)(m+2)} + \frac{1}{(m+2)(m+3)} + \frac{1}{(m+3)(m+4)} + \cdots$$

(a) Verify that

$$\frac{1}{(m+1)(m+2)} = \frac{1}{m+1} - \frac{1}{m+2}, \quad \frac{1}{(m+2)(m+3)} = \frac{1}{m+2} - \frac{1}{m+3}, \quad etc$$

(b) Explain why

$$\sum_{k=m+1}^{n} a_k = \frac{1}{m+1} - \frac{1}{n-1}$$

(c) By first letting $n \to \infty$ find a formula for the tail $\sum_{k=m+1}^{\infty} a_k$. Then use it to verify that $\sum_{k=m+1}^{\infty} a_k$ converges to 0 as $m \to \infty$.

If the infinite series does not converge then a limit s does not exist. The tail of a divergent series does not vanish no matter how much we take out of it. Consider the following two series

$$1 + 1 + 1 + 1 + \cdots$$
$$1 - 1 + 1 - 1 + \cdots$$

The first will look exactly the same no matter how many 1's you take out and the second will look essentially the same (it may start with a 1 or a -1). These divergent infinite series have no identifiable tails, let alone shrinking ones!

6.3 The geometric series: beauty in simplicity

The series in (6.3) is a special case of the following series, known as the *geometric series*:

$$a + ar + ar^2 + \cdots \qquad a \neq 0 \qquad (6.14)$$

Two numbers completely characterize this series. They are the *first term* a and the number r, called the *common ratio* because it is the fixed ratio of each term over the one right before it: $ar/a = r$ and $ar^2/ar = r$ and so on. The series in (6.1) is geometric with $a = r = 1/2$.

Here is a useful interpretation for (6.14): drop a rubber ball from a height a meters, or feet, and let it bounce on solid flat ground. In the first bounce it rises straight up a fraction of a by a factor r which in this case is a fixed number between 0 and 1. After the ball reaches the height ar it goes down again. When it bounces a second time, assume that it goes up by the same fraction r of the height ar. So in the second bounce the ball goes up to ar^2. This happen repeatedly, each time the height reached is smaller than the height of the previous bounce by the same percentage r. Allowing this to go on indefinitely, the total distance that the ball moves is the series (6.14). Figure 6.1 shows the bouncing process plotted against time; the ball's motion in space is one dimensional, of course.

The geometric series is important because it occurs frequently in calculus; it is also nice because there is a formula for its exact sum, a rarity for infinite series. The derivation of the formula is essentially the same as that for (6.3) *but this time we don't need to assume that the series converges*. The n-th partial sum of the series (6.14) is

$$s_n = a + ar + ar^2 + \cdots + ar^{n-1} \qquad (6.15)$$

It ends with power $n-1$ because we want s_n to be the sum of n terms. Now we apply the procedure used for (6.3) to (6.15); the important difference is that this time we are adding a finite number of terms so we need not worry about infinity. Multiply by r and subtract to obtain:

$$\begin{aligned} s_n - rs_n = {} & a + ar + ar^2 + \cdots + ar^{n-1} \\ & - ar - ar^2 - ar^3 - \cdots - ar^n \end{aligned}$$

Infinite Series

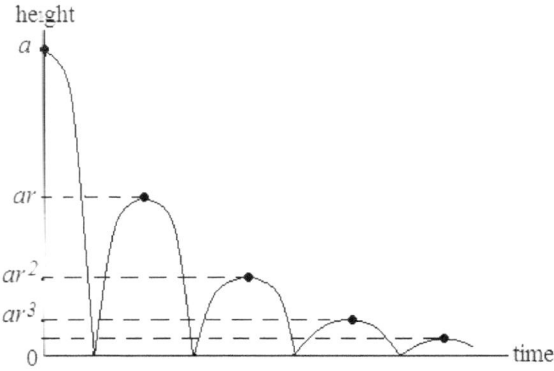

Figure 6.1: Height of the bouncing ball as a function of time

After cancelling all the like-terms in the middle we are left with

$$s_n - rs_n = a - ar^n$$
$$(1-r)s_n = a(1-r^n)$$

At this point, if $r \neq 1$ we divide by $1-r$ to obtain the following useful formula for s_n:

The geometric partial sums formula.

$$s_n = \frac{a(1-r^n)}{1-r} \quad \text{if } r \neq 1 \tag{6.16}$$

If $r = 1$ then we have the easy case where we insert 1 for r in (6.15) to get

$$s_n = a + a(1) + a(1)^2 + \cdots + a(1)^{n-1} = \underbrace{a + a + \cdots + a}_{n \text{ times}} = an \tag{6.17}$$

The n-th partial sum (6.15) of the series (6.14) is also called the n-term or finite geometric sum.

Exercise 65 *There is also something called the "arithmetic sum" that is defined as*

$$s_n = a + (a+d) + (a+2d) + \cdots + [a + (n-1)d]$$

Let's derive a formula for this finite sum. Rewrite s_n backwards right under the above to make two rows. Then add the two rows to get $2s_n$ on the left and a simplified expression on the right. Solve this equation to get the formula[5]

$$s_n = \frac{n}{2}[2a + (n-1)d] \tag{6.18}$$

---

[5]This derivation may be called the "accountant's method" since it is similar to the way accountants used to double check the sum of entries in a list in the pre-spreadsheet days.

Use (6.18) to find a formula for each of the following:

(a) $1 + 2 + 3 + \cdots + n$ (b) $1 + 3 + 5 + \cdots + 2n - 1$

The sum of a geometric series.

Now that we have a simple formula for s_n it is an easy matter to see what happens as $n \to \infty$. The situation here is similar to that of the telescoping series that we met earlier, although the formula for s_n now is a little more complex. The only part of (6.16) that changes with n is r^n. It is necessary to determine what happens to r^n as $n \to \infty$. The answer depends on the value of r so I consider the various possible cases:

$$r < -1, \quad r = -1, \quad -1 < r < 1, \quad r > 1$$

The case $r = 1$ uses the different formula (6.17) so I excluded it from the above list; $s_n = an$ goes to ∞ if $a > 0$ and to $-\infty$ if $a < 0$ so (6.14) diverges if $r = 1$.

To simplify the calculations for the other values of r, I use the absolute value to write the cases of interest above as the following three:

$$|r| < 1, \quad r = -1, \quad |r| > 1$$

It is easier to examine what happens to the size or magnitude of r^n using $|r|^n$ which is never negative. If $|r| < 1$ then repeated multiplication gives a decreasing sequence

$$|r| > |r| \cdot |r| = |r|^2 > |r|^2 |r| = |r|^3 > \cdots$$

For instance, if $r = -1/2$ then

$$\left|-\frac{1}{2}\right| = \frac{1}{2}, \quad \left|-\frac{1}{2}\right|^2 = \frac{1}{4}, \quad \left|-\frac{1}{2}\right|^3 = \frac{1}{8} \quad \text{and so on}$$

Ultimately *this sequence approaches zero* because each time that we multiply by $|r|$ we cut the value down by a fixed fraction, namely, $|r|$. In this case r^n approaches 0 as n goes to infinity so from 6.16 we infer that the partial sums s_n converge to the ratio $a/(1-r)$.

This proves the following result about the convergence of infinite geometric series:

$$a + ar + ar^2 + \cdots = \frac{a}{1-r} \quad \text{if } |r| < 1, \text{ i.e. } -1 < r < 1 \qquad (6.19)$$

Next, consider $|r| > 1$; this is the opposite case since repeated multiplication gives an increasing sequence

$$|r| < |r| \cdot |r| = |r|^2 < |r|^2 |r| = |r|^3 < \cdots$$

In this case the numbers go up without bound, or go to infinity, since each time that we multiply by $|r|$, we bump the value up by the fixed amount $|r|$. Finally, let $r = -1$. Then $r^n = (-1)^n$ is either 1 when n is even or -1 when n is odd; so there is no definite answer. From 6.16 we conclude that s_n diverges. In summary,

Infinite Series 133

> The geometric series $a + ar + ar^2 + \cdots$ diverges if $|r| \geq 1$.

As an example, let us calculate the value of
$$0.9 + 0.9^2 + 0.9^3 + \cdots \tag{6.20}$$

To use the formula (6.19) we must match this series up with the left hand side in (6.19). We see that $a = 0.9$; also $r = 0.9$ since we get a common ratio of 0.9 when dividing successive pairs of terms. Because $|r| < 1$ in this case we may substitute $r = a = 0.9$ in (6.19) to obtain
$$0.9 + 0.9^2 + 0.9^3 + \cdots = \frac{0.9}{1 - 0.9} = \frac{0.9}{0.1} = 9$$

So the series in (6.20) adds up to 9. A similar argument applies to the series
$$0.9 - 0.9^2 + 0.9^3 - \cdots = 0.9 + 0.9(-0.9) + 0.9(-0.9)^2 + \cdots \tag{6.21}$$

Here you notice that with $r = -0.9$ it is still true that $|r| = 0.9 < 1$ so we get the following result:
$$0.9 - 0.9^2 + 0.9^3 - \cdots = \frac{0.9}{1-(-0.9)} = \frac{0.9}{1.9} = \frac{9}{19}$$

The series in (6.21) adds up to 9/19. On the other hand, the series
$$1 + \frac{4}{3} + \left(\frac{4}{3}\right)^2 + \cdots$$

does not add up to a real number since $r = 4/3 > 1$. Let's check some of the partial sums of this series using formula (6.16) for the finite geometric sum; for example, with 10 terms a calculator gives:
$$s_{10} = \frac{1[1 - (4/3)^{10}]}{1 - 4/3} = \frac{1 - 1048576/59049}{-1/3} \simeq -3(1 - 17.76) = 50.28$$

Similarly:
$$s_{20} = \frac{1[1 - (4/3)^{20}]}{1 - 4/3} \simeq 943, \quad s_{100} = \frac{1[1 - (4/3)^{100}]}{1 - 4/3} \simeq 9353947230620.83$$

As expected, the partial sums are getting large without bounds. By contrast, when $a = r = 0.9$ we have
$$s_{10} = \frac{0.9[1 - (0.9)^{10}]}{1 - 0.9} \simeq \frac{0.586}{0.1} = 5.86$$
$$s_{20} = \frac{0.9[1 - (0.9)^{20}]}{1 - 0.9} \simeq \frac{0.791}{0.1} = 7.91$$
$$s_{100} = \frac{0.9[1 - (0.9)^{100}]}{1 - 0.9} \simeq \frac{0.89998}{0.1} = 8.9998$$

As expected, these partial sums are indeed approaching 9.

Exercise 66 *Which of the geometric series in (a) and (c) converge? What number do they add up to, if any?*
(a)
$$\frac{1}{4} + \left(-\frac{3}{4}\right)^2 + \left(-\frac{3}{4}\right)^3 + \cdots$$

Here you can tell easily what r is but be careful about a.
(b) *Calculate s_{10} and s_{100} for the series in (a) using the formula in (6.16).*
(c)
$$-\frac{1}{3} + \left(-\frac{4}{3}\right)^2 + \left(-\frac{4}{3}\right)^3 + \cdots$$

(d) *Calculate s_{10} and s_{100} for the series in (c) using the formula in (6.16).*

6.4 Testing for convergence (without calculating the sum)

Most infinite series of interest in pure and applied mathematics are not geometric or telescoping and we can't tell what their exact sums are. This loss of computational precision is not really something to worry about in practice if we can tell for sure whether a given series of interest converges or not.

If we know that $\sum_{n=1}^{\infty} a_n$ converges then we can estimate its value to a desired degree of accuracy, essentially by adding enough terms a_n of the series. Knowing the precise limit is not really necessary; we just need to be certain that we are not estimating a non-existent quantity by adding terms of a divergent series. But telling whether a series converges for sure is usually not obvious. In this section, we cover a few methods that are often used to tell whether a series converges or not.

The material in this section contains a few basic "convergence tests" that we can use to tell whether a series converges *without knowing (or caring about) what quantity it converges to*.[6]

Cauchy sequences of real numbers and the Cauchy criterion.

Now that we have a working definition of infinite series which reduces the problem of dealing with a series to one where we examine a sequence, some of the concepts that we mentioned about sequences will prove useful.

Let's begin by extending to all real numbers the definitions of convergent and Cauchy sequence that we saw earlier given for rational numbers.

A sequence x_n of real numbers *converges* to a real number x if $|x_n - x| \to 0$ as $n \to \infty$. In this case, the notation $x_n \to x$ is also used. Similarly, a sequence of real numbers is *Cauchy* if $|x_n - x_m| \to 0$ as $m, n \to \infty$.

We showed that every sequence of real numbers that converges is Cauchy. This, together with the completeness property of real numbers implies the following important fact about sequences of real numbers:

---
[6] Not every standard convergence test is discussed here; several tests that help simplify dealing with infinite series of constants are omitted. But such tests can be found in most standard calculus textbooks.

Infinite Series

> **The Cauchy convergence criterion:** *Every Cauchy sequence of real numbers converges to a real number and every convergent sequence is Cauchy.*

For infinite *series*, the Cauchy criterion is applied to the sequence of partial sums and its chief importance is that it enables us to assert that a series converges without knowing what it converges to. This fact is often useful in proving statements about series where finding a formula for the partial sums s_n is difficult or even impossible. As I show in the next section, this means that it applies to series for which, unlike telescoping series, there are no known formulas for s_n.

You may ask what good it is to know that a series converges without knowing what the value of the sum is; this is a good question and it has a good answer: *if we know that a series converges then we may estimate its value to any desired accuracy by using enough numbers in the series!*

First, check the n-th term!

There is a useful corollary of the Cauchy criterion that is sometimes called a *divergence test*,[7] implying that a series does *not* converge (or the series *diverges*). If this strikes you as an ironic consequence of the Cauchy criterion (which characterizes the *convergence* of series) then check the equivalent statement that follows it.

> **A divergence test:** *If the sequence of terms a_n does not converge to 0 then the infinite series $\sum_{k=1}^{\infty} a_k$ does not converge (to any real number).* The contrapositive equivalent of the above statement is: *If the series $\sum_{k=1}^{\infty} a_k$ converges then the sequence of its terms a_n converges to 0.*

Let's prove the contrapositive statement. If $\sum_{k=1}^{\infty} a_k$ converges then the sequence of partial sums converges and according to the Cauchy criterion, this means that the sequence of partial sums is Cauchy, i.e. $|s_n - s_m| \to 0$ as $m, n \to \infty$. Now, consider the special case that $m = n - 1$. In this case,

$$|s_n - s_{n-1}| = |(a_1 + a_2 + \cdots + a_n) - (a_1 + a_2 + \cdots + a_{n-1})| = |a_n|$$

so $|a_n| = |s_n - s_{n-1}| \to 0$ as $n \to \infty$. But if $|a_n|$ goes to zero then a_n must approach zero too and we are done!

As a quick application, consider the series (6.6) where we have $a_n = 1$ for every n. Since the n-th term does not converge to zero as $n \to \infty$ the series in (6.6) does not converge. The divergence test also implies that the series

$$\sum_{n=1}^{\infty} (-1)^{n-1} = 1 - 1 + 1 - 1 + \cdots$$

diverges because the n-th term $a_n = (-1)^{n-1}$ does not converge to any number, let alone to 0. Here is a more interesting example:

$$\sum_{n=1}^{\infty} \frac{100-n}{2n} = \frac{99}{2} + \frac{98}{4} + \frac{97}{6} + \cdots \qquad (6.22)$$

---
[7] The name *n-th term test* is also often used. This emphasizes the fact that the test is applied to the individual or generic term of the series (for which a formula is given) rather than to the partial sums (for which we don't have a formula).

The sequence of terms of (6.22) is:

$$a_n = \frac{100-n}{2n} = \frac{100}{2n} - \frac{n}{2n} = \frac{50}{n} - \frac{1}{2}$$

As $n \to \infty$ the first fraction $50/n$ approaches 0 so $a_n \to -1/2$. Since this is not 0, the series in (6.22) diverges by the divergence test even though the sequence of numbers a_n converges (to $-1/2$).

To see what actually happens to the series in (6.22) as $n \to \infty$ consider Figure 6.2 that shows a plot of the first 1000 partial sums from s_1 to s_{1000} (every 20th sum is plotted to avoid a misleading continuous appearance).

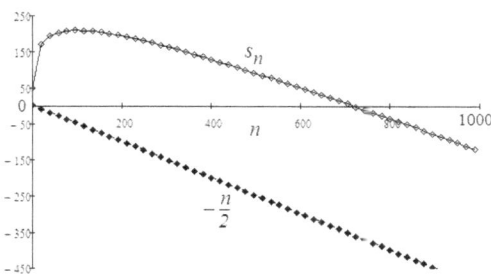

Figure 6.2: The eventual behavior of the partial sum sequence

After an initial rise (to overcome the effect of positive 100) the partial sums start following the sequence $-n/2$, nearly in parallel.

Why $-n/2$?

Look back at what the numbers a_n are doing as $n \to \infty$; if n is very large then $a_n \simeq -1/2$; see Figure 6.3. If I started adding $-1/2$ repeatedly from the start then I would get $(-1/2)n = -n/2$ after adding n times. This is what the series in (6.22) prefers to do but the added 100 delays its eventual destiny. In the end, both series will acquire infinitely negative magnitude, or diverge to $-\infty$.

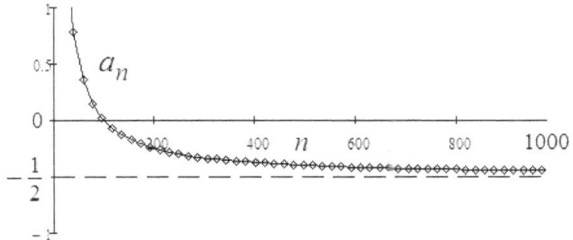

Figure 6.3: The eventual behavior of the n-th term

The series in (6.22) has another important message for us; to see what it is, I have magnified the initial part of Figure 6.2 and shown it separately in Figure 6.4. If I was content to decide on whether the series in (6.22) converges or diverges by just examining the first 100 partial sums then Figure 6.4 might lead me to the wrong conclusion, namely, that the series converges to some number near 210.

Therefore, in calculus numerical investigations alone are usually not enough to base conclusions on,[8] although they can often be helpful in *pointing to* the correct conclusion or at least, discarding some incorrect ones.

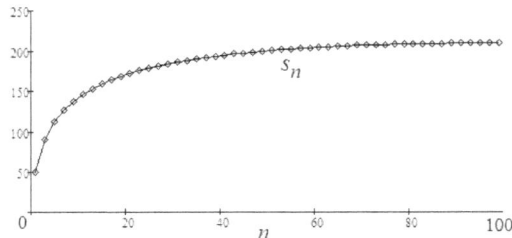

Figure 6.4: An insufficient convergence check

One more point about the divergence test is worth emphasizing:
It is for divergence only!
Remember that to get the divergence test we used only one possibility in the Cauchy criterion, namely, $m = n - 1$. But this does not take care of all other possible values of m.

What about the converse? *We looked at the divergence test and its contrapositive above. Now we consider the converse statement (not contrapositive).*

The converse of the divergence test states that: If a_n converges to 0 then the series $\sum_{k=1}^{\infty} a_k$ converges to a real number.

This converse statement is false. We discussed an example earlier that refutes it: in the harmonic series

$$\sum_{n=1}^{\infty} \frac{1}{n} = 1 + \frac{1}{2} + \frac{1}{3} + \frac{1}{4} + \frac{1}{5} + \cdots \qquad (6.23)$$

the sequence of terms $1, 1/2, 1/3, \ldots, 1/n, \ldots$ converges to zero, but the series as a whole diverges.

For a series to converge it is *necessary* that the sequence of numbers a_n converge to zero but this property is *not sufficient*; more restrictions on the terms of the series are typically required that ensure rapid convergence of the sequence of terms to 0.

The Comparison Test: Does your series pass?

There are two things that are common to all the series in this section: every number a_n is non-negative, i.e. $a_n \geq 0$ and $a_n \to 0$ as $n \to \infty$. The latter is a necessary feature of every convergent series (check the divergence test above) even if not explicitly stated. But why non-negative? The main reason is the following simple observation:

---

[8] A friend of yours says that she is generally convinced by what a million partial sums do. She has access to computational devices that easily calculate and plot millions of partial sums. You can then ask her to examine the series
$$\sum_{n=1}^{\infty} \frac{1,000,000 - n}{2n}$$
After plotting a million partial sums she will get a result that looks like that in Figure 6.4.

(Increasing partial sums) *For a series with positive terms, i.e. $a_n \geq 0$ for every $n = 1, 2, 3, \ldots$ the sequence of partial sums is non-decreasing.*

This is easy to understand: if $a_n \geq 0$ then $s_n = s_{n-1} + a_n \geq s_{n-1}$ for all n, which is to say that the partial sum sequence is non-decreasing. Stated more explicitly:

$$\underbrace{a_1}_{s_1} \leq \underbrace{a_1 + a_2}_{s_2} \leq \underbrace{a_1 + a_2 + a_3}_{s_3} \leq \cdots \leq \underbrace{a_1 + a_2 + \cdots + a_n}_{s_n} \leq \cdots$$

This simple observation when used together with Theorem (BC) and the least upper bound property has some far-reaching consequences based on the next important result.

> **(PSC) The positive series criterion**: *If all the numbers in an infinite series $\sum_{k=1}^{\infty} a_k$ are non-negative then the series converges if and only if the sequence of partial sums is bounded from above. Thus a series with non-negative terms either converges to the supremum of its sequence of partial sums or else, it has infinite sum, i.e. it diverges to ∞.*

As stated the (PSC) is quite general but not easy to use in specific cases because it requires checking the partial sums. There being no formula generally for the partial sums, it is more expedient to translate the (PSC) into conditions directly on the numbers a_n, the terms of the series. I discuss some of these conditions in the rest of this section.

> **The comparison test (CT)**: *Consider two infinite series, $\sum_{k=1}^{\infty} a_k$ and $\sum_{k=1}^{\infty} b_k$ and assume that $0 \leq a_k \leq b_k$ for every $k = 1, 2, 3, \ldots$ If the series $\sum_{k=1}^{\infty} b_k$ converges then so does $\sum_{k=1}^{\infty} a_k$ and if $\sum_{k=1}^{\infty} a_k$ diverges so does $\sum_{k=1}^{\infty} b_k$.*

You may have noticed two things about the above statement: first, it is both a convergence and a divergence criterion and second, there is no explicit mention of partial sums; the "comparison" is based on the inequalities $a_k \leq b_k$ between the corresponding numbers, or terms, of each series.

Why is (CT) true? The proof is quite simple: first, if $\sum_{k=1}^{\infty} b_k$ converges then there is a real number B such that $\sum_{k=1}^{\infty} b_k = B$. Since $a_k \leq b_k$ for all k we can write:

$$a_1 + a_2 + \cdots + a_n \leq b_1 + b_2 + \cdots + b_n \leq B \quad \text{for every } n = 1, 2, 3, \ldots$$

This shows that the sequence of partial sums of $\sum_{k=1}^{\infty} a_k$ is bounded from above so by the (PSC) the series must converge to some number that is less than or equal to B. On the other hand, suppose that $\sum_{k=1}^{\infty} a_k$ diverges. Then the sequence of partial sums of $\sum_{k=1}^{\infty} a_k$ is not bounded from above. Because

$$b_1 + b_2 + \cdots + b_n \geq a_1 + a_2 + \cdots + a_n \quad \text{for every } n = 1, 2, 3, \ldots$$

you see that the sequence of partial sums of $\sum_{k=1}^{\infty} b_k$ is also not bounded from above. The PSC then implies that $\sum_{k=1}^{\infty} b_k$ cannot converge and we are done!

Let's use the (CT) to prove the convergence of the following series:

$$\sum_{n=1}^{\infty} \frac{1}{n^2} = \frac{1}{1^2} + \frac{1}{2^2} + \frac{1}{3^2} + \cdots \tag{6.24}$$

Infinite Series 139

Let this be the series $\sum_{k=1}^{\infty} a_k$ in (CT) and recall the telescoping series (6.11). If we add a 1 to that series to get

$$1 + \sum_{n=1}^{\infty} \frac{1}{n(n+1)} = 1 + \frac{1}{(1)(2)} + \frac{1}{(2)(3)} + \frac{1}{(3)(4)} + \cdots$$

and use this as the $\sum_{k=1}^{\infty} b_k$ in (CT) then we can make the following comparisons:

$$\underbrace{\frac{1}{1^2} = 1}_{a_1} \leq \underbrace{1}_{b_1}, \quad \underbrace{\frac{1}{2^2}}_{a_2} \leq \underbrace{\frac{1}{(1)(2)}}_{b_2}, \quad \underbrace{\frac{1}{3^2}}_{a_3} \leq \underbrace{\frac{1}{(2)(3)}}_{b_3}, \quad \underbrace{\frac{1}{4^2}}_{a_4} \leq \underbrace{\frac{1}{(3)(4)}}_{b_4}, \cdots$$

From earlier discussion we know that $\sum_{k=1}^{\infty} b_k$ converges because:

$$1 + \sum_{n=1}^{\infty} \frac{1}{n(n+1)} = 1 + 1 = 2$$

So by the (CT) the series (6.24) also converges, and moreover, its value cannot exceed 2.[9] Here is an example in which the geometric series is used for comparison:

$$\sum_{n=0}^{\infty} \frac{\cos^2 n}{2^n} = \frac{\cos^2 0}{2^0} + \frac{\cos^2 1}{2^1} + \frac{\cos^2 2}{2^2} + \cdots \tag{6.25}$$

Here $2^0 = 1$ when $n = 0$ and the cosine values are calculated in radians. You may remember from past exposition to algebra or trigonometry the basic fact that

$$-1 \leq \cos x \leq 1$$

so after squaring:

$$0 \leq \cos^2 x \leq 1$$

Now we can compare the series in (6.25) with the geometric series (candidate for $\sum_{k=1}^{\infty} b_k$)

$$\sum_{n=0}^{\infty} \frac{1}{2^n} = \frac{1}{2^0} + \frac{1}{2^1} + \frac{1}{2^2} + \cdots \tag{6.26}$$

and notice that for every index n,

$$\frac{\cos^2 n}{2^n} \leq \frac{1}{2^n}$$

In (6.26) we have the first term $a = 1$ and common ratio $r = 1/2$ so

$$\sum_{n=0}^{\infty} \frac{1}{2^n} = \frac{1}{1 - (1/2)} = 2$$

by the (CT) the series in (6.25) converges and its value is less than 2.

---

[9]The series in (6.24) is the value of the Riemann "zeta function" $\zeta(s)$ at $s = 2$; this is known to be $\pi^2/6 \simeq 1.645$. This was proved by Euler in 1734 in his solution to the "Basel problem". We solve this problem in a different way than Euler's in the section on Fourier series below.

Exercise 67 *Use the (CT) and an appropriate series that we have discussed earlier to show that each of the following series converges:*

$$(a) \sum_{n=1}^{\infty} \left(\frac{\cos n}{n}\right)^2 \qquad (b) \sum_{n=0}^{\infty} \frac{3}{2^n + 1}$$

In (a) it helps to first distribute the power 2 to the numerator and the denominator. In both (a) and (b) you should be able to specify their upper bounds, i.e. numbers that the values of each of the given series cannot exceed.

Exercise 68 *The argument that we used to prove the convergence of the series in (6.25) can be extended to series of type*

$$\sum_{n=0}^{\infty} \frac{c_n}{2^n} \qquad (6.27)$$

where c_n is any bounded sequence of positive real numbers whatsoever with $c_n \leq C$ for all index values n where C is some positive constant. In (6.25) we had $c_n = \cos^2 n$ so $c_n \leq 1$. Prove that (6.27) converges and its sum is no greater than C.

Use your new result to prove that the infinite series

$$\sum_{n=0}^{\infty} \frac{9n^2}{2^n(3n^2 + 1)}$$

converges and its sum does not exceed 3.

Here is an example where we use (CT) to prove *divergence*.
Consider the series

$$\sum_{n=1}^{\infty} \frac{n+1}{n^2} = \frac{2}{1^2} + \frac{3}{2^2} + \frac{4}{3^2} + \cdots \qquad (6.28)$$

Here seeing n^2 in the denominator and recalling the convergent series (6.24) may flash "convergence" in mind. But we also have the $n + 1$ in the numerator that must be taken into account. We do this using inequalities:

$$\frac{n+1}{n^2} > \frac{n}{n^2} = \frac{1}{n}$$

so the proper series for comparison is the harmonic series (6.23), which diverges. Taking the harmonic series as our compare-to series $\sum_{k=1}^{\infty} a_k$ and the series (6.28) as $\sum_{k=1}^{\infty} b_k$ we quickly reach the conclusion that the latter series diverges.

Exercise 69 *Explain why each of the following series diverges; you do not have to use the (CT) for both series!*

$$(a) \sum_{n=1}^{\infty} \frac{2 - \cos n}{n} \qquad (b) \sum_{n=1}^{\infty} \frac{2n - 1}{6n}$$

Infinite Series

I must add that to apply the comparison test (CT) it is *not* necessary that *every* term of either or both series be non-negative; recalling the finite equivalence (F) above, each or both can have a finite number of negative terms without seriously affecting the use of (CT). This is because we may drop finitely many numbers from a series (change their values to 0) without affecting its convergence (or divergence). For example, you can use the (CT) to prove that the series

$$\sum_{n=1}^{\infty} \frac{2n-9}{n^2} \tag{6.29}$$

diverges by comparing this to the harmonic series even though $2n-9$ is negative for $n=1,2,3,4$. You can use the (CT) by taking note of the fact that $2n-9 \geq n$ if $n \geq 9$, then drop the first 9 terms from both the above series and the harmonic series and complete the argument using the (CT).

Exercise 70 *Fill in the details of the above argument showing that the series in (6.29) diverges.*

The two series $\sum_{n=1}^{\infty} 1/n$ and $\sum_{n=1}^{\infty} 1/n^2$ that we discussed above are special cases of the following series:

$$\sum_{n=1}^{\infty} \frac{1}{n^p} = \frac{1}{1^p} + \frac{1}{2^p} + \frac{1}{3^p} + \cdots \quad (p > 0) \tag{6.30}$$

This is often called the *p-series*, because the power function n^p is involved.[10]

It is easy to prove that the p-series converges for $p \geq 2$ using the comparison test: since $n^p \geq n^2$ for all integers $n \geq 1$ it follows that

$$\frac{1}{n^p} \leq \frac{1}{n^2} \quad \text{for all real values of } p \geq 2$$

We have already seen that $\sum_{n=1}^{\infty} 1/n^2$ converges by the (CT) upon comparison with the convergent telescoping series, so the p-series (6.30) also converges when $p \geq 2$ (in particular, if p is any integer greater than 1). We can also prove in a similar way that the p-series diverges if $p \leq 1$; because in this case, $n^p \leq n$ for all integers $n \geq 1$ so

$$\frac{1}{n^p} \geq \frac{1}{n}$$

We have already shown that $\sum_{n=1}^{\infty} 1/n$ diverges by the (CT), so the p-series (6.30) also diverges when $p \leq 1$.

After we introduce the *integral test* for series later, we can also cover the range $1 < p < 2$ and end up with the following useful criterion:

> **The p-series test:** *The p-series converges for all real values $p > 1$ and diverges if $p \leq 1$.*

---

[10] For $p > 1$ this series coincides with the value of the Riemann Zeta function $\zeta(s) = \sum_{n=1}^{\infty}(1/n^s)$ where the complex number $s = p$ is restricted to the interval $(1, \infty)$ on the real axis in the complex plane.

For instance the series
$$\sum_{n=1}^{\infty} \frac{1}{\sqrt{n}} = \frac{1}{\sqrt{1}} + \frac{1}{\sqrt{2}} + \frac{1}{\sqrt{3}} + \cdots$$
diverges as a p-series with $p = 1/2 < 1$ because $1/\sqrt{n} = 1/n^{1/2}$. On the other hand,
$$\sum_{n=1}^{\infty} \frac{1}{n\sqrt[3]{n}} = \frac{1}{1\sqrt[3]{1}} + \frac{1}{2\sqrt[3]{2}} + \frac{1}{3\sqrt[3]{3}} + \cdots$$
is a convergent p-series with $p = 4/3 > 1$ since
$$n\sqrt[3]{n} = nn^{1/3} = n^{1+1/3} = n^{4/3}$$

The p-series test is often used most fruitfully in conjunction with the comparison test (CT). Here is a quick example: the series
$$\sum_{n=1}^{\infty} \frac{2}{2n-1} = 2 + \frac{2}{3} + \frac{2}{5} + \cdots$$
is not a p-series. But it diverges because if you check its n-th term
$$\frac{2}{2n-1} > \frac{2}{2n} = \frac{1}{n}$$
compares correct way to a the n-th term of a divergent p-series, namely, the harmonic series ($p = 1$):
$$\sum_{n=1}^{\infty} \frac{1}{n}$$

The ratio test: extending the geometric series method.

As you explore the comparison test you come across series for which it may be hard to find a sister series with familiar properties, or it may not be easy to analyze a potential sister series. Here are a couple of examples:
$$\sum_{n=1}^{\infty} \frac{n^2}{2^n}, \quad \sum_{n=1}^{\infty} \frac{10^n}{n!}$$

Take the first series above and consider the ratios of consecutive terms for the series:
$$\sum_{n=1}^{\infty} \frac{n^2}{2^n} = \frac{1^2}{2^1} + \frac{2^2}{2^2} + \frac{3^2}{2^3} + \frac{4^2}{2^4} + \frac{5^2}{2^5} + \frac{6^2}{2^6} + \cdots$$
with the generic or n-term $a_n = n^2/2^n$. Dividing the first two terms gives
$$\frac{a_2}{a_1} = a_2 \frac{1}{a_1} = \frac{2^2}{2^2} \frac{2^1}{1^2} = 2$$

Infinite Series

Similarly,

$$\frac{a_3}{a_2} = \frac{3^2}{2^3}\frac{2^2}{2^2} = \frac{9}{8}, \quad \frac{a_4}{a_3} = \frac{4^2}{2^4}\frac{2^3}{3^2} = \frac{8}{9}$$

$$\frac{a_5}{a_4} = \frac{5^2}{2^5}\frac{2^4}{4^2} = \frac{25}{32}, \quad \frac{a_6}{a_5} = \frac{6^2}{2^6}\frac{2^5}{5^2} = \frac{36}{50} \cdots$$

Notice that whatever the actual values of the ratios, after the second ratio *they all seem to be less than* 1 and getting smaller as I keep dividing.

Why is this true? You may write a program to calculate many more ratios and that may be sufficiently convincing; alternatively, consider a generic ratio with unspecified n and do a little algebra:

$$\frac{a_{n+1}}{a_n} = a_{n+1}\frac{1}{a_n} = \frac{(n+1)^2}{2^{n+1}}\frac{2^n}{n^2}$$

$$= \frac{(n+1)^2}{n^2}\frac{2^n}{2^n(2)} = \left(\frac{n+1}{n}\right)^2\frac{1}{2} = \frac{1}{2}\left(1+\frac{1}{n}\right)^2$$

The very last term above gets smaller as n gets larger because and reduces the value of $1/n$ inside the parentheses. The value of the expression at the very end is always greater than $1/2$ so I can pick any number r between $1/2$ and 1, say, $r = 8/9$ the value of the third ratio, and knowing that all later ratios are no greater than r, write down

$$\frac{a_{n+1}}{a_n} \leq r \quad \text{so multiplying by } a_n: \quad a_{n+1} \leq ra_n \quad \text{for } n = 3, 4, 5, \ldots$$

This implies that

$$a_4 \leq ra_3, \quad a_5 \leq ra_4 \leq r(ra_3) = r^2 a_3, \quad a_6 \leq ra_5 \leq r^3 a_3, \ldots$$

The numbers ra_3, $r^2 a_3$, $r^3 a_3$ etc now look familiar as numbers in a geometric series. In fact, the above calculations let us compare the series that we started with to a geometric one as follows:

$$a_1 + a_2 + a_3 + a_4 + a_5 + a_6 + \cdots \leq a_1 + a_2 + (a_3 + ra_3 + r^2 a_3 + r^3 a_3 + \cdots)$$

The series inside the parenthesis is a convergent geometric series because its common ratio is $r < 1$. Now by the comparison test (CT) we conclude that the series on the left, which is the series that we started with, converges too!

You can see that this argument did not require anything substantial beyond a series in which the ratio of consecutive numbers fell below 1 after some index n and stayed below 1 thereafter. Given these features, the comparison test takes care of convergence, not only for the above series but also for all others that satisfy the same conditions!

Going back to the geometric series, you may also recall that that series converges even if r is negative, as long as $|r| < 1$; further, a geometric series with $|r| > 1$ always diverges. Of course, in a geometric series the common ratio r of consecutive numbers is fixed but in the series that we discussed above it varies. But as long as this variation stays below 1 in magnitude after some index, we can expect to get convergence and similarly, if it stays above 1 we expect to get divergence.

This argument (generalizing the idea behind geometric series) is essentially how the ratio test below is proved.[11] We discover the full power of this test later when discussing infinite series of functions.

> **(RT) The ratio test**: *Define $r_n = a_{n+1}/a_n$, $n = 1, 2, 3, \ldots$ for a series $\sum_{k=1}^{\infty} a_k$ with $a_k \neq 0$ for all k. If there is a real number r and an index N such that $0 < r < 1$ and $|r_n| \leq r$ for all $n \geq N$ then $\sum_{k=1}^{\infty} a_k$ converges. On the other hand, if r and N are such that $r > 1$ and $|r_n| \geq r$ for all $n \geq N$ then $\sum_{k=1}^{\infty} a_k$ diverges.*

To illustrate the application of (RT) consider the other series that I mentioned earlier, that is

$$\sum_{n=1}^{\infty} \frac{10^n}{n!} = \frac{10}{1!} + \frac{10^2}{2!} + \frac{10^3}{3!} + \frac{10^4}{4!} + \cdots \qquad (6.31)$$

The notation $n!$ defines the *factorial*

$$n! = n(n-1)(n-2) \cdots (3)(2)(1)$$

Thus,

$$1! = 1, \quad 2! = (2)(1) = 2, \quad 3! = (3)(2) = 6, \quad 4! = (4)(3)(2) = 24$$

and so on. We will encounter factorials later in our discussion of infinite series of functions where we also define

$$0! = 1.$$

Does the series in (6.31) converge or diverge? Let's check a few terms of the series:

$$\sum_{n=1}^{\infty} \frac{10^n}{n!} = 10 + 50 + 166.67 + 416.67 + 833.33 + 1388.89 + \cdots$$

From the few numbers listed above, it looks as if the series is diverging. But in an infinite series the first few terms are not indicative of convergence or divergence. Let's use the ratio test (RT) with

$$a_n = \frac{10^n}{n!}$$

Then

$$r_n = \frac{a_{n+1}}{a_n} = a_{n+1} \frac{1}{a_n} = \frac{10^{n+1}}{(n-1)!} \frac{n!}{10^n}$$

The definition of factorial gives

$$(n+1)! = (n+1)(n)(n-1) \cdots (3)(2)(1) = (n+1)n!$$

Using this break-down we get

$$r_n = \frac{10^n 10}{(n+1)n!} \frac{n!}{10^n} = \frac{10}{n+1}$$

---

[11] The ratio test first appeared around 1750 in a work of the French mathematician and physicist Jean-Baptiste le Rond d'Alembert (1717-1783).

Infinite Series

The cancellations left us with the simple fraction $10/(n+1)$ which approaches 0 as $n \to \infty$. In particular, r_n goes below 1 for all large values of n (in fact, for every $n \geq 10$) as it moves towards 0, so *any number between 0 and 1 will do as a valid r* (say, $r = r_{10} = 10/11$). We conclude that the series in (6.31) actually converges in spite of the increasing behavior of its initial segment.

Exercise 71 *Use (RT) to show that the following series diverges:*

$$\sum_{n=1}^{\infty} (-1)^n \frac{2^n}{n^2} = (-1)^1 \frac{2^1}{1^2} + (-1)^2 \frac{2^2}{2^2} + (-1)^3 \frac{2^3}{3^2} + (-1)^4 \frac{2^4}{4^2} + \cdots$$

$$= -2 + 1 - \frac{8}{9} + 1 - \frac{32}{25} + \cdots$$

When the ratio test is silent. Let's apply (RT) to the harmonic series

$$\sum_{n=1}^{\infty} \frac{1}{n} \qquad (6.32)$$

Here $a_n = 1/n$ so

$$\frac{a_{n+1}}{a_n} = a_{n+1} \frac{1}{a_n} = \frac{1}{n+1} n = \frac{n}{n+1}$$

Since $n/(n+1) < 1$ for all n we see that $r_n < 1$ for every n. Does this mean that the harmonic series converges, contrary to what we showed earlier? Not exactly. Let's take a closer look at what (RT); it requires that $r_n \leq r$ where $r < 1$ not just $r_n < 1$. *The distinction is essential in making the comparison with the geometric series work!* For the harmonic series

$$r_n = \frac{a_{n+1}}{a_n} = \frac{n}{n+1} \to 1 \quad \text{as } n \to \infty$$

so it is impossible to squeeze a real number r between 1 and every r_n. Therefore, (RT) is not applicable in this case. You will also run into the same problem if you try to use the (RT) on the series

$$\sum_{n=1}^{\infty} \frac{1}{n^2} \qquad (6.33)$$

that we earlier showed to *converge* by comparing it to the telescoping series. For this series r_n also approaches 1 and it is not possible to squeeze the r in as you discover for yourself in the next exercise.

Exercise 72 *Work out the details of why the ratio test cannot be used for the series in (6.33).*

Exercise 73 *Try the (RT) on the series in Exercise 69(b). Any conclusions?*

What if there are zeros in the series? The (RT) cannot be directly applied to a series whose terms include zeros because we have to divide consecutive terms. However, zeros do not contribute to the value of the series so we can take out all the nonzero numbers and add those as a new series. If this new series converges then so does the original infinite series! Only the nonzero terms matter in a series. This observation may not seem important but it is easy to overlook when working with series of functions. I should emphasize that zeros generally do matter in *sequences* that are not partial sum sequences. For example, the sequence

$$0, 1, 0, 2, 0, 1, 0, 2, \ldots$$

where the pattern $0, 1, 0, 2$ is repeated is a periodic sequence with period 4. But if I drop the zeros the resulting sequence is

$$1, 2, 1, 2, \ldots$$

which is a sequence of period 2.

On boxes and paradoxes!

The fact that the harmonic series (6.32) diverges while its squares series (6.33) converges leads to a peculiar thought experiment involving infinity that is worth discussing here.

Consider a collection of boxes, the largest of which has all sides one unit (meter or yard, say) long, the second has a length of one unit but its height and width are $1/2$ unit, etc as illustrated in Figure 6.5.

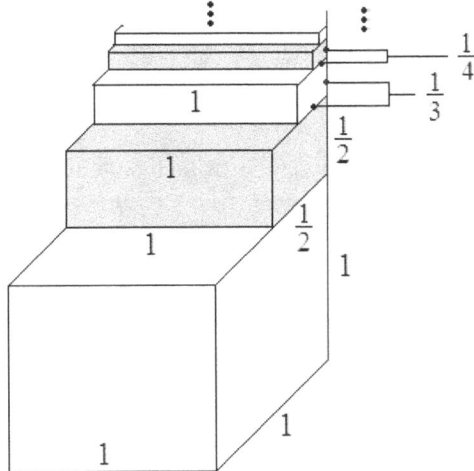

Figure 6.5: A paradoxical Tower of Boxes

The n-th box has dimensions $1 \times 1/n \times 1/n$ so its volume is $1/n^2$, which means that the total volume of all the boxes is finite and given by the sum of the convergent series (6.33); recall that this sum is less than 2 cubic units.

Now suppose that we stack the boxes up, each on top of the other as in Figure 6.5. The total height of the tower of boxes is given by the harmonic series (6.32) which has infinite value; this tower

Infinite Series

of boxes has finite volume that is less than 2 cubic units but it has infinite height! Even if we put them in open-top storage, the height would go past the farthest galaxies in the known universe.

An even more startling observation has to do with the total surface area: because some of the faces (four out of six) of the n-th box have area $(1)(1/n) = 1/n$ the total surface area of all the boxes is infinite. But if we filled all of the boxes with paint then a finite amount of paint (less than 2 cubic units) is enough to cover the infinite surface area of the interiors of all the boxes![12]

We may wonder if the infinite surface area be attributed in part to the infinite height of the tower; to see that *it is not*, let's modify our tower a little bit.

Suppose that the height of box n is $1/n^2$ for every n rather than $1/n$ with all other dimensions kept the same as those shown in Figure 6.5. Then the total height is finite because the series

$$\sum_{n=1}^{\infty} \frac{1}{n^2}$$

converges by the p-series test; in fact, we discover in Chapter 9 while discussing the "Basel Problem" that this series converges to the number $\pi^2/6$ Therefore, the height of the box tower is now finite. Its volume is thus also finite because the entire stack of boxes fits within a box with dimensions $1 \times 1 \times \pi^2/6$, so the total volume of all the boxes in the Tower is less than $\pi^2/6 \simeq 1.645$.[13] However, the total surface area of all the boxes is still infinite because the total are of all the horizontal sides (top or bottom of each box) is given by the divergent harmonic series.

Notice that the single larger box of volume $\pi^2/6$ has a total surface area of $2 + 4(\pi^2/6) \simeq 8.58$!

This paradoxical situation is a higher dimensional version of the area under the staircase curves that we mentioned in Chapter 1. In this case, suppose that the unit of length is centimeters or inches so the entire stack of boxes can fit in the palm of a hand. If we now inject paint into the stack we need less than 8.6 cubic centimeters or inches to fill the entire stack and yet, this amount of paint covers an infinite surface area!

The limit ratio test.

The variable ratios $r_n = a_{n+1}/a_n$ are in fact a sequence of real numbers. A restricted version of the ratio test that is typically discussed in calculus textbooks assumes that the ratios sequence approaches or converges to a fixed number. This version, which is usually called the "ratio test" in introductory calculus is often adequate for most computational purposes.[14]

> **(LRT) The limit ratio test:** *In a series $\sum_{k=1}^{\infty} a_k$ with $a_k \neq 0$ for all k, assume that the sequence $|a_{n+1}/a_n|$ converges to a real number R as $n \to \infty$. Then the series converges if $R < 1$ (including $R = 0$) and it diverges if $R > 1$. If $R = 1$ then no conclusion can be drawn as the series may converge or diverge in this case.*

---

[12] Here we have a discrete version of "Gabriel's Horn" that we discussed earlier in Chapter 1; also see the discussion of improper Riemann integration in Chapter 8 below.

[13] Alternatively, note that since the volume of the n-th box is $(1)(1/n)(1/n^2) = (1/n^3)$ the total volume of the Tower is given by the p-series

$$\sum_{n=1}^{\infty} \frac{1}{n^3}$$

which is a finite number by the p-series test.

[14] There is also a limit comparison test that I did not mention above but is found in most standard calculus textbooks. I discuss the limit ratio test in this book because it is useful in the discussion of series of functions, specifically, the power series.

It is easy to see why (LRT) is true; recall that $|a_{n+1}/a_n|$ converges to R if its distance to R approaches 0 as $n \to \infty$. With this in mind, first assume that $R < 1$. In this case, there is an index N large enough that if $n \geq N$ then the ratios a_{n+1}/a_n are so close to R that they are all less than 1 *by some margin*; that is they are less than some number $r < 1$ as shown in Figure 6.6.

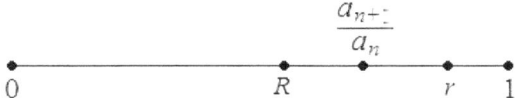

Figure 6.6: Illustrating the ratio test's convergence argument

This number is an r that works in (RT) and implies the convergence of the series. For the case $R > 1$ a similar argument works. As you see in Figure 6.7 there is some $r > 1$ such that after some index N the ratios $|a_{n+1}/a_n|$ are so close to R that $|a_{n+1}/a_n| \geq r$ for all $n \geq N$. We can use this r in (RT) to reach the conclusion that the series diverges.

Figure 6.7: Illustrating ratio test's divergence argument

But if $R = 1$ then the above argument breaks down because the ratios $|a_{n+1}/a_n|$ may appear on either side of 1 and that could imply convergence or divergence. So in this case we need to use some other method to find out whether the series converges or diverges. For instance, for the harmonic series (6.32) above, $R = 1$ since $n/(n+1)$ approaches 1 as shown and this series diverges. Similarly, $R = 1$ also for the infinite series in (6.33) that actually converges; see Exercise 72.

Let us see how the (LRT) can be used to prove the divergence of the series

$$\sum_{n=1}^{\infty} (-1)^n \frac{2^n}{n^2} = (-1)^1 \frac{2^1}{1^2} + (-1)^2 \frac{2^2}{2^2} + (-1)^3 \frac{2^3}{3^2} + (-1)^4 \frac{2^4}{4^2} + \cdots$$

We calculate R first:

$$\frac{a_{n+1}}{a_n} = \frac{(-1)^{n+1} 2^{n+1}}{(n+1)^2} \frac{n^2}{(-1)^n 2^n} = \frac{(-1)^n (-1) 2^n (2)}{(-1)^n 2^n} \frac{n^2}{(n+1)^2} = -\frac{2n^2}{n^2 + 2n + 1}$$

Dividing top and bottom of the fraction at the right end by n^2, simplifying and taking the absolute value gives

$$\left|\frac{a_{n+1}}{a_n}\right| = \frac{2n^2/n^2}{n^2/n^2 + (2n)/n^2 + 1/n^2} = \frac{2}{1 + 2/n + 1/n^2}$$

Infinite Series 149

Now it is easy to see that as $n \to \infty$ the reciprocals $2/n$ and $1/n^2$ both go to zero and leave us with $R = 2$. So the series diverges by the (LRT).

6.5 The effects of sign changes and Riemann's rearrangement theorem

From the beginning of this chapter and earlier discussion remember that series having infinitely many positive and negative numbers, or equivalently, series with infinitely many sign changes may converge to different numbers, or not at all, depending on how their terms are rearranged (by contrast, series with only finitely many sign changes give the same answer for every rearrangement of their numbers).

In this section I clarify some important issues involving two types of convergence that characterize this facet of infinite series.

The alternating series.

Consider the *alternating harmonic* series in which the positive and negative terms alternates:

$$\sum_{n=1}^{\infty} \frac{(-1)^{n-1}}{n} = 1 - \frac{1}{2} + \frac{1}{3} - \frac{1}{4} + \cdots \qquad (6.34)$$

Does this series diverge because the harmonic series does, or do the negative numbers reduce the partial sums enough for it to converge? If you review the criteria and tests that we discussed earlier you quickly realize that none of them apply to this series!

We may as well begin by checking the partial sums and it is helpful to do this pictorially. Figure 6.8 shows how it starts:

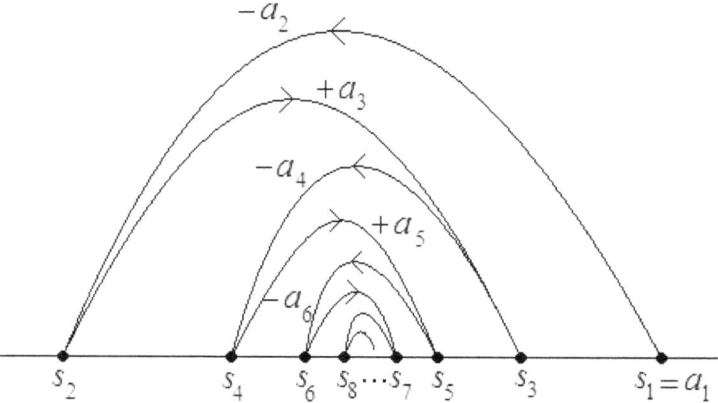

Figure 6.8: A convergent alternating series

In Figure 6.8 we can think of a_1, a_2, etc as 1, 1/2, etc from the series (6.34). We start with a_1

which is also the first partial sum s_1. Then

$$s_2 = 1 - \frac{1}{2} = a_1 - a_2 = s_1 - a_2$$
$$s_3 = 1 - \frac{1}{2} + \frac{1}{3} = a_1 - a_2 + a_3 = s_2 + a_3$$
$$s_4 = 1 - \frac{1}{2} + \frac{1}{3} - \frac{1}{4} = a_1 - a_2 + a_3 - a_4 = s_3 - a_4$$

and so on. Looking at the numbers on the right ends of the above equalities we see how the jumps back and forth in Figure 6.8 come about (follow the arrows). Further, because a_{n+1} is less than a_n for every n the jumps get smaller in length and result in a sequence of numbers (dots in Figure 6.8) that mover closer to some number in the middle. The sequence of partial sums converges to this number, so it must be the sum of the series!

Let's call the sum s. In general, it is not easy to find a formula for s (its exact value) although for the specific series (6.34) $s = \ln 2$ (the natural logarithm of 2); we demonstrate this later when discussing series of functions. The essentials of the above discussion are summarized in the following statement.[15]

> **(AST) Alternating series test**: *If a_n is a decreasing sequence of positive real numbers that converges to zero ($a_n \to 0$) then the alternating series below converges (to some real number):*
> $$\sum_{n=1}^{\infty} (-1)^{n-1} a_n = a_1 - a_2 + a_3 - a_4 + \cdots \qquad (6.35)$$

The two key concepts in the statement of (AST) to remember are "decreasing" and "converges to zero". The first is necessary to ensure that the jumps in Figure 6.8 have decreasing lengths; the second is needed to make the diminishing lengths to actually decrease all the way to 0. Indeed, if the sequences of numbers inside the series (6.35) did not converge to zero then the series would diverge by the divergence test!

Let us apply the (AST) to a more complicated series:

$$\sum_{n=1}^{\infty} \frac{(-1)^n}{n^2 + 1} = -\frac{1}{1^2 + 1} + \frac{1}{2^2 + 1} - \frac{1}{3^2 + 1} + \cdots \qquad (6.36)$$

In (6.36) the numbers a_n in the series are $1/(n^2+1)$. These decrease in value as the denominator increases with increasing n. In fact, as $n \to \infty$ the numbers a_n decrease all the way to 0. This is all that we need in the (AST) to conclude that the series in (6.36) converges; to what number exactly, the (AST) doesn't say any more than the (RT) did earlier.

Exercise 74 *A geometric series with common ratio $r < 0$ is also an alternating series. First, use the (AST) to prove that the series*

$$\sum_{n=0}^{\infty} \frac{(-1)^n 3^n}{4^{n+1}} = \frac{1}{4} - \frac{3}{4^2} + \frac{3^2}{4^3} - \frac{3^3}{4^4} + \cdots$$

---

[15] Leibniz was the first person known to have come up with the alternating series test.

Infinite Series

converges. Then use the geometric series formula to calculate its exact sum s. It may help to notice that

$$\frac{(-1)^n 3^n}{4^{n+1}} = \frac{1}{4}\left(\frac{-3}{4}\right)^n$$

Exercise 75 *Explain why an alternating p-series converges for all $p > 0$ and not just for $p > 1$*

$$\sum_{n=1}^{\infty} \frac{(-1)^{n-1}}{n^p} = 1 - \frac{1}{2^p} + \frac{1}{3^p} - \frac{1}{4^p} + \cdots \qquad (6.37)$$

Conclude that the following series converges:

$$\sum_{n=1}^{\infty} \frac{(-1)^{n-1}}{\sqrt{n}} = 1 - \frac{1}{\sqrt{2}} + \frac{1}{\sqrt{3}} - \frac{1}{\sqrt{4}} + \cdots$$

Error estimation in alternating series. *In general we rarely know what the exact sum s of a converging alternating series is but the nice thing about alternating series is that we have a simple error estimate when truncating a series at some finite term. Take another look at Figure 6.8 and notice that the difference between consecutive partial sums is given by the numbers in the alternating series. Specifically,*

$$|s_n - s_{n-1}| = a_n$$

If the series converges to a number s then s is tucked in between the consecutive jumps in the partial sum values, so for every n

$$|s_n - s| < |s_{n+1} - s_n| = a_{n+1}$$

So if we want to get an estimate of s that is accurate to, say, 4 decimal places then we want an index N large enough that $|a_{N+1}| < 0.00005$ and N is also the least such index.

For the series in (6.36) we find N by finding the least n that satisfies the inequality

$$\left|\frac{(-1)^{n+1}}{(n+1)^2 + 1}\right| \leq \frac{1}{(n+1)^2 + 1} < 0.00005 = \frac{5}{100000} = \frac{1}{20000}$$

This inequality can be solved with the help of a little algebra: flip the fractions (and do not forget to do the same to the inequality!) to get

$$(n+1)^2 + 1 > 20000$$
$$(n+1)^2 > 19999$$
$$n > \sqrt{19999} - 1 \simeq 140.4$$

So $N = 141$ will do the job! This means adding 141 terms of the series guarantees that the partial sum s_{141} gives the actual sum s of the series in (6.36) accurate to 4 decimal places *even though we do not know what the exact value of s is*!

Using a calculator, $s \simeq s_{141} = -0.363961$ is an estimate of the unknown exact sum of the series that we know is correct to at least 4 decimal places.

Exercise 76 *(a) Estimate the sum s of the alternating harmonic series in (6.35) that is accurate to 3 decimal places. How many terms of the series do you need for this accuracy? Use a calculator or computer to compute this approximation and check it against your calculator's value for $\ln 2$.*

(b) The index N that we get in (a) "guarantees" accuracy to 3 decimal places but it is not necessarily the smallest index to do the job. Use a calculator, and the value of $\ln 2$ as a check, to show that smaller values of N than that in (a) also give the same level of accuracy.

Absolute convergence.

Alternating series are among the simplest types of infinite series with infinitely many sign changes. In general, the sign changes may not occur in alternate fashion one after the other. Here is an example:

$$\sum_{n=1}^{\infty} \frac{\cos n}{n^2} = \frac{\cos 1}{1^2} + \frac{\cos 2}{2^2} + \frac{\cos 3}{3^2} + \cdots \tag{6.38}$$

The cosine in the numerator changes sign, going back and forth from positive to negative but not in an alternating fashion. Here is a list of the first 14 values of $\cos n$ rounded to two decimal places, where the cosine is calculated in radians:

n	1	2	3	4	5	6	7
$\cos n$	0.54	-0.42	-0.99	-0.65	0.28	0.96	0.75
n	8	9	10	11	12	13	14
$\cos n$	-0.15	-0.91	-0.84	0.004	0.84	0.91	0.14

The (AST) does not apply to the series in (6.38). On the other hand, the n^2 in the denominator is reminiscent of a convergent series, and you may also remember that $\cos n$ is always between -1 and 1. The comparison test (CT) now comes to mind; but unfortunately this test does not work if there are infinitely many sign changes!

The (CT) *would* work if we replaced $\cos n$ with its absolute value for then we end up with a series made of non-negative numbers:

$$\sum_{n=1}^{\infty} \frac{|\cos n|}{n^2} \tag{6.39}$$

Since

$$\frac{|\cos n|}{n^2} \leq \frac{1}{n^2} \quad \text{for } n = 1, 2, 3, \ldots$$

and we have already seen that the series $\sum_{n=1}^{\infty} 1/n^2$ converges by the p-series test, the (CT) implies that the series in (6.39) also converges.

Infinite Series

I have yet to show that the series in (6.38) converges but I am closer to the proof. To see how the series in (6.38) relates to its absolute value series in (6.39) is a general argument that applies to all infinite series because it is based on the *Cauchy convergence criterion* that we discussed earlier in this chapter.

Consider an infinite series $\sum_{k=1}^{\infty} a_k$ whose absolute value series $\sum_{k=1}^{\infty} |a_k|$ we know converges. Consider the partial sum sequences in each case:

$$s_n = \sum_{k=1}^{n} a_k = a_1 + a_2 + \cdots + a_n \quad \text{and} \quad s'_n = \sum_{k=1}^{n} |a_k| = |a_1| + |a_2| + \cdots + |a_n|$$

Because s'_n converges, the Cauchy criterion says that s'_n is a Cauchy sequence which means that $|s'_n - s'_m| \to 0$ as $m, n \to \infty$. Assuming without loss of generality that $m < n$ we have

$$|s'_n - s'_m| = (|a_1| + |a_2| + \cdots + |a_n|) - (|a_1| + |a_2| + \cdots + |a_m|)$$
$$= |a_{m+1}| + |a_{m+2}| + \cdots + |a_n|$$

Next, let's look at the partial sums of the original series $\sum_{k=1}^{\infty} a_k$ and notice that

$$|s_n - s_m| = |a_{m+1} + a_{m+2} + \cdots + a_n| \leq |a_{m+1}| + |a_{m+2}| + \cdots + |a_n|$$

where the inequality above comes from the triangle inequality. I have now shown that

$$|s_n - s_m| \leq |s'_n - s'_m|$$

Since the right hand side of the above inequality converges to 0 as $m, n \to \infty$ the even smaller left hand side does the same, which means that the partial sum sequence of $\sum_{k=1}^{\infty} a_k$ is Cauchy too. Now, once again the Cauchy criterion (applied in reverse) implies that the partial sums of $\sum_{k=1}^{\infty} a_k$ converge to a real number. And this is exactly what we mean by saying that the series $\sum_{k=1}^{\infty} a_k$ converges!

We have shown the useful fact that if the series of absolute values converges then so does the original series. In this case, we define:

> The series $\sum_{k=1}^{\infty} a_k$ *converges absolutely* if the series of absolute values $\sum_{k=1}^{\infty} |a_k|$ converges.

With this terminology, what we showed above can be stated as:

> **Absolute-convergence test**: *If an infinite series converges absolutely then it converges.*

Now our task regarding the series in (6.38) is complete: this series converges because it converges absolutely. But along the way we also found a new and general criterion that can be used for many different types of series. Here is another example: consider the series

$$\sum_{n=1}^{\infty} \frac{n^2 \cos n}{2^n} = \frac{1^2 \cos 1}{2^1} + \frac{2^2 \cos 2}{2^2} + \frac{3^2 \cos 3}{2^3} + \cdots \quad (6.40)$$

This series is a non-alternating version of that in Exercise 71. It is difficult to apply (LRT) or (RT) to this series directly; if you are skeptical then I encourage you to try it and maybe notice some interesting things about the variation of the ratios of discrete cosine values.

So let's check if the series in (6.40) converges absolutely. We know that $|\cos n| \leq 1$ which we can use to simplify the calculation. The series of absolute values of (6.40) is

$$\sum_{n=1}^{\infty} \left| \frac{n^2 \cos n}{2^n} \right| = \sum_{n=1}^{\infty} \frac{n^2}{2^n} |\cos n| \qquad (6.41)$$

Since the factors n^2 and 2^n are positive they are not affected by the absolute value and so they came out of the absolute value symbol. Now, since

$$\frac{n^2}{2^n} |\cos n| \leq \frac{n^2}{2^n}$$

the comparison test (CT) implies that the series (6.41) converges if the following series does:

$$\sum_{n=1}^{\infty} \frac{n^2}{2^n} \qquad (6.42)$$

Now, this is a series to which (LRT) applies easily. Set $a_n = n^2/2^n$ and calculate:

$$\frac{a_{n+1}}{a_n} = \frac{(n+1)^2}{2^{n+1}} \frac{2^n}{n^2} = \frac{1}{2} \frac{(n+1)^2}{n^2} = \frac{1}{2} \left(\frac{n+1}{n} \right)^2 = \frac{1}{2} \left(1 + \frac{1}{n} \right)^2$$

Since $1/n \to 0$ as $n \to \infty$ the quantity at the right end above approaches $1/2$. With this as R in the (LRT) we conclude that the series in (6.42) converges. Then, like dominos, the series in (6.41) converges also by the (CT) so the series in (6.40) converges absolutely, and finally by the absolute-convergence test, the original series in (6.40) also converges!

Exercise 77 *Show that the series below converges absolutely, and therefore, it converges:*

$$\sum_{n=1}^{\infty} \frac{n^3 \sin n}{2^n + 1} = \frac{1^3 \sin 1}{2^1 + 1} + \frac{2^3 \sin 2}{2^2 + 1} + \frac{3^3 \sin 3}{2^3 + 1} + \cdots$$

In this exercise, make a slightly greater use of the (CT) than I did in my discussion of the series in (6.40).

Exercise 78 *Explain why the alternating p-series (6.37) is absolutely convergent only for $p > 1$.*

Guessing convergence or divergence based on rates.

At this stage you may be wondering what test to use to prove whether an infinite series converges or diverges. This is a skill that is refined by experimentation and reflection; in short, getting your hands dirty with technical details. As you know, skill refinement is not a goal of this book, but even with a limited toolkit of concepts and functions it is possible to explore a couple of points about series with positive terms or more generally, series that converge absolutely.

To begin with, when dealing with series whose terms are fractions containing algebraic type functions (consisting of integer or real powers of n) it is useful to remember that the p-series test and the (CT) imply:

Infinite Series 155

If the highest power in the denominator (say, p) exceeds that in the numerator (p') by more than 1, i.e. $p > p' + 1$ then the series converges, otherwise, i.e. if $p \leq p' + 1$ then the series diverges.

For instance, if you are looking at a series like:

$$\sum_{n=1}^{\infty} \frac{n+6}{n^2 + 3n}$$

then $p = 2$ and $p' = 1$ so this series is going to diverge and you can formally prove it using the (CT) and the p-series test. Similarly, for the series

$$\sum_{n=1}^{\infty} \frac{\sqrt{n}+6}{n^2 + 3n}$$

$p = 2$ and $p' = 1/2$ so this series converges.

The ratio test is especially useful when there are exponential functions like 2^n or the factorial $n!$ present (as well as possibly some other functions such as the power functions like n^2). When dealing with these functions it is useful to remember that:

Factorials overwhelm (grow much more rapidly than) exponentials, and exponentials overwhelm power functions.

This statement can be made more precise using mathematical expressions or ideas related to L'Hôpital's rule that we encounter in Chapter 9 but I interpret it in the series context using examples. For instance, the following series converges because the exponential in the denominator "overwhelms" the power function in the numerator:

$$\sum_{n=1}^{\infty} \frac{n^{20}}{2^n}$$

This means that the fraction $n^{20}/2^n$ approaches 0 as $n \to \infty$ fast enough to make the series converge, even though it takes a rather large n for 2^n to overtake n^{20}. This observation is not a trivial one and I think you will gain much insight if you use a calculator or computer software to figure out the value of n where 2^n overtakes n^{20} and further, what happens as you increase the value of n further beyond the overtaking threshold to see the overwhelming in action.

For the same reason, the following series diverges:

$$\sum_{n=1}^{\infty} \frac{2^n}{n^{20}}$$

You can check these convergence claims (recommended) using the ratio tests (RT) or (LRT) like the examples discussed earlier. For a similar reason, all four of the series

$$\sum_{n=1}^{\infty} \frac{n^{100}}{n!}, \quad \sum_{n=1}^{\infty} \frac{10^n}{n!}, \quad \sum_{n=1}^{\infty} \frac{n!}{n^n}, \quad \sum_{n=1}^{\infty} \frac{n^n}{2^{n^2}}$$

converge, while all of the following series diverge:

$$\sum_{n=1}^{\infty} \frac{n!}{n^{100}}, \quad \sum_{n=1}^{\infty} \frac{n!}{10^n}, \quad \sum_{n=1}^{\infty} \frac{n^n}{n!}, \quad \sum_{n=1}^{\infty} \frac{2^{n^2}}{n^n}$$

Finally, you may find it necessary to use the comparison test (CT), in addition to the (LRT) or (RT) if required, when you see more "cluttered" expressions inside the series like

$$\sum_{n=1}^{\infty} \frac{5n^{100}}{2^n + 1}, \quad \sum_{n=1}^{\infty} \frac{10^n \cos n}{n! + 1}, \quad \sum_{n=1}^{\infty} \frac{(-1)^n n}{n^3 + 1}$$

where, as in the examples discussed earlier, you can take advantage of the inequalities below to show that all three series converge absolutely:

$$\frac{5n^{100}}{2^n + 1} < 5\left(\frac{n^{100}}{2^n}\right), \quad \left|\frac{10^n \cos n}{n! + 1}\right| \leq \frac{10^n}{n! + 1} < \frac{10^n}{n!},$$

$$\left|\frac{(-1)^n n}{n^3 + 1}\right| = \frac{n}{n^3 + 1} < \frac{n}{n^3} = \frac{1}{n^2}$$

If you find these concepts interesting enough to do more then you have a natural talent for calculation! You might find it rewarding to follow up and work your way through a standard calculus textbook for fun.

Conditional convergence: rearrangements give different answers!

It so happens that *the converse of the absolute convergence criterion is false*: if a series converges then it may not do so absolutely. The alternating harmonic series (6.34) is a prime example of a series that converges but not absolutely. So the following definition is meaningful:

> **Conditional convergence.** An infinite series *converges conditionally* if it converges but not absolutely; or more precisely, the infinite series $\sum_{k=1}^{\infty} a_k$ converges conditionally if it converges but the series of absolute values $\sum_{k=1}^{\infty} |a_k|$ diverges.

Keep in mind that the existence of infinitely many sign changes in a series is not the same thing as conditional convergence; for example, an alternating series diverge and fail the (AST).

The fact that conditionally convergent series will diverge if the negative signs are all changed to positive is related to some of the most startling effects of infinity that sequences and series allow us to explore at a deep level.

We saw in Chapter 1 that the value of the alternating harmonic series (6.34):

$$1 - \frac{1}{2} + \frac{1}{3} - \frac{1}{4} + \cdots$$

changes if we add its terms in a different order where we rearrange infinitely many of its terms.

This observation leads to one of the most amazing results concerning infinite series that I discuss now.[16]

---

[16] This result appeared in a 1866 paper of the German mathematician Bernhard Riemann (1826-1866) published after his death; but based on his correspondences, he likely proved it around 1853.

Infinite Series

> **The Rearrangement Theorem (conditional convergence):** *if an infinite series $\sum_{n=1}^{\infty} a_n$ converges conditionally then for **every** real number r there is a rearrangement $a_{k_1}, a_{k_2}, a_{k_3}, \ldots$ of the numbers in the series such that the new, rearranged series $\sum_{n=1}^{\infty} a_{n_k}$ converges to r; i.e.*
> $$a_{k_1} + a_{k_2} + a_{k_3} + \cdots = r$$

In Chapter 1 the rearranged sequence of fractions in the alternating harmonic series was

$$a_{k_1}, a_{k_2}, a_{k_3}, \ldots = a_1, a_3, a_2, a_5, a_7, a_4, \ldots$$

This corresponds to the following rearrangement of indices:

$$k_1 = 1,\ k_2 = 3,\ k_3 = 2,\ k_4 = 5,\ k_5 = 7,\ k_6 = 4, \ldots$$

Of course, to get a result other than $3s/2$ a different rearrangement of the fractions is needed. Surprisingly, it is not hard to explain how to find the correct rearrangement. The key idea is a sequence of over-estimates and under-estimates that is similar to what we saw earlier with the alternating series test (AST). Rather than write the technical details of the proof, I will proceed as in the discussion of (AST) and use the alternating harmonic series to illustrate the key idea behind the process.

Let's rearrange the fractions in the alternating harmonic series so that the rearranged series converges to zero. Since the series starts with a 1 which is greater than 0, we go to $-1/2$ or the first negative number and add that to 1. The result is $1/2$ which is still greater than 0, so on to the next negative number and continue until we reach or pass 0:

$$1 - \frac{1}{2} - \frac{1}{4} - \frac{1}{6} - \frac{1}{8} \simeq -0.042 < 0$$

Now add the first positive number after 1 which is $1/3$ to get

$$1 - \frac{1}{2} - \frac{1}{4} - \frac{1}{6} - \frac{1}{8} + \frac{1}{3} \simeq 0.292 > 0$$

This is back to positive (past 0 again) but notice that it is closer to 0 than 1 was; this is a gain!

Now go the next available negative number $-1/10$ which is too small when added to the above sum to reach 0. So continue adding negative numbers till we reach or pass 0:

$$1 - \frac{1}{2} - \frac{1}{4} - \frac{1}{6} - \frac{1}{8} + \frac{1}{3} - \frac{1}{10} - \frac{1}{12} - \frac{1}{14} - \frac{1}{16} \simeq -0.026 < 0$$

We have overshot 0 but notice that this time the total is closer to zero than the first time we over-shot 0 and got about -0.042. We have made more progress.

To get back towards 0, I add positive fractions like before:

$$1 - \frac{1}{2} - \frac{1}{4} - \frac{1}{6} - \frac{1}{8} + \frac{1}{3} - \frac{1}{10} - \frac{1}{12} - \frac{1}{14} - \frac{1}{16} + \frac{1}{5} \simeq 0.174 > 0$$

Again, though I have overshot 0 the total is closer than it was in the previous step where I got a value of about 0.292.

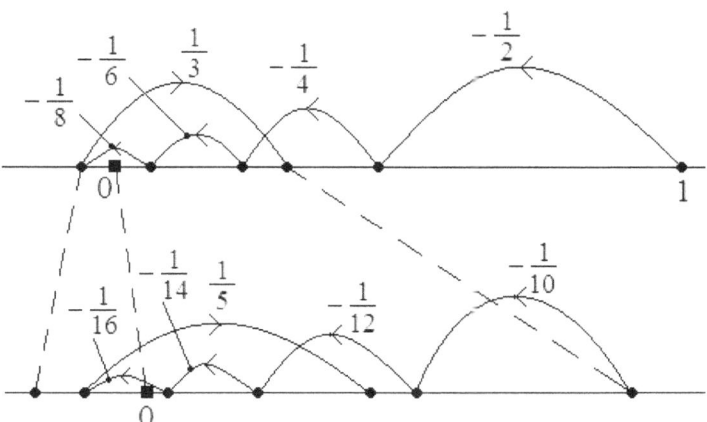

Figure 6.9: Illustrating Riemann's rearrangement argument

Figure 6.9 captures the above steps in the process; as you can see it looks like Figure 6.8 though a bit more complex since the back and forth jumps may take several steps in this case. The question now is: if the above process is continued indefinitely will the total approach 0?

Recall that at each step of the process when a change in direction (or sign) occurs, the numbers to the right of zero and those to the left get smaller:

$$-0.042 < -0.026 < -0.018 < \cdots < 0 < \cdots < 0.174 < 0.292 < 1$$

This observation shows that the overshoot sequence to the left of 0 and the one to its right both converge to 0 and the key reason for this is that the individual numbers in the series ($1/n$ or $-1/n$) approach 0. For example, in the first step overshooting 0 occurred by adding $1/3$ but in the second step we added the smaller number $1/5$. Later steps involve even smaller numbers. Going in the other direction, the negative number that just gets the sum past 0 is smaller in each step: first, $-1/8$ then $-1/16$ and so on.

Another important question that is not very transparent in the above discussion is this: *how do we know that we will in fact overshoot 0 in each direction?*

To see why this is a relevant question, consider a number that is farther from 1 than 0 was; say $\pi \simeq 3.1416$. Although this is not much farther away, starting from 1, we must add enough positive fractions to reach (or pass) π. This is accomplished with no fewer than 76 fractions:

$$1 + \frac{1}{3} + \frac{1}{5} + \cdots + \frac{1}{151} \simeq 3.1471$$

The next step requires just one negative term to pass π again:

$$1 + \frac{1}{3} + \frac{1}{5} + \cdots + \frac{1}{151} - \frac{1}{2} \simeq 2.6471$$

Although the number 2.6471 is closer to π than 1 was, we must add smaller available positive fractions to get back to the neighborhood of π; in fact, 128 more fractions are needed to pass π

Infinite Series

again:
$$1 + \frac{1}{3} + \frac{1}{5} + \cdots + \frac{1}{151} - \frac{1}{2} + \frac{1}{153} + \frac{1}{155} + \cdots + \frac{1}{409} \simeq 3.1433$$

And so on. The farther the chosen number r is from 1, the more terms (positive or negative) are needed to reach or pass r. For example, if $r = 6$ then starting from 1, nearly 23,000 odd fractions are needed to pass it!

So, again, *how can we be sure that we will reach or pass r in each direction?*

The key observation that answers this question requires the hypothesis that the series does not converge absolutely.

> In every series that converges conditionally, like the alternating harmonic series, *the sum of all negative numbers diverges to* $-\infty$ and *the sum of all positive numbers diverges to* ∞.
>
> Why is this true? We use a contradiction argument: If the negative numbers added to a finite value, say, $-A$ then the positive numbers would have to add to ∞ because the absolute value series diverges to ∞. But if the positive numbers add up to ∞ and the negative numbers add up to $-A$ then the series would not converge in the first place! This is a contradiction that came about by the assumption that the negative numbers add up to $-A$; therefore, this assumption must be false. A similar reasoning rules out the possibility that the positive numbers add to a finite value.

In a conditionally convergent series the half that diverges to ∞ and the half that diverges to $-\infty$ balance each other. There are always enough positive numbers and negative numbers available to reach or pass r, no matter what number r we pick. The fact that every real number is covered may be attributed to the coarseness of infinity; that is, the difference $\infty - \infty$ can be any number.

The above discussion is the essence of the proof of the rearrangement theorem; it shows how the process works and why the theorem is true.

Exercise 79 *Do the first 4 steps of the rearrangement process where you reach or pass $r = \sqrt{2}$ in different directions. List the calculated terms of your alternating harmonic series for each step.*

What about rearranging an *absolutely* convergent series?

You might guess (correctly) that those series are not affected by the rearrangements of their terms. The next result validates this conclusion;[17] it is best explained in general terms since rather than finding a particular rearrangement as in the previous theorem, now we want to show that *all rearrangements have no effect* on the sum of the series.

You won't miss anything essential to the rest of the book if you skip the explanation after the statement of the theorem below, but I explain just in case you are curious and want to dig in, especially following Riemann's theorem.

---

[17]This result first appeared in 1837 in a paper by the German mathematician Peter Gustav Lejeune Dirichlet (1805-1859). It was likely proved by Riemann also.

> **The Rearrangement Theorem (absolute convergence):** *If $\sum_{n=1}^{\infty} a_n$ converges absolutely then its sum does not change by rearranging its terms.*

First recall that if all the numbers a_n in the series are positive then rearranging the terms does not affect any of the upper bounds of the series, and so the *least upper bound* or *supremum* is not affected either (recall the section on the positive series criterion). Since the sequence of partial sums is increasing (each partial sum s_n is calculated by adding a positive number a_n to the last partial sum s_{n-1}) the supremum is in fact the sum of the positive series.

Next, suppose that $\sum_{n=1}^{\infty} a_n$ converges absolutely but has infinitely many sign changes. Still, its absolute value series $\sum_{n=1}^{\infty} |a_n|$ converges and since this latter is a positive terms series, rearranging its terms gives the same supremum. So if $\sum_{n=1}^{\infty} a_{n_k}$ is the rearranged series then $\sum_{n=1}^{\infty} |a_{n_k}|$ converges and therefore, $\sum_{n=1}^{\infty} a_{n_k}$ converges absolutely. Now let's see why $\sum_{n=1}^{\infty} a_{n_k}$ adds up to the same number as $\sum_{n=1}^{\infty} a_n$.

Let $s = \sum_{n=1}^{\infty} a_n$ and consider the partial sums

$$s_n = a_1 + a_2 + \cdots + a_n \quad \text{and} \quad s'_n = a_{k_1} + a_{k_2} + \cdots + a_{n_k}$$

By assumption, $s_n \to s$ as $n \to \infty$ so $|s_n - s| \to 0$. As a result, if an index N is large enough then $|s_N - s|$ is arbitrarily small. By the triangle inequality

$$|s'_n - s| = |s'_n - s_N + s_N - s| \leq |s'_n - s_N| + |s_N - s|$$

If I show that $|s'_n - s_N|$ can be made arbitrarily small also for all n and N large enough then I will have shown that $|s'_n - s| \to 0$ as $n \to \infty$ and complete the proof. Because N is finite while $n \to \infty$ it is possible to take n so large that all of the numbers $1, 2, 3, \ldots, N$ are among the indices $k_1, k_2, \ldots, n_k$ (this is a key observation). These integers also include a finite number of indices that are larger than N. Let M be the set of all such indices, namely, those among $k_1, k_2, \ldots, n_k$ that are not one of $1, 2, \ldots, N$. Then

$$M \subset \{N+1, N+2, N+3, \ldots\}$$

and

$$|s'_n - s_N| = |a_{k_1} + \cdots + a_{n_k} - a_1 - \cdots - a_N| = \left| \sum_{j \in M} a_j \right| \leq \sum_{j \in M} |a_j| \leq \sum_{j=N+1}^{\infty} |a_j| \quad (6.43)$$

Finally, assuming absolute convergence, $\sum_{n=1}^{\infty} |a_n|$ is a finite quantity, so the last sum in (6.43) approaches 0 as N gets larger and larger (recall the *shrinking tail detail* above) which is what we needed to show to complete the proof. *The key part of the proof was the observation about the set M which made the inequalities in (6.43) possible.*

6.6 The real numbers revisited ... rational, irrational and transcendental

Rational numbers are easy to understand because each one is just a ratio of two integers. When we divide these integers using long division, we inevitably reach a stage where digits begin to repeat in

Infinite Series 161

a periodic fashion. But irrational numbers are more opaque; there is no analog of the long division method that generates every irrational number.

Consider the decimal expansions of familiar irrational numbers, like

$$\pi = 3.14159265358979323846264338332\cdots$$
$$\sqrt{2} = 1.41421356237309504880168887242\cdots$$

Where do the digits in each case come from? How do we generate more digits? Must the decimal expansions of all irrational numbers look unpredictable?

Obviously, these questions involve infinity in a substantial way and we use infinite series to answer them. Earlier we constructed the real numbers as equivalence classes of Cauchy sequences of rational numbers. In this section we show that every real number can be represented as an infinite series of integer powers of 10.

Series expansions of real numbers.

The value of π shown above can be written as the infinite series

$$\pi = \frac{3}{10^0} + \frac{1}{10^1} + \frac{4}{10^2} + \frac{1}{10^3} + \frac{5}{10^4} + \frac{9}{10^5} + \frac{2}{10^6} + \cdots \qquad (6.44)$$

Of course, we do not have a formula that gives the numerator of the fraction having 10^n in the denominator for every value of n. But we know that the infinite series on the right converges because the number in the numerator of each fraction is just an integer between 0 and 9 while the number in the denominator is an increasing power of 10.

The natural question now is:

Can we write all real numbers as an infinite series like the one in (6.44)?

The answer is *yes* though this takes a little explaining...

Before proceeding, I point out that the same infinite series representation works in any (integer) base b, not just base 10. If b is a fixed integer, $b \geq 2$, and $d_1, d_2, \ldots$ is a sequence of integers where $0 \leq d_n \leq b-1$ for every $n \geq 1$ then the infinite series:

$$\sum_{n=1}^{\infty} \frac{d_n}{b^n} = \frac{d_1}{b^1} + \frac{d_2}{b^2} + \frac{d_3}{b^3} + \cdots \qquad (6.45)$$

converges to a real number between 0 and 1. If $b=2$ then the above series is the *binary* expansion; the only digits allowed are 0 and 1. If $b=3$ then the series gives the *ternary* expansion with digits 0,1,2; and so on.

Exercise 80 *Use the comparison test (CT) to prove that the series in (6.45) converges to a number $r < 1$. Recall that $d_n \leq b-1$ and if you replace all the numerators with $b-1$ (the largest possible) then you get a geometric series.*

The *converse* statement of the above observation is more important:

> **Representation by infinite series.** *Every real number r between 0 and 1 can be written as the series in (6.45) with an appropriate sequence $d_1, d_2, \ldots$ This representation is unique (one infinite series for each real number r) except for rational numbers of type m/b^k where k, m are natural numbers. Each one of these rational numbers has two series representations.*

The numbers $d_1, d_2, \ldots$ are the *digits* of r and the series in (6.45) is the *base b expansion* of r. If $b = 10$ and $0 \le d_n \le 9$ for every $n \ge 1$ then the *decimal expansion* of r is given by the series

$$\sum_{n=1}^{\infty} \frac{d_n}{10^n} = \frac{d_1}{10^1} + \frac{d_2}{10^2} + \frac{d_3}{10^3} + \cdots \tag{6.46}$$

We usually abbreviate this series as:

$$0.d_1 d_2 d_3 \ldots$$

With regard to the two ways of representing certain rational numbers in the above theorem, consider an example: $r = 0.38$ in base 10 can be written in two ways as an infinite series:

$$0.38 = \frac{38}{100} = \frac{3}{10} + \frac{8}{100} \quad \text{(the digits can only range from 0 to 9)}$$

$$0.38 = \frac{3}{10} + \frac{7+1}{100} = \frac{3}{10} + \frac{7}{100} + \frac{\sum_{i=1}^{\infty} 9/10^i}{100} = \frac{3}{10} + \frac{7}{100} + \sum_{i=1}^{\infty} \frac{9}{10^{i+2}}$$

One expansion ends in all zeros: $0.3800\ldots = 0.38\bar{0}$ and the other in all 9's: $0.3799\ldots = 0.37\bar{9}$.

Real numbers less than 0 or greater than 1 are obtained by simply adding an integer to the series in (6.46), or to (6.45) if the integer is expressed in base b also.

It is not difficult to explain why the infinite series representation theorem is true. Basically, we need to calculate a sequence of digits from a given r and show that there can be only one such sequence. The argument is straightforward though a bit lengthy. You may skip to the next section without losing anything that is essential to understanding the rest of the book.

Consider an arbitrary real number r between 0 and 1. We can exclude 0 and 1 from consideration because the former has a unique expansion with all digits zeros and the latter has the unique expansion $1 = 0.\bar{9}$ which is a series with all digits equal to 9 (I will focus on the decimal expansion $b = 10$ since this is the most familiar base; the argument is essentially the same for any base b).

In order to illustrate the process in a concrete way, first consider a specific real number, say, $r = 0.3861\cdots$ Each time I multiply r by 10 the decimal point moves one digit to the right. So $10r = 3.861\cdots$ Now if I truncate the digits after the decimal point I get the first digit 3. For every real number x this truncation is denoted $[x]$ which means the *greatest integer that is less than or equal to x*. In this notation, the first digit is

$$d_1 = 3 = [3.861\cdots] = [10r]$$

The second digit of r is 8 and it can be extracted from r in a similar way once we eliminate the first digit, 3. So define

$$r_1 = 0.861\cdots = 3.861\cdots - 3 = 10r - d_1$$

Infinite Series

Here r_1 simply is r with the first digit removed. Now the second digit can be calculated as follows:
$$d_2 = 8 = [8.61\cdots] = [10r_1] = [10^2 r - 10 d_1]$$

In this way, the *same process* of multiplication by 10 and truncation can be repeated to extract more digits from r. Define $r_2 = 10 r_1 - d_2 = 0.61 \cdots$ which is r without the first two digits, and repeat the process to get
$$d_3 = 6 = [6.1\cdots] = [10 r_2] = [10^2 r_1 - 10 d_2] = [10^3 r - 10^2 d_1 - 10 d_2]$$

And so on. Since the process is now *recursive*, we can generate the sequence of digits $d_1, d_2, \ldots$ by simple iteration. Further, the above process applies to any real number r between 0 and 1, not just the specific one that I used above.

Even better, the pattern is simple enough that for any r we can guess that
$$d_4 = [10^4 r - 10^3 d_1 - 10^2 d_2 - 10 d_3]$$

and so on.

Exercise 81 *Can you guess what d_5 is for any r between 0 and 1, based on the pattern for d_1, d_2, d_3, d_4?*

Deriving the decimal representation. Here is how we construct the infinite series (6.46) from a given real number r and show that the series converges to r. Let's start with the fact that for every real number x, rational or irrational, we have $[x] \leq x < [x]+1$. Now, with $x = 10 r$ and $d_1 = [10 r]$ note that
$$d_1 \leq 10 r < [10 r] + 1 = d_1 + 1$$
so dividing by 10 we get
$$\frac{d_1}{10} \leq r < \frac{d_1}{10} + \frac{1}{10} \tag{6.47}$$
Next, recalling that $d_2 = [10^2 r - 10 d_1]$ I have
$$d_2 \leq 10^2 r - 10 d_1 < [10^2 r - 10 d_1] + 1 = d_2 + 1$$
so that
$$10 d_1 + d_2 \leq 10^2 r < 10 d_1 + d_2 + 1$$
Now, if I divide by 10^2 then
$$\frac{d_1}{10} + \frac{d_2}{10^2} \leq r < \frac{d_1}{10} + \frac{d_2}{10^2} + \frac{1}{10^2} \tag{6.48}$$

Notice that in going from (6.47) to (6.48) we have come closer to r from both sides and the approximation error is lowered ten times, from $1/10$ to $1/100$. With the sequence of digits $d_1, d_2, \ldots$ calculated recursively using the multiplication and truncation algorithm above, I can further refine the sets of inequalities (6.47) and (6.48) using more of the digits and higher powers of 10. After n steps, we will have obtained:
$$\frac{d_1}{10} + \frac{d_2}{10^2} + \cdots + \frac{d_n}{10^n} \leq r < \frac{d_1}{10} + \frac{d_2}{10^2} + \cdots + \frac{d_n}{10^n} + \frac{1}{10^n}$$

The sum on the left hand side above is exactly the n-th partial sum of the series in (6.46); if I call it s_n and subtract it from r then I get

$$0 \le r - s_n < \frac{1}{10^n}$$

Since $1/10^n \to 0$ as $n \to \infty$ you see that the sequence s_n increases to reach r which means that r is given by the infinite series in (6.46) as claimed in the theorem, i.e.

$$r = \frac{d_1}{10^1} + \frac{d_2}{10^2} + \frac{d_3}{10^3} + \cdots \qquad (6.49)$$

It remains to show the *uniqueness* of the decimal expansion by the infinite series in (6.46). Suppose that somebody comes along and claims that she has *another* decimal expansion, say, $r = 0.d'_1 d'_2 d'_3 \cdots$ which can be written as the infinite series

$$r = \frac{d'_1}{10^1} + \frac{d'_2}{10^2} + \frac{d'_3}{10^3} + \cdots \qquad (6.50)$$

If I subtract the series in (6.49) from that in (6.50) I must get 0 since both series equal r. Combining like terms

$$0 = \frac{d'_1 - d_1}{10^1} + \frac{d'_2 - d_2}{10^2} + \frac{d'_3 - d_3}{10^3} + \cdots$$

Some of the digits in (6.49) may be the same as the corresponding ones in (6.50) but suppose that not all of them are the same so we have a claim to consider! Maybe the first index where the digits differ is the integer $k \ge 1$ so that $d'_k - d_k \ne 0$. Then

$$0 = \frac{d'_k - d_k}{10^k} + \frac{d'_{k+1} - d_{k+1}}{10^{k+1}} + \frac{d'_{k+2} - d_{k+2}}{10^{k+2}} + \cdots$$

Multiplying the last equality by 10^k and simplifying gives

$$0 = d'_k - d_k + \frac{d'_{k+1} - d_{k+1}}{10^1} + \frac{d'_{k+2} - d_{k+2}}{10^2} + \cdots \qquad (6.51)$$

If we take a close look at what we have here, we notice that the *magnitude* of the numerator of each of the fractional terms is at most 9:

$$\left|d'_{k+1} - d_{k+1}\right|, \left|d'_{k+2} - d_{k+2}\right|, \ldots \le 9$$

so the fractional terms can add up to at most

$$\frac{9}{10^1} + \frac{9}{10^2} + \cdots = \frac{9/10}{1 - 1/10} = 1$$

On the other hand, $|d'_k - d_k|$ is *at least* 1 since d'_k and d_k are integers. So the only way that (6.51) can hold is if every one of the numerators $d'_{k+1} - d_{k+1}, d'_{k+2} - d_{k+2}, \ldots$ is 9 (if $d'_k - d_k = -1$) or -9 (if $d'_k - d_k = 1$). Therefore, the only way that the decimal expansion in (6.46) is not unique is if r has one of the following forms:

$$r = \frac{d_1}{10^1} + \frac{d_2}{10^2} + \cdots + \frac{d_k}{10^k} \qquad (6.52)$$

or

$$r = \frac{d_1}{10^1} + \frac{d_2}{10^2} + \cdots + \frac{d_k}{10^k} + \frac{9}{10^{k+1}} + \frac{9}{10^{k+2}} + \cdots \qquad (6.53)$$

$$r = \frac{d_1}{10^1} + \frac{d_2}{10^2} + \cdots + \frac{d_k + 1}{10^k}$$

This completes our derivation of the decimal expansion.

Infinite Series 165

In either case, (6.52) or (6.53), r is a rational number of type $m/10^k$ as stated in the statement of the theorem, where m is found by combining the fractions on the right hand side of (6.52):

$$m = 10^{k-1}d_1 + 10^{k-2}d_2 + \cdots + d_k$$

This is a positive integer with decimal expansion $m = d_1 d_2 \cdots d_k$.

The cardinality of the set of all real numbers.

The infinite series representation theorem is not a tool for calculating or approximating irrational numbers but it does confirm that *every real number has a unique decimal expansion* (with the exception of certain rational numbers that have just two decimal expansions, one finite and one ending in all 9's).

This fact has an important consequence about the size of the set of real numbers!

Earlier we used Cantor's diagonal argument to prove that $\mathbb{R}$ is uncountable. This differentiates $\mathbb{R}$ from smaller sets like $\mathbb{N}$ or $\mathbb{Q}$ but it does not say *how large* $\mathbb{R}$ is; we now show that $\mathbb{R}$ is a "small" uncountable set.

Every real number r where $0 \leq r \leq 1$ has a unique expansion as the infinite series in (6.46); for the exceptional rational numbers we just take one of the two possibilities, say, the one that ends in all 9's and discard the other. Now we have a bijection between infinite series of type (6.46) and all sequences $d_1, d_2, d_3, \cdots$ of integers between 0 and 9, excluding those sequences that end is zeros (eventually 0 sequences). We showed earlier in our discussion of infinite products of sets that the set of all such sequences has cardinality $2^{\aleph_0}$ (see Theorem (PS) above). Therefore we conclude that:

The set $\mathbb{R}$ of all real numbers has cardinality $2^{\aleph_0}$.

An interesting by-product of this fact is that there is a bijection between $\mathbb{R}$ and the power set $\mathcal{P}(\mathbb{N})$ so every subset of $\mathbb{N}$ uniquely corresponds to a real number. Further, the set of all transcendental numbers has cardinal number $2^{\aleph_0}$.

Irrational numbers with predictable decimal expansions.

We have seen that the decimal expansions of irrational numbers can be unpredictable, like those of π or $\sqrt{2}$ above. But is this true of all irrational numbers? Is it a characteristic property?

Let's observe that non-periodic or non-repetitive decimal expansion is not the same thing as unpredictable. For instance, here is an irrational number that is very orderly and predictable:

$$0.123..89101112..1819202122..2829303132\cdots$$

This number, which is simply a concatenation of all non-negative integers, is obviously irrational since its digits do not appear in a periodic fashion. [18]

With the help of infinite series, we can show more: the set of all irrational numbers that have orderly or predictable decimal expansions also has cardinal number $2^{\aleph_0}$, the same as the set of all real numbers! And in a historically remarkable twist, this fact was known three decades before Cantor's work on infinite sets (not in terms of cardinalities, of course).

---

[18]The number appeared in 1933 in a thesis by D.G. Champernowne. More significantly, it later shown to be actually transcendental by the mathematician K. Mahler.

Back in 1844, the French mathematician Joseph Liouville (1809-1882) discovered certain irrational numbers that turned out to have some remarkable features. These *Liouville numbers* are defined (in base 10) by infinite series as

$$\sum_{n=1}^{\infty} \frac{d_n}{10^{n!}} = \frac{d_1}{10^1} + \frac{d_2}{10^2} + \frac{d_3}{10^6} + \frac{d_4}{10^{24}} + \frac{d_5}{10^{120}} + \cdots \qquad (6.54)$$

In particular, when $d_n = 1$ for every n we get the so-called *Liouville's constant*:

$$L = 0.11000100000000000000000100\cdots \qquad (6.55)$$

where we see the in-between zeros grow in number quite rapidly with the fast growth of the factorial. Since the decimal expansions of Liouville numbers are not eventually periodic, all of them are irrational. Furthermore, since the numbers $d_1, d_2, d_3, \ldots$ can be any integers from 0 to 9, using the same argument that showed the cardinality of $\mathbb{R}$ is $2^{\aleph_0}$, we can prove the following:

$\boxed{\textit{The set of all Liouville numbers is uncountable with cardinality } 2^{\aleph_0}.}$

I emphasize: this does not mean that the irrational numbers with *unpredictable* decimal expansions are few; proper subsets of infinite sets can be in bijective correspondence with the entire set so it is entirely possible to have two disjoint subsets of $\mathbb{R}$ each having the same cardinality $2^{\aleph_0}$.

We can easily locate Liouville's constant in (6.55) on the number line; writing it as an infinite series:

$$L = \frac{1}{10^1} + \frac{1}{10^2} + \frac{1}{10^6} + \frac{1}{10^{24}} + \frac{1}{10^{120}} + \cdots$$

we can see that L is greater than the rational number $1/10 = 0.1$ but less than $0.\bar{1}$ which is a convergent geometric series

$$0.\bar{1} = \frac{1}{10^1} + \frac{1}{10^2} + \frac{1}{10^3} + \cdots = \frac{1/10}{1 - 1/10} = \frac{1}{9}$$

So L is sandwiched between two rational numbers

$$\frac{1}{10} < L < \frac{1}{9}$$

When you look at the Liouville constant you notice that there is nothing unpredictable about its decimal expansion; if a digit is indexed by a factorial then it is 1; otherwise, it is 0 and that is all there is to it! Given that the set of all Liouville numbers is equipollent to the set of all real numbers, we see that there are lots of irrational numbers with very predictable decimal expansions.

Additional properties of Liouville numbers. Liouville numbers have many other interesting properties; I list two below that relate to what we have been doing in this book with going into the (substantial) technical details. Recalling that well-known transcendental numbers like π have unpredictable decimal expansions, it seems natural to conclude that irrationals with well-behaving decimal expansions can't be transcendental. But Liouville proved otherwise:

$\boxed{\textit{All Liouville numbers are transcendental.}}$

Infinite Series

In other words, Liouville numbers are not roots of polynomials with rational coefficients. Historically, Liouville numbers were the first transcendental numbers to be discovered, coming ahead of e which was shown to be transcendental in 1873 by Hermite. Prior to Liouville numbers the existence of transcendental numbers was suspected (prominent candidates were π and e) but not yet proved. Another interesting property of Liouville numbers is the following:

$\boxed{\textit{The set of all Liouville numbers is dense in } \mathbb{R}.}$

In other words, between any pair of real numbers there is a Liouville number. Thus Liouville numbers form an uncountable dense subset of the real numbers, just like the irrational numbers. However, when it comes to density, the smaller the set the more useful it is; so the rational numbers are the most important dense subset of $\mathbb{R}$.

Exercise 82 *Consider the number*

$$0.1010010001000010000010\cdots \qquad (6.56)$$

Here we start with a 1 to the right of the decimal point and every occurrence of 1 is followed by a corresponding number of zeros: the first appearance of 1 is followed by one 0, the second appearance by two zeros, the third by three zeros and so on.

(a) Explain why this number is irrational and show that is it smaller than Liouville's constant but still between $1/10$ and $1/9$;

(b) Explain why there are $2^{\aleph_0}$ real numbers of type

$$0.d_1 0 d_2 00 d_3 000 d_4 0000 d_5 00000 d_6 00 \cdots$$

where $d_1, d_2, d_3, \ldots$ are integers ranging from 0 to 9 that replace the 1's in the constant in (6.56);

(c) Write the numbers in (b) as infinite series (just replace the $10^{n!}$ in (6.54) with a suitable power of 10).

Chapter 7

Derivatives: *changing by infinitely little*

Velocity, acceleration, slope, rate of change: these are concepts associated with a single idea in mathematics, namely, the "derivative". Since its birth in the 17th century this concept been associated with the infinitely small (or infinitesimal) magnitudes, a concept whose proper understanding requires substantial abstraction, although its use in calculus is often described in an intuitive way.

The standard definition uses function limits based on the ε and δ definition. This approach is precise and avoids infinitesimals, but it is somewhat of an acquired taste; I discuss it in a separate section below to give you a flavor of it, and to also explain its equivalence to the sequence-based approach that we discuss here.

In this chapter, we work with convergent sequences instead because they use the infinity more explicitly than the so-called ε-δ approach and further, learning to work with sequences is important when dealing with the infinite dimensional spaces of modern analysis.

7.1 Measuring and calculating the velocity - without speedometers

Average velocity.

The *average velocity* $\bar{v}$ of any moving object is the distance it travels (the change in its position) over a given interval of time:

$$\bar{v} = \frac{\text{change in position}}{\text{change in time}} = \frac{\Delta x}{\Delta t}$$

In this notation x denotes the *position* or location of the object relative to a point of reference and t denotes the time as measured from a suitable starting point, like when we start measuring positions and times. Velocity can be a negative quantity when the object is moving in the opposite direction. The absolute value of velocity is the *speed* of the object.

The average velocity is the actual velocity of a moving object only when the object moves with constant velocity, which is rare in the real world. How then do we define or measure velocity that is changing?

Suppose that we have a set of data points on the locations of a freely falling pebble at various times and want to calculate the velocity of the pebble. The following table shows how far the pebble has fallen each second after it was dropped from a known height. The units of time and distance can be anything (nanoseconds, seconds, minutes, and meters, feet, miles, etc) but for the sake of concreteness here let us agree that distance is measured in meters and time in seconds. So in the table, the pebble has fallen 4.9 meters one second after being released, 19.6 meters two seconds later and so on.

t	0	1	2	3	4	5	6	7	8	9
x	0	4.9	19.6	44.1	78.4	122.5	176.4	240.1	313.6	396.9

Figure 7.1 shows a set up where the pebble's position as it falls down alongside a measuring stick is recorded perhaps by photographs taken once every second.

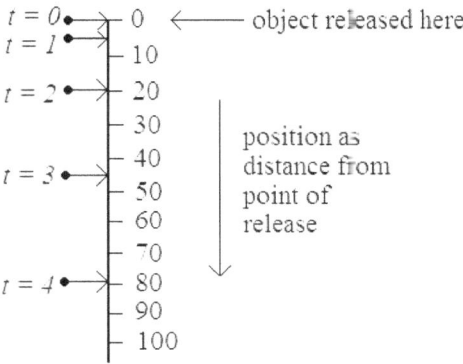

Figure 7.1: Free fall measurements

We can calculate the average velocity over various time intervals using the above table. For example, the average velocity from the start time $t = 0$ to $t = 1$ is

$$\bar{v} = \frac{4.9 - 0}{1 - 0} = 4.9 \text{ meters per second, or m/s}$$

Similarly, the average velocity over the next one-second interval from time $t = 1$ to $t = 2$ is

$$\bar{v} = \frac{19.6 - 4.9}{2 - 1} = 14.7 \text{ m/s}$$

The next table lists the values of average velocity for five 1-second time intervals:

time interval	$[0, 1]$	$[1, 2]$	$[2, 3]$	$[3, 4]$	$[4, 5]$
$\bar{v}$	4.9	14.7	24.5	34.3	44.1

We see in this table that the average velocity is increasing as time goes by from one interval to the next. But what about the velocity of the pebble at a specific instant of time, say, $t = 3$? Can we find a specific number that indicates how fast it is moving 3 seconds after being dropped?

Derivatives

Instantaneous velocity.

We can guess from the last table above that the pebble's speed at $t = 3$ is going to be faster than 24.5 m/s and slower than 34.3 m/s since it is moving slower before $t = 3$ and faster after $t = 3$. In order to narrow things down a bit more, let's measure the position of the pebble every *tenth of a second*; so $\Delta t = 0.1$ second. The next table lists the position measured shortly before and after $t = 3$:

t	2.8	2.9	3	3.1	3.2
x	38.42	41.21	44.1	47.1	50.18

Using these numbers we get

$$\bar{v} = \frac{44.1 - 41.21}{3 - 2.9} = 28.91 \text{ m/s}$$

Notice that this value is larger than 24.5 m/s that was measured over the time interval $[2,3]$ ($\Delta t = 1$) because the interval $[2.9, 3]$ is at the end of $[2, 3]$ and the pebble is moving faster. The graphs of data points for $\Delta t = 0.1$ are shown in Figure 7.2, in which time values are plotted on the x-axis in both panels and the measured position, or the calculated average velocity, on the y-axis.

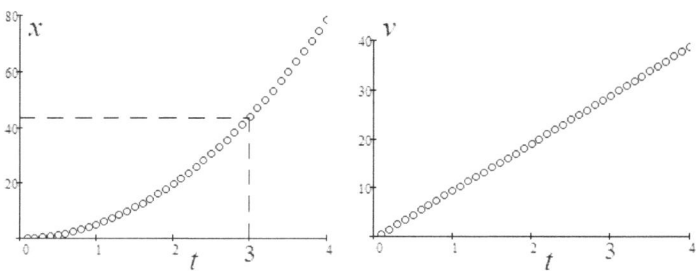

Figure 7.2: Data plots of position (left) and velocity vs time

But still the pebble's velocity at $t = 3$ is greater than the average velocity over $[2.9, 3]$. So we go ahead and refine our measurements by measuring the position every *hundredth of a second*; so $\Delta t = 0.01$ second. The next table lists the positions shortly before and after $t = 3$:

t	2.98	2.99	3	3.01	3.02
x	43.51	43.81	44.1	44.39	44.69

Using these numbers we get

$$\bar{v} = \frac{44.1 - 43.81}{3 - 2.99} = 29.35 \text{ m/s}$$

If we keep refining the position measurements then we can calculate the average velocity of the pebble over shorter and shorter intervals ending with $t = 3$. The next table summarizes the results for the first few steps:

Δt	1	0.1	0.01	0.001	0.0001
$\bar{v}$	24.5	28.91	29.35	29.395	29.3995

We see in the above table that the velocity changes very little as we shrink the time interval further. So we may reasonably conclude that the velocity of the pebble at the *time instant* $t = 3$ is 29.4 m/s. This number corresponds to velocity at an instant of time so we call it the *instantaneous velocity* of the pebble.

These observations suggest that the instantaneous velocity is the limit of a sequence of average velocity numbers that are calculated over a sequence of shrinking time intervals. More precisely, if we take a sequence of time intervals whose lengths Δt_n converge to 0 (say, $\Delta t_n = 10^{-n}$ as in the above discussion) and for each interval, measure the distance Δx_n traveled by the object then the instantaneous velocity at time $t = 3$ is the limit

$$\lim_{n \to \infty} \frac{\Delta x_n}{\Delta t_n}$$

We discuss and calculate limits like the one above in great detail in this chapter. But can this number be actually *measured*? Even if there were an exact value for instantaneous velocity in nature, we could not measure that value precisely.[1] In a continuous model *we assume that such a value exists* based on our intuitive dispositions, so whether we can actually measure a continuous quantity like the instantaneous velocity is not a scientific issue. We only require that our measurements *closely approximate* the expected continuous values. Such an agreement, when it is found, verifies the model. In modern science, especially physics, the reality is much more illusive than we think. But if we are not far off then we may claim that we understand it in some sense; that is, through mathematical modeling, at least to the extent that the model is valid.

Galileo's equations of motion.

In the case of the freely falling pebble, things are simple enough that years before Newton and Leibniz, Galileo Galilei (1564-1642) came up with a continuous model for it. He derived the equations that produce the values of the position x and velocity v using common sense assumptions and some experimentation (you have likely heard stories, true or not, about him dropping various items from the leaning tower of Pisa). In modern notation, these equations are:[2]

$$x = \frac{1}{2}gt^2 \quad \text{and} \quad v = gt \tag{7.1}$$

where g is the near-Earth value of gravitational acceleration whose value in the metric system is approximately 9.8. For instance, if $t = 3$ then Galileo's equations give

$$x = \frac{1}{2}(9.8)(3^2) = 44.1 \quad \text{and} \quad v = (9.8)(3) = 29.4$$

---

[1] It is worth mentioning that to get the value of the velocity in the last column we would need to measure the location of the pebble every 1/10000 of a second (a tenth of a millisecond). The *practical value* of doing such labor (even electronically) versus the greater accuracy obtained is certainly debatable for macroscopic objects like our pebble.

[2] These equations also assume negligible air friction, which is the case for pebbles but not for, say, feathers! The air friction is entirely responsible for causing pebbles (or bowling balls) to fall to the ground much faster than feathers. This was dramatically tested on video; check later in this section!

Derivatives

We will later derive both of Galileo's equations from Newton's laws of motion, when discussing integration in Chapter 8. Notice that Galileo's equations are continuous functions; you may recognize a parabola in the equation for x and a straight line with slope g in the equation for v; Figure 7.2 illustrates a discrete estimation of these continuous curves.

Next, consider Galileo's equations with $g = 9.8$

$$x = 4.9t^2 \quad \text{and} \quad v = 9.8t$$

Following the same line of reasoning used in the above discussion involving data without equations, we can derive the equation for velocity *mathematically* from the equation for position using the average velocity concept and a sequential approximation that involves letting $\Delta t \to 0$ (this time we are not dealing with actual physical measurements). Take the time instant $t = 3$ (or any positive number) and *any sequence t_n* of time instants that converges to 3; for example, we may take the sequence

$$2, 2.9, 2.99, \ldots$$

that we considered earlier, or any other such time sequence. The important thing is that $t_n \to 3$ as $n \to \infty$; other than this, *we do not impose any further restrictions on t_n*. If x_n is the position measured at time t_n then according to Galileo's equations, the average velocity for step n is

$$\frac{\Delta x_n}{\Delta t_n} = \frac{4.9 t_n^2 - 4.9(3^2)}{t_n - 3} \tag{7.2}$$

As $n \to \infty$ the time instants t_n approach 3 and the instantaneous velocity is defined by the limit

$$v = \lim_{n \to \infty} \frac{\Delta x_n}{\Delta t_n}$$

Let us take a closer look at the equation in (7.2); a little algebra gives

$$\frac{\Delta x_n}{\Delta t_n} = \frac{4.9(t_n^2 - 3^2)}{t_n - 3} = \frac{4.9(t_n + 3)(t_n - 3)}{t_n - 3} = 4.9(t_n + 3)$$

Now, if we let $n \to \infty$ then $t_n \to 3$ and by the limit theorems in Chapter 4

$$v = \lim_{n \to \infty} \frac{\Delta x_n}{\Delta t_n} = 4.9(3 + 3) = 29.4$$

There are a few points worth making about this calculation: first, the particular value 3 is not a decisive part of the calculation. You notice that it enters the calculation only marginally. Suppose that we choose *any unspecified (positive) real number t* and pick *any sequence t_n whatsoever that converges to t* (but of course, $t_n \neq t$ for all n). Then repeating each step of the same calculation but with 3 changed to t gives:

$$v = \lim_{n \to \infty} \frac{\Delta x_n}{\Delta t_n} = \lim_{n \to \infty} \frac{4.9(t_n^2 - t^2)}{t_n - t} = 4.9(t + t) = 9.8t$$

This is just Galileo's velocity equation!

On using sequences to define the velocity. You may have noticed my not so subtle mentions of the fact that t_n must be an *arbitrary* sequence, besides converging to t (and of course, $t_n \neq t$ so we don't divide by 0). If the particular choice of a sequence t_n were to affect the value of velocity v then different measurement data would result in different values for instantaneous velocity, which would be crazy where motion is concerned. But it is mathematically possible for the choice of the converging sequence to affect the limit; we will discuss these types of examples below and find that in these cases we cannot define the instantaneous rate of change, or put differently, *the derivative does not exist* (even when the path is continuous). We will see later in this chapter that sequentially defined limits (including the derivative) are equivalent to the more standard functional limits that we typically see in calculus textbooks. The sequence-based approach has the advantage of being more closely associated with the actual measurement idea; the only thing that is added that goes beyond actual measurements is the infinity and this is done in a direct manner with sequences through $n \to \infty$ rather than by appealing to infinitesimals.

7.2 Derivative and the tangent line: here comes infinity

Suppose that we have a general equation $y = f(x)$ defining y as a function of a variable x, like the position being given as a function of time. Both x and y represent real numbers in this setting. We define the derivative of $f(x)$ as follows.

> **The derivative.** Let a be any fixed real number in the domain of the function $y = f(x)$. The *derivative* of $f(x)$ at $x = a$ is a number $f'(a)$ that is defined as the following limit:
>
> $$f'(a) = \lim_{n \to \infty} \frac{f(x_n) - f(a)}{x_n - a} = \lim_{n \to \infty} \frac{\Delta y_n}{\Delta x_n} \qquad (7.3)$$
>
> provided that the limit exists and has the same value $f'(a)$ for *every* sequence x_n in the domain of $f(x)$ that converges to a and $x_n \neq a$ for all n.

One-sided derivative. When we talk about the derivative of a function on an interval like $[a, b)$ or $[a, b]$ that contain one or both of their end-points, we take the arbitrary sequences x_n only in the interval of interest. So the derivative at an end-point a or b involves sequences that converge to the end-point from one side only. These end-point derivatives are often called "one-sided derivatives".

For functions that are defined by simple formulas, like the position function in Galileo's equation, (7.3) is easy to apply.

Exercise 83 *Show that the derivative of $y = x^2 - 7$ at $x = a$, where a is any unspecified real number, is $2a$; consider using a method similar to the one discussed above for deriving Galileo's speed equation from the equation for position.*

Derivatives

> **Differentiable function.** If a function $f(x)$ has a derivative at $x = a$ then we say that $f(x)$ is *differentiable at $x = a$*. If $f(x)$ is differentiable at every point of the domain of $f(x)$ then we just say that $f(x)$ is *differentiable*.
> If a function is differentiable on an interval I then its derivative defines another function $f'(x)$ on I. We call $f'(x)$ the *derivative function* of $f(x)$.

For instance, the function $f(x) = x^2 - 7$ in Exercise 83 has a derivative function $f'(x) = 2x$ on the entire set of real numbers, namely, the largest possible interval $(-\infty, \infty)$. Similarly, the derivative of Galileo's position function $x = 4.9t^2$ is the velocity function $v = 9.8t$.

The derivative notation. The notation used to indicate a derivative tends to vary according to the context. There are two common notations, one being the prime notation that we discussed above; the other is the *differential notation* involving infinitesimal symbols of Leibniz. It goes as follows: if $y = f(x)$ then
$$\frac{dy}{dx} = f'(x)$$
This fractional notation is not a ratio of real numbers or quantities, and we usually think of it as just a symbol. However, in a number of cases such as when dealing with the "chain rule" for derivatives or some integration formulas it is helpful to work with this notation as if it were a fraction (albeit not of real magnitudes). One more notation that is largely confined to physics applications is Newton's "dot notation" that is typically used for time derivatives; so if $x(t)$ is the position of an object at time t then
$$\dot{x} = x'(t) = \frac{dx}{dt}$$
Thus $\dot{x}$ is just the velocity v.

In the next exercise, using simple calculations you can derive the following useful derivative formulas:
$$\frac{d(x)}{dx} = 1 \quad \text{and:}$$
$$\frac{d(c)}{dx} = 0 \quad \text{for each fixed real number } c$$

Exercise 84 *Use (7.3) to show that the identity function $f(x) = x$ is differentiable for all real x and show that $f'(x) = 1$. Do the same for a constant function $f(x) = c$ where c is any fixed real number and show that $f'(x) = 0$. Your calculation should make it clear that the specific value of c doesn't matter.*

The derivative is the slope of the tangent line.

The definition of the derivative involves the ratio or fraction:
$$\frac{\Delta y_n}{\Delta x_n} = \frac{y_n - f(a)}{x_n - a} = \frac{f(x_n) - f(a)}{x_n - a} \tag{7.4}$$

This ratio has a familiar geometric meaning. You may recall from elementary algebra that the *slope* of any straight line with two points (x_1, y_1) and (x_2, y_2) specified on it is

$$\frac{\text{rise}}{\text{run}} = \frac{y_2 - y_1}{x_2 - x_1}$$

The ratio in (7.4) gives the slope of a straight line through two points

$$(x_n, f(x_n)) \quad \text{and} \quad (a, f(a))$$

on the graph of the function $y = f(x)$; see Figure 7.3. Because this line cuts through the graph at distinct points, it is called a *secant line*.

The slope of the secant line is the geometric meaning of the ratio in (7.4).

It follows that as $n \to \infty$ the slopes of the secant lines approach the derivative of the function $f(x)$ at $x = a$.

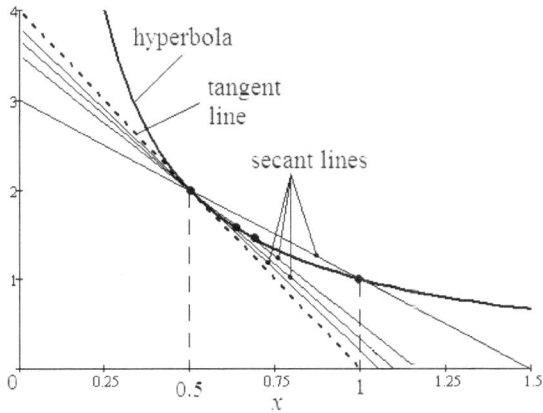

Figure 7.3: Secant lines approaching the tangent line

Now let's see what happens to the secant lines themselves as $n \to \infty$. To make the presentation more concrete, consider a specific function, say,

$$f(x) = \frac{1}{x} \quad x \neq 0$$

The graph of this function is a hyperbola and a portion of it for $x > 0$ is shown in Figure 7.3, together with a number of secant lines that pass through the point $(1/2, 2)$ on the graph.

Let $a = 1/2$ and pick any sequence x_n that approaches $1/2$ as $n \to \infty$. The secant lines shown in Figure 7.3 are some of those corresponding to the specific sequence $x_n = (n+1)/2n$. As $n \to \infty$ we use the ideas in Chapter 4 to calculate

$$\lim_{n \to \infty} x_n = \lim_{n \to \infty} \frac{n+1}{2n} = \lim_{n \to \infty} \left(\frac{1}{2} + \frac{1}{2n}\right) = \frac{1}{2} + 0 = a$$

You notice that the secant lines shown in Figure 7.3 approach the dashed line, which is just *the tangent line to the graph of $f(x)$*. This observation is valid for any function $f(x)$; simply replace

Derivatives

$1/x$ and its graph with another formula $f(x)$ and the corresponding graph. So we may conclude the following:

> The derivative $f'(a)$ of a function $f(x)$ at $x = a$ is the slope of the tangent line to the graph of $f(x)$ at $x = a$.

Of course, once we know the slope of the tangent line we can find its equation using the familiar equation $y - f(a) = m(x - a)$ from algebra, which with slope $m = f'(a)$ gives the formula

$$y = f(a) + f'(a)(x - a) \tag{7.5}$$

for the *equation of the tangent line at $x = a$*.

Let's find this equation for the reciprocal function $f(x) = 1/x$ at $x = a$ for any given $a > 0$. First, we calculate the slope:

$$f'(a) = \lim_{n \to \infty} \frac{f(x_n) - f(a)}{x_n - a} = \lim_{n \to \infty} \frac{1/x_n - 1/a}{x_n - a} \tag{7.6}$$

where x_n is now an arbitrary sequence of real numbers that converges to a and $x_n \neq a$ for all n. Doing a little algebra changes the fraction on the right above to

$$\frac{(a - x_n)/ax_n}{x_n - a} = \frac{-(x_n - a)}{(x_n - a)ax_n} = -\frac{1}{ax_n} \tag{7.7}$$

By assumption, $\lim_{n \to \infty} x_n = a$ so using the property of limits that allows distribution of the limit over the fraction,[3] we get

$$f'(a) = -\frac{\lim_{n \to \infty} 1}{\lim_{n \to \infty} ax_n} = -\frac{1}{a \lim_{n \to \infty} x_n} = -\frac{1}{a^2} \tag{7.8}$$

In particular, when $a = 1/2$ we find $f'(1/2) = -4$, which is the slope of the (dashed) tangent line in Figure 7.3. Now we are ready to find the equation of the tangent line: if $x = a$ then $f(a) = 1/a$ so from (7.5) we get

$$y = \frac{1}{a} - \frac{1}{a^2}(x - a) = \frac{1}{a} - \frac{1}{a^2}x + \frac{1}{a} \quad \text{or:}$$

$$y = -\frac{1}{a^2}x + \frac{2}{a}$$

For $a = 1/2$ the equation of the tangent line in Figure 7.3 is

$$y = -4x + 4$$

Exercise 85 *(a) Calculate the derivative of the following function at $x = a$ where $a > -1$*

$$f(x) = \frac{1}{x + 1}$$

(b) Find the equations of the tangent lines to the graph of $f(x)$ in (a) for each of the values of a below:

$$a = 0, \quad a = 1, \quad a = 9$$

---

[3]The problematic quantity $x_n - a$ in (7.6) is cancelled in (7.7) so division by zero is avoided in (7.8) when we distribute the limit.

7.3 Derivative formulas and higher derivatives

Calculating derivatives using its definition in (7.3) is fine as long as the function in question is simple enough, but it gets cumbersome and impractical even for functions that are not very complex. To help with calculations, I list a number of basic formulas for derivatives. If you have had any exposure to calculus then you have likely come across these useful facts most of which are derived in the typical calculus textbook. I discuss most of the proofs in Appendix 11.7, using only elementary arguments.

The power rule.

Remember that a function of type $f(x) = x^p$ where p is a real number is a *power function*.

These functions may be undefined for negative real numbers x unless p is a positive integer (natural number). For example, if $p = -1$ then $f(x) = x^{-1} = 1/x$ is not defined for $x = 0$; or if $p = 1/2$ then $f(x) = x^{1/2} = \sqrt{x}$ is not defined (as a real number) for $x < 0$. If p is irrational then x^p may be defined using the exponential and logarithmic functions (we discuss these later in Chapter 8); this requires that the domain of x^p be $x > 0$, which is the same as the domain of the logarithm.

> **The power rule for derivatives.** The derivative of a power function is given by the formula:
> $$f'(x) = px^{p-1} \tag{7.9}$$
> The power rule may be stated in the differential notation:
> $$\frac{d(x^p)}{dx} = px^{p-1} \tag{7.10}$$

We assume in the above rule that the value of x is properly restricted so that x^{p-1} is defined as discussed above; for instance, if $p = -1/2$ then (7.9) is valid for $x > 0$ only whereas if p is a positive integer then (7.9) holds for all real values of x.

Let's verify the power rule in a special case, say, $f(x) = x^3$. For any fixed real number a and any sequence x_n of real numbers that converges to a and $x_n \neq a$, (7.3) gives:

$$f'(a) = \lim_{n \to \infty} \frac{x_n^3 - a^3}{x_n - a}$$

We use a standard algebraic identity for the "difference of cubes"

$$u^3 - v^3 = (u - v)(u^2 + uv - v^2)$$

to factor the numerator out:

$$f'(a) = \lim_{n \to \infty} \frac{(x_n - a)(x_n^2 + x_n a + a^2)}{x_n - a} \tag{7.11}$$

Now we see that the difference $x_n - a$ can be cancelled out, leaving us with a simple expression that can be calculated using the basic limit theorems that we discussed in Chapter 4:

$$f'(a) = \lim_{n \to \infty} (x_n^2 + x_n a + a^2) = \lim_{n \to \infty} x_n^2 + a \lim_{n \to \infty} x_n + \lim_{n \to \infty} a^2 = a^2 + a^2 + a^2 = 3a^2$$

Here a is an arbitrary value of x so for $f(x) = x^3$ we may write: $f'(x) = 3x^2$. In the formula, if $p = 3$ then $px^{p-1} = 3x^2$ so we have proved the formula for $p = 3$.

Derivatives

The key step in this calculation was the factorization in (7.11) that made it possible to cancel out $x_n - a$ from the fraction, thus avoiding division by 0. The difference of powers formula can be extended to all rational values of p so the above argument may be used to prove the power rule for all rational powers; see Appendix 11.7.[4] For irrational powers, we derive the power rule from formulas for exponential functions in Chapter 8.

Derivatives of sums, products and quotients.

We now discuss formulas for finding the derivatives of sums, products and quotients. Among other things, they allow us to speed up calculations considerably.

Suppose that a given pair of function $f(x)$ and $g(x)$ have derivatives $f'(a)$ and $g'(a)$ at $x = a$. Then the following are true:

> **The Sum Rule:** *The derivative of the sum $(f+g)'(a)$ exists and*
> $$(f+g)'(a) = f'(a) + g'(a)$$
>
> **The Product Rule:** *The derivative of the product $(fg)'(a)$ exists and*
> $$(fg)'(a) = f'(a)g(a) + f(a)g'(a)$$
>
> **The Quotient Rule:** *The derivative of the quotient $(f/g)'(a)$ exists provided that $g(a) \neq 0$ and*
> $$\left(\frac{f}{g}\right)'(a) = \frac{f'(a)g(a) - f(a)g'(a)}{[g(a)]^2}$$
>
> *These rules are more often written in the differential notation that turns out to be more convenient to use. A common way would be to set $u = f(x)$ and $v = g(x)$ so that the derivatives at an arbitrary point are written as du/dx and dv/dx. With these symbols, we can write the above three rules as follows:*
>
> $$\frac{d}{dx}(u+v) = \frac{du}{dx} + \frac{dv}{dx}$$
> $$\frac{d}{dx}(uv) = \frac{du}{dx}v + u\frac{dv}{dx}$$
> $$\frac{d}{dx}\left(\frac{u}{v}\right) = \frac{1}{v^2}\left[\frac{du}{dx}v - u\frac{dv}{dx}\right]$$

Why do the product and quotient rules involve unexpected combinations of functions and derivatives? After all, derivatives are limits and the *limit* theorems of Chapter 4 did not contain such surprises. But remember that the derivative involves the limit of a *quotient* namely, $\frac{\Delta y}{\Delta x}$ so there is really no reason for us to think that derivative rules must look like the limit rules.

Let's examine this issue using simple examples. Consider $f(x) = x^3$ and $g(x) = x^2 + 1$. Then
$$(fg)(x) = f(x)g(x) = x^3(x^2+1) = x^5 + x^3$$

---
[4]An alternative derivation is based on the binomial theorem but the two approaches are equivalent. We discuss the binomial expansion in Chapter ??.

so the power rule and sum rule imply that $(fg)'(x) = 5x^4 + 3x^2$. However, if we take the derivatives of f and g separately using the same rules and then multiply we get: $f'(x)g'(x) = (3x^2)(2x+0) = 6x^3$ which is not the correct answer! On the other hand, if we use the product rule then

$$(fg)'(x) = f'(x)g(x) + f(x)g'(x) = (3x^2)(x^2+1) + (x^3)(2x+0) = 3x^4 + 3x^2 + 2x^4 = 5x^4 + 3x^2$$

which is the same as the answer that we got by first multiplying the functions to find the product function. A similar example with simple functions, say, $f(x) = x^3$ and $g(x) = x^2$ shows that the quotient rule really does need to be as stated above.

The sum rule is easy to prove using the definition of derivative (7.3) and some simple algebra; the proofs of the product and quotient rules are in see Appendix 11.7. As usual, let x_n be an arbitrary sequence that approaches a but $x_n \neq a$ for all n. Then by (7.3)

$$f'(a) = \lim_{n \to \infty} \frac{f(x_n) - f(a)}{x_n - a} \quad \text{and} \quad g'(a) = \lim_{n \to \infty} \frac{g(x_n) - g(a)}{x_n - a} \tag{7.12}$$

are both well-defined numbers; we make use of this fact as follows:

$$\begin{aligned}
(f+g)'(a) &= \lim_{n \to \infty} \frac{(f+g)(x_n) - (f-g)(a)}{x_n - a} \\
&= \lim_{n \to \infty} \frac{f(x_n) + g(x_n) - f(a) - g(a)}{x_n - a} \\
&= \lim_{n \to \infty} \frac{f(x_n) - f(a) + g(x_n) - g(a)}{x_n - a} \\
&= \lim_{n \to \infty} \left[\frac{f(x_n) - f(a)}{x_n - a} + \frac{g(x_n) - g(a)}{x_n - a} \right]
\end{aligned}$$

Each of the two fractions is a separate sequence and we can distribute the limit over the sum as discussed in Chapter 4 to get:

$$(f+g)'(a) = \lim_{n \to \infty} \frac{f(x_n) - f(a)}{x_n - a} + \lim_{n \to \infty} \frac{g(x_n) - g(a)}{x_n - a} = f'(a) + g'(a)$$

Here is a useful consequence of the product rule in which we take $g(x) = c$ to be a constant function with c being an arbitrary fixed real number. Then

$$(cf)'(a) = 0f(a) + cf'(a) = cf'(a)$$

The 0 above comes from the fact mentioned earlier that the derivative of a constant is always 0.

The constant multiple rule. If $f(x)$ has a derivative $f'(a)$ at $x = a$ then for every real constant:
$$(cf)'(a) = cf'(a)$$
This rule can be written in the differential notation with $u = f(x)$ as:
$$\frac{d}{dx}(cu) = c\frac{du}{dx}$$

Derivatives

When used together with the power rule, the above derivative rules make it easy to find derivatives of a variety of important algebraic functions, including polynomials, rational functions and more. For example, to find the derivative of the rational function

$$f(x) = \frac{x^5 - x^2}{x^6 - 7x - 8}$$

we use the quotient rule along with the sum rule, the constant multiple rule and the power rule to obtain

$$f'(x) = \frac{(5x^4 - 2x)(x^6 - 7x - 8) - (x^5 - x^2)(6x^5 - 7)}{(x^6 - 7x - 8)^2}$$
$$= \frac{-x^{10} + 4x^7 - 28x^5 - 40x^4 + 7x^2 + 16x}{(x^6 - 7x - 8)^2}$$

With this derivative function at hand, we may calculate the derivative at every point in the domain of the original function $f(x)$. For instance, $f'(0) = 0$ and $f'(1) = -6/343$ are calculated by just setting $x = 0$ or $x = 1$ in $f'(x)$ rather than performing a separate calculation in each case using a difference quotient and a limit.

Exercise 86 *Calculate the derivative of $f(x)$ below then find $f'(1)$:*

$$f(x) = \frac{x^2 - x + 2}{3x + 1}$$

The chain rule.

Recall that another basic way of combining functions is by composing them. Now let's see how to calculate the derivative of a function that can be written as a composition two or more other functions. In symbols:

$$(g \circ f)(x) = g(f(x))$$

provided that the function $g(x)$ is defined over the range of $f(x)$. In the order shown above, the composition operation essentially inserts one function, $f(x)$ into another, $g(x)$. This operation, like addition and multiplication, generates a new function from the two given ones.

Now, suppose that both $f(x)$ and $g(x)$ are differentiable functions; does it follow that their composition $(g \circ f)(x)$ is also differentiable? If so then what is its derivative?

It is fortunate that not only the answer is yes, but there is also a simple formula for calculating the derivative of the composition in terms of the derivatives of the constituent functions. As we discover in the course of our trek in this book, the resulting formula is one of the most useful and important derivative formulas in calculus!

> **The Chain Rule:** If $f(x)$ is differentiable at $x = a$ and $g(x)$ is differentiable at $x = f(a)$ then the composition $(g \circ f)(x)$ is differentiable at $x = a$ and
>
> $$(g \circ f)'(a) = g'(f(a))f'(a) \tag{7.13}$$
>
> The notation $g'(f(a))$ in (7.13) means: calculate the derivative of $g(x)$ at $x = f(a)$. The chain rule is usually written in the differential notation: If $u = f(x)$ and $y = g(u)$ then
>
> $$\frac{dy}{dx} = \frac{dy}{du}\frac{du}{dx}$$

The differential notation is especially useful and intuitive in multivariable calculus where the presence of two or more variables leads to different versions of the chain rule.

A proper proof of the chain rule requires certain technical details that we will not discuss here. But to get a sense of where the expression in 7.13 comes from, here is an oversimplified argument:

Let x_n be a sequence of real numbers that converges to a (but $x_n \neq a$ for all n and note that

$$(g \circ f)'(a) = \lim_{n \to \infty} \frac{g \circ f(x_n) - g \circ f(a)}{x_n - a}$$
$$= \lim_{n \to \infty} \frac{g(f(x_n)) - g(f(a))}{f(x_n) - f(a)} \frac{f(x_n) - f(a)}{x_n - a}$$

In the last step we divided and multiplied by the quantity $f(x_n) - f(a)$ and rearranged the fractions.[5] Now, $f(x_n) \to f(a)$ as $x_n \to a$ because $f'(x)$ exists by assumption at $x = a$. Let $b = f(a)$ and $y_n = f(x_n)$ for each n so that $y_n \to b$ as $n \to \infty$. Rewriting the last limit above, we get

$$(g \circ f)'(a) = \lim_{n \to \infty} \frac{g(y_n) - g(b)}{y_n - b} \frac{f(x_n) - f(a)}{x_n - a}$$
$$= \lim_{n \to \infty} \frac{g(y_n) - g(b)}{y_n - b} \lim_{n \to \infty} \frac{f(x_n) - f(a)}{x_n - a}$$

Now each of the limits on the right hand side is a derivative by hypothesis:

$$(g \circ f)'(a) = g'(b)f'(a) = g'(f(a))f'(a)$$

which is the expression that we see in (7.13).

Let's explore the usefulness of the chain rule by discussing some examples. First, consider the function

$$h(x) = (x^3 + 1)^2$$

One way of finding the derivative $h'(x)$ is to first square out the expression on the right hand side to get $x^6 + 2x^3 + 1$ and then use the rules that we previously discussed, namely, sum rule and constant multiple rule to get:

$$6x^5 + 2(3x^2) + 0 = 6x^5 + 6x^2$$

---

[5] In dividing by $f(x_n) - f(a)$ we ignored the possibility that this quantity might be zero, since it is entirely possible that $f(x_n) = f(a)$ for some (or even all) values of n even if $x_n \neq a$ for every n. A proper proof avoids this issue and can be found in most books in advanced calculus or real analysis that cover the development of the derivative.

Derivatives

A second way is to first write $h(x)$ as:

$$h(x) = (x^3 + 1)(x^3 + 1)$$

and then apply the product rule to obtain:

$$h'(x) = 3x^2((x^3 + 1) + (x^3 + 1)3x^2 = 6x^2(x^3 + 1) = 6x^5 + 6x^2$$

matching the answer that we got earlier.

Alternatively, we may use the chain rule; define $u = f(x) = x^3 + 1$ and $y = g(u) = u^2$ so that

$$h'(x) = \frac{dy}{dx} = \frac{dy}{du}\frac{du}{dx} = 2u(3x^2) = 6x^2(x^3 + 1)$$

Using the chain rule above does not seem to have made the calculations any easier, not to mention requiring conceptual work like the substitutions involving y and u. But the true value and power of the chain rule shows when we consider higher powers or non-integer powers on the parenthesis. The chain rule applies in essentially the same way and no additional effort to a function like $h(x) = (x^3 + 1)^{100}$ as to $h(x)$ above when the power was just 2. We simply set $u = x^3 + 1$ and $y = u^{100}$ in the chain rule to get

$$h'(x) = \frac{dy}{du}\frac{du}{dx} = 100u^{99}(3x^2) = 300x^2(x^3 + 1)^{99}$$

It would be practically infeasible to expand $(x^3 + 1)^{100}$ by raising to the power 100. Also, to use the product rule to find the derivative of this function we must apply the product rule 100 times (not just twice as we did above); again not a practical approach!

Here is another example where the chain rule is the only formula that works: consider the function $h(x) = \sqrt{x^2 + 1}$. The derivative formulas that we discussed earlier in this section simply do not apply to this function. However, we can think of it as a composition of two functions to which those formulas *do* apply. By visual inspection, we pick $u = x^2 + 1$ and $y = \sqrt{u} = u^{1/2}$ so that

$$h'(x) = \frac{dy}{du}\frac{du}{dx} = \frac{1}{2}u^{-1/2}(2x) = x(x^2 + 1)^{-1/2}$$

These examples also illustrate how the chain rule can be used to derive a general version of the power rule.

The general power rule: Let p be a rational number and $f(x)$ a differentiable function that satisfies the isolation property for some real number a. Then the derivative of $h(x) = [f(x)]^p$ at $x = a$ is

$$h'(a) = p[f(a)]^{p-1}f'(a)$$

This formula says that to find the derivative of a function $f(x)$ to some power p, first use the power rule thinking of $f(x)$ as a new variable u and then multiply the result of the power rule by the derivative of $f(x)$. In the differential notation this comes out just as described, where if $u = f(x)$ then

$$\frac{d(u^p)}{dx} = pu^{p-1}\frac{du}{dx}$$

Exercise 87 *Prove the general power rule using the power rule and the chain rule.*

Exercise 88 *Find the derivatives of the following functions and then calculate the derivatives at $x = 1$. Note that you may need to use some of the previous rules like the product rule, etc. It is a good idea to list each derivative rule that you use to keep track of things:*

$$\text{(a)} \quad f_1(x) = \frac{x^2 + 1}{(3x - 2)^7} \qquad \text{(b)} \quad f_2(x) = 2x^3 \sqrt{x^2 + x}$$

Derivatives of trigonometric functions.

We now consider the derivatives of trigonometric functions.

There are a variety of equations that relate the familiar trigonometric functions $\sin x$, $\cos x$, $\tan x$, etc (see Appendix 11.2). Once we know the derivative of, say, $\sin x$ we can use these equations together with some of the derivative rules that we discussed above to find the derivatives of all the other trigonometric functions. In Appendix 11.6. we prove that if $f(x) = \sin x$ then for every real number a

$$f'(a) = \cos a$$

In the differential notation this is written in the somewhat easier to remember form:

$$\frac{d}{dx} \sin x = \cos x$$

Let's use this to find the derivatives of $\cos x$ and $\tan x$. First, we know that $\cos x = \sin(x + \pi/2)$ (see Appendix 11.2); so let $h(x) = \cos x$, $g(x) = \sin x$ and $f(x) = x + \pi/2$ and use the chain rule and the above formula to get:

$$h'(a) = g'(f(a))f'(a) = \cos(f(a))(1) = \cos\left(a + \frac{\pi}{2}\right) = -\sin a$$

In the differential notation this derivative formula for the cosine function is expressed as:

$$\frac{d}{dx} \cos x = -\sin x$$

Similarly, if $f(x) = \tan x = (\sin x)/(\cos x)$ where $\cos a \neq 0$, i.e., $a \neq \pi k + \pi/2$ for all integers k then applying the quotient rule gives:

$$f'(a) = \frac{(\cos a)\cos a - (\sin a)(-\sin a)}{(\cos a)^2} = \frac{\cos^2 a + \sin^2 a}{\cos^2 a} = \frac{1}{\cos^2 a}$$

or in the differential notation,

$$\frac{d}{dx} \tan x = \frac{1}{\cos^2 x}$$

The derivatives of other trigonometric functions or combinations of trigonometric functions can be found using the sum, product, quotient and chain rules.

Derivatives

Exercise 89 *Find the derivatives of each function below using both the prime notation and the differential notation:*

(a) $f_1(x) = \sec x = \dfrac{1}{\cos x}$ (b) $f_2(x) = \sin(2x+3)$ (b) $f_3(x) = x^2 \tan x$

Higher order derivatives.

We saw earlier in this chapter that the rate of change of position $x(t)$ of a moving object is its (instantaneous) velocity $v(t) = x'(t)$. The derivative $v'(t)$ of the velocity function gives the rate of change of velocity and is called the *acceleration* of the object, which we denote by $a(t)$. Since the velocity itself is a derivative we can then write, using the differential notation (because it is more clear in this case):

$$a(t) = v'(t) = \frac{dv}{dt} = \frac{d}{dt}\left(\frac{dx}{dt}\right)$$

The last expression on the right hand side shows that we may alternatively calculate the acceleration by taking the derivative of the position function *twice in a row*. This suggests the alternative primes notation

$$a(t) = x''(t)$$

In some cases, is convenient to use this derivative of order 2 or "the second derivate" directly. An example is Newton's second law of motion that is often written as

$$ma = F$$

In words, the net of all forces acting on a moving object at any time is proportional to its acceleration, with the mass m of the object being the constant of proportionality. This law can be stated in two equivalent ways, each of which stresses a different aspect of motion. In some cases, we write

$$mv' = F$$

in which case the left hand side is interpreted as the rate of change of the "momentum", which is defined as mv.[6]

In other cases, the expression

$$mx'' = F$$

renders itself naturally to the derivation of a differential equation for the motion. Each solution of such an equation gives the position of the particle at all times subject to specified initial values (for position and velocity).

To illustrate this latter interpretation, recall that Galileo's equation for velocity was $v(t) = 9.8t$. The derivative of this function is $v'(t) = 9.8$, or equivalently,

$$x''(t) = 9.8 \qquad (7.14)$$

This is simply a mathematical expression of the experimental fact that acceleration due to gravity near the surface of Earth is the constant 9.8. In 7.14 we have a differential equation of order 2 whose

---

[6] In post-Newtonian physics it is often more convenient to define "force" as the rate of change of momentum and consider the momentum as a fundamental parameter of motion, along with the position.

solutions determine Galileo's equations of motion. A substantial mathematical theory for solving differential equations of order 2 and greater has been developed over the centuries and this version of Newton's law has been a prominent model building tool in physics and engineering since Newton's time.

For Galileo's simple equations it is not necessary to study differential equations. We discuss how to go backward and derive the position function from 7.14 in Chapter 8 where we discuss the concept of integral.

Now we formally define derivatives of order 2 and higher.

> **Higher order derivatives.** Suppose that $y = f(x)$ is differentiable and has a derivative function $f'(x)$. If $f'(x)$ is also differentiable then its derivative is the *second derivative* of $f(x)$ and denoted, in our usual two notations,
> $$f''(x) = \frac{d^2y}{dx^2}$$
> The second derivative of $f(x)$ is also called the *derivative of order 2*. We also refer to $f'(x)$ as the derivative of order 1 or the first derivative of $f(x)$.
> More generally, for each positive integer k the *derivative of order k* or the *kth derivative* of $f(x)$ is obtained by taking the derivative k times (assuming that we can) and it is denoted
> $$f^{(k)}(x) = \frac{d^ky}{dx^k}$$

Finding higher order derivatives is usually just a matter of repeated differentiation. Let's look at some examples.

Consider the power function $f(x) = x^3$. We can take derivatives of this function repeatedly using the rules and formulas that we discussed earlier.

$$f'(x) = 3x^2 \quad f''(x) = 2(3x^1) = 6x \quad f'''(x) = 6, \quad f^{(4)}(x)) = 0$$

It is common practice to write $f'''(x)$ instead of $f^{(3)}(x)$); either notation is fine but clearly we cannot keep adding primes indefinitely and it is rare to see derivatives of order 4 or greater written using a sequence of prime symbols.

Notice that for the cube function x^3 derivatives of order 4 or greater are all zeros and repeated differentiation does not produce any new functions.

Exercise 90 *Find all higher order derivatives of the polynomial function $f(x) = x^4 - x^2 + 7$.*

For non-polynomial functions, it may be possible to take the derivative infinitely often and generate new functions at each step. For instance, consider $y = f(x) = 1/x$. To find the derivatives of this function it is more convenient to write it as a power function: $y = x^{-1}$. Then (using the differential notation):

$$\frac{dy}{dx} = -x^{-2}, \quad \frac{d^2y}{dx^2} = -(-2)x^{-3} = 2x^{-3}, \quad \frac{d^3y}{dx^3} = 2(-3x^{-4}) = -6x^{-4},$$

Derivatives

and so on. We may also write these answers as fractions:

$$\frac{dy}{dx} = -\frac{1}{x^2}, \quad \frac{d^2y}{dx^2} = \frac{2}{x^3}, \quad \frac{d^3y}{dx^3} = -\frac{6}{x^4}, \ldots$$

It is clear that if we continue to take derivatives then we just get power functions of greater negative degree; this can go on indefinitely and generate an infinite sequence of distinct power functions.

For some functions the process does not generate distinct derivative functions but also does not end in the zero function. The obvious examples are the trigonometric functions $\sin x$ and $\cos x$. For each of these functions only the first three derivatives are distinct; the fourth cycles back to the original function.

Exercise 91 *Find all distinct higher order derivatives of $\sin x$ (only a few of these derivatives are distinct, or non-repeating).*

Repeated differentiation may not be possible for all functions, at least, not on their entire domains. For example, consider $f(x) = x^{5/3}$ which is well defined for all real values of x, that is, its domain is $(-\infty, \infty)$. The first and second derivatives of this function are

$$f'(x) = \frac{5}{3}x^{2/3} \quad f''(x) = \frac{5}{3}\left(\frac{2}{3}x^{-1/3}\right) = \frac{10}{9x^{1/3}}$$

We see that while the first derivative is also defined on the entire set $(-\infty, \infty)$, the second derivative is not defined at $x = 0$. We say that $x^{5/3}$ is differentiable but not twice differentiable on $(-\infty, \infty)$.

Here is a more extreme example:[7] let $f(x) = \cos^{1/3} x$, that is, the cube root of $\cos x$. The domain of this function is $(-\infty, \infty)$ but its derivative

$$f'(x) = -\frac{1}{3}\sin^{-2/3} x = -\frac{1}{3\sin^{2/3} x}$$

is undefined at all values of x where $\sin x = 0$. This is the infinite set of numbers $n\pi$ for all integers n.

We discuss various case where the derivative fails to exist in the next section. I would like to close this section by briefly discussing the importance of higher derivatives.

There are specific occurrences of derivatives of order 2 or greater in physics. For example, if $x(t)$ denotes the position of a moving object at time t then the third derivative $\frac{d^3x}{dt^3}$ gives the rate of change of acceleration and is know as the "jerk" function. For instance, if a freely falling pebble with constant downward acceleration 9.8 received a puff of air directly from below that causes it to momentarily slow down then its acceleration is impacted negatively and a jerk in the motion occurs.

Derivatives of even higher order also occur; for example, the Euler-Bernoulli equation for the deflection of a beam under a load involves a derivative of order 4.

---
[7]This is far from the worst possible; there are functions whose domain is all of $(-\infty, \infty)$ but fail to have a derivative at even one point in $(-\infty, \infty)$. We discuss the famous example found by K. Weierstrass in Chapter 9.

Possibly the most important use of higher order derivatives occurs in expansions of functions in the form of Taylor series that we discuss in some detail in Chapter 9. These infinite series, which usually involve derivatives of all possible orders, are among the most useful tools in applied mathematics.

7.4 When derivatives fail to exist: more often than may seem

The derivative of a function $f(x)$ may not exist at every point of the function's domain. If a is a point in the domain of $f(x)$ then $f(a)$ is a well-defined number. But for $f'(a)$ to also exist, *the limit in (7.3) must exist and have the same value for every sequence x_n that converges to a.*

Let's focus on the last statement; assuming that $x_n \neq a$ for all n, the fraction

$$\frac{f(x_n) - f(a)}{x_n - a} \qquad (7.15)$$

is a well-defined number for every n. But the *limit* of this fraction as $n \to \infty$ is another matter; here is what can go wrong:

1. *The limit may not exist for some numbers a regardless of the choice of x_n;*

2. *The limit may exist for some sequences x_n that converge to a but not for all such sequences;*

3. *The limit may exist for all sequences that converge to some a but may have different values for different sequences.*

I illustrate each of the above possibilities using an example.

Since graphs help visualize each case, let's also remember that derivatives are slopes of tangent lines. So each of the above three possibilities is a case where it is not possible to draw a unique tangent line with a well-defined slope.

Let's start with the cube-root function

$$f_1(x) = x^{1/3} = \sqrt[3]{x}$$

The graph is shown in Figure 7.4

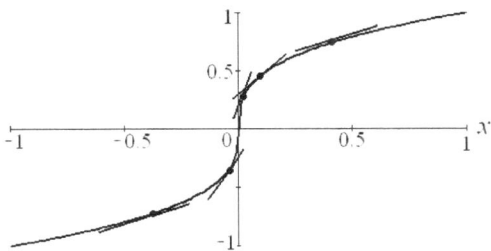

Figure 7.4: No derivative at the origin: Vertical tangent line

Although the y-axis appears to be a unique tangent line at the origin $x = 0$, it does not have a well-defined slope (or has infinite slope) there. If we draw some tangent lines to the graph of the

Derivatives

function both to the left of 0 and to the right, the lines appear to be getting vertical and approaching the y-axis as we choose the points of tangency closer to 0. Let's take a look at (7.15) in this case:

$$\frac{\sqrt[3]{x_n} - \sqrt[3]{0}}{x_n - 0} = \frac{x_n^{1/3}}{x_n} = \frac{1}{x_n^{1-1/3}} = \frac{1}{x_n^{2/3}}$$

Since $x_n \to 0$ it follows that $x_n^{2/3}$ does too.[8] This implies that $1/x_n^{2/3} \to \infty$ (the sequence diverges to infinity) so the limit in (7.3) does not exist, regardless of what sequence x_n we choose that approaches 0.

Exercise 92 *Study the fraction in (7.15) for the square of the cube-root function*

$$f(x) = x^{2/3} = \left(\sqrt[3]{x}\right)^2$$

Show that if $x_n \to 0$ and x_n is negative for all n then the slope fraction goes to $-\infty$ while if $x_n \to 0$ and x_n is positive then the slope fraction goes to ∞. So the situation here is worse than the cube-root function! Consider graphing $x^{2/3}$ over an interval that contains 0, say, $[-1,1]$ and draw tangent lines at various points both to the left and the right of 0. What happens to the slopes of the tangent lines at points closer and closer to 0 from either side? The origin in this example is called a "cusp" and considered a type of singularity.

To illustrate Item 2 above, consider the two-piece function

$$f_2(x) = \begin{cases} \sqrt{x}, & x \geq 0 \\ x^2, & x \leq 0 \end{cases}$$

A graph of this function is shown in Figure 7.5.

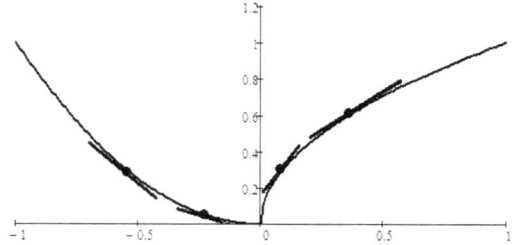

Figure 7.5: No derivative at the origin: different tangents from opposite sides

Notice that $f_2(0) = 0$. If we draw tangent lines to the graph at various points to the *right* of 0 then they appear to be getting vertical and their slopes become infinitely large; but if we draw the tangent lines to the *left* of 0 the tangent lines get more horizontal as I approach 0 so the slopes

---

[8]You can check this easily using a calculator. It is easy to prove this fact using a general result that we discuss in the section on continuity.

approach 0. We can verify this dual behavior analytically by examining (7.15). Choose a sequence x_n that converges to 0 and $x_n \neq 0$. If $x_n > 0$ for all n then

$$\frac{f_2(x_n) - f_2(0)}{x_n - 0} = \frac{\sqrt{x_n}}{x_n} = \frac{1}{\sqrt{x_n}}$$

If $x_n \to 0$ then so does its $\sqrt{x_n} \to 0$ as $n \to \infty$ so the above slope fraction gets infinitely large. Clearly, there is no limit when x_n is positive for all n. But if $x_n < 0$ for all n then the limit does exist because:

$$\lim_{n \to \infty} \frac{f_2(x_n) - f_2(0)}{x_n - 0} = \lim_{n \to \infty} \frac{x_n^2}{x_n} = \lim_{n \to \infty} x_n = 0$$

consistent with the tangent lines from the left of 0 getting horizontal when approaching 0.

Of course, this does not mean that the derivative exists because that requires the limit to exist and be the same number for all sequences x_n that approach 0. In this case, the derivative does not exist because some of the sequences fail to converge.[9] Geometrically, both the x-axis and the y-axis are tangent to the graph at $x = 0$.

Finally, to illustrate Item 3 above consider the absolute value function $f_3(x) = |x|$, or equivalently,

$$f_3(x) = \begin{cases} x, & x \geq 0 \\ -x, & x \leq 0 \end{cases}$$

The v-shaped graph of $f_3(x)$ appears in Figure 7.6.

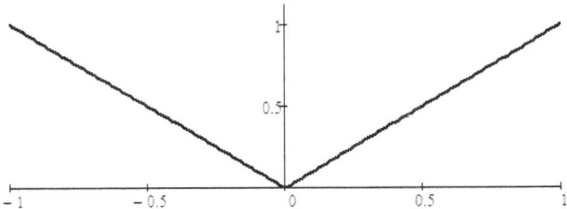

Figure 7.6: No derivative at the origin: different slopes from opposite sides

In this case, the graph is just a pair of lines, each being its own tangent line; the line on the right has slope 1 while the one on the left has slope -1. As in the discussions for the previous two items, this duality indicates that $f_3(x)$ does not have a derivative at 0. Let's examine (7.15) in this case:

$$\frac{f_3(x_n) - f_3(0)}{x_n - 0} = \begin{cases} \frac{x_n}{x_n} = 1, & x_n > 0 \\ \frac{-x_n}{x_n} = -1, & x_n < 0 \end{cases}$$

So if the sequence x_n is chosen to the right of zero then the slope fraction approaches 1; but if x_n is chosen to the left of zero then the slope fraction approaches -1. For the derivative to exist, the limit must always be the *same* number (namely, the derivative).

---

[9] I should point out that an arbitrary sequence x_n may also dance around 0 by jumping back and forth. It maybe positive for an infinite number of indices and negative for also an infinite number of indices. For instance, if $x_n = (-1)^n/n$ then all odd terms are negative and the slopes converge to zero, whereas all even terms are positive and the slopes diverge to infinity. But we don't need to consider such sequences here because we already found two sequences that converged to different limits, which is bad enough!

7.5 Continuity and singularities

One way of estimating the value of $\sqrt{2}$ is to graph the polynomial function $f(x) = x^2 - 2$ and find where it crosses the x-axis, because $\sqrt{2}$ is a number where $x^2 - 2 = 0$. We take for granted that the curve representing the graph of $x^2 - 2$ does in fact cross the x-axis because:

(a) The x-axis has no gaps or holes (the set of real numbers is complete);

(b) The curve $x^2 - 2$ also has no holes or cuts, rather like a piece of string or a strand of thin noodle.

We might imagine sketching the graph of such a curve without lifting the tip of the pen from the paper (or white board) so when the tip reaches the number line and has to go above it, then the pen's tip *has to* meet, or intersect the x-axis.

We call a curve like that in (b) above "continuous". But how do we define this so that we can answer basic questions with precision?

We certainly can't sketch the graph of every function by hand and track the tip of the proverbial pen over an infinite domain. Suppose that we carefully sketch x^2 from say, $x = -10$ to $x = 10$; we find that we can do this without lifting our pen so we may declare that x^2 is continuous from $x = -10$ to $x = 10$. But this does not show that x^2 is continuous for $x > 10$ or $x < -10$. We would need new sketches that cover the entire x-axis; failing that, which is certain, we can only claim that x^2 is continuous over the largest interval that we can cover.

If this limitation doesn't inspire considering a more clever way of dealing with the continuity issue then consider the fact that to show that slight variations like $x^2 + 1$ or $-x^2$ over *any interval* we must sketch each of these all over again to make sure that our pen doesn't leave the page. And this says nothing about other functions that we suspect must be continuous too, like x^3, $\sqrt{x}$ or $\sin x$.

Adding to the uncertainty is the fact that sketches done on digital computers or calculators are inconclusive since they have finite resolution both in calculations (due to finite memory) and in display (device screens have finitely many pixels). The image that we see on the screen (after zooming if necessary) is just a sequence of tiny squares that are stitched together by some algorithm.

For these reasons and many others, we need a more precise understanding of what makes a function continuous; and as you might guess, the infinity plays a critical role here. Earlier in our study of the set $\mathbb{R}$ of real numbers we saw that the number line is "continuous" (technically, complete) because the irrational numbers filled in all the gaps among the rational numbers. In this section we study the continuity of functions whose domains and ranges are contained in $\mathbb{R}$. Similarly to how we defined the continuity of $\mathbb{R}$ using convergent sequences, we define continuity of functions using convergent sequences. In the process we identify where and how the infinity comes into play.

Continuity through limits.

Consider the function $f(x) = x^2 - 2$ again. From the sequence limit theorems in Chapter 4 it follows that

$$\lim_{n \to \infty} f(x_n) = \lim_{n \to \infty} (x_n^2 - 2) = \lim_{n \to \infty} (x_n^2) - \lim_{n \to \infty} 2 = (\lim_{n \to \infty} x_n)^2 - 2$$

If x_n is an arbitrary sequence on the x-axis that converges to $\sqrt{2}$ so that $\lim_{n \to \infty} x_n = \sqrt{2}$ then

$$\lim_{n \to \infty} f(x_n) = (\sqrt{2})^2 - 2 = 0$$

Therefore, the numbers $f(x_n)$ on the y-axis converge to 0. Now remember that the points

$(x_n, f(x_n))$ are on the graph of $f(x) = x^2 - 2$ which means that the graph must be approaching the point $(\sqrt{2}, 0)$ on the x-axis; see Figure 7.7.

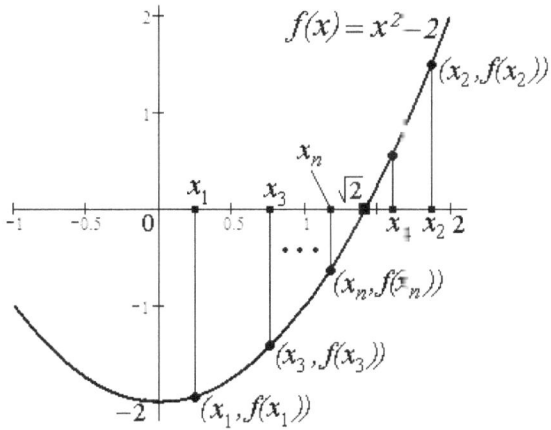

Figure 7.7: Explaining continuity using sequences

The point $(\sqrt{2}, 0)$ is also on the graph of $f(x)$ because the y-coordinate at $x = \sqrt{2}$ is

$$f(\sqrt{2}) = (\sqrt{2})^2 - 2 = 0$$

Therefore, the graph of $f(x)$ *must* approach the x-axis and intersect it at the point $(\sqrt{2}, 0)$!

This proves our intuitive guess based on the image of a pen drawing without being lifted; not only that, it suggests a general definition of continuous functions.

> ***Continuous function.*** A function $f(x)$ is *continuous* at $x = a$ if for *every* sequence x_n of real numbers that approaches a as $n \to \infty$ the image sequence $f(x_n)$ approaches the image $f(a)$. This definition may be stated more precisely as follows:
> ***Continuous functions and limits.*** $f(x)$ is *continuous* at a point a if $f(a)$ exists and for *every* sequence x_n of real numbers
>
> $$\lim_{n \to \infty} x_n = a \quad \text{implies} \quad \lim_{n \to \infty} f(x_n) = f(a). \qquad (7.16)$$
>
> If $f(x)$ is continuous at every point of its domain then we say that $f(x)$ is a *continuous function*. On the other hand, $f(x)$ is *discontinuous* at $x = a$ if the first equality in (7.16) holds but *not the second*, for at least one sequence x_n that converges to a.

To further illustrate the consistency of this definition with our common sense understanding of continuity let's look at an example in which it fails and we see that the resulting "discontinuity" is what we would expect. Consider the following simple two-piece function

$$f(x) = \begin{cases} 1, & x \geq 0 \\ -1, & x < 0 \end{cases} \qquad (7.17)$$

The graph of this simple function is shown in Figure 7.8; notice that $f(0) = 1$ in this definition and I highlight this in the graph by placing a full circle on the y-axis at 1; the hollow circle at -1 indicates that this point on the y-axis is *not* on the graph of $f(x)$.

Derivatives

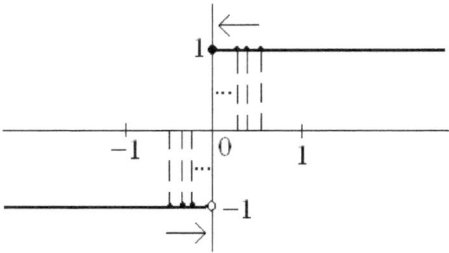

Figure 7.8: Illustrating a jump discontinuity: different sequences have different limits

It is clear that there is a break in the graph of $f(x)$ at $x = 0$ that is characterized by the jump in the value of $f(x)$ at the origin. Now, let's go back to the definition of continuity above and pick a sequence that converges to 0 from the left, say, $x_n = -1/n$. Then, as we see in Figure 7.8,

$$\lim_{n \to \infty} x_n = 0 \quad \text{and} \quad \lim_{n \to \infty} f(x_n) = \lim_{n \to \infty} (-1) = -1$$

But $f(0) = 1$ which is not equal to the limit -1 above! So our definition of continuity does imply that this function is discontinuous at $x = 0$ as we would expect it to do.

This example also illustrates that to show a function is continuous at a point a, the second equality in (7.16) must hold for **every** sequence that converges to a. Indeed, for the function in (7.17) the condition in (7.16) does hold for *some* sequences that converge to 0; if we pick $x_n = 1/n$ then

$$\lim_{n \to \infty} x_n = 0 \quad \text{and} \quad \lim_{n \to \infty} f(x_n) = \lim_{n \to \infty} (1) = 1 = f(0)$$

So if some sequences satisfy the limit condition in (7.16) but *there is even one sequence that doesn't* then the function is discontinuous!

The next exercise illustrates a function with a hole in its graph, which if we fill with just the right value then the function can be made continuous.

Exercise 93 *Consider the function*

$$f(x) = \frac{x^2 - 1}{x - 1} \quad \text{if } x \neq 1 \tag{7.18}$$

This function needs to be defined at $x = 1$. Explain why if we define $f(1) = 2$ then $f(x)$ is continuous at $x = 1$ but if $f(1) \neq 2$ then the function is discontinuous at $x = 1$. Note that the function simplifies algebraically after factoring the numerator; then make a quick sketch of the (simple) graph, making sure that you put a gap or hole above $x = 1$.

It is worth emphasizing at this stage if there are no points at which (7.16) fails for a function $f(x)$ then $f(x)$ is continuous. To illustrate this subtle aspect of the definition, consider the reciprocal function $f(x) = 1/x$. We commonly state that the domain of this function is the set of all nonzero real numbers, which we may write as $D = (-\infty, 0) \cup (0, \infty)$ Now, since $1/x$ is continuous at every point of this set D we would conclude that $1/x$ is a continuous function. This seems at odds with what we see in a typical graph of the reciprocal function, which jumps to infinity as the value of x approaches 0. But there are really no conflicts; the discontinuity occurs at 0 which is not in D.

To conclude that $1/x$ is a discontinuous function we must consider the fact that there is a break in the graph at $x = 0$. The proper way of handling this issue is through the concept of singularity that we consider next.

Discontinuity and singularities.

The hole in the graph of the function in (7.18) could be filled by a carefully chosen value at $x = 1$, i.e. $f(1) = 2$. This type of discontinuity can literally be erased or "removed" by choosing the right value for the function at the point in question. Other functions, like $1/x$ are not like this; the singularity of this function at $x = 0$ cannot be removed because the function shoots up (and down from the other side) to infinity. Still other types of functions may remain perfectly bounded but possess rather complicated singularities.

These considerations lead us to the following definition.

> **Singularities.** Consider a function $f(x)$ that is not continuous at $x = a$. Then:
> (a) $f(x)$ has a *removable singularity* at $x = a$ if there is a real number b such that when we set (or reset) $f(a) = b$ then $f(x)$ will be continuous at $x = a$.
> (b) If there are no real numbers that satisfy (a) then the singularity at $x = a$ is *non-removable*.

Removable singularities are far from being insignificant; notice that the derivative itself can exist at $x = a$ only when the quotient (7.15) has a removable singularity there! More on this later.

There are different types of non-removable singularities. For some functions a hole or gap cannot be patched up with any possible new value for the function; in this situation the function clearly has a non-removable singularity at the point in question. A common occurrence is where the function value jumps a certain distance that may be finite or infinite. A finite jump case is illustrated in Figure 7.8.

The function $1/x$ presents a case where there is an infinite jump at $x = 0$. No real number can fill in the gap at 0 in such a way that the function becomes continuous (conditions in (7.16) are satisfied). So we arrive at the conclusion:

$$f(x) = \frac{1}{x} \quad \text{has a non-removable singularity at } x = 0$$

Let's call this easy to spot singularity an *infinite singularity*.[10]

A singularity may occur more subtly as in the case of:

$$f(x) = \sin\frac{\pi}{x} \quad \text{also has a non-removable singularity at } x = 0$$

See Figure 7.9 for a graph of $\sin(\pi/x)$ near the origin.

As you can see in Figure 7.9 the only candidates that might bridge the gap at 0 are the numbers on the y-axis between -1 and 1; but none of these infinitely many numbers (in fact, $2^{\aleph_0}$) can satisfy the conditions in (7.16)!

---

[10] In complex analysis this type of singularity is called a "pole".

Derivatives

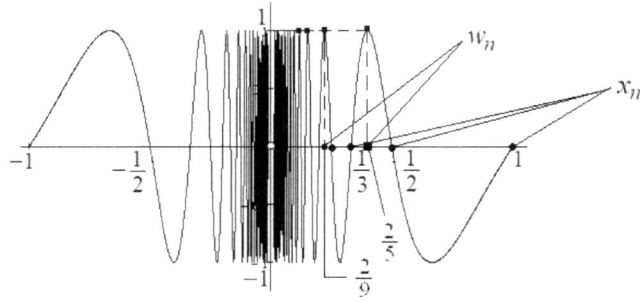

Figure 7.9: An extreme type of discontinuity at the origin

To see why not, first let's pick the sequence $x_n = 1/n$ that converges to 0 and calculate

$$\sin\frac{\pi}{x_n} = \sin \pi n = 0 \quad \text{for all } n$$

Evidently, (7.16) may hold for this particular sequence if we define $f(0) = 0$. But now suppose that we pick another sequence that also converges to 0 on the x-axis but has nonzero function values; based on Figure 7.9 we consider $w_n = 1/(2n + 1/2) = 2/(4n + 1)$ where the peaks occur. Then

$$\sin\frac{\pi}{w_n} = \sin\frac{\pi(4n+1)}{2} = \sin\left(2\pi n + \frac{\pi}{2}\right) = \sin\frac{\pi}{2} = 1 \quad \text{for all } n$$

This requires defining $f(0) = 1$ in order to satisfy (7.16). It follows that (7.16) cannot be satisfied in this case because as we have just seen, different sequences that converge to 0 on the x-axis generate sequences of function values on the y-axis that converge to different numbers.

The singularity at the origin has a complicated nature in this case; let's think of the graph of this function as an imaginary snaking road that winds infinitely often as it reaches a city whose right edge is the y-axis. As we drive on this winding road we find that that we are *approaching every point on the y-axis between -1 and 1 infinitely often*, each time getting a little closer. We get so close that we can touch fences, buildings, shrubbery, etc on the outer edge of the city over and over again but not yet having arrived there.

It is worth mentioning here that the curve in Figure 7.9 has infinite length over any interval on the x-axis that includes the origin. This is a simple consequence of the fact that the function is continuous and its graph stretches, from a minimum value of -1 to a maximum of 1, infinitely often. Each such stretch has a length that is greater than 2 so the length of the curve is greater than $2n$ for every large enough positive integer n.

Both the curve in Figure 7.9 and the graph of $1/x$ have infinite length on every interval of type $(0, a)$ where $a > 0$. In the case of $1/x$ the infinite length is due to unbounded magnitude. For the function shown in Figure 7.9 and similar rapidly oscillating functions whose magnitudes remain quite bounded, the infinite length comes from *unbounded variation*.

In the next exercise we look at a function whose graph approaches *all real numbers* infinitely often!

Exercise 94 *Consider the function*

$$f(x) = \frac{1}{x} \sin \frac{\pi}{x}$$

(a) Use a computing device to sketch a graph of this function for $-2 \le x \le 2$; be sure to set a high resolution on the x-axis when graphing the portion near the origin. You should be able to see that this function oscillates rapidly near the origin and further, its variation near the origin covers the entire y-axis, because $1/x$ becomes infinitely large as x approaches 0.

(b) Calculate $f(x_n)$ and $f(w_n)$ where x_n and w_n are the sequences defined in the preceding discussion for $\sin(\pi/x)$.

Additional complicated singularities can be obtained by combining the above ideas. Consider the function

$$f(x) = \frac{1}{\sin(\pi/x)}$$

Each of the numbers $s_n = 1/n$ is an infinite singularity of $f(x)$ because

$$\sin \frac{\pi}{s_n} = \sin \pi n = 0 \quad \text{for all } n$$

Therefore, this function $f(x)$ has a sequence of infinite singularities that converges to 0. The origin itself is a complicated singularity because it is a limit of infinite singularities. It is not easy to plot any significant stretch of this function close to 0 but a part of it that is not too far from 0 is shown in Figure 7.10.

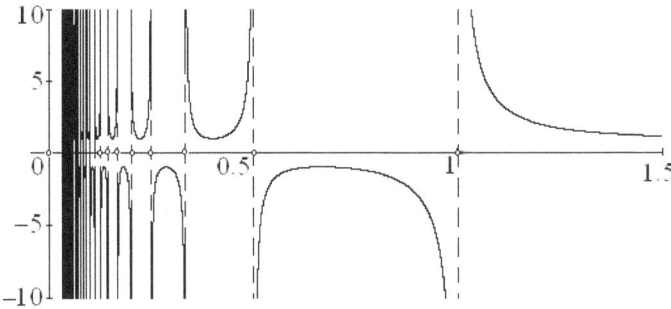

Figure 7.10: A function with a converging sequence of infinite discontinuities

Extreme discontinuity: Dirichlet's function.

The functions that we encounter in standard calculus textbooks, or in typical engineering or scientific models, may be discontinuous at many points but they are continuous in between those points. It is difficult to visualize a function that is not continuous at every point of a given interval, but it is surprisingly easy to define such a function mathematically. Consider the following function, named after the German mathematician J.P.G.L. Dirichlet (1805-1859)

$$D(x) = \begin{cases} 1, & \text{if } x \text{ is rational} \\ 0, & \text{if } x \text{ is irrational} \end{cases}$$

Derivatives

Recall that each of the sets of rational numbers and irrational numbers is dense in the set of real numbers. So for each real number a there is a sequence q_n of rational numbers that converges to a and we have:

$$\lim_{n\to\infty} q_n = a \quad \text{and} \quad \lim_{n\to\infty} D(q_n) = \lim_{n\to\infty} 1 = 1$$

There is also a sequence x_n of irrational numbers that converges to a and for this sequence we have:

$$\lim_{n\to\infty} x_n = a \quad \text{and} \quad \lim_{n\to\infty} D(x_n) = \lim_{n\to\infty} 0 = 0$$

Therefore, (7.16) cannot be satisfied in this case because as we have just seen, it cannot hold for *every* sequence that converges to a; the function values converge to 1 for some sequences that converge to a and to 0 for others. It follows that D is discontinuous at a, and since this argument is valid for every real number a we conclude that $D(x)$ is discontinuous everywhere!

It is impossible to draw an actual graph of $D(x)$ but you can imagine it as consisting of the x-axis with all the rational numbers moved up one unit to form a grainy straight line, a fine mist of tiny dots, that hangs above and parallel to the x-axis. This image is not strictly correct but it is intuitively helpful.

Exercise 95 *Consider the following Dirichlet-type function*

$$D_0(x) = \begin{cases} x, & \text{if } x \text{ is rational} \\ 0, & \text{if } x \text{ is irrational} \end{cases}$$

Can you explain why $D_0(x)$ is continuous at $x = 0$ but at no other real number? Try to imagine what the graph of this function may look like; then proceed by using (7.16) as in the discussion above for $D(x)$.

Every time you take a derivative you erase a removable singularity.

Let's see precisely how the derivative is related to continuity. Consider the *difference quotient* of $f(x)$ at $x = a$:

$$F(x) = \frac{f(x) - f(a)}{x - a} \tag{7.19}$$

This function obviously has a singularity at $x = a$ but according to the definition of derivative, this singularity is removable when $f(x)$ has a derivative at $x = a$.

> A function $f(x)$ has a derivative $f'(a)$ if and only if its difference quotient $F(x)$ has a removable singularity at $x = a$. By defining the value of $F(x)$ at $x = a$ as:
>
> $$F(a) = f'(a)$$
>
> we ensure that the difference ratio $F(x)$ is a continuous function.

This is essentially the definition of derivative reworded without the limit notation; the limit is there implicitly through the removable singularity.

This way of looking at the derivative highlights the fact that *in finding the derivative of $f(x)$ we are not taking the limit of $f(x)$ but of its difference quotient $F(x)$.*

As a bonus, we also discover the precise manner in which the infinity is incorporated in the derivative concept; specifically, the derivative exists as a finite quantity when the singularity in $F(x)$ is removable; which is like saying that the infinity is removable from the difference quotient!

To illustrate the above discussion visually, let's go back to the functions $f_1(x), f_2(x)$ and $f_3(x)$ that we discussed earlier and examine them in this light. For $f_1(x)$ that is graphed in Figure 7.4 if we draw secant lines that pass *through the origin* and a point x on the left half of the curve then the slopes of these secant lines jumps to infinity as x reaches the origin. As soon as x moves past the origin and goes on to the right half, the slope of the corresponding secant lines drop down to a finite number again. The difference quotient function in this case is

$$F_1(x) = \frac{x^{1/3} - 0}{x - 0} = \frac{1}{x^{-1/3}x} = \frac{1}{x^{2/3}}$$

This gives the numerical values of the slopes of secant lines mentioned above. This function does not have a removable singularity at the origin.

For $f_2(x)$ the difference quotient function has a two-piece graph: the left half is $y = x$ and the right half is $y = 1/\sqrt{x}$. If you graph this two-piece curve using a graphics device then you see that its singularity is non-removable also. Similarly, the difference quotient of $f_3(x)$ is easy to draw (see Figure 7.8) and again you see that the jump discontinuity is not a removable singularity in this case either.

Beware: continuous doesn't imply differentiable!

Suppose that $f(x)$ is a function that has a derivative at $x = a$. Then we can approximate the slope $f'(a)$ of the tangent line by the slope of a secant line that is nearly aligned with the tangent; so for n large,

$$\frac{f(x_n) - f(a)}{x_n - a} \simeq f'(a)$$

By assumption, $x_n \to a$ as $n \to \infty$, so $x_n - a \to 0$. In order that the fraction on the left hand side above stay near the finite number $f'(a)$ it is also necessary that its numerator $f(x_n) - f(a)$ converge to zero as $n \to \infty$; otherwise, the fraction would become infinitely large and there would be no hope of approaching the number $f'(a)$. We conclude that:

> If $f'(a)$ exists and is a real number then $f(x_n) \to f(a)$ for every sequence x_n of real numbers that approaches a as $n \to \infty$. This is a necessary condition for the existence of derivatives; if it is not true then $f'(a)$ cannot exist.

In other words,

If $f(x)$ is differentiable at $x = a$ then $f(x)$ is continuous at $x = a$.

The converse of the above statement is not true, that is, *continuity is not a sufficient condition* for differentiability; when talking about how derivatives may fail to exist, we discussed three examples of continuous functions that are not differentiable at $x = 0$; see Figures 7.4-7.6.

Derivatives

Continuity is a weaker condition than the existence of derivative; so corners, sharp points or vertical tangent lines are all possible for continuous functions but not for differentiable ones.

The points at which a continuous function $f(x)$ is not differentiable may be singularities of the derivative function $f'(x)$. By switching from $f(x)$ to $f'(x)$ in our earlier definitions of singularity we see that new definitions are not necessary. However, certain singularities of the derivative have been given names for historic and semantic reasons.

For example, the origin in Figure 7.5 is sometimes referred to as a "corner" because the different slopes from opposite sides introduce a sharp corner. The singularity of the type that we discussed earlier in $x^{2/3}$ at $x = 0$ is called a "cusp".

It is easy to visualize continuous functions with a large number of corners, cusps etc. We can even imagine continuous curves that have infinitely many such singularities of the derivative that occur in the form of a well defined sequence, like having a cusp at every integer. But it is quite difficult to visualize a continuous curve whose derivative is nowhere defined.

If visualizing was how we decided on the existence of things in math then so such continuous curve could exist. But it so happens that such curves do exist; not only do they exist but they are in fact more abundant than curves having derivatives at some point! We present one such curve in Chapter 9 (Weierstrass's function) where we discuss the issue of differentiable, continuous functions a bit more.

Continuity and algebraic operations.

Continuity is essentially the existence of limits so the properties of limits that allow us to distribute limits over sums, products and quotients translate into similar properties for continuous functions. We summarize these properties as follows:

> **Addition and multiplication:** If the functions $f(x)$ and $g(x)$ are continuous at $x = a$ then so are their sum $f(x) + g(x)$, their product $f(x)g(x)$ and if $g(a) \neq 0$ then also their quotient $f(x)/g(x)$.

You may recall that another way to combine functions is to compose them: $g \circ f(x) = g(f(x))$. The following result is often useful in calculus.

> **Composition:** If the functions $f(x)$ is continuous at $x = a$ and $g(x)$ is continuous at $x = f(a)$ then their composition $g \circ f(x)$ is continuous at $x = a$. Therefore, the composition of two continuous functions is a continuous function.

This is easy to prove using the definition in (7.16). Start with an arbitrary (unspecified) sequence x_n that converges to a. Since $f(x)$ is continuous at $x = a$, it follows that $\lim_{n \to \infty} f(x_n) = f(a)$. Now relabel: $f(x_n) = y_n$ and $f(a) = b$. Since $g(x)$ is continuous at $x = b$ and we just showed that $\lim_{n \to \infty} y_n = b$ it follows that $\lim_{n \to \infty} g(y_n) = g(b)$ or equivalently, $\lim_{n \to \infty} g(f(x_n)) = g(f(a))$. In other words,

$$\lim_{n \to \infty} x_n = a \quad \text{implies} \quad \lim_{n \to \infty} g \circ f(x_n) = g \circ f(a)$$

which means that $g \circ f(x)$ is continuous at $x = a$.

> A continuous function has the nice operational property that
> $$\lim_{n\to\infty} f(x_n) = f(\lim_{n\to\infty} x_n) \qquad (7.20)$$
> as long as $\lim_{n\to\infty} x_n$ is a real number. An extremely useful way to interpret (7.20) is as *a rule that enables us to take a limit in or out of the function $f(x)$.*

We can do this only when $f(x)$ is continuous!

For instance, $f(x) = \sqrt{x}$ is continuous for all $x > 0$ (it is actually differentiable as a power function $x^{1/2}$). So if $\lim_{n\to\infty} x_n = 2$, say, then we can use this fact and (7.20) to calculate

$$\lim_{n\to\infty} \sqrt{x_n} = \sqrt{\lim_{n\to\infty} x_n} = \sqrt{2}.$$

You see that "we moved the limit inside the square root". We can't do this with discontinuous functions like the function $f(x)$ in (7.17), because as we saw earlier, the equality in (7.20) may or may not hold if $\lim_{n\to\infty} x_n = 0$ depending on the particular sequence x_n that is chosen (holds only if $x_n \geq 0$ for all n).

For the function $f(x)$ in (7.18), using the limit theorems for sequences we find that if $\lim_{n\to\infty} x_n = 1$ then

$$\lim_{n\to\infty} f(x_n) = \lim_{n\to\infty} \frac{x_n^2 - 1}{x_n - 1} = \lim_{n\to\infty} \frac{(x_n - 1)(x_n + 1)}{x_n - 1} = \lim_{n\to\infty} (x_n + 1) = 2$$

Therefore, if $f(1) \neq 2$ then the equality in (7.20) *never* holds for *any* sequence x_n that converges to 1.

The Intermediate Value Theorem: missing no points in-between!

At the beginning of this section we considered the function $f(x) = x^2 - 2$ and how the continuity of this function guarantees that its graph crosses the x-axis at $x = \sqrt{2}$. But the x-axis is not really special; consider this function on the interval $[0, 2]$ and note that $f(0) = -2$ and $f(2) = 2$. Looking back at Figure 7.7, if we draw a horizontal line $y = d$ where d is any number between -2 and 2 (not only $d = 0$) then we expect that this line crosses the graph of $x^2 - 2$ at some point because the graph is continuous. At that point, $f(x) = d$ that is, $f(x)$ takes the value d. This observation holds for all continuous functions defined on closed intervals.

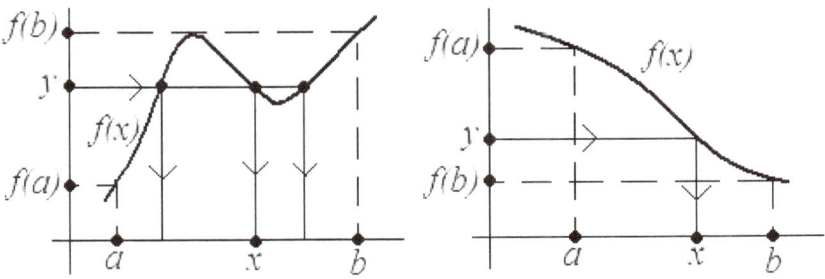

Figure 7.11: The intermediate value property of continuous functions

Derivatives

Figure 7.11 displays the graphs of two *continuous* functions. Each function takes the values $f(a)$ and $f(b)$ at each end of the interval and if $f(a) \neq f(b)$ then $f(x)$ also takes *every value $y = d$* between $f(a)$ and $f(b)$ for at least one value of the variable x.

Figure 7.11 illustrates the following fundamental property of continuous functions.

> **Intermediate Value Theorem (IVT):** *If a function $f(x)$ is continuous at every point of the interval $[a, b]$ and $f(a) \neq f(b)$ then for every number y between $f(a)$ and $f(b)$ there is a number c between a and b such that $y = f(c)$.*

You may be wondering why in Figure 7.11 the arrows go from the numbers on the y-axis to numbers on the x-axis rather than the other way around. To see why, consider the graphs in Figure 7.12; notice that in both graphs every value of x on the x-axis between a and b does correspond to at least one number between $f(a)$ and $f(b)$ on the y-axis and yet, both functions are discontinuous. Notice that there numbers on the y-axis that don't come from any numbers on the x-axis through $f(x)$.

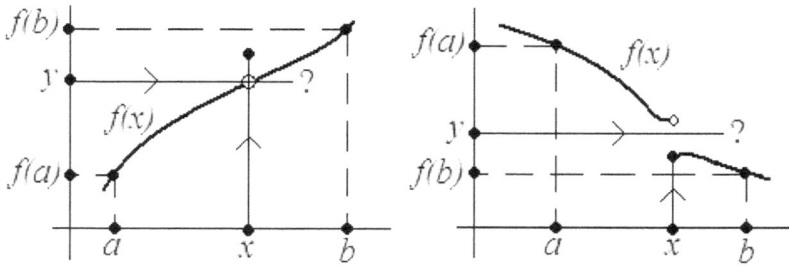

Figure 7.12: The intermediate value property does not hold if there is a discontinuity

In applied mathematics, the IVT is often used in conjunction with Newton's fast method for approximating roots or solutions of nonlinear equations. We discuss this application (which also involves the derivative) later in this chapter.

Although the Intermediate Value Theorem is intuitively obvious, its proof is not very clear;[11] As you may have gathered from the earlier discussion, the proof requires two important facts: the continuity of $f(x)$ and the completeness of the set $\mathbb{R}$ of real numbers. To see how the latter property comes into play, note that if we were using $\mathbb{Q}$ instead of $\mathbb{R}$ as the x-axis then the graph of $x^2 - 2$ would slip right through the hole where $\sqrt{2}$ would be without hitting any point of $\mathbb{Q}$; so the IVT would fail even though $x^2 - 2$ is continuous!

The Extreme Value Theorem: let there be max and min!

Imagine that we have a bounded function $f(x)$ on a set S so $f(x)$ does not grow infinitely large in magnitude. Let m be the absolute minimum of $f(x)$ so that the graph of $f(x)$ comes down as low as m as x ranges over S but no lower; see Figure 7.13.

---

[11] The Intermediate Value Theorem is a simple corollary of a topological result that *continuous functions map (or transform) connected sets into connected sets.* In the one-dimensional space $\mathbb{R}$ with the Euclidean (or absolute value) metric that we are working with here, connected sets are precisely the intervals.

202   Infinite Resolution

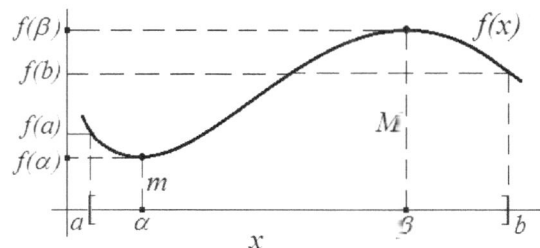

Figure 7.13: Illustrating the extreme value property

Similarly, let M be the absolute maximum of $f(x)$ so that the graph reaches up to M as x ranges over S but goes no higher. The numbers m and M are the extreme values of $f(x)$. It is reasonable to conclude that the lowest value m must be associated with some number α in S and the highest M to some β in S so that $f(\alpha) \leq f(x) \leq f(\beta)$ for all x in S see Figure 7.13.

This scenario is intuitively plausible but not always correct. To see what may be missing consider the functions shown in Figure 7.14.

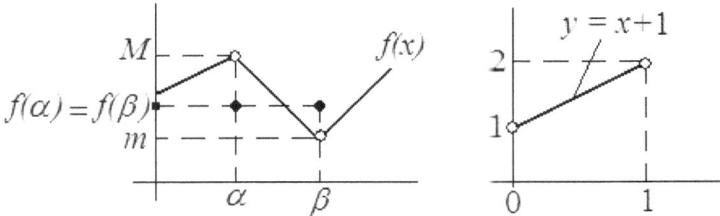

Figure 7.14: Cases where the Extreme Value Theorem does not hold

The left panel in Figure 7.14 shows a simple, bounded function over an interval. This function has both a maximum value M and a minimum value m. But *there are no points in* the interval that correspond to its lowest and highest values! The values of the function shown at the numbers labeled α and β in Figure 7.14 are different from both M and m. This discrepancy was caused by the discontinuities at the two points shown.

It seems from Figure 7.14 that the highest value M is never really attained by $f(x)$ since there is a hole at the peak. But this high value is reached in the limit. To see how, imagine a sequence x_n on the x-axis that approaches α from either the left or the right and track the function values $f(x_n)$ on the graph. We see that these numbers do in fact approach M in the sense that

$$\lim_{n \to \infty} f(x_n) = M$$

If, say, $M = 12$ then there are index values n so large that $f(x_n)$ is larger than $11.999\ldots 9$ for any number of 9's after the decimal point. A similar observation holds for m.

The assumption of continuity alone is also not enough; consider the line $y = x + 1$ for $0 < x < 1$ that is shown in the right panel of Figure 7.14 over the open interval $(0, 1)$. This function is

Derivatives

continuous for all x, its lowest value $m = 1$ occurs (in the limit) at $x = 0$ and its highest value $M = 2$ at $x = 1$. But neither 0 nor 1 is in $(0,1)$; here the problem is with the missing endpoints. These observations explain the assumptions in the next theorem.

> **Exterme Value Theorem (EVT):** If $f(x)$ is continuous on a closed interval $[a, b]$ then $f(x)$ achieves it maximum and minimum values at points in $[a, b]$. More precisely, there are α, β in $[a, b]$ such that for all x
> $$f(\alpha) \leq f(x) \leq f(\beta)$$

The EVT as stated above can be proved using either the Nested Intervals Property (NIP) or the Least Upper Bound Property for Sets (LUBS). Note that both of these results require the completeness property.[12]

The EVT is an important property of continuous functions on closed intervals. It is used to establish the existence of maxima and minima of functions and we also use it in our study of the integral.

7.6 Newton's method: fast convergence with a risk of singularities

When studying mathematical models in science and engineering we often find it necessary to solve nonlinear equations. In a one-dimensional context, the equation looks like

$$f(x) = 0 \qquad (7.21)$$

where $f(x)$ is some *nonlinear* function, i.e. its graph is *not* a straight line. We will be looking for a value of x that makes the value of the function on the left hand side zero, so we call the solution a *zero* (or a *root*) of $f(x)$. Here is a concrete example of a nonlinear equation:

$$\cos x = x \quad \text{or equivalently,} \quad x - \cos x = 0 \qquad (7.22)$$

The zeros of this equation are the x-coordinates of points where $f(x) = x - \cos x = 0$ (equivalently, where $y = \cos x$ intersects the line $y = x$).

Note that $f(x) = x - \cos x$ is continuous (everywhere) since it is a difference of two continuous functions (the power function and the cosine). Further,

$$f(0) = 0 - \cos 0 = -1 < 0, \qquad f\left(\frac{\pi}{2}\right) = \frac{\pi}{2} - \cos\left(\frac{\pi}{2}\right) = \frac{\pi}{2} > 0$$

so the IVT implies that there is a zero z between 0 and $\pi/2 \simeq 1.57$; see Figure 7.15.

Can we find the *exact value* of the solution z?

The answer is generally *no* because z is usually irrational. This is true typically of nonlinear equations not just for the one in (7.22). When faced with infinitely many digits and no connection to familiar numbers like π or $\sqrt{2}$ our theories are as powerless as our computing devices in helping us scale this infinitely tall barrier!

---

[12]The EVT is also a simple corollary of a general topological fact about *compact sets*. Closed intervals are compact sets.

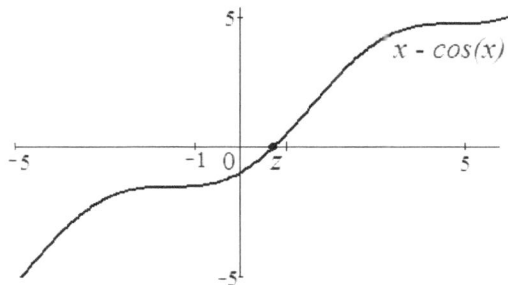

Figure 7.15: The x-intercept of a smooth curve

But although we cannot find the exact solutions *we can approximate them as closely as we want* by finding as many decimal figures as may be feasible, depending on what we want to do with the results. There is a great deal of information that our theories, enhanced by our technology do provide in this direction.

Newton's method[13] is not only a powerful and practical calculation tool, it also happens to be a topic where the infinity may occur in a hidden and unexpected way. We see in this section that infinite singularities may arise in Newton's method even though none is present in the function $f(x)$, which is why I call them *hidden*. We also explore the potentially adverse, chaotic effects of these infinities on the results of calculation if we are not careful enough to avoid them.

In this section we discuss both the basics of the method and the role of infinity in it. Although there are approximation methods without infinite singularities, most of them are slower to converge to a solution than Newton's method. Working with the infinity may have some risks but it comes with nice rewards in this case.

The Newton-Raphson recursion.

Let's go back to (7.21). We will be using tangent lines, so we assume that $f(x)$ is differentiable, at least near z. Pick a point x_1 near z and calculate the equation of the tangent line at $x = x_1$ as we discussed earlier in this chapter,

$$y = f(x_1) + f'(x_1)(x - x_1)$$

This line crosses the x-axis when $y = 0$; we label this crossing point x_2 so that

$$f(x_1) + f'(x_1)(x_2 - x_1) = 0$$

See Figure 7.16.

When we solve this equation for x_2 we get

$$x_2 = x_1 - \frac{f(x_1)}{f'(x_1)}$$

---

[13] The method was introduced by Newton in 1669 but was not published by him. John Wallis did that for him in 1685. The improved version that we discuss here is due to Joseph Raphson (1648-1715) who published it in 1690. Although this is an old result, special cases of it were known much earlier, like the divide and average rule for estimating $\sqrt{2}$ that we discussed before. The method of Newton and Raphson itself is a special type of one-dimensional *discrete dynamical system*, one among many such systems that are collectively known as *fixed point methods*.

Derivatives

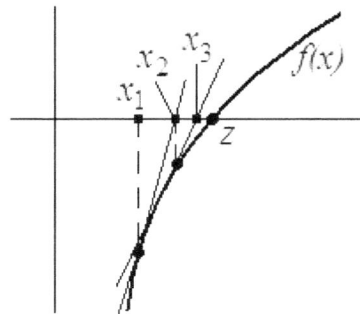

Figure 7.16: Illustrating the Newton-Raphson method

Now that we have found x_2 which is closer to z than our initial guess x_1 was, we discard x_1 and repeat the above process with x_2 instead. Find the tangent line $x = x_2$ and calculate the point x_3 where this line crosses the x-axis which comes out to be

$$x_3 = x_2 - \frac{f(x_2)}{f'(x_2)}$$

Repeating n times, we get

$$x_{n+1} = x_n - \frac{f(x_n)}{f'(x_n)}, \quad n = 1, 2, 3, \ldots \quad (7.23)$$

This is the *Newton-Raphson recursion*.

The most basic question about using (7.23) is *where to begin*, or what to pick as x_1 as our initial guess. There is no general rule of thumb for selecting this initial value; I will discuss some helpful facts a little later, but for now let's use (7.23) to find the solution z of (7.22) correct to 6 decimal places.

First, we need to calculate $f'(x)$. Using the derivative formulas discussed above, we get

$$f'(x) = 1 + \sin x$$

Now, let $x_1 = 0$, for no particular reason other than the fact that it is a nice round number that is not far from z in Figure 7.15. Then

$$x_2 = 0 - \frac{f(0)}{f'(0)} = -\frac{0 - \cos 0}{1 + \sin 0} = 1$$

This in turn gives

$$x_3 = 1 - \frac{1 - \cos 1}{1 + \sin 1} \simeq 0.75036387$$

Here I used a basic calculator and kept one or more decimal places beyond the desired level of accuracy, which we set as 6 decimal places. The chart below shows the values that I generated by carrying out a couple of additional steps of this process:

n	1	2	3	4	5	6
x_n	0	1	0.75036387	0.73911289	0.73908513	0.73908513

Since the first 6 decimal places do not change going from step 5 to step 6, we conclude that to 6 decimal places,[14]

$$z \simeq 0.739085$$

is the solution of the equation in (7.22) which was obtained in just 5 steps starting from $x_1 = 0$. Feel free to check this in (7.22) with your own calculator or computer (in radian mode). If you do the check then try using different initial values also; for example, $x_1 = 1/2$.

In the next exercise you can find the zeros of a nonlinear function $f(x)$ both analytically and using Newton's method.

Exercise 96 *(a) Write the nonlinear equation $x^3 = 2x + 1$ as $f(x) = 0$; consider sketching a graph of your $f(x)$ between $x = -2$ and $x = 2$ to see that it has three crossings with the x-axis, namely, the zeros of $f(x)$.*

(b) Calculate the derivative of your $f(x)$ in (a) and figure out the Newton-Raphson recursion;

(c) Use the recursion in (b) to calculate the positive zero of $f(x)$ accurate to 6 decimal places with $x_1 = 1$.

(d) Verify by direct substitution that $f(-1) = 0$ so that -1 is a zero of $f(x)$ too. If you remember how to divide polynomials then divide $f(x)$ by $x + 1$ to get the quotient $x^2 - x - 1$; so $f(x) = (x+1)(x^2 - x - 1)$. Now use the quadratic formula from algebra to find the roots of the quotient and get the exact values of the other two zeros of $f(x)$ as:

$$\frac{1 \pm \sqrt{5}}{2}$$

The positive one of these is the "golden ratio" φ, one of the most famous irrational numbers known. The negative one is the conjugate of φ that we denote by $\bar{\varphi}$. A calculator can now independently verify the accuracy of your result in (c).[15]

We will return to the above exercise later for additional interesting ideas and to uncover a hidden infinity that is responsible for unpredictable outcomes!

Estimating roots very fast!

A practical and simple application of Newton's method is to estimate roots of real numbers to a high degree of accuracy in a few iterations. Let r be any positive real number and m an integer greater than 1. The m-th root of r, that is, $\sqrt[m]{r} = r^{1/m}$ is then a root of the polynomial

---

[14] Strictly speaking this is not always a valid criterion for ending the iteration process because it is possible that after several steps the numbers will shift again and approach some other value. But given the estimated location of z in this example, we can be confident that we have reached the desired accuracy in this case.

[15] You can also do it without using calculators. If you estimate $\sqrt{5}$ to seven decimal places as a solution of $x^2 - 5 = 0$ using the Newton-Raphson method then you can find φ accurate to at least 6 decimal places without needing a calculator! Remember that computing devices use algorithms like the Newton-Raphson method or infinite series; they don't know irrational numbers!

Derivatives

$f(x) = x^m - r$. The derivative is $f'(x) = mx^{m-1}$ so the Newton-Raphson recursion for this function is:

$$x_{n+1} = x_n - \frac{x_n^m - r}{mx_n^{m-1}} = \frac{(m-1)x_n^m + r}{mx_n^{m-1}} \tag{7.24}$$

In the special case $m = r = 2$ (7.24) can be written as:

$$x_{n+1} = \frac{x_n^2 + 2}{2x_n} = \frac{1}{2}\left(x_n + \frac{2}{x_n}\right)$$

The right hand side of the above recursion is easy to remember as the divide and average rule for estimating $\sqrt{2}$ that we discussed earlier.

Let's use (7.24) to estimate $\sqrt[3]{2}$ correct to 6 decimal places. Here $m = 3$ and $r = 2$. Pick $x_1 = 1$ (a convenient number not far from the target $\sqrt[3]{2}$ but any positive value works). Then

$$x_2 = \frac{2x_1^3 + 2}{3x_1^2} = \frac{4}{3} \simeq 1.3333333$$

I rounded the answer to 7 decimal places. Next,

$$x_3 \simeq \frac{2(1.3333333)^3 + 2}{3(1.3333333)} \simeq 1.2638889$$

Using similar computation, we find that

$$x_4 \simeq 1.2599335, \qquad x_5 \simeq 1.2599211, \qquad x_6 \simeq 1.2599211$$

As a check we see that multiplying the last number above three times yields the following result:

$$1.2599211^3 = 2.00000024$$

which differs from 2 by less than $0.000001 = 10^{-6}$. It is worth mentioning that you can do the above calculations without a calculator, or at most, using a basic calculator that can only add, multiply and divide numbers. Further, all of the approximations are rational numbers.

Exercise 97 *Use the recursion in (7.24) with the initial value $x_1 = 1$ to estimate each of the following roots correct to at least 6 decimal places:*

$$\sqrt{5}, \qquad \sqrt[5]{2}$$

Where to begin?

If we sketch the graph of a function $f(x)$ using a computing device then we can often pick a point that is close to a zero of $f(x)$ as the starting point of iteration. Newton's method then usually gives great accuracy in a relatively small number of iterations.

If $f(x)$ is continuous then a zero exists by the Intermediate Value Theorem (IVT) provided that there are numbers a and b such that $f(a)$ *and* $f(b)$ *have opposite signs*. For instance, in the case of (7.22) as we saw previously,

$$f(0) = -1 < 0, \qquad f\left(\frac{\pi}{2}\right) = \frac{\pi}{2} > 0$$

so there is a zero of $f(x)$ between $x = 0$ and $x = \pi/2$. Likewise, for $f(x) = x^3 - 2x - 1$ in Exercise 96

$$f(1) = -2 < 0, \qquad f(2) = 3 > 0$$

so there is a zero of $f(x)$ between $x = 1$ and $x = 2$.

The bisection algorithm. The use of IVT above is not sufficient to ensure the existence of a *unique* solution unless a and b are so close that there are no other zeros of $f(x)$ between them.[16] But it *is* possible to narrow things down without graphing $f(x)$. For example, getting back to (7.22) we calculate

$$f\left(\frac{\pi}{4}\right) = \frac{\pi}{4} - \cos\frac{\pi}{4} = \frac{\pi}{4} - \frac{1}{\sqrt{2}} \simeq 0.08 > 0$$

which means that the zero z is between 0 and $\pi/4 \simeq 0.79$. This idea is sufficiently useful to be called the *bisection method* or *bisection algorithm* for solving equations! Further, the bisection method does not require that $f(x)$ have a derivative and it always works if $f(x)$ is just continuous. Why then bother with Newton's method? Because the bisection method is much slower in estimating the zero z to a desired accuracy than Newton's method is. Also, during the bisection process, we may inadvertently pass by a very close estimate of the zero; for instance, in the above discussion, $\pi/4$ is rather close to z but if we continue the bisection process by going to half of the interval from 0 to $\pi/4$ we get $\pi/8 \simeq 0.45$ which has moved away from z. In practice the bisection method is sometimes used to just get a number close enough to z to serve as x_1 in the Newton-Raphson recursion.

So, why do we need to get close to z to use Newton's method?

Beware of the hidden infinity!

For the Newton-Raphson method to work it is essential that x_1 be "close enough" to z. But how close is that? In practice, a "good" sketch of the graph of $f(x)$ using a computing device is usually all that we need to guess a suitable value for x_1. But the graph of $f(x)$ sometimes doesn't show all that is relevant!

Looking at the graph of $f(x) = x - \cos x$ in Figure 7.15 nothing jumps out as odd; all is nice and smooth with not a trace of abnormalities. But looking at the wider picture, in the Newton-Raphson recursion itself we spot a division by the derivative $f'(x_n)$. If this quantity gets close to 0 for some n while $f(x_n)$ is not near 0 then the next value x_{n+1} spikes and goes off track.

There are a number of places on the graph of $f(x) = x - \cos x$ in (7.22) where this occurs; in Figure 7.15 the places where the curve is flat (with horizontal tangent line) are where the derivative

---

[16] For instance, recall that the function $\sin(\pi/x)$ is continuous if $x > 0$ and has an infinite number of tightly packed zeros between 0 and 1.

Derivatives

is near 0 or equal to it. We can be more precise in this case: since $f'(x) = 1 + \sin x$ we calculate that $f'(x) = 0$ where $\sin x = -1$, that is,

$$x = -\frac{\pi}{2} + 2k\pi = (4k-1)\frac{\pi}{2}, \quad \text{all } k \text{ in } \mathbb{Z} \tag{7.25}$$

We see two of these points in Figure 7.15. *These points are infinite singularities of the Newton-Raphson recursion; flatness in $f(x)$ may be a sign of trouble!*

Figure 7.17 illustrates what happens to the recursion function

$$F_{NR}(x) = x - \frac{f(x)}{f'(x)}$$

near these two points. We call $F_{NR}(x)$ the *Newton-Raphson recursion function (or recursion map)*.

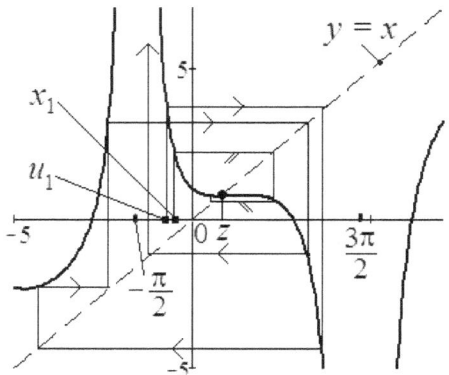

Figure 7.17: Exposing hidden infinities in the Newton-Rhaphson method

In Figure 7.17 we see a part of the graph of the recursion function, that is,

$$F_{NR}(x) = x - \frac{x - \cos x}{1 + \sin x} = \frac{x \sin x + \cos x}{1 + \sin x} \tag{7.26}$$

that consists of three separate pieces and contains z and two singularities of $F_{NR}(x)$. The sequence generated by the iteration of $F_{NR}(x)$ may jump from one piece to another, if it does not actually land on one of the singularities. If we start from x_1 shown in Figure 7.17 then $x_2 = F(x_1)$ is obtained by moving vertically from x_1 to the curve then horizontally until we reach the identity line $y = x$. This is where x_2 is on the x-axis. Starting from this point, we get to $x_3 = F(x_2)$ again by moving vertically to the curve, then horizontally to $y = x$ and so on. Repeating this process gives the successive values

$$x_2 = F_{NR}(x_1)$$
$$x_3 = F_{NR}(x_2) = F_{NR}(F_{NR}(x_1))$$
$$x_4 = F_{NR}(x_3) = F_{NR}(F_{NR}(F_{NR}(x_1)))$$
$$\vdots$$

Figure 7.17 shows that we get close enough to z in just three iterations that it is hard to show further approximations x_n distinctly from z.

By contrast, the iterations starting from the initial point u_1 that is near x_1 result in a series of numbers that do not approach z; if you follow the cobweb of arrows $\longrightarrow$ from u_1 on the x-axis to the curve, then to $y = x$ to get u_2, then back to the curve again and repeat then you see that the path can move far from z and there is no reason to think it will ever approach it.

The following table lists the results obtained using the Newton-Raphson method with a few starting points from -0.75 to -0.8, about halfway between 0 and the nearest infinite singularity at $-\pi/2 \simeq -1.57$. As you can see there is significant unpredictability about the outcome even though the graph of $x - \cos x$ is perfectly smooth and showing no hint of infinities of any kind in this range! *This curve is not even flat between -0.75 and -0.8!*

x_1	Outcome
-0.75	reaching 0.739 in about 215 iterations
-0.76	reaching 0.739 in about 7 iterations
-0.77	iteration values exceed 10^{12} in 76 iterations
-0.78	iteration values exceed 10^{12} in 31 iterations
-0.79	reaching 0.739 in about 9 iterations
-0.80	iteration values exceed 10^{12} in 110 iterations

This table shows that for Newton-Raphson method to work it is important to avoid potential singularities in $F_{NR}(x)$ by starting sufficiently close to the solution. When starting close enough, Newton's method converges quite rapidly to the solution owing to the fact that $F_{NR}(x)$ has a slope of zero at $x = z$.

Exercise 98 *You can uncover the hidden infinities of Newton's method and explore their effects by yourself in this simple exercise that continues the study that we began in Exercise 96. If you worked out that exercise then you already have a good deal of information that you can use here too.*

(a) Set the derivative of $f(x) = x^3 - 2x - 1$ equal to zero and solve $f'(x) = 0$ to calculate the exact values of the infinite singularities of the Newton-Raphson function $F_{NR}(x)$.

(b) Find $F_{NR}(x)$ and sketch its graph carefully so you can visually identify its two infinite singularities as well as its crossings with the identity line $y = x$, i.e. the zeros -1, $\bar{\varphi}$ and φ of $f(x)$.

(c) Use your figure in (b) to pick a good initial value to estimate the negative zero $\bar{\varphi}$ accurately to 6 decimal places. Why did you pick that particular initial value? Why can you not use -1 as an initial value? Consider using the same reasoning to find a range of possible initial values that will always lead to φ.

(d) It would seem reasonable that an initial value between -1 and $\bar{\varphi}$ should generate a sequence that converges to either -1 or to $\bar{\varphi}$. But the initial value $x_1 = -0.81$ quickly takes you to the positive zero φ of $f(x)$! You can check this out by iterating $F_{NR}(x)$ a few times; how would you explain this unexpected outcome?

It must be said that *not all functions hide infinite singularities* in the Newton-Raphson recursion. For instance, a slightly altered version of the function in the above Exercise is $f(x) = x^3 + 2x - 1$ whose derivative $f'(x) = 3x^2 + 2$ is always positive (in fact, never less than 2) so the associated

Derivatives

function $F_{NR}(x)$ has no infinite singularities. The IVT shows that this function has a zero between $x = 0$ and $x = 1$. You can figure out the zero of this function ($z \simeq 0.45$) very quickly.

Things can go off track even when there are no infinite singularities. Take a slightly different version of the above function, that is, $f(x) = x^3 + ax - 1$ where a is positive but tiny; say, 0.001. Then the derivative $f'(x) = 3x^2 + a$ is still positive but its value is a when $x = 0$ so there is a large spike at the origin, even if not an infinite one. With the initial value $x_1 = 0$ the next value comes out as

$$x_2 = 0 - \frac{f(0)}{f'(0)} = -\frac{-1}{a} = \frac{1}{a} = 1000$$

Since the actual zero z of $x^3 + 0.001x - 1$ is between $x = 0$ and $x = 1$ choosing $x_1 = 0$ is not outlandish; yet, $x_2 = 1000$ is nowhere near z. The graph of $f(x)$ in this case is nearly flat near $x = 0$, a warning to expect a little fireworks if not careful with the initial guess.

7.7 The Mean Value Theorem and the shapes of functions

When discussing the measurement of speed for a freely falling body, we saw that the real data is always discrete and allows only computing the average velocity. Our measurements show that the body is moving ever faster as it goes down but only through snapshots. But we believe that the falling body is moving *continuously* in-between the measurements, so if the *average* velocity from, say, $t = 1$ to $t = 2$ is 14.7 m/s then its *instantaneous* velocity is smaller at $t = 1$ than at $t = 2$. Based on these criteria, we conclude that *at some point in time between 1 second and 2 seconds the object is moving at exactly* 14.7 *m/s*. For example, if in Galileo's equations we set the velocity equal to 14.7 then $14.7 = 9.8t$. So the time instance at which the object has this exact velocity is $t = 14.7/9.8 = 1.5$ seconds.

The same reasoning applies to any time interval as long as the motion of the falling body is not altered. How do we know this? And more importantly, how do justify it for other, more complex types of motion than free fall?

The Mean Value Theorem.

Suppose that a function $f(x)$ is differentiable (has a derivative) at every point of some interval $a < x < b$. Then $f(x)$ is continuous over the same interval, since differentiability is a stronger requirement than continuity. Let's also assume that $f(x)$ is continuous at the two points a and b so that its value does not suddenly jump up or down when $x = a$ or $x = b$. Figure 7.18 illustrates such a configuration.

The variation, or the change in $f(x)$ as x goes from a to b is $f(b) - f(a)$ and of course, the change in x is $b - a$. If we divide these two changes then we get the ratio

$$\frac{f(b) - f(a)}{b - a}$$

You may recall that this ratio is the slope of the secant line that passes through the points $(a, f(a))$ and $(b, f(b))$ on the graph of $f(x)$ as shown in Figure 7.18.

It is not hard to see that there must be at least one point on the arc of the curve in Figure 7.18 at which the tangent line has the same slope as the secant line (two such points exist in Figure 7.18

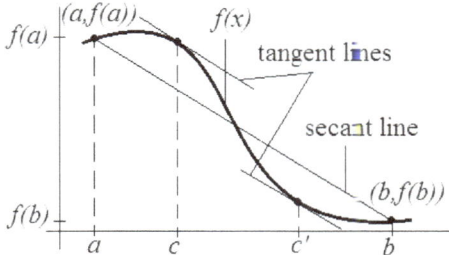

Figure 7.18: Illustrating the mean value property

that are indicated as c and c'). Having the same slopes at some point means that *the tangent line is parallel to the secant line* at that point. If the x-coordinate of the point is $x = c$ then

$$\frac{f(b) - f(a)}{b - a} = f'(c) \qquad (7.27)$$

You can see why we need differentiability: there must be a tangent line for the above discussion to make sense! I summarize the above discussion as follows:

> **The Mean Value Theorem (MVT):** *If a function $f(x)$ is differentiable for $a < x < b$ and also continuous for $a \leq x \leq b$ then there is a number c between a and b such that (7.27) holds.*

This verifies our earlier hunch that the speed 14.7 m/s being attained at a time instant c between $t = 1$ and $t = 2$. Specifically, if $x(t)$ is the position function of the freely falling body then

$$\frac{x(2) - x(1)}{2 - 1} = v(c)$$

The name "mean value" is rooted in this connection between average or mean velocity and the instantaneous velocity.

The proof of the MVT is not difficult but requires some rather lengthy preliminaries that you can find in texts on real analysis or advanced calculus. Like the proof of the IVT and other fundamental results involving the real numbers, this proof requires the completeness of the set of real numbers; otherwise, there might be gaps in the interval $a < x < b$ where the sought-after c might fall through the cracks, so to speak.

It is easy to verify the MVT in simple examples; let's find the point (or points) in the interval $-2 < x < 2$ where the function $f(x) = x^3 - 2x - 1$ satisfies the MVT. We are looking for one or more points that satisfy (7.27), that is,

$$\frac{f(2) - f(-2)}{2 - (-2)} = f'(c), \quad -2 < c < 2$$

$$\frac{3 - (-5)}{4} = 3c^2 - 2$$

Derivatives

The last equality above reduces to the equation $3c^2 = 4$ which is easy to solve; we get:

$$c = \pm \frac{2}{\sqrt{3}} \simeq \pm 1.155$$

See Figure 7.19 which shows a sketch of the graph of $f(x)$, the secant line through the points $(-2, -5)$ and $(2, 3)$ and the two tangent lines that are parallel to it. Both $c_+ = 2/\sqrt{3}$ and $c_- = -2/\sqrt{3}$ are between $x = -2$ and $x = 2$ so both of them are valid choices for c in (7.27).

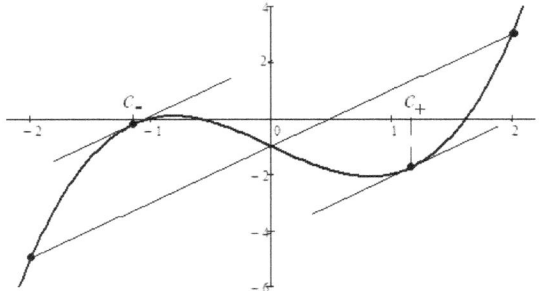

Figure 7.19: Verifying the mean value property in a specific case

---

Exercise 99 *For the function $f(x) = x^3 - 2x - 1$ find the (only) number c that satisfies (7.27) for $0 < x < 3$. Note that your c will not be the same as c_+ above.*

If the derivative of $f(x)$ fails to exist *even at a single point* between $x = a$ and $x = b$ then the MVT may be totally false!

Consider the function $f(x) = |x|$ for $-1 < x < 2$. Recall that this function has a derivative at every real number x as long as $x \neq 0$; in fact,

$$f'(x) = \begin{cases} 1 & \text{if } x > 0 \\ -1 & \text{if } x < 0 \end{cases}$$

But since $f'(0)$ does not exist we cannot say that "$f(x)$ is differentiable on the interval $-1 < x < 2$". With $a = -1$ and $b = 2$ in (7.27) we get

$$\frac{f(2) - f(-1)}{2 - (-1)} = \frac{|2| - |-1|}{3} = \frac{1}{3}$$

This slope value is neither 1 nor -1 so there is no number c between -1 and 2 that satisfies (7.27); the MVT fails in this case because $f(x) = |x|$ does not have a derivative at the single point $x = 0$.

Is your function increasing or decreasing?

In a standard calculus textbook, or course, we discover that the derivative provides a good deal of information about the shapes of functions and their behaviors. This information often comes courtesy of the MVT. In this book we need not get into such details, fascinating as they are; but I discuss one important case to illustrate the role of the MVT in such results.

Recall that a function $f(x)$ is *increasing* (or *non-decreasing*) from $x = a$ to $x = b$ if for every pair of numbers u, v between a and b

$$u < v \quad \text{implies} \quad f(u) < f(v) \quad (\text{or } f(u) \leq f(v)) \tag{7.28}$$

Similarly, $f(x)$ is *decreasing* (or *non-increasing*) from $x = a$ to $x = b$ if for every pair of numbers u, v between a and b

$$u < v \quad \text{implies} \quad f(u) > f(v) \quad (\text{or } f(u) \geq f(v)) \tag{7.29}$$

In short, a function $f(x)$ is increasing (or decreasing) if an *increase* in the value of x causes the value of $f(x)$ to increase (or decrease). The graph of $f(x)$ in the xy-plane rises (or falls) as x moves *to the right*. If a function is always non-increasing or always non-decreasing then it is a *monotone function*.

It is often possible by sketching the graph to tell whether a function is increasing or decreasing. If a function is not monotone then a graph often tells us for what values of x the function is increasing and for what values it is decreasing, as in, say, Figure 7.19. But sometimes graphs are not convincing; for instance, in Figure 7.20 we see the graphs of the cubic polynomial $f(x) = x^3 + ax - 1$ for two different values of a. Both of the curves in this figure seem to be increasing or at least non-decreasing, with a flat portion in the middle. What is the difference, if any?

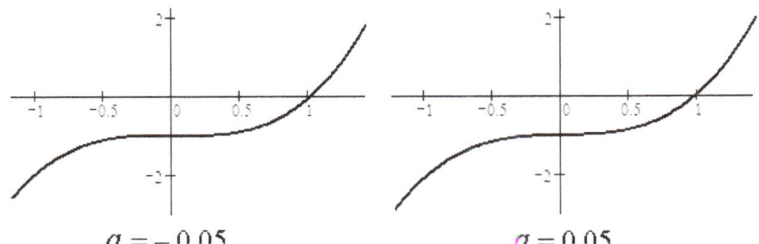

Figure 7.20: Are both of these curves increasing?

And how about the parts of the graphs *outside* the range shown in Figure 7.20? How can we be certain that neither curve changes direction if we enlarge the view?[17]

To answer these questions, let's use the definitions in (7.28) and (7.29). To check whether $f(x)$ is increasing we need to show that

---

[17] With a graphing device we might consider zooming in on each graph; any subtle differences would show up if we zoom in enough. But remember that the value of a can be arbitrarily small in principle; so small in fact that even the most sophisticated devices cannot resolve the difference between the graphs, since physical devices have finite resolution. Also, zooming out would cover larger intervals on the x-axis but since the x-axis has infinite extent we can never cover its entirety using a physical device.

Derivatives

$$u < v \quad \text{implies} \quad u^3 + au - 1 < v^3 + av - 1$$

for every pair of real numbers u, v. With a little algebra we find:

$$u^3 + au < v^3 + av$$
$$0 < v^3 - u^3 + av - au$$
$$0 < v^3 - u^3 + a(v - u)$$

Since by assumption $u < v$ it follows that $v - u > 0$ so the right hand side of the last inequality above is positive if $a \geq 0$. We see that $f(x)$ is increasing if $a \geq 0$.

The above argument shows that the curve in the right hand panel of Figure 7.20 is indeed increasing; but it says nothing useful about the curve in the left hand panel. Also the argument itself was cumbersome and it gets worse for more complicated functions.

But there is a clever way of dealing with this problem using the MVT. Write (7.28) in the equivalent form

$$u < v \quad \text{implies} \quad f(v) - f(u) > 0$$

By the MVT there is a number c between u and v such that

$$\frac{f(v) - f(u)}{v - u} = f'(c), \quad \text{or:} \quad f(v) - f(u) = f'(c)(v - u)$$

Since $v - u > 0$ we see that $f(v) - f(u) > 0$ if $f'(c) > 0$. A similar argument using (7.29) shows that $f(v) - f(u) < 0$ if $f'(c) < 0$.

For the function $f(x) = x^3 + ax - 1$ we have $f'(x) = 3x^2 + a$ so $f'(c) = 3c^2 + a$. Notice that $3c^2 \geq 0$ for any real number c so if $a > 0$ then $f'(c) > 0$. On the other hand, if $a < 0$ then for the values of c small enough it is possible to have $3c^2 + a < 0$.

For instance, if $a = -0.05$ as in the panel in the left hand side of Figure 7.20 then for small values of c like 0, -0.1, 0.1 we calculate $3c^2 - 0.05$ to get, respectively, 0, -0.02, -0.02. So if $u < 0 < v$ and the difference $v - u$ is smaller than 0.1 then a portion of $f(x)$ is decreasing when x is near 0.

This solves the problem, not only for the specific values of a in Figure 7.20 but for *all* values of a.

We also see that both functions in Figure 7.20 are increasing far from the origin since the derivative $f'(x) = 3x^2 + a$ is positive if the magnitude of x is large enough that $3x^2 + a > 0$, regardless of the sign of x. Therefore, we can be sure that the graph will not change direction at some far away point.

The part of the above argument using the MVT did not really depend on a specific function $f(x)$. So the same argument applies to any differentiable function $f(x)$ and gives us the following useful result.

> **Derivatives and the increasing and decreasing behavior:** Suppose that $f(x)$ is differentiable over an interval $a < x < b$. If $f'(x) > 0$ (or $f'(x) < 0$) for all x between a and b then $f(x)$ is increasing (respectively, decreasing) over the interval $a < x < b$.

A word of caution! To see how the above derivative criterion can be misunderstood, consider the reciprocal function $f(x) = 1/x$, $x \neq 0$. The derivative $f'(x) = -1/x^2$ is always negative so does it follow that $1/x$ is a decreasing function everywhere? Clearly $1/x$ is not decreasing everywhere; for example, $-2 < 1$ but $f(-2) = -1/2$ is less than $f(1) = 1$, *not greater* as required of a decreasing function. Can you explain this? We must read the statement of the criterion carefully: it says that $f(x)$ must be differentiable on *all* of the interval; $1/x$ is not differentiable on any interval that contains 0.

Also the *converse* of the above criterion is not true. For example, $f(x) = x^3$ is increasing on all of $(-\infty, \infty)$ but $f'(0) = 0$. A more severe example that we discussed earlier (see Figure 7.15) is $f(x) = x - \cos x$ which is also increasing on $(-\infty, \infty)$ but $f'(x) = 1 + \sin x = 0$ for all $x = -\pi/2 + 2\pi k$ where k is any integer.

7.8 What about the ε and δ?

When you browse through a standard calculus textbook, there is usually a chapter or a section devoted to limits. But the limits that you find there are different from what we discussed above. There is no mention of sequences usually until the chapter on infinite series where, of course, converging *sequences* of partial sums need to be defined as we have already seen.

In this section we discuss the type of limit that is used in a standard coverage of calculus for defining continuity and derivatives. We will also see that this concept is entirely equivalent to the one involving sequences that we discussed earlier: each one implies the other. You may skip this section without loss of continuity (no pun intended).

In order to allow taking limits of functions such as $F(x)$ in (7.19) we need to slightly generalize the concept beyond (7.16) and continuity. Consider the function $f(x)$ in (7.18) again. As long as $x \neq 1$

$$\frac{x^2 - 1}{x - 1} = \frac{(x-1)(x+1)}{x - 1} = x + 1 \tag{7.30}$$

The graph is that of the straight line $y = x + 1$ with a hole or gap where the point $(1, 2)$ would be. If x_n is any sequence that approaches 1 as $n \to \infty$ (but $x_n \neq 1$ for all n) then

$$\lim_{n \to \infty} \frac{x_n^2 - 1}{x_n - 1} = \lim_{n \to \infty} (x_n + 1) = 2$$

Note that this limit exists even though the fraction itself is not defined at $x = 1$. So we may define the limit of a function, whether it is continuous or not, as follows:

> A function $f(x)$ has a *limit* L at $x = a$ if for *every* sequence x_n such that $x_n \neq a$ for all n,
> $$\lim_{n \to \infty} x_n = a \quad \text{implies} \quad \lim_{n \to \infty} f(x_n) = L. \tag{7.31}$$

In the above definition you do not see any mention of $f(a)$ and indeed, the function $f(x)$ does not have to be defined at $x = a$. This is important not only for the function in (7.30) but also even more for the difference quotient $F(x)$ in (7.19) where $F(a)$ is not defined by (7.19). And as before, the term "every" is important when using sequences to present a fact about functions.

The ε again, but now paired with a δ.

Converging sequences convey a sense of nearness through the idea of approaching or approximation. But this is not the only way of quantifying "nearness". Let us take a closer look at the two

Derivatives

limits in (7.31). Going back to our limit definition in (7.31), these statements say that for every real number $\varepsilon > 0$ there is a positive integer N such that

$$|f(x_n) - L| < \varepsilon \quad \text{for all } n > N$$

provided that x_n is "close enough" to the point a; or, more precisely, provided that there is some (small enough) real number $\delta > 0$ [18] such that

$$|x_n - a| < \delta \quad \text{for all } n > N$$

We can streamline the above description by dropping any mention of sequences and just define:

> **The function limit.** The function $f(x)$ approaches the number L as x approaches the number a if for every real number $\varepsilon > 0$ there is a real number $\delta > 0$ such that
>
> $$|f(x) - L| < \varepsilon \quad \text{whenever} \quad 0 < |x - a| < \delta \tag{7.32}$$
>
> The customary notation that summarizes (7.32) is
>
> $$\lim_{x \to a} f(x) = L \tag{7.33}$$

This notation dispenses with sequences altogether and efficiently combines the two limits in (7.31) into a single limit.

You see that there is no explicit occurrence of the infinity in this definition. This does not mean that the infinity is not involved, only that its presence is concealed for the sake of efficiency.[19]

The side inequality $0 < |x - a|$ in (7.32) simply says that $x - a \neq 0$ or $x \neq a$; which says that we do not need to require $f(x)$ be defined at $x = a$. As we saw earlier, this is necessary for defining derivatives through the difference quotient (7.19). In particular, the derivative can be defined (as in the standard calculus books) using function limits:

$$f'(a) = \lim_{x \to a} \frac{f(x) - f(a)}{x - a}$$

In this definition it is essential that $x \neq a$ so that the quotient on the right hand side is defined. But when defining the continuity of a function $f(x)$ at $x = a$ where $f(a)$ is actually defined, we can relax and allow $x = a$. We say that $f(x)$ is continuous at $x = a$ if:

$$\lim_{x \to a} f(x) = f(a)$$

Or more precisely, for every real number $\varepsilon > 0$ there is a real number $\delta > 0$ such that

$$|f(x) - f(a)| < \varepsilon \quad \text{whenever} \quad |x - a| < \delta$$

You may want to carefully compare this statement with (7.32) above!

---

[18] We can use any other symbol instead of δ (other than ε of course, which is already used). The symbols ε and δ have simply become standardized; beyond that there is nothing special about them.

[19] There are more profound reasons beyond efficiency for using the function limit definition; for example, it is more intuitive geometrically, especially when dealing with spaces of dimension two or three. The idea also renders itself to an abstract topological definition of continuity that we can't discuss here because it leads us away from our main goals in this book.

The absolute value notation may not adequately convey the sense of proximity or "neighborhood" involved in the function limit idea, so let us look at the inequalities in (7.32) in a more geometric way, via intervals and diagrams. First, let's use the definition of absolute value to translate the inequalities as

$$-\varepsilon < f(x) - L < \varepsilon \quad \text{whenever} \quad -\delta < x - a < \delta \quad \text{and} \quad x - a \neq 0, \quad \text{or:}$$
$$L - \varepsilon < f(x) < L + \varepsilon \quad \text{whenever} \quad a - \delta < x < a + \delta \quad \text{and} \quad x \neq a \quad (7.34)$$

We read (7.34) as:

"$f(x)$ is between $L - \varepsilon$ and $L + \varepsilon$ whenever x is between $a - \delta$ and $a + \delta$ but $x \neq a$".

Now "between $L-\varepsilon$ and $L+\varepsilon$" simply means "in the interval $(L-\varepsilon, L+\varepsilon)$". Similarly, "between $a - \delta$ and $a + \delta$" means "in the interval $(a - \delta, a + \delta)$". The inequalities (7.34) are illustrated in Figure 7.21.

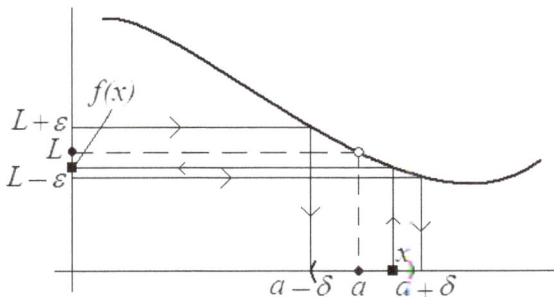

Figure 7.21: Illustrating the function limit

The arrows going from the y-axis to the x-axis in Figure 7.21 highlight the important fact that *ε is given and δ is determined by it*; further, if you look closely then you see that the symmetric interval $(L - \varepsilon, L + \varepsilon)$ on the y-axis is reflected onto a non-symmetric interval on the x-axis; this is typically the case. We can choose any δ small enough that the symmetric interval $(a - \delta, a + \delta)$ fits inside the non-symmetric interval, as shown in Figure 7.21.

We may now translate (7.34) in words as:

$f(x)$ is guaranteed to be in the (small) interval $(L-\varepsilon, L+\varepsilon)$, or neighborhood, of the number L if x is in the (small) interval $(a - \delta, a + \delta)$ or neighborhood of the number a. In other words, we can bring $f(x)$ as close to the number L as we may want simply by bringing the variable x close enough to a.

Function limits are equivalent to sequence limits.

Our next task is to show that the two concepts of limit, one using ε and δ and the other using ε and N do in fact say the same thing.

The function limit definition in (7.32) or (7.33) is just a different way of phrasing the sequential definition (7.31). The two definitions are equivalent.

Derivatives

Technically, we must show that (7.33) holds *if and only if* (7.31) does.

Let's first suppose that (7.33) is true and take x_n to be an arbitrary sequence that converges to a but $x_n \neq a$. We need to explain why x_n satisfies (7.31).

Since the distance between x_n and a is decreasing as $n \to \infty$ the sequence x_n will enter very interval neighborhood $(a - \delta, a + \delta)$ of a no matter how small δ is; so there is a large enough positive integer N such that $a - \delta < x_n < a + \delta$ for all $n > N$. Now (7.34) guarantees that $L - \varepsilon < f(x_n) < L + \varepsilon$ for all $n > N$, or in absolute value notation, $|f(x_n) - L| < \varepsilon$ for all $n > N$. This means that $\lim_{n \to \infty} f(x_n) = L$ so (7.31) is true!

Now let's prove the converse, by supposing that (7.31) is true and proving that (7.33) must be true; that is, we must show that for every $\varepsilon > 0$ there is $\delta > 0$ such that (7.32) holds. *We use a proof by contradiction*; suppose that no matter how small a $\delta > 0$ we choose, the distance $|f(x) - L|$ will exceed a positive number ε_0 for *some* x that satisfies $0 < |x - a| < \delta$.[20] I show that this assumption leads to a contradiction by constructing a sequence u_n that converges to a but the numbers $f(u_n)$ do not go near L.

Let's start with some convenient initial $\delta > 0$, say, $\delta = 1$. By our "suppose not" assumption, there is some real number u_1 in the interval $(a - 1, a + 1)$ such that

$$|f(u_1) - L| \geq \varepsilon_0 \quad \text{where } 0 < |u_1 - a| < 1$$

Next, pick a strictly smaller δ, say, $\delta = 1/2$ and find u_2 in the interval $(a - 1/2, a + 1/2)$ such that

$$|f(u_2) - L| \geq \varepsilon_0 \quad \text{where } 0 < |u_2 - a| < \frac{1}{2}$$

Note that u_2 may be equal to u_1; the important thing is that u_2 is in a smaller neighborhood of a, thus closer to it. We keep going in this direction; there is a number u_3 in the interval $(a - 1/3, a + 1/3)$ such that

$$|f(u_3) - L| \geq \varepsilon_0 \quad \text{where } 0 < |u_3 - a| < \frac{1}{3}$$

You may see where this is going now: for every index n there is a number u_n such that

$$|f(u_n) - L| \geq \varepsilon_0 \quad \text{where } 0 < |u_n - a| < \frac{1}{n}$$

The last inequality implies that $u_n \to a$ as $n \to \infty$ since $1/n \to 0$. So $\lim_{n \to \infty} u_n = a$. But no matter how large the index n gets, the inequality $|f(u_n) - L| \geq \varepsilon_0$ translates into

$$f(u_n) \geq L + \varepsilon_0 \quad \text{or} \quad f(u_n) \leq L - \varepsilon_0$$

Therefore, the distance between $f(u_n)$ and L is never less than ε_0. Of course, this contradicts our converse assumption that (7.31) is true and $\lim_{n \to \infty} f(x_n) = L$ for *every* sequence x_n that converges to a, *including* u_n. This conflict came about by assuming that (7.33) was false; it follows that (7.33) must be true if (7.31) is.

Which limit definition is preferable?

---

[20] This may be a good time to take a look back at the negations of statements that we discussed in the logic review section in Chapter 2.

Which of the equivalent definitions (7.31) or (7.33) we use depends on what our purpose is. In standard calculus texts function limits are usually preferred for a variety of reasons. In addition to the one that I already mentioned above, the function limit definition is easier to use for studying continuity in the Euclidean spaces of dimensions two and three that arise in multivariable calculus. Since our brains see these spaces as continuous, our intuition comes in handy in visualizing the function's behavior (as a patch of a surface, a cross section, a contour map, etc) in the neighborhood of a point. In two or three dimensions it is not easy to visualize or classify all possible sequences that converge to a point or specify the function's behavior over *all* such sequences.

On the other hand, the sequences approach to limits is easier and more natural to use when proving discontinuity since all it takes to do that is to *find just one sequence that violates* (7.16); and of course, sequence convergence arises in completely different contexts, like in the infinite series and their partial sums where sequences are essential.

In spaces of dimension greater than three our visual intuition is less availing so the function limit idea loses its intuitive, geometrical advantage. In spaces of infinite dimension, such as those encountered when studying the solutions of differential equations, sequences are indispensable as long as the space is not "too crowded".[21] In this book our stated goal is to highlight the occurrences of infinity as explicitly as possible; for this purpose, the sequence approach is preferable.

---

[21] In topological terms, the space must be "first countable" which basically means that at each point there is a countable collection of open sets (called neighborhoods) that characterize the local properties of space near the point. All finite dimensional Euclidean spaces and all infinite dimensional *metric* spaces, like the Hilbert spaces of square integrable functions encountered in quantum theory are first countable. In spaces without this property, as in some topological spaces of functions, there are points with so many neighborhoods that a sequence, being a countable set, is just not "infinite enough" to enter *every* neighborhood, as required for convergence. To characterize convergence and continuity in such spaces suitable abstract concepts are needed that we cannot discuss here. Suffice it to say that such spaces are locally too complex to be metrizable; i.e. we cannot define distances between distinct points.

Chapter 8

Integrals: *not just areas*

When introducing derivatives, we discussed how we might calculate the velocity of a moving object from its position data (or position function, where available). For freely falling bodies, this is likely how Galileo guessed his equations of motion from experiments that he conducted. Later on, Newton managed to derive the same equations from his laws of motion by essentially going backwards, in the opposite direction. Instead of calculating the velocity as the rate of change of position, he would derive the velocity from acceleration. Illustrating Newtonian ideas on free fall motion makes for an ideal start to this chapter because it shows how the idea of integration can come up naturally in studying an ordinary-life situation.

8.1 From acceleration to velocity to position to ... *area?*

Newton's gravitational force (of attraction) between two bodies was found to be

$$F = \frac{GMm}{R^2} \tag{8.1}$$

where G is a universal constant that is determined by experimental measurements, M and m are masses of the two bodies and R is the distance between the bodies.

For objects falling to the surface of the Earth, M is the mass of the Earth and R is Earth's radius plus the height h of the object above Earth's surface. Typically, h is much smaller than R, which is about 6,371 km (3,959 mi) so $R + h \simeq R$ (mass of the nearly spherical Earth can be assumed to be concentrated at its center). On the other hand, according to Newton's second law $F = ma$ where a is the acceleration of the object moving under the action of F so with this in (8.1) we get

$$ma = \frac{GMm}{R^2}$$
$$a = \frac{GM}{R^2}$$

The quantities G, M and R being well-known constants in physics, the quantity on the right hand side works out to about 9.8 meters per second squared (or 32 feet per second squared).[1] It

---

[1] In metric units, $G \simeq 6.674 \times 10^{-11}$ Nm^2/kg^2, mass of the Earth is $M \simeq 5.972 \times 10^{24}$ kg and the radius of the

represents the near-Earth acceleration and is denoted by the letter g so

$$g = \frac{GM}{R^2} \simeq 9.8 \text{ m/s}^2 = 32 \text{ ft/s}^2$$

Although not strictly constant, changes in the value of g are negligible near the Earth's surface.

Deriving the velocity function from acceleration.

Recall that acceleration is generally the rate of change of the velocity of a moving body. This is true for g too, so

$$g = \frac{\text{change in velocity}}{\text{change in time}} = \frac{\Delta v}{\Delta t}$$

In this case, we know what acceleration is; but how do we use this knowledge to find the velocity?

To find the velocity of a falling object at a certain time, say, t seconds after release, let's write $v(t)$ for velocity at time t and note that $v(0) = 0$ since we just drop the object starting from rest at time $t = 0$. Therefore,

$$g = \frac{v(t) - v(0)}{t - 0} = \frac{v(t)}{t}$$

If we solve this simple equation for $v(t)$ we get

$$v(t) = gt \tag{8.2}$$

This is Galileo's equation for the velocity of a falling object, derived from Newton's second law.

Deriving the position function from velocity.

In our discussion of derivatives we saw how to calculate the velocity if the position function were known. But in the previous subsection we derived the velocity from acceleration because the position function was not given. Why not use the same idea to calculate the position from the velocity function in (8.2) that we just figured out?

As above, we start with the fact that

$$v(t) = \frac{\text{change in position}}{\text{change in time}} = \frac{\Delta x}{\Delta t}$$

But unlike the constant acceleration g, velocity is *not* constant so the left hand side changes with time. To track the changes in position consider the following idea:

If we measure the velocity $v(t)$ over a small enough time period then its value does not change significantly from the start of the period to the end.

This observation takes us back to the situation in the last section where acceleration was constant. Here *the velocity is (almost) constant over a small time interval,* so we can repeat the above calculation; of course, we must do it over many successive short time intervals, one interval at a time, till we reach the desired time instant.

Earth is $R \simeq 6.371 \times 10^6$ m (averaged out over a comparatively small range of variations due to mountains, valleys, etc).

Integrals

So let's divide up (or partition) the time interval from 0 to t into N equal time-subintervals (each t/N seconds in duration) to get

$$t_0 = 0, \quad t_0 + \frac{t}{N} = t_1, \quad t_1 + \frac{t}{N} = t_2, \ldots, t_{N-1} + \frac{t}{N} = t_N = t \tag{8.3}$$

For example, if time is measured in seconds and $N = 100$ then each time unit is a hundredth of a second (the accuracy of a typical stopwatch). So if the object falls for 10 seconds then $t = 10$ and each subinterval has duration $t/N = 10/100 = 0.1$ second, so that

$$t_0 = 0, \quad t_1 = t_0 + 0.1 = 0.1, \quad t_2 = t_1 + 0.1 = 0.2, \ldots, t_{100} = t_{99} + 0.1 = 10 = t$$

For each $n = 1, 2, 3, \ldots, N$ let $v(t_n)$ be the velocity of the object n time-subintervals after the object is released. Now, over the time-subinterval $[t_{n-1}, t_n]$ the object moves a little bit, from $x(t_{n-1})$ to $x(t_n)$. If the duration $\Delta t = t_n - t_{n-1}$ is small (that is, *if N is large*) then the velocity changes very little from $v(t_{n-1})$ at the beginning of the short subinterval $[t_{n-1}, t_n]$. This reduces the approximation error in finding the velocity when using the formula:

$$\frac{x(t_n) - x(t_{n-1})}{t_n - t_{n-1}} \simeq v(t_{n-1}) \tag{8.4}$$

Over the brief time period $[t_{n-1}, t_n]$ the velocity is essentially constant and given by the number $v(t_{n-1})$. Let's remove the fraction and write this as

$$x(t_n) - x(t_{n-1}) \simeq v(t_{n-1})(t_n - t_{n-1}) \tag{8.5}$$

For $n = 1, 2, 3, \ldots, N$ this gives us a set of differences:

$$x(t_1) - x(t_0) = v(t_0)(t_1 - t_0)$$
$$x(t_2) - x(t_1) = v(t_1)(t_2 - t_1)$$
$$x(t_3) - x(t_2) = v(t_2)(t_3 - t_2)$$
$$\vdots$$
$$x(t_N) - x(t_{N-1}) = v(t_{N-1})(t_N - t_{N-1})$$

Adding all of the above, and canceling all equal terms with opposite signs on the left hand side gives

$$x(t_N) - x(t_0) \simeq v(t_0)(t_1 - t_0) + v(t_1)(t_2 - t_1) + \cdots + v(t_{N-1})(t_N - t_{N-1})$$

Since $t_N = t$ and $x(t_0) = x(0)$ we can rewrite the above as (using the summation symbol):

$$x(t) \simeq x(0) + \sum_{n=1}^{N} v(t_{n-1}) \Delta t \tag{8.6}$$

Further, we have $\Delta t = t/N$ and by (8.2),

$$v(t_{n-1}) = gt_{n-1} = g\frac{(n-1)t}{N} = gt\frac{n-1}{N}$$

and if we insert these quantities into (8.6) then

$$\begin{aligned}x(t) &\simeq x(0) + \sum_{n=1}^{N} gt\frac{n-1}{N}\frac{t}{N} \\ &= x(0) + \sum_{n=1}^{N} gt^2\frac{n-1}{N^2} \\ &= x(0) + gt^2(0) + gt^2\left(\frac{1}{N^2}\right) + gt^2\left(\frac{2}{N^2}\right) + \cdots + gt^2\left(\frac{N-1}{N^2}\right) \\ &= x(0) + gt^2\frac{1+2+3+\cdots+N-1}{N^2}\end{aligned} \qquad (8.7)$$

The sum in the numerator is an arithmetic sum and easy to calculate as follows: Let S be its total and notice that

$$\begin{array}{ccccccccccccc}2S &=& 1 &+& 2 &+& 3 &+& \cdots &+& N-2 &+& N-1 \\ &+& N-1 &+& N-2 &+& N-3 &+& \cdots &+& 2 &+& 1 \\ 2S &=& N &+& N &+& N &+& \ldots &+& N &+& N\end{array}$$

Therefore, $2S$ is just N added to itself $N-1$ times, or:

$$2S = N(N-1)$$
$$1 + 2 + 3 + \cdots + N - 1 = \frac{N(N-1)}{2}$$

Inserting this in (8.7) and setting $x(0) = 0$ (so that we measure the distance from the point where the object is dropped) we get:

$$x(t) \simeq gt^2\frac{N-1}{2N} \qquad (8.8)$$

We can improve the above approximation by choosing N larger, because larger N makes the duration $\Delta t = t/N$ between t_{n-1} and t_n smaller, and this in turn reduces the change in velocity from $v(t_{n-1})$ to $v(t_n)$. Taking the limit as $N \to \infty$ reduces Δt to zero and gives us a precise value for the position in (8.6) as:

$$x(t) = \lim_{N\to\infty} gt^2\frac{N-1}{2N} = gt^2\lim_{N\to\infty}\left(\frac{1}{2} - \frac{1}{2N}\right) = gt^2\left(\frac{1}{2} - \lim_{N\to\infty}\frac{1}{2N}\right) = \frac{1}{2}gt^2$$

We have just derived Galileo's equation of motion for falling bodies that I had promised in the previous chapter! In doing so, we only needed Newton's second law and his formula for the gravitational force; and the useful idea of starting by partitioning the time interval, calculating the distance (change in position) for each short time period, then adding our results and finally taking a limit. We have much more to say about this interesting three-step process below.

Caught on video! Experimentally verifying Galileo's equations.

You may have noticed that Galileo's equations do not contain the masses of objects. This is not intuitively obvious: consider that if dropped from, say, the top of the leaning tower of Pisa, then a

Integrals

bowling ball takes the same amount of time to hit the ground as a small rock. In the absence of air resistance, a feather falls just as fast as the bowling ball. Counter-intuitive as this may sound, given the mental conditioning that we receive from our daily experiences here on Earth, there have been experiments done in vacuum that verify this fact–*visually*!

One memorable test was actually done on *the Moon*! Astronaut David Scott, commanding the Apollo 15 mission in 1971, held a hammer in one hand and a feather in the other, then released them at the same time. A (grainy) video is available on the Internet that shows both objects hitting the ground at the same time.

A more recent experiment was performed dramatically by the physicist Brian Cox with the help of researchers at the NASA Glenn Research Center's Space Power Facility in Ohio where the world's largest vacuum chamber is located. There, a bowling ball and some feathers were released in a special chamber after the air was pumped out. Captured clearly on video (in slow motion!), the ball and the feathers are seen falling down *side by side* all the way to the bottom. It is well worth watching these amazing videos which may be accessible online; Galileo would have been overjoyed!

Area versus slope.

There is a geometric way of looking at the derivation of position from velocity that will prove to be very useful. In Figure 8.1 you see a graph of the velocity function $v(t) = gt$.

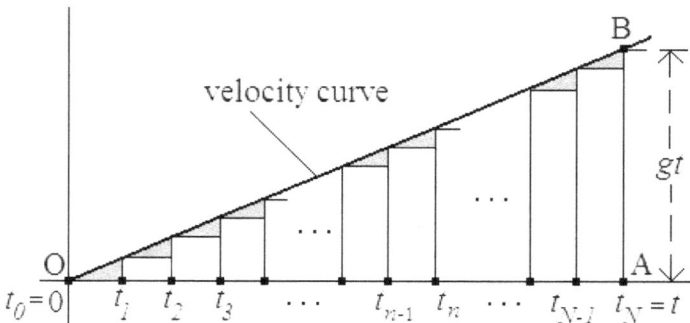

Figure 8.1: A partition of the area under the velocity curve

On the horizontal (time) axis in Figure 8.1 we see a partition of the interval from 0 to t into N subintervals of equal length. The n-th vertical strip of width Δt has area

$$v(t_{n-1})\Delta t = gt_{n-1}\Delta t = g\frac{(n-1)t}{N}\frac{t}{N} = gt^2\frac{n-1}{N^2}$$

This is recognizable as one term of the sum in (8.7). We see in Figure 8.1 that *this sum is the total area of all the strips, so it approximates (under-estimates in this case) the area of triangle OAB*:

$$\sum_{n=1}^{N} v(t_{n-1})\Delta t \simeq \text{area of triangle OAB}$$

The error of this approximation is the sum of the areas of all the small shaded regions on top of the strips.

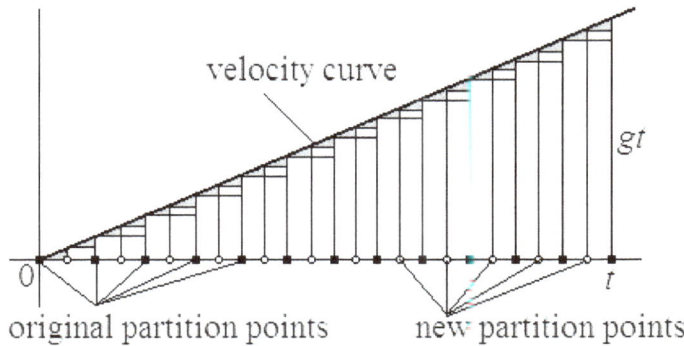

Figure 8.2: Refining a partition reduces the approximation error

In Figure 8.2 I have doubled N, the number of strips and with that, reduced the approximation error. We see in this figure that the sum of the little shaded areas on top of each strip is clearly less than the similarly shaded area in Figure 8.1.

As we increase N we reduce the error. Letting $N \to \infty$ the width Δt of each strip approaches zero and the strips thin out to line segments. The error goes to zero so we end up with the area of triangle OAB in the limit. The area of OAB is simply half of its height gt multiplied by the length t of the base, so

$$\text{area of triangle OAB:} \quad \frac{(gt)(t)}{2} = \frac{1}{2}gt^2$$

Going back to (8.6) with $x(0) = 0$ we conclude that

$$x(t) = \lim_{N \to \infty} \sum_{n=1}^{N} v(t_{n-1}) \Delta t = \frac{1}{2} gt^2$$

which is what we derived in the previous subsection using arithmetic rather than geometry.

Notice that the arithmetic calculation did not require *the assumption* that the approximation error goes to zero. This observation is important for two reasons: one is that it proves (rigorously) that our assumption was valid: the error does vanish as $N \to \infty$. Further, by using a similar calculation we can calculate some areas that we don't have a geometric formula for, as in the next exercise.

Exercise 100 *Let's calculate the area $A(t)$ under the graph of the position function $x(t) = gt^2/2$.[2] The analog of the approximating sum (8.6) is*

$$A(t) = A(0) + \sum_{n=1}^{N} x(t_{n-1}) \Delta t$$

---

[2]There is no physics concept associated with this area so just think of it as practice; we will soon develop more powerful methods for finding areas.

Integrals

Start from here and proceed as before to find the equation for $A(t)$. At some point along the way you will need the following summation formula:

$$1^2 + 2^2 + 3^2 + \cdots + (N-1)^2 = \frac{N(N-1)(2N-1)}{6}$$

The above discussion contains two important points relevant to our purpose in this chapter:

(a) The area bounded by a curve may be approximated by adding the areas of thin strips set up over partitions of the interval on the x-axis;
(b) The position function $x(t)$ is the area under the graph of the velocity function $v(t)$, which is just the derivative $x'(t)$. In fact, the same is true about velocity and acceleration: velocity $v(t)$ is the area under the graph of (constant) acceleration g.

Both of these facts are highlighted in Figure 8.3.

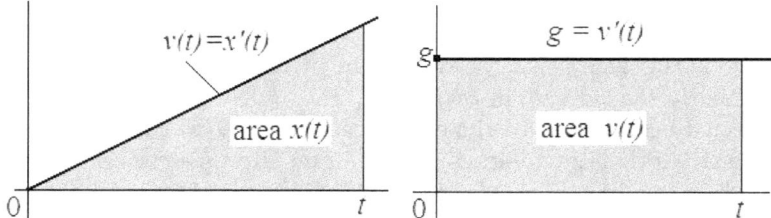

Figure 8.3: From areas to curves!

Let's take a closer look at Figure 8.3. In the right panel we see that the *area* of the shaded rectangle is gt which is none other than the velocity *function* in the left panel; the area in the right panel increases with t as a linear function of t that shows up as the straight line of slope g in the left panel. The area of the shaded triangle in the left panel also increases with t but does so as a quadratic function $gt^2/2$ that we recognize as the equation of a parabola.

We can think of the *velocity as the "area function" of acceleration* and of *position as the "area function" of velocity*.

These observations show that *area* is a geometric representation of "backward" derivatives, better known as *antiderivatives*. Just as the slope of the tangent line gives the rate of change of a function, the function itself gives the rate of change of the area under its graph; so every function is an antiderivative of its "slope function".

The term "area" makes sense only as a positive quantity and to ensure this, we may stipulate that the function is non-negative. But this is like putting a straight jacket on functions; for example, both velocity and acceleration functions meaningfully take on negative values, as in an object that is slowing down (negatively accelerating, or decelerating) or an object that is moving back and forth (velocity changes sign).

We might consider defining "negative area" to allow for a greater range of functions; but rather than doing that, mathematicians coined a new term to deal with this issue: the *integral*. So the position function $x(t)$ is an integral of the velocity $v(t)$, just as the velocity is the integral of acceleration.

8.2 The integral: partition, add, take limit!

Integrals and antiderivatives that I mentioned above are not the same things; the latter is essentially a tool that is used to calculate the former in simple enough cases, using the so-called *Fundamental Theorem of Calculus*. In introductory calculus textbooks this result is often cited as the key to calculating integrals; although true from a historical perspective, calculating integrals via antiderivatives is limited to a very small selection of functions and of limited practical value in modern applications. Also, the common tricks for finding antiderivatives shed no light on the role of infinity in calculus, which is our focus here.

The importance of the Fundamental Theorem lies not in its potential as a tool for calculation, but in the link that it provides between derivatives and integrals; it completes the discussion that we started in the last section about the relationship between areas and slopes. But before we discuss the precise nature of this relationship we need to have a clear definition of the integral.

The trio of steps mentioned earlier, namely:

partitioning an interval,
forming a sum based on the partition, and
taking the limit of the sum

has proven to be amazingly successful in solving a variety of scientific and engineering problems, from finding areas and centers of mass to calculating electric and magnetic fields, to finding probabilities in statistics.

Step 1: Interval partitions.

It all starts with an interval $[a,b]$ of real numbers and a finite set of numbers $P = \{x_0, x_1, x_2, \ldots, x_N\}$ in this interval, *not necessarily equally spaced*, such that

$$a = x_0 < x_1 < x_2 < \cdots < x_{N-1} < x_N = b$$

Partition. The set P of $N+1$ points above is a *partition* of $[a,b]$. If the points in P are equally spaced then P is a *regular partition*.

Figure 8.4 illustrates a partition visually.

Figure 8.4: A partition of the interval $[a,b]$

There are practical as well as theoretical reasons for allowing the partition points to be arbitrarily placed between a and b. I will discuss the theory below, but for a practical reason consider that if we were using rectangular strips to estimate the area under the graph of a function that changes slowly over a part of $[a,b]$ and rapidly over another, as in Figure 8.5, then we might save some time and effort by taking fewer subintervals under the flat portion and more under the steep portion so as to get roughly the same errors (shaded) per strip in approximating the area without unnecessary added calculations.

Integrals

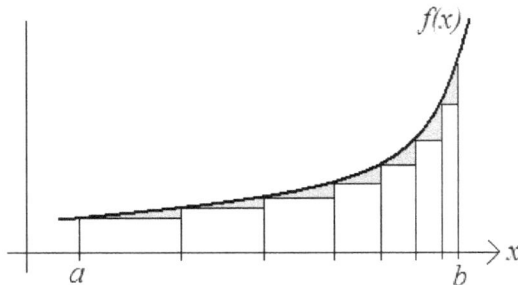

Figure 8.5: When a non-regular partition is more efficient

Subinterval. The set of all real numbers in between each pair of numbers in a partition is a *subinterval* of $[a,b]$.

The $N+1$ points of a partition P generate N of these subintervals:

$$[x_0, x_1], [x_1, x_2], \ldots, [x_{N-1}, x_N]$$

These intervals do not overlap except at their endpoints; we allow their lengths to be different in general so the i-th subinterval $[x_{i-1}, x_i]$ has length

$$x_i - x_{i-1} = \Delta x_i \quad i = 1, 2, \ldots, N$$

The mesh. The length of the *largest* of subintervals is often called the the *mesh (or norm) of the partition* which is commonly denoted by $\|P\|$. If the partition is regular then

$$\|P\| = \frac{b-a}{N} \quad \text{(mesh of a regular partition)}$$

The above fraction is evidently the minimum possible value for the mesh of any partition with N subinterval given that the total must be the length $b-a$ of the whole interval:

$$\sum_{i=1}^{N} \Delta x_i = (x_1 - x_0) + (x_2 - x_1) + \cdots + (x_{N-1} - x_{N-2}) + (x_N - x_{N-1}) = x_N - x_0$$

Refinements. By adding one or more partition points to an existing partition P of $[a,b]$ we *refine* the partition.

For example, for the interval $[0,1]$ a refinement of

$$P = \left\{0, \frac{1}{2}, 1\right\} \tag{8.9}$$

is the partition

$$P' = \left\{0, \frac{1}{2}, \frac{2}{3}, \frac{3}{4}, 1\right\}$$

Note that both of these partitions have mesh 1/2.

> **Union of partitions.** If P and P' are two partitions of $[a,b]$ then their *union* $P \cup P'$ is just the ordinary union of the two finite sets P and P'.

If P is a refinement of P' then clearly $P \cup P'$ doesn't add new points to P so $P \cup P' = P$. Otherwise, $P \cup P'$ refines P and P'. For example, if P is the partition in (8.9) and

$$P' = \left\{0, \frac{1}{4}, \frac{3}{4}, 1\right\}$$

then their union is

$$P \cup P' = \left\{0, \frac{1}{4}, \frac{1}{2}, \frac{3}{4}, 1\right\}$$

This partition refines both P and P'; also $P \cup P'$ is a regular partition in this case with mesh $1/4$ whereas each of P and P' has mesh $1/2$.

Exercise 101 *Consider the interval $[-1, 2]$.*
(a) Plot the partition P defined by the following set on the x-axis:

$$P = \{-1, 0, 1, 2\}$$

What is the mesh of this partition?
(b) Plot the partition P' defined by the following set on the x-axis:

$$P' = \left\{-1, -\frac{1}{2}, \frac{2}{3}, 1, \ 1.25, \ 1.3, \ 2\right\}$$

What is the mesh of this partition?
(c) Is P' a refinement of P? Or P of P'?
(d) Write down the union $P \cup P'$ explicitly as a partition. What is its mesh?

Step 2: the Riemann sums.

Suppose that $f(x)$ is any *bounded* function that is defined over an interval $[a, b]$ and let

$$P = \{x_0, x_1, x_2, \cdots, x_{N-1}, x_N\}$$

be a partition of $[a, b]$.

> A *Riemann sum* of $f(x)$ relative to P is the quantity
>
> $$\sum_{i=1}^{N} f(x_i^*) \Delta x_i = f(x_1^*)(x_1 - x_0) + f(x_2^*)(x_2 - x_1) + \cdots + f(x_N^*)(x_N - x_{N-1}) \qquad (8.10)$$
>
> where the numbers $x_1^*, x_2^*, \ldots, x_N^*$ are arbitrarily chosen in the corresponding subintervals of P; more precisely,
>
> $$x_{i-1} \leq x_i^* \leq x_i \quad \text{for } i = 1, 2, \ldots, N$$

Integrals

You may recognize that the sums of thin strips under the velocity curve that we discussed earlier were examples of Riemann sums with $x_i^* = x_{i-1}$ for $i = 1, 2, \ldots, N$ (the left hand endpoints of the subintervals $[x_{i-1}, x_i]$). We could alternatively choose the right hand endpoints $x_i^* = x_i$. There are practical and theoretical reasons for being flexible here; from a practical of view we may find it more useful in some cases to pick points other than the left or right hand endpoints. For example, we might pick the *midpoints*

$$x_i^* = \frac{x_{i-1} + x_i}{2}$$

because these often result is smaller errors per partition. For example, consider $f(x) = 2x$ for $0 \leq x \leq 2$ and take the regular partition $P = \{0, 1, 2\}$. To approximate the area under the graph of this $f(x)$ we calculate three Riemann sums below for the two subintervals $[0, 1]$ and $[1, 2]$ of P as follows:

$$\text{Left hand endpoints:} \quad 2(0)(1-0) + 2(1)(2-1) = 0 + 2 = 2$$
$$\text{Right hand endpoints:} \quad 2(1)(1-0) + 2(2)(2-1) = 2 + 4 = 6$$
$$\text{Midpoints:} \quad 2\left(\frac{1}{2}\right)(1-0) + 2\left(\frac{3}{2}\right)(2-1) = 1 + 3 = 4$$

The area in question is that of a right triangle with a base of length 2 and a height of length 4 which give a total area

$$\frac{1}{2}(2)(4) = 4$$

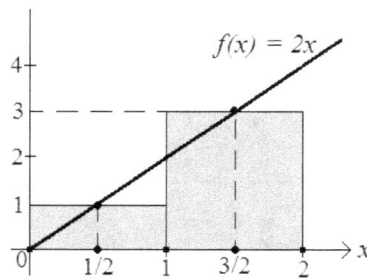

Figure 8.6: Riemann sum using midpoints

See Figure ??. In this case where the function is a straight line the midpoints give the exact answer because the over-estimation errors (dark triangles above the line) cancelled the under-estimation errors (clear triangles below the line).

Exercise 102 *Let $f(x) = x^2$ and consider the partition $P = \{0, 1, 2\}$ of the interval $[0, 2]$.*

(a) Calculate the three Riemann sums for the two subintervals $[0, 1]$ and $[1, 2]$ of P as was done above. Which answer do you think is a better approximation of the area under the graph of this function? A careful sketch of the graph is helpful here.

(b) Using essentially the same calculation as in Example 100 (or by just setting $g/2 = 1$ in that example) you can find the exact value of the area. Use this answer to tell how far each of your numbers in (a) is from the exact value.

(c) Repeat (a) for the exponential function $f(x) = 2^x$.[3]

Upper and lower Riemann sums. A *theoretically* important reason for allowing x_i^* to be flexible is that it is a useful option to have when we try to answer a question like the following about approximation of area:

Given any function $f(x)$ and any partition P, can we choose the x_i^* so that the Riemann sum under-estimates (or over-estimates) the area?

If $f(x)$ is *continuous* then there is a satisfying answer to the above question, thanks to the Extreme Value Theorem (EVT):

> For a continuous function, in every partition of the interval there is always a set of x_i^* that give an under-estimating Riemann sum and another set of x_i^* that give an over-estimating Riemann sum!

It is not hard to see why. If $f(x)$ is continuous on the interval $[a, b]$ then it is continuous at every point of this interval. In particular, $f(x)$ is continuous on every subinterval $[x_{i-1}, x_i]$, regardless of how we partition $[a, b]$.

Apply the EVT to pick a point x_i^* in $[x_{i-1}, x_i]$ for each index i to be where $f(x)$ achieves its minimum value m_i. Then the *Lower Riemann sum*

$$R_L(P) = \sum_{i=1}^{N} f(x_i^*)\Delta x_i = \sum_{i=1}^{N} m_i \Delta x_i$$

clearly under-estimates the area; see Figure 8.7 where $R_L(P)$ is shaded.

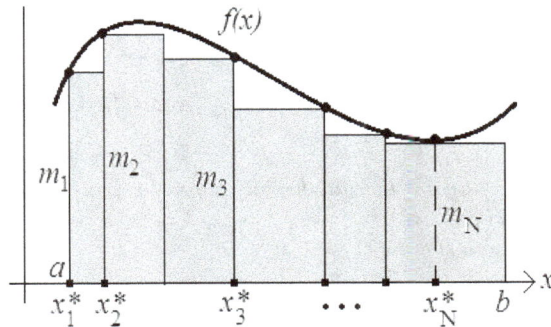

Figure 8.7: A lower Riemann sum

---

[3] In case you are curious, the exact value is
$$\frac{3}{\ln 2}$$
This number is not hard to determine but requires a little preparation that we discuss later in this chapter.

Integrals 233

Similarly, by the EVT there is x_i^* where $f(x)$ has its maximum value M_i in $[x_{i-1}, x_i]$ so that the *Upper Riemann sum*

$$R_U(P) = \sum_{i=1}^{N} f(x_i^*)\Delta x_i = \sum_{i=1}^{N} M_i \Delta x_i$$

over-estimates the area; see Figure 8.8 where $R_U(P)$ is shaded.

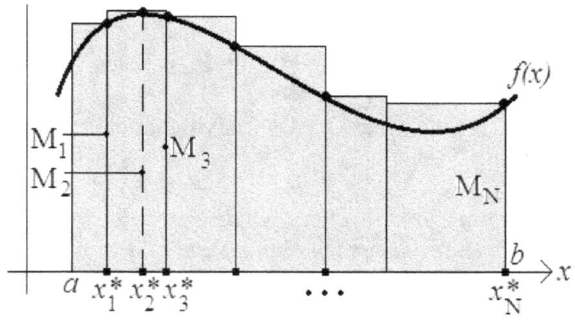

Figure 8.8: An upper Riemann sum

Exercise 103 *Consider $f(x) = x^2 + 1$ and the partition $P = \{-2, -1, 0, 1, 2\}$ of the interval $[-2, 2]$. Sketch graphs like those in Figures 8.7 and 8.8 for this case and calculate both $R_L(P)$ and $R_U(P)$. For comparison, also find the Riemann sum with x_i^* being the midpoint of each subinterval.*

Exercise 104 *Suppose that $f(x)$ is an **increasing** function on an interval $[a, b]$. Explain why the following equalities with the indicated values of x_i^* are valid:*

$$R_L(P) = \sum_{i=1}^{N} f(x_{i-1})\Delta x_i \quad \text{and} \quad R_U(P) = \sum_{i=1}^{N} f(x_i)\Delta x_i$$

*Sketching a graph may be helpful here. How do you think $R_L(P)$ and $R_U(P)$ change for a **decreasing** function $f(x)$?*

In practical terms, the upper and lower Riemann sums $R_L(P)$ and $R_U(P)$ are rarely used for approximating areas. The significance of $R_L(P)$ and $R_U(P)$ lies in the properties that make them theoretically useful.

Notice that *every Riemann sum relative to the same partition P is between $R_L(P)$ and $R_U(P)$*:

$$R_L(P) \leq \sum_{i=1}^{N} f(x_i^*)\Delta x_i \leq R_U(P) \tag{8.11}$$

regardless of where in $[x_{i-1}, x_i]$ we choose x_i^*, for every $i = 1, 2, \ldots, N$. Figure 8.9 illustrates these inequalities: the areas of strips with dashed tops are the terms of the Riemann sum; their areas, individually and collectively, are larger than the areas of the strips with heights m_1, m_2 etc (terms of the lower sum) but smaller than the areas of the strips with heights M_1, M_2 etc (terms of the upper sum).

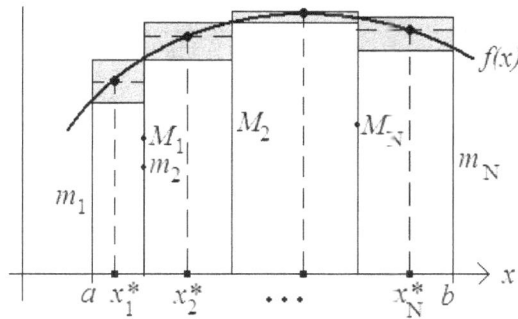

Figure 8.9: A generic Riemann sum together with the upper and lower sums

These lower and upper sums also have nice properties that make them suitable for defining limits, which we turn to next to complete our definition of the integral.

Step 3: take limits (invoking infinity)

You may recall that the extreme Riemann sums $R_L(P)$ and $R_U(P)$ were defined for continuous functions where we could use the EVT. But as long as a function $f(x)$ is bounded, we can define an upper and a lower sum regardless of whether $f(x)$ is continuous or not. Let's do so now:[4]

> **Upper and lower Darboux sums:** Let $f(x)$ be a bounded function on the interval $[a, b]$ and P be a partition of $[a, b]$. The *lower sum* $L(P)$ and the *upper sum* $U(P)$ of $f(x)$ relative to P are defined as
>
> $$L(P) = \sum_{i=1}^{N} m_i \Delta x_i \qquad U(P) = \sum_{i=1}^{N} M_i \Delta x_i$$
>
> where m_i and M_i are the minimum and maximum values of $f(x)$ on the subinterval $[x_{i-1}, x_i]$ for $i = 1, 2, \ldots, N$, i.e.
>
> $$m_i \leq f(x) \leq M_i \quad \text{for all } x \text{ in } [x_{i-1}, x_i]$$

It is a subtle fact that $L(P)$ and $U(P)$ are not necessarily Riemann sums because we cannot generally write them in the form (8.10) unless $f(x)$ is continuous. But they still satisfy the analog

---

[4] These are known as Darboux sums, after they were introduced by the French mathematician Jean-Gaston Darboux (1842-1917).

Integrals

of (8.11), that is, regardless of the choice of x_i^*

$$L(P) \le \sum_{i=1}^{N} f(x_i^*)\Delta x_i \le U(P) \qquad (8.12)$$

because $m_i \le f(x) \le M_i$ for every $i = 1, 2, \ldots, N$; see Figure 8.9. Therefore, for each partition P, every Riemann sum is between $L(P)$ and $U(P)$; and of course, thanks to the EVT:

> If $f(x)$ is continuous then $L(P) = R_L(P)$ and $U(P) = R_U(P)$.

The lower and upper sums have the following important *monotone* property.

> **The Monotone Sums Property (MSP):** *Suppose that P and P' are partitions of $[a, b]$ and $P \supset P'$ (P refines P'). Then*
> $$L(P') \le L(P) \le U(P) \le U(P') \qquad (8.13)$$

The inequalities in (8.12) and (8.13) ensure that *by refining a partition we narrow the range of variation of the Riemann sums.*

Let's see why the MSP is true. Because every partition point in P' is also in P we see that each subinterval $[x_{j-1}, x_j]$ of P' is a union of one or more subintervals of P; that is,

$$[x_{j-1}, x_j] = [x_{i-1}, x_i] \cup [x_i, x_{i+1}] \cup \cdots \cup [x_{i+k-1}, x_{i+k}]$$

Figure 8.10 illustrates the situation for the lower sums ($k = 2$ in the figure).[5]

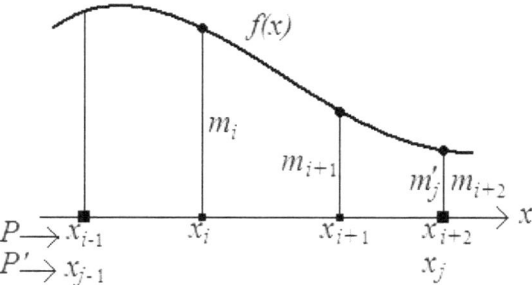

Figure 8.10: Illustrating the monotone property: lower sums

Because the subintervals do not overlap except at their endpoints, we have

$$\begin{aligned}
\Delta x_i + \Delta x_{i+1} + \cdots + \Delta x_{i+k} &= (x_i - x_{i-1}) + (x_{i+1} - x_i) + \cdots + (x_{i+k} - x_{i+k-1}) \\
&= x_{i+k} - x_{i-1} \\
&= x_j - x_{j-1} \\
&= \Delta x_j
\end{aligned}$$

---

[5]The situation is more complicated if there are discontinuities in the graph of $f(x)$, but the argument is not affected.

236   Infinite Resolution

Looking at the lower sum relative to P'

$$L(P') = \sum_{j=1}^{N'} m'_j \Delta x_j$$

we see that for each term j

$$\begin{aligned} m'_j \Delta x_j &= m'_j \Delta x_i + m'_j \Delta x_{i+1} + \cdots + m'_j \Delta x_{i+k} \\ &\leq m_i \Delta x_i + m_{i+1} \Delta x_{i+1} + \cdots + m_{i+k} \Delta x_{i+k} \end{aligned} \qquad (8.14)$$

This is true because *the lowest value of $f(x)$ on the set $[x_{j-1}, x_j]$ is the minimum of the lowest values of $f(x)$ on each of the subsets $[x_{i-1}, x_i], \ldots, [x_{i+k-1}, x_{i+k}]$*. Since each term of $L(P')$ consists of a sum of terms of $L(P)$ as in (8.14) it follows that $L(P') \leq L(P)$, as claimed in (8.13).

The argument for the upper sums is entirely similar (see Figure 8.11).

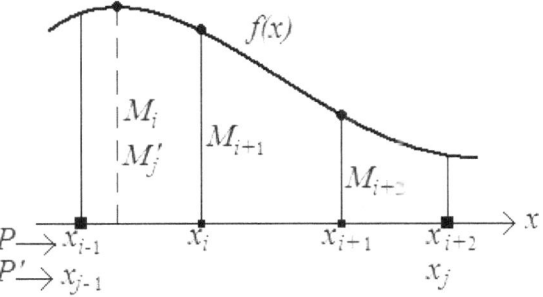

Figure 8.11: Illustrating the monotone property: upper sums

We have

$$U(P') = \sum_{j=1}^{N'} M'_j \Delta x_j$$

where for each term j

$$M'_j \Delta x_j \geq M_i \Delta x_i + M_{i+1} \Delta x_{i+1} + \cdots + M_{i+k} \Delta x_{i+k}$$

because *the largest value of $f(x)$ on the set $[x_{j-1}, x_j]$ is the maximum of the largest values of $f(x)$ on each of the subsets $[x_{i-1}, x_i], \ldots, [x_{i+k-1}, x_{i+k}]$*. It follows that $U(P') \geq U(P)$. This completes the proof of (8.13) because we can infer from (8.12) that $L(P) \leq U(P)$ for every partition P.

It is worth mentioning that there is a maximum possible value for all the upper sums and a minimum possible value for all the lower sums. Since $f(x)$ is bounded there are numbers m and M such that

$$m \leq f(x) \leq M \quad \text{for all } x \text{ in } [a, b]$$

Integrals 237

Clearly, $m \leq m_i$ and $M \geq M_i$ for all i so

$$U(P) = \sum_{i=1}^{N} M_i \Delta x_i \leq \sum_{i=1}^{N} M \Delta x_i = M \sum_{i=1}^{N} \Delta x_i = M(b-a)$$
$$L(P) = \sum_{i=1}^{N} m_i \Delta x_i \geq \sum_{i=1}^{N} m \Delta x_i = m \sum_{i=1}^{N} \Delta x_i = m(b-a)$$

To summarize:
$$m(b-a) \leq L(P) \leq U(P) \leq M(b-a) \tag{8.15}$$

We may identify $m(b-a)$ and $M(b-a)$ as the lower and upper sums of the coarsest possible partition of $[a, b]$, namely the two-point partition $\{a, b\}$:

$$L(\{a,b\}) = m(b-a) \quad \text{and} \quad U(\{a,b\}) = M(b-a)$$

The following corollary of the Monotone Sums Property is another important step towards the definition of the integral.

Let $f(x)$ be a bounded function on the interval $[a, b]$. For every pair of partitions P and P' of $[a, b]$ (P and P' are not assumed to be refinements of each other) it is true that

$$L(P) \leq U(P') \tag{8.16}$$

This is easy to prove: define the partition $P'' = P \cup P'$ so that P'' is a refinement of both P and P'. Then by the MSP
$$L(P) \leq L(P'') \leq U(P'') \leq U(P')$$
which shows that (8.16) is true.

Exercise 105 *All of the above inequalities are actually equalities for constant functions. Suppose that $f(x) = c$ for all x in an interval $[a, b]$. Show that for every partition P of the interval,*

$$L(P) = \sum_{i=1}^{N} f(x_i^*) \Delta x_i = U(P) = c(b-a)$$

The upper and lower integrals. We are now almost ready to define the integral of a bounded function. One more observation makes it easier to understand the definition and that is the following: for every (bounded) function $f(x)$ consider all possible partitions of $[a, b]$. Suppose that we collect all the possible lower sums and upper sums for *every partition* P and create two sets of real numbers:

$$\mathcal{L} = \{L(P) : P \text{ is a partition of } [a,b]\}$$
$$\mathcal{U} = \{U(P) : P \text{ is a partition of } [a,b]\}$$

By (8.16) every number in $\mathcal{L}$ is bounded above by some (any) number in $\mathcal{U}$ so $\mathcal{L}$ has a least upper bound , or supremum which we denote by a real number I_L. Similarly, every number in $\mathcal{U}$ is bounded below by some (any) number in $\mathcal{L}$ so $\mathcal{U}$ has a greatest lower bound, , or infimum that we denote by the real number I_U. Therefore,

$$I_L = \sup(\mathcal{L}) = \sup\{L(P) : P \text{ is a partition of } [a,b]\}$$
$$I_U = \inf(\mathcal{L}) = \inf\{U(P) : P \text{ is a partition of } [a,b]\}$$

Let's call I_L the *lower integral* of $f(x)$ and I_U the *upper integral*.[6] You may suspect that:

$$I_L \leq I_U \tag{8.17}$$

This is clearly true by (8.16) if $I_L = L(P)$ and $I_U = U(P')$ for some partitions P and P' of $[a,b]$. But if $\mathcal{L}$ and $\mathcal{U}$ are *infinite* sets (as is typically the case) then I_L and I_U need not equal $L(P)$ or $U(P)$ for *any* one partition P of the interval; we can only say that there are partitions P_n such that the *increasing* sequence of numbers $L(P_n)$ in $\mathcal{L}$ approaches I_L and likewise, there are partitions P'_n such that the *decreasing* sequence of numbers $U(P'_n)$ in $\mathcal{U}$ approaches I_U; see Figure 8.12.

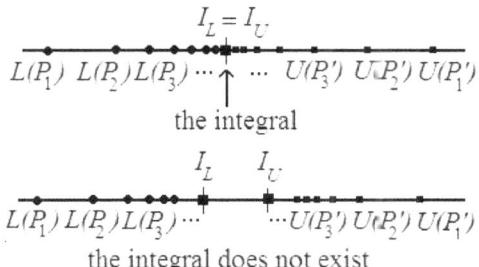

Figure 8.12: Existence and non-existence of integrals

To see that (8.17) is still true, first note that for every fixed positive integer m (8.16) implies that

$$U(P'_m) \geq L(P_n) \quad \text{for all } n$$

Therefore, $U(P'_m)$ is an upper bound for all $L(P_n)$ and as such, it can be no less that the *least* upper bound I_L; that is,

$$U(P'_m) \geq I_L \quad \text{for all } m.$$

Now we have established that I_L is a lower bound for $U(P'_m)$ and as such, it can be no greater that the *greatest* lower bound I_U; so (8.17) is true!

Feel free to pause for a minute to unravel this in your mind to your satisfaction; it is not obvious but easy to grasp when it sinks in. Also the think-over can help you follow some of the details ahead more easily.

We are finally ready to define the integral.[7] Since the existence of the numbers I_L and I_U requires the completeness of the set of real numbers, this definition makes rather explicit the role of completeness (and thus of infinity) in the concept of the integral.

---

[6]These are just convenient names; these two half-integrals do not make a whole integral unless they are equal to each other.

[7]This definition is actually due to Darboux. Riemann's definition, which naturally involves Riemann sums gives an equivalent concept (easy to verify when $f(x)$ is continuous). But the notation and various expressions are more elaborate with Riemann's definition so I use Darboux's here.

Integrals								239

> **The integral:** A bounded function $f(x)$ is *(Riemann) integrable on the interval* $[a,b]$ if its lower and upper integrals are equal: $I_L = I_U$. This common value is the *(Riemann) integral of* $f(x)$ and is denoted
> $$\int_a^b f(x)dx$$
> If $I_L < I_U$ then $f(x)$ is not (Riemann) integrable; its integral does not exist.

Figure 8.12 illustrates the above definition. This definition may not seem like the conventional definition involving limits of Riemann sums as "the mesh of the partitions approaches zero" that is typically given in calculus textbooks. But the above definition says the same thing more precisely because (8.12) implies that for each partition P of the interval all possible Riemann sums are confined between $L(P)$ and $U(P)$. The length of each subinterval $[x_{i-1}, x_i]$ of P is no greater than the mesh $\|P\|$, so $\|P\| \to 0$ means that we are adding more in-between points and refining P over and over. So the Monotone Sums Property implies that the upper sums move down and approach I_U while the lower sums move up and approach I_L.

Naturally, the upper and lower sums squeeze all the Riemann sums that are sandwiched between them. So if $f(x)$ is integrable as we defined above then the confined Riemann sums have nowhere to go but approach the common value $\int_a^b f(x)dx$ of I_L and I_U. So we can say that the Riemann sums approach the integral as $\|P\| \to 0$, just like we see in the calculus books!

I should highlight the fact that in the definition of the integral above it is assumed that $a < b$. It is convenient however to have a meaning for integrals such as

$$\int_a^a f(x)dx \quad \text{or} \quad \int_b^a f(x)dx$$

The integral on the left can be accounted for by extending the concept of partition just slightly to say that the only partition of the degenerate, *point interval* $[a,a] = \{a\}$ is $P_0 = \{a\}$ which has a single "subinterval" of length $\Delta x = 0$. Then $L(P_0) = 0 = R(P_0)$ and since P_0 is the only partition involved, $I_L = I_U = 0$. So we may conclude that

$$\int_a^a f(x)dx = 0$$

The second integral cannot be defined if $a < b$ but for operational reasons that we encounter later we define

$$\int_b^a f(x)dx = -\int_a^b f(x)dx \tag{8.18}$$

To illustrate the definition of integral let's find the integral of a very simple *point function*:

$$f(x) = 0 \quad \text{if } 0 \leq x \leq 2,\ x \neq 1$$
$$f(1) = d \quad d > 0$$

If P is any partition of $[0,2]$ then there is a subinterval $[x_{j-1}, x_j]$ that contains 1. Since every subinterval also contains points other than 1 we see that the smallest values m_i of $f(x)$ are all zeros for every i, including $i = j$. See Figure 8.13.

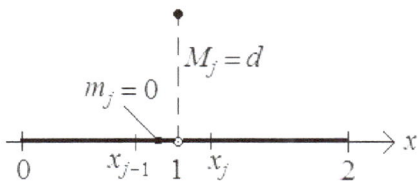

Figure 8.13: A point function

On the other hand, although the largest values M_i of $f(x)$ are zeros for $i \neq j$ it is true that $M_j = d > 0$. Therefore,
$$L(P) = 0, \quad U(P) = d(x_j - x_{j-1})$$
Since P is an arbitrary partition we conclude that $I_L = 0$ and
$$0 \leq I_U \leq U(P) \leq d\|P\|$$

So by letting $\|P\| \to 0$ we see that I_U has nowhere to go but to 0. It follows that $\int_0^2 f(x)dx = 0$.

Intuitively, if you think of the point function as a strip of zero width then you expect that such a "strip" will have zero area too. But of course, there is no such thing as a "zero-width strip"; the technical machinery developed above is designed to deal with the thorny issues about the infinity that intuition conveniently glosses over.

Exercise 106 (a) Explain how to modify the above argument to show that $\int_0^2 f(x)dx = 0$ also if $d < 0$.

(b) Explain how to modify the above argument to show that $\int_0^2 f(x)dx = 0$ if $f(x) = 0$ for $0 \leq x < 2$ but $f(2) = d \neq 0$.

(c) Consider a generalized version of $f(x)$ on an arbitrary interval $[a,b]$ that is nonzero at two numbers c_1, c_2 in $[a,b]$ and $f(x) = 0$ if $x \neq c_1, c_2$. What modifications are needed to the above argument to show that $\int_a^b f(x)dx = 0$?

(d) What do you expect to get if $f(x) = 0$ except for a finite number of points $c_1, c_2, \ldots, c_m$ in $[a,b]$? Do you see why?

Monotone functions.

In this subsection I show that every bounded *monotone function* (either always non-decreasing or non-increasing) is integrable. We won't put any further restrictions on $f(x)$ beyond boundedness. A key idea here is the *telescoping sums* that we discussed in Chapter 6. An important step in getting there was also discussed in Exercise 104.

Every monotone function on an interval $[a,b]$ is integrable.

Let's assume that $f(x)$ is non-decreasing (you can work out the non-increasing case similarly). If $P = \{x_0, x_1, \cdots, x_N\}$ is any partition of $[a,b]$ then the minimum m_i and maximum M_i of $f(x)$

Integrals

on each subinterval $[x_{i-1}, x_i]$ occur at an endpoint:

$$m_i = f(x_{i-1}), \qquad M_i = f(x_i)$$

Hence the lower and upper sums are in fact Riemann sums in this case:

$$L(P) = \sum_{i=1}^{N} f(x_{i-1})\Delta x_i, \qquad U(P) = \sum_{i=1}^{N} f(x_i)\Delta x_i$$

Let's subtract these numbers and collect the like terms on the right hand side to get:

$$U(P) - L(P) = \sum_{i=1}^{N} [f(x_i) - f(x_{i-1})]\Delta x_i$$

Recall that $\Delta x_i = x_i - x_{i-1} \leq \|P\|$ for every index i so we have

$$U(P) - L(P) \leq \sum_{i=1}^{N} [f(x_i) - f(x_{i-1})]\|P\| = \|P\| \sum_{i=1}^{N} [f(x_i) - f(x_{i-1})]$$

The last sum is a telescoping sum that collapses to just a single difference:

$$[f(x_1) - f(x_0)] + [f(x_2) - f(x_1)] + [f(x_3) - f(x_2)] + \cdots [f(x_{N-1}) - f(x_{N-2})]+$$
$$+ [f(x_{N-1}) - f(x_{N-2})] + [f(x_N) - f(x_{N-1})] = f(x_N) - f(x_0) = f(b) - f(a)$$

Therefore,

$$U(P) - L(P) \leq \|P\|\,[f(b) - f(a)]$$

Next, we refine the partitions so that $\|P\| \to 0$. Then $U(P) - L(P) \to 0$ and since

$$0 \leq I_U - I_L \leq U(P) - L(P)$$

we conclude that $I_U - I_L = 0$ which means that $I_U = I_L$ and $f(x)$ is integrable.

Exercise 107 *Prove the above result when $f(x)$ is a non-increasing function (you need to make only one essential change to the above argument).*

The above result tells us that we can integrate every monotone function but *it does not tell us what the integral is*, or even how to integrate the function efficiently. Calculating the numbers I_L and I_U directly is usually not easy even in simple cases. The definition of the integral above is not well-suited to direct calculations so it is glossed over in introductory calculus textbooks (and courses) in favor of the more computational topics. On the other hand, the monotone functions in question are arbitrary other than being bounded on a closed interval; there is no result beyond the above that tells us they must be integrable. An example that gives the flavor of the generality but still is not very complicated is the following.

Example: an integrable function with infinitely many points of discontinuity. In typical calculus textbooks functions are usually taken to be continuous, or at worst have a few points where there may be a gap or a tear in the graph (such functions are called *piece-wise continuous*). But the definition of Riemann integrability given above does not make this simplifying assumption. Consider the function below that is defined on the interval $[0,1]$

$$f(x) = \frac{1}{2^{n-1}} \quad \text{for } \frac{1}{2^n} < x \leq \frac{1}{2^{n-1}}, \quad f(0) = 0 \tag{8.19}$$

This is a non-decreasing function with infinitely many points of discontinuity; a rough sketch is shown in Figure 8.14. By the Monotone Functions Integrals theorem above, this function is integrable.

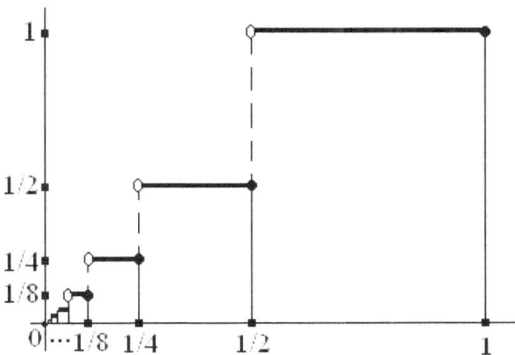

Figure 8.14: An increasing function with infinitely many discontinuities

The integral of this function cannot be found using the Fundamental Theorem of Calculus or other similar results that are true for piece-wise continuous functions.

The integral is just the total area of all the strips in Figure 8.14. The area of the largest rectangular strip is "base times height" or: $(1/2)(1)$. Similarly, the strip next to it has area

$$\left(\frac{1}{2} - \frac{1}{2^2}\right)\frac{1}{2} = \left(\frac{1}{2^2}\right)\frac{1}{2} = \frac{1}{2^3}$$

and so on; adding the areas of all the strips gives the value of the integral as an infinite series:

$$\int_0^1 f(x)dx = \frac{1}{2} + \frac{1}{2^3} + \frac{1}{2^5} + \cdots$$

The infinite series on the right hand side is geometric with first term $1/2$ and common ratio $1/2^2 = 1/4$ so as we saw earlier, its sum is easy to find:

$$\int_0^1 f(x)dx = \frac{1/2}{1 - 1/4} = \frac{2}{3}$$

Example: A non-integrable function. In general, non-monotone functions that are discontinuous at an infinite number of points are typically not Riemann integrable. An example is

Integrals

Dirichlet's function $D(x)$; let's consider this function over any closed interval, say, $[-1, 1]$ and let P be any partition of $[-1, 1]$. Since rational numbers are dense every subinterval of P contains a rational number at which $D(x) = 1$, that is, $M_i = 1$ for all i. Similarly, the irrational numbers are dense so every subinterval contains an irrational number where $D(x) = 0$; therefore, $m_i = 0$ for all i. It follows that

$$L(P) = \sum_{i=1}^{N} m_i \Delta x_i = 0$$

$$U(P) = \sum_{i=1}^{N} M_i \Delta x_i = \sum_{i=1}^{N} \Delta x_i = x_N - x_0 = 1 - (-1) = 2$$

Since the above two values come out for every partition of $[-1, 1]$ it follows that the sets $\mathcal{L}$ and $\mathcal{U}$ each contain just one number

$$\mathcal{L} = \{0\}, \quad \mathcal{U} = \{2\}$$

Therefore, $I_L = 0$ and $I_U = 2$; these two numbers not being equal, we conclude that $D(x)$ is not Riemann integrable.[8]

Exercise 108 *$D(x)$ is defined over the set $\mathbb{R}$ of all real numbers but it is not integrable over any closed interval $[a, b]$. Prove this generalization by extending the above argument for $[-1, 1]$ to $[a, b]$ (simply figure out what I_L and I_U are).*

8.3 The Fundamental Theorem of Calculus: opposites annihilate!

Now that we have a clear definition of what the integral is, let's proceed to study its properties and its relationship to the other major calculus concepts, namely, derivatives and continuity. In this section we also discuss a way of calculating the integrals of some familiar functions; although this method of calculation is of limited practical value, it is still much easier to use than the definition of the integral (calculating I_U and I_L).

Antiderivatives.

Recall that the derivative of x^3 is $3x^2$. We may think of x^3 as an antiderivative of $3x^2$. More generally:

> **Antiderivative.** Let $f(x)$ be a function on an interval I of real numbers. A function $F(x)$ is an *antiderivative* of $f(x)$ on I if $F'(x) = f(x)$ for all x in I. Since the derivative of a constant C is zero we see that if $F(x)$ is an antiderivative of $f(x)$ then so is $F(x) + C$.

---

[8] The Dirichlet function may be integrable via a different integral. For example, it is "Lebesgue integrable" and its integral over the entire set $\mathbb{R}$ is zero. The reason it is Lebesgue integrable is simply that the set of rational numbers is so small compared to the set of all real numbers that it has "measure zero"; whatever a function does on this set contributes nothing to the (Lebesgue) integral.

For example, x^3, $x^3 - 2$, $x^3 + \pi$ are all antiderivatives of $3x^2$. Further, if $F_1(x)$ and $F_2(x)$ are both antiderivatives of the *same function* $f(x)$ then for all x

$$\frac{d}{dx}[F_1(x) - F_2(x)] = F_1'(x) - F_2'(x) = f(x) - f(x) = 0$$

It follows that $F_1(x) - F_2(x) = C$ must be a constant. We conclude that all possible antiderivatives of $f(x)$ are accounted for by the functions $F(x) + C$ where $F(x)$ is any particular antiderivative of $f(x)$. Because every possible real value of C is admissible, C often called the "(arbitrary) constant of integration".

Each derivative formula that we discussed above yields an antiderivative formula; for example, since the derivative of $\sin x$ is $\cos x$ we conclude that the antiderivatives of $\cos x$ are $\sin x + C$. To calculate the antiderivative of x^p where p is a non-zero real number we first recall that by the Power Rule:

$$\frac{d}{dx} x^p = px^{p-1}$$

This shows that an antiderivative of x^{p-1} is x^p/p. By adding 1 to p we get:

All the antiderivatives of x^p are given by

$$\frac{x^{p+1}}{p+1} + C \quad \text{if } p \neq -1$$

For instance, the antiderivatives of x^2 are $x^3/3 + C$. The above formula leaves out one value of p, namely, $p = -1$. The reason is, of course, because we can't divide by 0. But the function whose derivative is $1/x$ is a special transcendental function that we know as the natural logarithm; we discuss it later in this chapter.

Exercise 109 *Find formulas for the antiderivatives of each of the following functions:*

$$\frac{1}{x^3}, \quad \sqrt{x}, \quad \frac{1}{\sqrt{x}}, \quad \sin x$$

A point worth emphasizing about antiderivatives is this:

If $F(x)$ is an antiderivative of $f(x)$ then by definition $F(x)$ is differentiable, that is, $F'(x)$ exists; in fact, $F'(x) = F(x)$.

Some functions simply do not have antiderivatives; for instance,

$$f(x) = \begin{cases} 1 & \text{if } x \geq 0 \\ -1 & \text{if } x < 0 \end{cases}$$

cannot have an antiderivative on any open interval containing 0.

To see why not, consider that at $x = 0$ we must have $F'(0) = f(0) = 1$. But for all $x < 0$ we have $F'(x) = f(x) = -1$ so that $F(x) = -x + C$. Now if we take a sequence x_n that converges to 0 and $x_n < 0$ then

$$\frac{F(x_n) - F(0)}{x_n - 0} = \frac{(-x_n + C) - (0 - C)}{x_n} = -1$$

Integrals

for all n so that
$$F'(0) = \lim_{n \to \infty} \frac{F(x_n) - F(0)}{x_n - 0} = -1$$
which is a contradiction!

You may have noticed that $|x|$ *would be* an antiderivative if not for the trouble at $x = 0$. No matter how we define $f(x)$ above at $x = 0$ (we defined $f(0) = 1$ arbitrarily) the antiderivative cannot be defined there. Evidently, the jump discontinuity of $f(x)$ at 0 is the cause of the problem here; we see in the next subsection that if a function $f(x)$ is continuous on an interval I then it has a *smooth* antiderivative $F(x)$; that is $F(x)$ has no corners or sharp points.

On the other hand, continuity is a sufficient condition that is not necessary for the existence of an antiderivative; for example, consider
$$F(x) = x^2 \sin \frac{1}{x} \quad \text{if } x \neq 0, \quad F(0) = 0$$

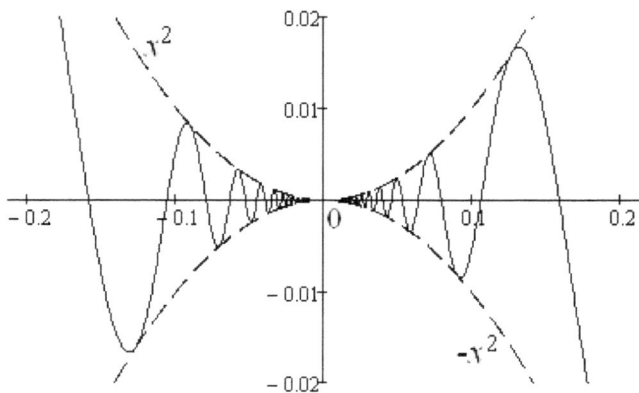

Figure 8.15: A differentiable function with discontinuous derivative

See Figure 8.15. This figure also shows that $F(x)$ is continuous everywhere, including at 0. Now, if $x \neq 0$ then using the derivative rules and formulas, we calculate
$$F'(x) = 2x \sin \frac{1}{x} - \cos \frac{1}{x} \doteq f(x) \tag{8.20}$$

The derivative function $f(x)$ is discontinuous at $x = 0$ where it has a non-removable singularity; see Figure 8.16 for a graph of $f(x)$.[9] But for every sequence x_n that converges to 0 and $x_n \neq 0$ we have
$$\lim_{n \to \infty} \frac{F(x_n) - F(0)}{x_n - 0} = \lim_{n \to \infty} x_n \sin \frac{1}{x_n} = 0$$

So $F'(0)$ exists and equals 0. So if we define $f(0) = 0$ then we see that the function $f(x)$ in (8.20) which is discontinuous with a nasty-looking singularity at 0 can very well have a *continuous* antiderivative!

---

[9] $F(x)$ here is an example of a function that is differentiable everywhere but not continuously; that is, its derivative exists but is discontinuous at some point ($x = 0$ in this case).

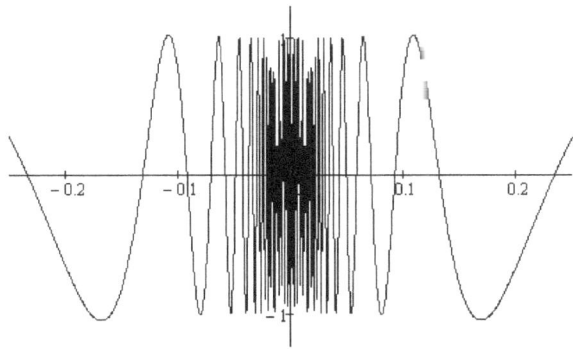

Figure 8.16: The discontinuous derivative of the function in Figure 8.15

The Fundamental Theorem of Calculus, Part 1.

Our discussion of antiderivatives above suggests that the existence of antiderivative is a weaker condition than continuity, whereas the existence of derivatives is stronger or more restrictive condition, in the following sense:

The derivative of $f(x)$ exists $\implies$ $f(x)$ is continuous $\implies$ the antiderivative of $f(x)$ exists

We saw earlier why the first implication is true, and the Fundamental Theorem of Calculus(Part 2) proves the second. There are two halves of the Fundamental Theorem that say different things, so they are usually presented separately in standard calculus textbooks and I follow the same practice here. We start with the first half which is based on the Mean Value Theorem.

The Fundamental Theorem, Part 1 (FT1). *Assume that $f(x)$ is an integrable function on the interval $[a, b]$. If $f(x)$ has an antiderivative $F(x)$ on $[a, b]$ then*

$$\int_a^b f(x)dx = F(b) - F(a) \tag{8.21}$$

It is not hard to see why this is true. Consider an arbitrary partition $P = \{x_0, x_1, \ldots, x_N\}$ of $[a, b]$. Since $F(x)$ is differentiable it has the Mean Value Theorem; so there is a c_i in each of the subintervals $[x_{i-1}, x_i]$ such that

$$F(x_i) - F(x_{i-1}) = F'(c_i)(x_i - x_{i-1}) = f(c_i)\Delta x_i$$

The special Riemann sum with $x_i^* = c_i$ telescopes nicely:

$$\begin{aligned}
R(P) &= \sum_{i=1}^{N} f(c_i)\Delta x_i \\
&= \sum_{i=1}^{N} [F(x_i) - F(x_{i-1})] \\
&= F(x_N) - F(x_0) \\
&= F(b) - F(a)
\end{aligned}$$

Integrals

We recall that this (and every other) Riemann sum is bounded by the upper and lower sums, so:

$$L(P) \leq F(b) - F(a) \leq U(P)$$

This is true for every partition P of $[a, b]$ so the first inequality extends to the least upper bound I_L of all lower sums, and the second inequality to the greatest lower bound I_U of all upper sums; that is,

$$I_L \leq F(b) - F(a) \leq I_U$$

Finally, $I_L = I_U = \int_a^b f(x)dx$ since by assumption $f(x)$ is integrable; so (8.21) is true!

This theorem makes calculating some integrals very easy. For example, an antiderivative of $f(x) = x^2$ is $F(x) = x^3/3$ so

$$\int_{-1}^{1} x^2 dx = F(1) - F(-1) = \frac{(1)^3}{3} - \frac{(-1)^3}{3} = \frac{1}{3} + \frac{1}{3} = \frac{2}{3}$$

More generally, over any interval $[a, b]$ we have

$$\int_a^b x^2 dx = F(b) - F(a) = \frac{b^3}{3} - \frac{a^3}{3} = \frac{b^3 - a^3}{3}$$

Still more generally, for every $p \neq -1$ using the power rule for the antiderivative of x^p above we obtain

$$\int_a^b x^p dx = \frac{b^{p+1} - a^{p+1}}{p+1} \qquad (8.22)$$

It would be quite difficult to obtain this integration formula directly from the definition of the integral (calculating I_L and I_U).

Notation. It is convenient to use a shorthand notation for the right hand side of (8.21) as follows:

$$F(b) - F(a) = F(x)\big]_a^b$$

With this notation we can write the power rule for integrals in (8.22) as follows:

Power rule for integrals.

$$\int_a^b x^p dx = \frac{x^{p+1}}{p+1}\bigg]_a^b \qquad (p \neq -1)$$

Exercise 110 *Find each of the following integrals:*

$$\text{(a)} \int_0^4 \sqrt{x}\, dx \qquad \text{(b)} \int_1^4 \frac{1}{\sqrt{x}}\, dx \qquad \text{(c)} \int_0^{\pi/2} \sin x\, dx$$

How do we know that $\sqrt{x}$ is integrable on $[0, 4]$? We check if it is monotone! The same is true of $1/\sqrt{x}$ on $[1, 4]$ and $\sin x$ on $[0, \pi/2]$.

Continuity and integrability.

In Theorem FT1 we needed to assume that $f(x)$ is integrable on an interval. In general, showing that a function $f(x)$ is integrable is not an easy task. We showed earlier that all monotone functions are integrable; this was enough to for Exercise 110 but what if we enlarge the interval of integration in (c) of that Exercise past $\pi/2$ where the sine function is no longer monotone? How do we know that the sine function is integrable over larger intervals?

Further, in our derivation of the integral formula for power functions we simply assumed that x^p is integrable on every interval. Again, some power functions are certainly not monotone.

You might think that since power functions and the sine function all have antiderivatives they must be integrable. But this assumption is not generally valid as we see in the next example.

The existence of an antiderivative for a function doesn't mean that the function is integrable!

A function that has an antiderivative but is not integrable. Consider the function $f(x) = 1/\sqrt{x} = x^{-1/2}$. This has an antiderivative $F(x) = 2\sqrt{x}$ as $F'(x) = f(x)$. This is defined for all $x \geq 0$ so in particular, on the interval $[0, 1]$. Further,

$$F(1) - F(0) = 2\sqrt{1} - 2\sqrt{0} = 2 - 0 = 2$$

But this *isn't* $\int_0^1 1/\sqrt{x}\,dx$ because $1/\sqrt{x}$ is *not* bounded on $[0,1]$ and thus, *not* integrable!

Thinking of the integral as the area of the shaded region under the graph does not help in this case because as we see in Figure 8.17, this region is unbounded (there is no end at the top).[10]

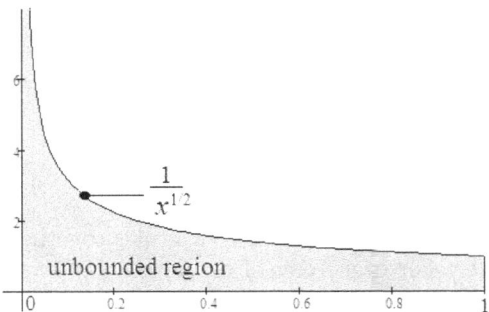

Figure 8.17: An unbounded function with an antiderivative

In standard calculus texts you don't run into this *integrability issue* because in those books functions are usually assumed to be continuous. The next result explains this choice.

---

[10]It so happens that the unbounded region in Figure 8.17 has a finite area. We came across this counterintuitive situation once before in the discussion of the infinite tower of boxes in Chapter 6. But interestingly, the area is 2 which is also the difference $F(1) - F(0)$. We clear up this issue in our discussion of the improper Riemann integral later in this chapter.

Integrals

> **Continuous implies integrable.** *If $f(x)$ is continuous at all x in $[a,b]$ then $f(x)$ is integrable on $[a,b]$.*

The proof of this theorem is available in standard real analysis texts. It is worth highlighting the following implications that we have now established:

The derivative of $f(x)$ exists $\Longrightarrow$ $f(x)$ is continuous $\Longrightarrow$ the integral of $f(x)$ exists

Or more succinctly,

> *Differentiable implies continuous and continuous implies integrable.*

We have already discussed examples of functions that are integrable but not continuous and also examples of continuous functions that are not differentiable.

Basic properties of the integral.

Before moving on to the second half of the Fundamental Theorem, we need to list a few basic properties of the integral. These are proved using the original definition of the integral and straightforward though technical reasoning.[11] They also help extend the range of functions that we can integrate.

Suppose that $f(x)$ and $g(x)$ are both integrable functions on an interval $[a,b]$.

> **1. The sum rule:** $f(x) + g(x)$ is integrable on $[a,b]$ and
> $$\int_a^b [f(x) + g(x)]dx = \int_a^b f(x)dx + \int_a^b g(x)dx$$
>
> **2. The constant-multiple rule:** for every fixed real number c the function $cf(x)$ is integrable on $[a,b]$ and
> $$\int_a^b cf(x)dx = c\int_a^b f(x)dx$$
>
> **3. The linear property:** The above two rules together state that the integral has the following property: For every pair of fixed real numbers,
> $$\int_a^b [cf(x) + dg(x)]dx = c\int_a^b f(x)dx + d\int_a^b g(x)dx$$

To illustrate the usefulness of these properties in basic calculations, consider the function $f(x) = (x^3 - 2)^2/x^3$ which is continuous, hence integrable on any interval that does not contain 0. We can

---

[11] The proofs are found in most introductory analysis textbooks, intended usually for college juniors or seniors in the mathematics curriculum. They do not shed new light on the role of infinities in calculus but are helpful in refining proof skills. Proofs of some of these results may also be found in many standard calculus textbooks.

calculate its integral using the above results and a little bit of algebra:

$$\int_1^2 f(x)dx = \int_1^2 \frac{(x^3-2)^2}{x^3}dx$$
$$= \int_1^2 \frac{x^6 - 4x^3 + 4}{x^3}dx$$
$$= \int_1^2 \left(x^3 - 4 + 4x^{-3}\right)dx$$
$$= \int_1^2 x^3 dx - \int_1^2 4dx + 4\int_1^2 x^{-3}dx$$

Did you notice where we used the sum rule and the constant multiple rule? Next, we calculate each of the three integrals on the right hand side using the FTC:

$$\int_1^2 f(x)dx = \left.\frac{x^4}{4}\right]_1^2 - 4x]_1^2 + 4\left.\frac{x^{-2}}{-2}\right]_1^2$$
$$= \frac{15}{4} - 4 - 2\left(\frac{1}{2^2} - 1\right)$$
$$= \frac{5}{4}$$

Note that we did not need a formula for integrating quotients in this case because the algebra simplified the fraction.

Exercise 111 *Explain why the functions within each integral sign are integrable on the indicated intervals and calculate the integrals:*

$$(a) \quad \int_1^3 \frac{2x - x^3}{x}dx \qquad (b) \quad \int_0^\pi (1 + 2\sin x)\,dx$$

The above rules say essentially that is okay to distribute the integral sign over a sum and to move it past a constant. Can we do the same for products and quotients?

If you recall we could not distribute the derivative over products or quotients; a similar issue exists for integrals: *the integral sign does not distribute over products or quotients* of functions. For example, $f(x) = x^2$ and $g(x) = x^3$ are integrable functions on any interval as is their product $f(x)g(x) = x^2 x^3 = x^5$ so using (8.22)

$$\int_0^1 f(x)g(x)dx = \int_0^1 x^5 dx = \left.\frac{x^6}{6}\right]_0^1 = \frac{1^6}{6} - \frac{0^6}{6} = \frac{1}{6}$$

whereas if we distribute the integral over the product then we get:

$$\int_0^1 f(x)dx \int_0^1 g(x)dx = \int_0^1 x^2 dx \int_0^1 x^3 dx = \left(\frac{1}{3}\right)\left(\frac{1}{4}\right) = \frac{1}{12}$$

Integrals

Although in general it is true that

$$\int_a^b f(x)g(x)dx \neq \int_a^b f(x)dx \int_a^b g(x)dx$$

it can be shown that:

> The product $f(x)g(x)$ is integrable if both $f(x)$ and $g(x)$ are integrable on the same interval.

This leads to something called the *integration by parts formula* which is analogous to the product rule for derivatives.

---

4. Integration by parts: Assume that $f(x)$ and $g(x)$ are both differentiable functions and their derivatives $f'(x)$ and $g'(x)$ are integrable on an interval $[a,b]$. Then

$$\int_a^b f(x)g'(x)dx = f(x)g(x)]_a^b - \int_a^b f'(x)g(x)dx \qquad (8.23)$$

It should not be surprising that the integration by parts formula can be used to find the integral of a product, albeit in a rather indirect way. For example, the function $2x \sin x$ is continuous, hence integrable on every interval of real numbers. To calculate its integral on, say, $[0, \pi/2]$ we identify it with the product inside the integral on the left hand side of (8.23). There are two natural ways of doing this:

$$f(x) = 2x, \quad g'(x) = \sin x, \quad \text{or:} \qquad (8.24)$$
$$f(x) = \sin x, \quad g'(x) = 2x \qquad (8.25)$$

Which of these works best depends entirely on whether the integral on the right hand side of (8.23) will be easier to calculate than the one we started with. Using (8.24) in (8.23) gives:

$$\int_0^{\pi/2} 2x \sin x \, dx = 2x \sin x]_0^{\pi/2} - \int_0^{\pi/2} 2(-\cos x)dx \qquad (8.26)$$

The last integral above is obtained by calculating $f'(x) = 2$ from $f(x) = 2x$ and calculating $g(x)$ from its derivative $g'(x) = \sin x$. The antiderivative of $\sin x$ is $-\cos x$ which we see inside the integral. Now the last integral is indeed easier to calculate than the one we started with so we continue:

$$\int_0^{\pi/2} 2x \sin x \, dx = 2\frac{\pi}{2} \sin \frac{\pi}{2} - 0 \sin 0 + 2 \int_0^{\pi/2} \cos x \, dx$$
$$= \pi + 2 \sin x]_0^{\pi/2}$$
$$= \pi + 2 \simeq 5.142$$

What would happen if we used (8.25) instead? This choice of functions would give us:

$$f'(x) = \cos x, \quad g(x) = 2\left(\frac{x^2}{2}\right) = x^2$$

and the integral on the right hand side of (8.23) would then be:

$$\int_0^{\pi/2} x^2 \cos x\, dx$$

This is not easier than the integral that we started with; so the selection (8.25) is not useful.

Integration by parts is a versatile and useful formula that is applied in a few different ways in calculus and you can find these applications in most standard calculus textbooks.

Another simple and useful integration method is based on the derivative *chain rule*; recall that this rule changes the differentiation variable from x to, say, u. A similar idea holds for the variable of integration.

> **5. Integration by substitution:** Suppose that $g(x) = u$ is a function whose derivative $g'(x)$ is continuous on an interval $[a, b]$ and that $f(x)$ is continuous on the range of $g(x)$. Then
>
> $$\int_a^b f(g(x))g'(x)dx = \int_{g(a)}^{g(b)} f(u)du \qquad (8.27)$$

We see that the integration variable changes from x on one side to u on the other side of (8.27). So if a hard integral can be translated properly to look like the one on the left hand side and the integral on the right hand side is easy then we manage to solve the originally hard problem. The key is to find a suitable part of the expression inside the integral on the left to call $g(x)$, or u for short. Although the formula in (8.27) seems somewhat complicated, it is often easy to use in solving the problems to which it applies. For example, consider

$$\int_0^{\pi/2} 2x \sin(x^2)dx \qquad (8.28)$$

This integral is similar to the one on the left hand side of (8.26) but cannot be calculated using integration by parts. Let's take a close look at the function inside; it is not tough to spot the u here: if we set $x^2 = g(x) = u$ then $g'(x) = 2x$ is indeed a part of the function inside the integral. If we make the identification:

$$\sin(x^2)(2x) = f(g(x))g'(x)$$

then it is clear that defining f to be the sine function gives us a true identity. Now with $a = 0$ and $b = \pi/2$ we find that

$$g(a) = \sin 0 = 0 \quad \text{and} \quad g(b) = \sin\left(\frac{\pi}{2}\right)^2 = \sin\frac{\pi^2}{4} \simeq 0.624$$

Now, using (8.27) we get:

$$\int_0^{\pi/2} 2x \sin(x^2)dx = \int_0^{\sin(\pi^2/4)} \sin u\, du \simeq -\cos(0.624) - (-\cos 0) \simeq 0.189$$

Here is another typical example of the use of integration by substitution:

$$\int_0^1 \frac{x}{\sqrt{2-x^2}} dx \qquad (8.29)$$

Integrals

After a moment's reflection, we spot a good candidate for u

$$2 - x^2 = g(x) = u \implies g'(x) = 0 - 2x = -2x$$

The expression $-2x$ does not appear in full in (8.29) but the x part does and the rest is just a number -2 that can be easily handled by the constant multiple rule. Here's then how we proceed:

$$\frac{x}{\sqrt{2-x^2}}dx = \frac{1}{\sqrt{2-x^2}}xdx = \frac{1}{\sqrt{u}}\frac{-2xdx}{-2} = -\frac{1}{2}\frac{1}{\sqrt{u}}du$$

I used the notation u directly so as to emphasize the change in variables from x on the far left to u on the far right. Now the function f is clearly

$$f(u) = \frac{1}{\sqrt{u}} = u^{-1/2}$$

which is easy to integrate as a power function. So with $a = 0$ and $b = 1$ we have $g(a) = 2 - 0 = 2$, $g(b) = 2 - 1 = 1$ and therefore,

$$\int_0^1 \frac{x}{\sqrt{2-x^2}}dx = \int_2^1 -\frac{1}{2}u^{-1/2}du = -\frac{1}{2}\int_2^1 u^{-1/2}du$$

Now, we integrate on the right hand side using the power rule to obtain:

$$\int_0^1 \frac{x}{\sqrt{2-x^2}}dx = -\frac{1}{2}\frac{u^{1/2}}{1/2}\Big|_2^1 = -1^{1/2} - (-2^{1/2}) = \sqrt{2} - 1 \simeq 0.414$$

Other important and useful properties of the integral that we need to be aware of are the following.

---

6. Splitting an interval: Assume that $f(x)$ is bounded on the interval $[a, b]$ and let c be any real number (typically $a < c < b$). Then $f(x)$ is integrable on $[a, b]$ if and only if $f(x)$ is integrable on the two intervals $[a, c]$ and on $[c, b]$. In this case,

$$\int_a^b f(x)dx = \int_a^c f(x)dx + \int_c^b f(x)dx \qquad (8.30)$$

7. Absolute value: If $f(x)$ is integrable on $[a, b]$ then so is $|f(x)|$ and

$$\left|\int_a^b f(x)dx\right| \leq \int_a^b |f(x)|\,dx \qquad (8.31)$$

It is worth pointing out that the *inequality in (8.31) is usually strict*; so the absolute value and the integral can be switched but at a cost. For example, using the power rule for integrals, you can easily show that

$$\left|\int_{-1}^1 xdx\right| = \left|\frac{1^2}{2} - \frac{(-1)^2}{2}\right| = \left|\frac{1}{2} - \frac{1}{2}\right| = 0$$

On the other hand, since $|x| = -x$ when $x \leq 0$ and $|x| = x$ when $x \geq 0$ we find that:

$$\int_{-1}^{1} |x|dx = \int_{-1}^{0} -xdx + \int_{0}^{1} xdx$$
$$= -\left(\frac{0^2}{2} - \frac{(-1)^2}{2}\right) + \left(\frac{1^2}{2} - \frac{0^2}{2}\right)$$
$$= \frac{1}{2} + \frac{1}{2}$$
$$= 1$$

As we see in this example, the integral inside the absolute value on the left hand side of (8.31) is subject to the cancellation of postive and negative quantities. Even if these numbers did not cancel out to 0 the result wouldn't be greater than what we would get from the right hand side of (8.31), since the absolute value inside the integral is always non-negative and it generates no cancellations.

The Fundamental Theorem of Calculus, Part 2.

Part 1 of the Fundamental Theorem requires that $f(x)$ be *both integrable and have an antiderivative* on the interval of interest. *These are independent properties* because as we have seen, one does not generally imply the other. But FT1 doesn't say anything if either property is missing. It is the second part of the Fundamental Theorem, FT2, that connects these two properties.

In particular, we discover the important fact that on closed and bounded intervals, continuous functions are not only integrable but they also have antiderivatives.[12]

The basic idea is very simple:

For the sake of illustration, assume for the moment that $f(x)$ is both continuous and positive so that $\int_a^b f(x)dx$ is just the area under the graph of $f(x)$. Pick an arbitrary c in $[a,b]$; the area $A(C)$ under the graph from $x = a$ to $x = c$ is $\int_a^c f(x)dx$ as we see in Figure 8.18.

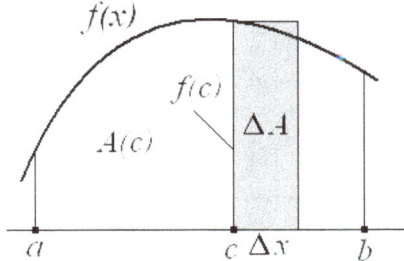

Figure 8.18: Illustrating a point: the second fundamental theorem

In Figure 8.18 we also see that the area from $x = c$ to $x = c + \Delta x$, call it ΔA, is approximately equal to the area of the shaded rectangular strip; so

$$\Delta A \simeq f(c)\Delta x$$

---
[12]FT2 does not help with the explicit calculation of antiderivatives, but it does provide a blueprint for estimating them using the integral.

Integrals

Dividing by Δx gives
$$\frac{\Delta A}{\Delta x} \simeq f(c)$$
This approximation gets better as $\Delta x \to 0$ so that
$$\frac{dA}{dx} = f(c)$$

This means that the area function $A(x)$, which is the integral of $f(x)$, is also an antiderivative of $f(x)$!

The above observation is FT2 in a nutshell; let's get to the details now, and discover a richer story about a much larger class of functions than the area function $A(x)$.

The Fundamental Theorem, Part 2 (FT2). *Assume that $f(x)$ is an integrable function on the interval $[a,b]$ and for each x in $[a,b]$ define*

$$F(x) = \int_a^x f(t)dt \tag{8.32}$$

(a) The function $F(x)$ above is continuous on $[a,b]$;
(b) If the original function $f(x)$ is continuous at a point c in $[a,b]$ then $F(x)$ is differentiable at $x = c$ and
$$F'(c) = f(c)$$
Therefore, if $f(x)$ is continuous on $[a,b]$ then $F(x)$ is an antiderivative of $f(x)$ with the property that $F(a) = 0$.

Inside the integral in (8.32) I used the symbol t to distinguish it from the variable x.[13]

Average values.

We now discuss the average values of functions over intervals. This is an interesting application of integration that does not require a background in science or engineering. First, we discuss the Mean Value Theorem for Integrals, which is the proverbial "low-hanging fruit" after the derivation of the Fundamental Theorem. It is a good application of that theorem because it uses both FT1 and FT2 thus illustrating the significance of the two parts working together.

The Mean Value Theorem for Integrals (MVTI): *If $f(x)$ is continuous on $[a,b]$ then there is a number c in $[a,b]$ such that*

$$\int_a^b f(x)dx = f(c)(b-a) \tag{8.33}$$

In terms of areas this statement says that (for a positive function) the area under the graph of $f(x)$ is equal to the area of a suitably chosen rectangle of height $f(c)$ and base-length $b - a$; see

---

[13] Any symbol is actually fine as long as it causes no confusion; $\int_a^b f(t)dt$ and $\int_a^b f(x)dx$ say exactly the same thing. The symbol that is used inside the integral is typically called the "dummy variable" because it is a temporary variable that is "integrated away". The answer is a number or a parameter that does not contain the dummy variable.

Figure 8.19 the shaded area over-estimate by the rectangle is equal in size to the also shaded area under-estimate.

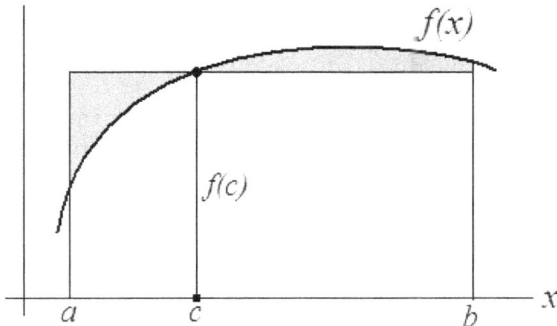

Figure 8.19: Illustrating the Mean Value Theorem for Integrals

The MVTI is generally false if $f(x)$ is *not* continuous; looking at Figure 8.19 again if there were a hole in the graph of $f(x)$ right above where c is located then $f(c)$ would have a higher or lower value, making it greater or less than required in (8.33).

It is easy to see why the MVTI is true. By FT2 the continuous function $f(x)$ has an antiderivative, namely, $F(x) = \int_a^x f(t)dt$. Being an antiderivative, $F(x)$ is differentiable so by the original MVT (for derivatives) there is a point c in $[a,b]$ such that

$$F(b) - F(a) = F'(c)(b-a)$$

But $F'(c) = f(c)$ and by FT1 the left hand side of the above equality is the integral $\int_a^b f(x)dx$. So (8.33) is true!

The number $f(c)$ in (8.33) is the *mean or average value of the function* $f(x)$ over the interval $[a,b]$ though this is not readily apparent. To see why the term "mean value" is apt let's go back to the usual idea of average value. The average of N numbers $y_1, y_2, \ldots, y_N$ is

$$\frac{y_1 + y_2 + \cdots + y_N}{N}$$

If $y = f(x)$ is a function defined on an interval $[a,b]$ then its average value involves all of its values, that is, *an (uncountable) infinity of numbers*. Obviously we can't add up so many numbers, but the average of any finite number of values of the function $f(x)$ at points $x_1, x_2, \ldots, x_N$ in an interval $[a,b]$ is just:

$$\frac{f(x_1) + f(x_2) + \cdots + f(x_N)}{N} \tag{8.34}$$

To go from here to the average of the whole function, suppose that the N points $x_1, x_2, \ldots, x_N$ are equally spaced so that with $x_1 = a$ and $x_N = b$ we have a regular partition of the interval $[a,b]$. Each subinterval of this partition has length

$$\Delta x = \frac{b-a}{N}$$

Integrals

Other than $b-a$ all the ingredients of a Riemann sum are present in (8.34). So to complete the picture, lets multiply and divide the fraction in (8.34) by $b-a$ and write it as

$$\frac{1}{b-a}\frac{b-a}{N}[f(x_1)+f(x_2)+\cdots+f(x_N)]=\frac{1}{b-a}\sum_{i=1}^{N}f(x_i)\Delta x$$

If $f(x)$ is integrable (for instance, if it is continuous) then the Riemann sum on the right hand side converges to the integral of $f(x)$ as we refine the partition by increasing N indefinitely. This leads to the following:

> **Average Value of a function:** If $f(x)$ is integrable over an interval $[a,b]$ then its average value over this interval is
> $$\frac{1}{b-a}\int_a^b f(x)dx$$

For example, let's calculate the average value of $f(x)=1-1/x^2$ over the interval $[1,3]$.

$$\frac{1}{3-1}\int_1^3\left(1-\frac{1}{x^2}\right)dx=\frac{1}{2}\left[x+\frac{1}{x}\right]_1^3=\frac{1}{2}\left[\left(3+\frac{1}{3}\right)-\left(1+\frac{1}{1}\right)\right]=\frac{2}{3}\simeq 0.67$$

It is worth mentioning that we can use the average value to find a number c that works in (8.33). For $f(x)=1-1/x^2$ over $[1,3]$ we have

$$f(c)=\frac{1}{3-1}\int_1^3\left(1-\frac{1}{x^2}\right)dx$$

$$1-\frac{1}{c^2}=\frac{2}{3}$$

We use a little algebra to solve the above equation for c:

$$\frac{1}{c^2}=\frac{1}{3}$$
$$c^2=3$$

The last equation has two possible solutions: $c=\pm\sqrt{3}$. We are looking for a c in the interval $[1,3]$ where only $c=\sqrt{3}\simeq 1.73$ works.

Exercise 112 *(a) Find the average value of $f(x)=x^2$ first over the interval $[-1,1]$ and then over $[0,2]$ which has the same length. In each case, find the point c and the height $f(c)$ of the rectangle that satisfies (8.33). Notice that more than one value of c works in $[-1,1]$; sketching the graph of x^2 sheds more light! Find all valid values of c for both intervals.*

(b) Find the average value of $\sin x$ over one of its humps, say, $0\leq x\leq \pi$. Can you tell what the average value is for two adjacent humps ($0\leq x\leq 2\pi$, say) without explicitly calculating it?

(c) The function $y=\sin x$ takes on every y value from 0 to 1 exactly twice (except for $y=1$ which occurs only once at $x=\pi/2$). Why is it then that the average of $\sin x$ over a hump is not $1/2$?

8.4 Logarithmic and exponential functions: from areas to exponents

We can define variable-power functions for integers rather easily: if a is a nonzero real number and n is a positive integer then

$$a^n = \underbrace{(aa\cdots a)}_{n \text{ times}}$$

If m is another positive integer then

$$a^{m+n} = \underbrace{(aa\cdots aa)}_{m+n \text{ times}} = \underbrace{(aa\cdots a)}_{m \text{ times}}\underbrace{(a\cdots aa)}_{n \text{ times}} = a^m a^n$$

and similarly,

$$(a^m)^n = \underbrace{(aa\cdots a)}_{n \text{ times}}\ldots\underbrace{(aa\cdots a)}_{m \text{ times}} = a^{m+n}$$

It is possible to build up the exponential function from this starting point and then to define logarithms as their inverse functions. But, given that we have already defined logarithms using the integral, it is easier for us to start with the logarithm and define the exponential function as the inverse function. In the process we also find interesting applications of the Fundamental Theorem and thus, also of integration.

The natural logarithm function.

Let's use the second half of the Fundamental Theorem to study a special case that we left out from the power rule for integrals (8.22). The function $f(x) = x^{-1} = 1/x$ is continuous for all $x > 0$ so its integral exists for every closed interval $[a, b]$ where $0 < a < b$. Let's consider the antiderivative of $1/x$ as the following function $L : (0, \infty) \to \mathbb{R}$

$$L(x) = \int_1^x \frac{1}{t} dt, \quad x > 0 \tag{8.35}$$

Note that $L(1) = 0$ and by FT2, $L'(x) = 1/x$ so $L(x)$ *is differentiable on its domain* $(0, \infty)$.

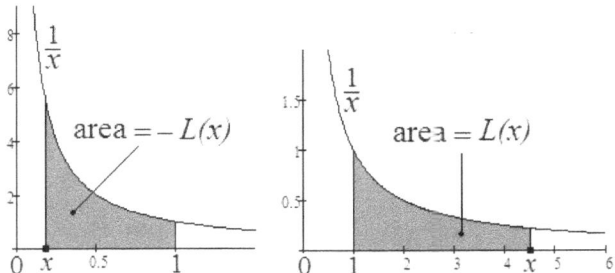

Figure 8.20: Defining the natural logarithm using an integral

Integrals

Further, since $x > 0$ we conclude that $L(x)$ *is an increasing function*. As we see Figure 8.20, for each fixed $x > 1$ the value of $L(x)$ is the area under the graph of $1/x$; if $x < 1$ then the area under the graph of $1/x$ is the negative of the value of $L(x)$ since for $0 < x < 1$

$$L(x) = \int_1^x \frac{1}{t} dt = -\int_x^1 \frac{1}{t} dt < 0$$

It is not hard to prove that $L(x) \to \infty$ as $n \to \infty$.

Consider a regular partition $P_n = \{1, 2, 3, \ldots, n\}$ of the interval $[1, n]$. Note that all subintervals of P_n have length $\Delta x = 1$. Let's calculate the lower Riemann sum $R_L(P_n)$ corresponding to P_n as an underestimation of $L(n)$:

$$L(n) \geq R_L(P_n) = \sum_{i=1}^n \frac{1}{x_i} \Delta x = \sum_{i=1}^n \frac{1}{i+1}$$

The lower sum $R_L(P_n)$ is shaded in Figure 8.21; we see that the total areas of the shaded strips is less than the actual area $L(n)$ under the graph of $1/x$ from $x = 1$ to $x = n$.

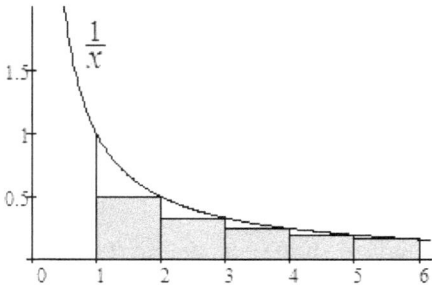

Figure 8.21: A Riemann sum approximation of the natural logarithm

Next, take a close look at the series for $R_L(P_n)$: we observe that this is exactly the n-th partial sum (minus 1) of the harmonic series that we earlier showed to be a divergent (unbounded) series. Therefore, $L(n) \to \infty$ as $n \to \infty$.

Now, let x be any positive real number, not necessarily an integer. If x goes to infinity then so does the *integer* $[x]$, namely, the greatest integer that is less than or equal to x. Because $L(x)$ is an increasing function $L(x) \geq L([x])$. We conclude that $L(x) \to \infty$ as $x \to \infty$, as was claimed.

An consequence of this observation that is important to our later work is that $L(x)$ must exceed 1 for every sufficiently large value of x, so the Intermediate Value Theorem and the increasing nature of $L(x)$ imply that there is a *unique* real number e such that

$$e > 1 \quad \text{and} \quad L(e) = 1.$$

This observation leads to the definition of the natural logarithm[14]

---

[14] The Scottish mathematician John Napier (1550-1617) is credited with the discovery of logarithm around 1614; he found it to be a powerful computational aid. The term "natural logarithm" is due to Nicholas Mercator (1620-1687), a German mathematician who should not be confused with the more famous cartographer Gerhard (Gerardus) Mercator who lived in the 1500's and created the familiar Mercator projection. The symbol e is due to Euler.

$L(x)$ is the *natural logarithm function, or the logarithm to the base e*. It is commonly denoted $\ln x$.

Numerical approximations of e to high levels of accuracy are known; for example, we discover the following infinite series formula later:

$$e = \sum_{n=0}^{\infty} \frac{1}{n!} = 1 + \frac{1}{1!} + \frac{1}{2!} + \frac{1}{3!} + \cdots$$

This series converges rapidly (as you can verify using the ratio test) so just a few terms give very good approximations of e. For instance, adding just 10 terms of the series gives:

$$e \simeq \sum_{n=0}^{9} \frac{1}{n!} \simeq 2.7182815$$

an estimate that is accurate to 6 decimal places.

The values of $\ln x$ can be accurately approximated using Riemann sums or other, more accurate sums that are not of Riemann type; we discuss some of these latter sums below. A graph of $\ln x$ is shown in Figure 8.22.

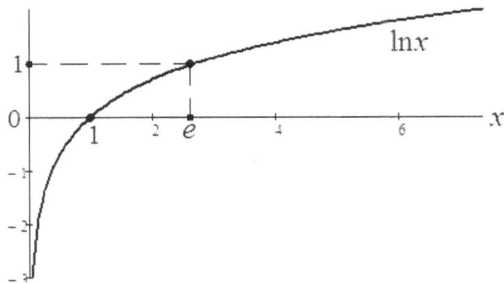

Figure 8.22: A graph of the natural logarithm function

To summarize, we have shown that the natural logarithm $\ln x$ that is defined by (8.35) is an unbounded, increasing and differentiable function whose derivative is $1/x$. We have also supplied the missing case in the power rule (8.22) by showing, via FT1, that

$$\int_a^b x^{-1} dx = \int_a^b \frac{1}{x} dx = \ln b - \ln a, \quad 0 < a < b \tag{8.36}$$

The logarithmic growth of the harmonic series.

We can use the logarithm to show not only that the old harmonic series $\sum_{n=1}^{\infty} 1/n$ diverges slowly, but also that it diverges at a logarithmic rate.

Let's go back to Figure 8.21 and observe that

$$\frac{1}{2} + \frac{1}{3} + \frac{1}{4} + \frac{1}{5} + \frac{1}{6} < \ln 6$$

Integrals

since the area of the dark strips adds up to less than the area under $1/x$. This observation is not limited to the number 6 and can be extended to any positive integer k, so

$$\sum_{n=2}^{k} \frac{1}{n} < \ln k$$

Therefore, adding 1 to both sides gives:

$$\sum_{n=1}^{k} \frac{1}{n} < 1 + \ln k \tag{8.37}$$

The strips in Figure 8.21 happen to be rectangles with heights $1/x_n$ where x_n is the right hand endpoint of the interval $[n, n+1]$ for $n = 1, 2, 3, \ldots$ The areas of these rectangles underestimate the area under $1/x$. Now, imagine using the strips whose heights are calculated using the left hand endpoints of the same integer intervals; the areas of these strips overestimate the area under $1/x$ so that

$$\sum_{n=1}^{k} \frac{1}{n} > \ln(k+1) \tag{8.38}$$

For instance, if $k = 5$ then the sum of the areas of the 5 overestimating strips exceeds the area under $1/x$ from 1 to 6. Now, from (8.37) and (8.38) it follows that

$$\ln(k+1) < \sum_{n=1}^{k} \frac{1}{n} < 1 + \ln k \tag{8.39}$$

We may conclude from these inequalities that:

For large values of k, *the partial sums of the harmonic series increase logarithmically.*

More precisely, since $\ln(k+1) \simeq \ln k$ when k is a large positive integer, the partial sum $\sum_{n=1}^{k} 1/n$ is between $\ln k$ and $1 + \ln k$ and if we use the average of these bounds as an approximation to the partial sum then we have the estimate:

$$\sum_{n=1}^{k} \frac{1}{n} \simeq \frac{1}{2} + \ln k \quad (\text{large } k)$$

For instance, if $k = 100,000$ then $\ln(k+1) \simeq 11.512935$, $\ln k \simeq 11.512925$ and

$$\sum_{n=1}^{k} \frac{1}{n} \simeq 12.090, \quad \frac{1}{2} + \ln k \simeq 12.013$$

The properties of the natural logarithm.

We can use our definition of the natural logarithm to discover some additional properties of $\ln x$. For instance, let's see why the following identity holds:

$$\ln(uv) = \ln u + \ln v \tag{8.40}$$

assuming, of course, that both u and v are positive real numbers. This is done using a simple trick that lets us use the derivative.

Let's fix v and think of u as a variable so that we are thinking of a function

$$g(u) = \ln(uv)$$

With v held constant, we use the chain rule to find[15]

$$g'(u) = \frac{d}{du}\ln(uv) = \frac{1}{uv}\frac{d}{du}(uv) = \frac{1}{uv}v = \frac{1}{u}$$

This means that $g(u)$ is an antiderivative of $1/u$, that is, $g(u) = \ln u + C$ where C is a constant of integration. Notice that

$$g(1) = \ln 1 + C = C$$

and

$$g(1) = \ln(1v) = \ln v$$

so

$$\ln(uv) = g(u) = \ln u + g(1) = \ln u + \ln v$$

and we have shown that (8.40) is true!

Exercise 113 *(a) Use the fact that $u(1/u) = 1$ and (8.40) to prove that for all $u > 0$*

$$\ln\left(\frac{1}{u}\right) = -\ln u \qquad (8.41)$$

(b) Use (a) to prove that

$$\ln\frac{u}{v} = \ln u - \ln v$$

(c) Why is the following not true?

$$\ln\frac{u}{v} = \frac{\ln u}{\ln v}$$

Consider testing it with some specific values, like $u = 1$ and any $v > 0$. What if $v = 1$ and $u > 0$?

(d) Explain why $\ln x \to -\infty$ as x approaches zero from the right; consider what happens to $u = 1/x$ as $x \to 0$ and use (a).

A property of the natural logarithm that, among other things, is important for the development of exponential functions is the following:

$$\ln u^p = p \ln u \qquad (8.42)$$

where p is any fixed *rational* number. The restriction of p to rational numbers is temporary and needed because *we have not yet defined what we mean by raising a real number u to an irrational power*, like $3^{\sqrt{2}}$ or $(\sqrt{2})^\pi$. This issue will be resolved when we develop a general concept of exponential function that is designed essentially for this purpose.

---

[15]The value of v is fixed but not specified; it can be any positive number.

Integrals

To prove (8.42) we start with $p = n$ a positive integer. Then repeated applications of (8.40) n times give

$$\ln u^n = \ln(u \cdots u) = \ln u + \cdots + \ln u = n \ln u \qquad (8.43)$$

Therefore, (8.42) is true when p is a positive integer. It is also true if $p = 0$ because[16]

$$\ln u^0 = \ln 1 = 0 = 0 \ln u$$

If $p = -n$ is a negative integer then by (8.41) and (8.43):

$$\ln u^{-n} = \ln \left(\frac{1}{u}\right)^n = n \ln \frac{1}{u} = -n \ln u$$

so (8.42) still holds. Next, let $p = 1/n$ where n is a positive integer and notice that since $u^{1/n} = \sqrt[n]{u}$ it follows that

$$n \ln \sqrt[n]{u} = \ln \left(\sqrt[n]{u}\right)^n = \ln u$$

which after dividing by n gives (8.42) again. Finally, if $p = m/n$ is any rational number then

$$\ln u^{m/n} = \ln \left(\sqrt[n]{u}\right)^m = m \ln \sqrt[n]{u} = \frac{m}{n} \ln u$$

so we see that (8.42) is true for all rational p, as claimed.

The natural exponential function.

The fact that $\ln x$ is an increasing function implies that it is one-to-one and therefore, has an inverse function. Let's go back to the L notation for a moment: $\ln x = L(x)$. The inverse function $L^{-1} : \mathbb{R} \to (0, \infty)$ satisfies the relations

$$L^{-1}(L(x)) = x \quad \text{and} \quad L(L^{-1}(y)) = y$$

for all x in $(0, \infty)$ and y in $\mathbb{R}$. See Figure 8.23.

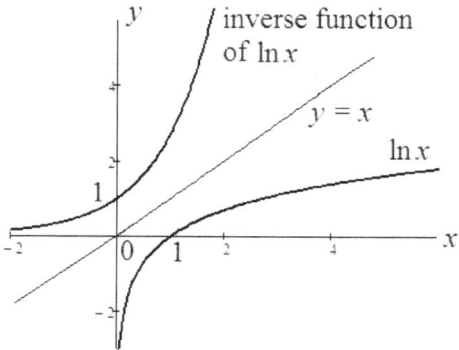

Figure 8.23: The natural logarithm and its inverse function

---

[16] To see why $u^0 = 1$ note that $u^0 = u^{1-1} = u^1 u^{-1} = u(1/u) = 1$. We need to exclude the value $u = 0$ because we are using the fraction $1/u$. In fact, 0^0 is an "indeterminate form" that can have no specific value.

Now let's define $E(y) = L^{-1}(y)$. Then we can rewrite the above two inverse function relations as
$$E(\ln x) = x \quad \text{and} \quad \ln E(y) = y \qquad (8.44)$$

These relations allow us to derive the properties of the inverse function $E(y)$. First, since the range of $E(y)$ is $(0, \infty)$ (the same as the domain of $\ln x$) we have
$$E(y) > 0 \quad \text{for all } y \text{ in } \mathbb{R}$$

Next, setting $x = 1$ and $x = e$ in the equality on the left in (8.44) gives
$$E(\ln 1) = 1, \qquad E(\ln e) = e$$

which we can rewrite as:
$$E(0) = 1, \qquad E(1) = e$$

I now show that $E(y)$ is differentiable and find its derivative using the inverse function relations (8.44).

Let y be a fixed real number and y_n any sequence of real numbers that converges to y with $y_n \neq y$ for all n. If $x = E(y)$ and $x_n = E(y_n)$ for each n then $y = \ln x$ and $y_n = \ln x_n$ and
$$\frac{E(y_n) - E(y)}{y_n - y} = \frac{x_n - x}{\ln x_n - \ln x} = \frac{1}{(\ln x_n - \ln x)/(x_n - x)}$$

Recall that $y_n \neq y$ so $\ln x_n \neq \ln x$ for all n; since $\ln x$ is a one-to-one function we see that $x_n \neq x$ for all n. Taking the limit as $n \to \infty$ we get:
$$E'(y) = \lim_{n \to \infty} \frac{E(y_n) - E(y)}{y_n - y} = \frac{1}{\lim_{n \to \infty}(\ln x_n - \ln x)/(x_n - x)} = \frac{1}{d/dx(\ln x)} = x$$

Since $x = E(y)$ we have derived the formula for the derivative of the exponential function:
$$E'(y) = E(y)$$

The derivative of $E(y)$ is itself; as a consequence we see that the slope of the tangent line at every point equals the value of the function.

Another important by-product of the derivative calculation above is that it shows $E(y)$ to be *continuous and (strictly) increasing on its domain* $\mathbb{R}$. These facts come in handy when studying the properties and applications of the inverse function $E(y)$, better known as the *natural exponential function*.

> **The exponential notation.** $E(y)$ is usually written as e^y. It is not hard to see why this is a valid interpretation, but first it is necessary to explain what we mean by raising the number e to a real number power y, which can certainly be irrational. Recall from (8.42) that if $y = m/k$ is a fixed rational number then e^y is defined as $e^{m/k} = \sqrt[k]{e^m}$ and
> $$\ln e^y = y \ln e = y$$
> So it follows that for each rational number y
> $$\ln e^y = \ln E(y)$$

Integrals 265

and because $\ln x$ is a one-to-one function we must have

$$e^y = E(y), \quad y \text{ rational}$$

Now recall that the rational numbers are *dense* in $\mathbb{R}$ so if y is in $\mathbb{R}$ then there is a sequence of rational numbers q_n that converge to y; that is, $y = \lim_{n \to \infty} q_n$. Therefore, remembering that $E(y)$ is continuous:

$$E(y) = E(\lim_{n \to \infty} q_n) = \lim_{n \to \infty} E(q_n) = \lim_{n \to \infty} e^{q_n}$$

This shows that if we define

$$e^y = \lim_{n \to \infty} e^{q_n}$$

then the interpretation of $E(y) = e^y$ as the "y-th power of e" is extended to all real y (rational or irrational). This will help us define general exponential functions below.

The properties of the natural exponential function.

Also as promised, with the help of the above extension we can extend (8.42) to all real p. Let $p = \lim_{n \to \infty} q_n$ and note that

$$\ln u^p = \ln \left(\lim_{n \to \infty} u^{q_n} \right)$$
$$= \lim_{n \to \infty} \ln(u^{q_n}), \quad \text{since } \ln x \text{ is continuous}$$
$$= \lim_{n \to \infty} q_n \ln u, \quad \text{using (8.42) on the rational powers } q_n$$
$$= p \ln u$$

We conclude that:

The power rule (8.42) holds for all real p.

We discovered earlier that $\ln x \to \infty$ as $x \to \infty$ and $\ln x \to -\infty$ as $x \to 0$. Since $\ln x$ is continuous the Mean Value Theorem implies that it is onto $\mathbb{R}$. So for arbitrary real numbers u and v there are positive real numbers s and t such that $u = \ln s$ and $v = \ln t$. Note that s and t are uniquely determined by u and v because $\ln x$ is a one-to-one function. Now, from (8.44) and using (8.40) we derive

$$E(u + v) = E(\ln s + \ln t) = E(\ln st) = st$$

Since $s = E(\ln s) = E(u)$ and likewise, $t = E(v)$ we have shown that

$$E(u + v) = E(u)E(v) \tag{8.45}$$

This identity translates into the more familiar form:

$$e^{u+v} = e^u e^v$$

Exercise 114 *Use an argument similar to the one above that we used to derive (8.45) to derive each of the following identities:*

$$E(-u) = \frac{1}{E(u)} \qquad E(pu) = [E(u)]^p$$

where p is a fixed real number. In the exponential notation, these identities translate into

$$e^{-u} = \frac{1}{e^u} \qquad e^{pu} = (e^u)^p$$

The power function and its derivative revisited.

Back in Chapter 7 we noted that the power function x^p is not well-defined for all real numbers if p is not a positive integer or 0. If p is negative then x^p is not defined at $x = 0$ (think of $x^{-1} = 1/x$). If p is a rational number that is not an integer the x^p may be undefined for all $x \leq 0$ (think of $x^{-1/2} = 1/\sqrt{x}$). But at least we were able to define x^p for all rational p using roots and fractions as necessary.

If p is irrational then it is not obvious that x^p can be defined even for $x > 0$. For instance, how do we define $x^{\sqrt{2}}$?

The above extension of (8.42) to all real values of p provides a way of defining x^p for all real numbers p as long as $x > 0$ (the domain of $\ln x$). Notice that

$$\ln(x^p) = p \ln x$$

is well-defined for all $x > 0$ and can be written in inverse form as an exponential function:

$$x^p = e^{p \ln x}$$

Thus, for example, $x^{\sqrt{2}} = e^{\sqrt{2} \ln x}$ and $x^{-\pi} = e^{-\pi \ln x}$.

Feeling empowered by this new idea, we may wonder if irrational powers of negative numbers can also be meaningful. The short answer is obvious: no, because $\ln x$ is not defined–as a real number. But if we extend our number system to the complex numbers then we can define real powers of negative numbers also.

This should not come as a surprise since we have already encountered a prominent example, even without irrational powers: $(-1)^{1/2} = i$ is the unit imaginary number. To go further and define, say, $(-1)^{\sqrt{2}}$ we need to define the complex versions of the logarithm and exponential functions, topics that belong to the field of complex analysis proper.

With the definition of power functions extended as above an important question follows: is the derivative of a power function still given by the formula in Chapter 7?

The answer is easy to figure out using the chain rule:

$$\frac{d}{dx}(x^p) = \frac{d}{dx}(e^{p \ln x}) = e^{p \ln x} \frac{d}{dx}(p \ln x) = e^{p \ln x} p \frac{1}{x} = p x^p \frac{1}{x} = p x^{p-1}$$

as expected!

The general exponential and logarithmic functions.

We now define the extensions of the natural exponential and logarithmic functions. It is more convenient to start with the general exponential functions since there is no "natural" interpretation of the general logarithmic functions in terms of an integral.

We want to extend the concept of "raising to a power" to all real numbers. The standard way is by defining a function a^x that we prove to be continuous. Then we can use this continuity and the

Integrals 267

fact that rational numbers are dense among the real numbers to bump up from rational powers and bases to all real powers and bases. We did the same thing when defining e^x.

Let a be any fixed real number, where $a > 0$ and $a \neq 1$.[17] If x is a real variable then how do we define a function a^x? We have already defined the function e^y; now consider $y = x \ln a$ and use the results of the last sub-section to derive

$$e^{x \ln a} = e^{\ln(a^x)} = a^x \tag{8.46}$$

> The function a^x is *the (general) exponential function in base a* where $a > 0$ and $a \neq 1$. If $a = e$ then we recover the natural exponential function (in base e) since $\ln e = 1$.

Notice that the domain of a^x is all of $\mathbb{R}$ since the quantity $x \ln a$ in $e^{x \ln a}$ ranges over $\mathbb{R}$ as x does. Further, $a^x > 0$ for all real x because $e^{x \ln a} > 0$. Figure 8.24 shows the graphs of a^x for three different values of a.

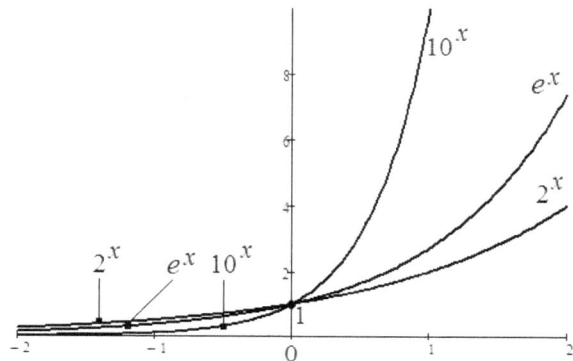

Figure 8.24: Exponential functions with different bases

By now you may be wondering what happens if $a = 1$; let's see. Since $1^n = 1$ for every positive integer n we can use arguments similar to what we have seen before to deduce that $1^q = 1$ for every rational number. Since rationals are dense in the set of all real numbers it follows that $1^x = 1$ for all real x. This is a constant function; you may think of it as a *degenerate* exponential function but in any case, it is an uninteresting case that we may ignore without any loss.

Using (8.46) we can derive the properties of a^x rather quickly. For example, the derivative formula is readily found using the chain rule:

$$\frac{d}{dx} a^x = \frac{d}{dx} e^{x \ln a} = e^{x \ln a} \frac{d}{dx}(x \ln a) = a^x \ln a \tag{8.47}$$

In particular, a^x is continuous since is differentiable. The algebraic properties of exponents are likewise a breeze; for instance, for all real numbers u, v

$$a^{u+v} = e^{(u+v) \ln a} = e^{u \ln a + v \ln a} = e^{u \ln a} e^{v \ln a} = a^u a^v$$

---

[17]It is necessary that $a > 0$ because we will be using the quantity $\ln a$. As for why $a \neq 1$, see below.

Exercise 115 *Derive the following properties of the general exponential function:*

$$a^{-u} = \frac{1}{a^u} \qquad a^{pu} = (a^u)^p \qquad (ab)^u = a^u b^u$$

Assume that $p \in \mathbb{R}$ and $b > 0$ are fixed real numbers.

We can quickly gain significant information from (8.47) about the function a^x. Specifically, the derivative is negative if $a < 1$ since $\ln a < 0$ and the derivative is positive if $a > 1$; see Figure 8.25. Therefore,

> *The general exponential function a^x is continuous and monotone on $\mathbb{R}$, decreasing if $a < 1$ and increasing if $a > 1$.*

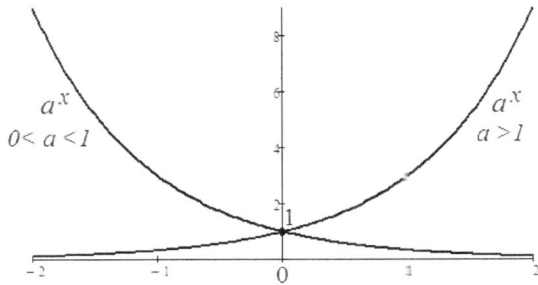

Figure 8.25: Exponential functions with bases larger and smaller than 1

The monotone feature of a^x implies that it has an inverse function whose domain is the range $(0, \infty)$ of a^x. Like the case $a = e$ the inverse function is a logarithmic function that we denote by $\log_a x$ and call it *the logarithm to the base a* (here it is necessary that $a \neq 1$; can you recall why?)

$$y = \log_a x \quad \text{if} \quad x = a^y$$

Of course, $\ln x = \log_e x$ is the logarithm to the base e. This definition gives us the useful inverse function relations:

$$\log_a(a^x) = x, \qquad a^{\log_a x} = x$$

From these we may calculate the various properties of the base a logarithm function, which are identical to those of the natural logarithm $\ln x$. Just as we defined $a^x = e^{x \ln a}$ we can define $\log_a x$ in term of the natural logarithm. From the inverse function relations we have:

$$x = a^{\log_a x} = e^{(\log_a x)(\ln a)}$$

Set $w = (\log_a x)(\ln a)$ so that the above reads: $x = e^w$. Therefore, $w = \ln x$ so that

$$(\log_a x)(\ln a) = \ln x$$

$$\log_a x = \frac{\ln x}{\ln a} \qquad (8.48)$$

It is worth noticing that if $a < 1$ then $\ln a < 0$ so $\log_a x$ is a flipped version of $\ln x$.

Integrals   269

Exercise 116 *Explain why* $\log_a x > \ln x$ *if* $1 < a < e$ *but* $\log_a x < \ln x$ *if* $a > e$.

All the properties of $\log_a x$ are quickly found using (8.48). For example, since $1/\ln a$ is a constant and $\ln x$ is differentiable it follows that $\log_a x$ is differentiable too and its derivative formula is:

$$\frac{d}{dx}\log_a x = \frac{1}{x \ln a}$$

Exercise 117 *Use (8.48) to show that if* $a > 0$, $a \neq 1$ *then*

$$\log_a 1 = 0, \qquad \log_a a = 1$$

Also verify the following identities:

$$\log_a(uv) = \log_a u + \log_a v, \qquad \log_a \frac{1}{u} = -\log_a u, \qquad \log_a(u^p) = p \log_a u$$

Assume that u, v *are positive real variables,* p *is any fixed real number.*

It worth mentioning here that since scientific and engineering calculations usually involve powers of 10 it is more convenient to use the base 10 logarithm in those contexts than the natural logarithm; we also usually drop the base to avoid notational clutter. Using (8.48) we have

$$\log_{10} x = \log x = \frac{\ln x}{\ln 10} \simeq 0.43 \ln x \qquad (8.49)$$

We can convert $\ln x$ to $\log x$ if need be; using (8.48) with $a = 10$ we calculate

$$(\log e)(\ln 10) = \ln e = 1$$

$$\ln 10 = \frac{1}{\log e}$$

So from (8.49) we derive

$$\ln x = (\log x)(\ln 10) = \frac{\log x}{\log e} \simeq 2.3 \log x$$

Let's end this section with a few computational exercises that are helpful in reviewing some of the topics that we discussed earlier.

Exercise 118 *(a) We defined* $\ln x$ *as an integral. But since it is a continuous function* $\ln x$ *is also integrable; show that:*

$$\int_1^e \ln x \, dx = 1$$

using integration by parts with $f(x) = \ln x$ and $g'(x) = 1$. Then use (8.48) to show that

$$\int_1^e \log x \, dx = \log e$$

(b) *Explain why the following functions are differentiable and use appropriate formulas to prove each of the following equalities:*

$$\frac{d}{dx}e^{-x^2} = -2xe^{-x^2}, \qquad \frac{d}{dx}e^{\sin x} = e^{\sin x} \cos x, \qquad \frac{d}{dx}\pi^{ex} = \pi^{ex} \ln(\pi^e)$$

$$\frac{d}{dx}(x \log x - x + e) = \log x + \log e - 1, \qquad \frac{d}{dx} \ln \sqrt{x^2 + 1} = \frac{x}{x^2 + 1}$$

To simplify calculations for the last formula use the fact that $\ln u^{1/2} = (1/2) \ln u$.

8.5 Numerical approximations of integrals: all you need is a computer

In most practical situations we can rarely find the precise value of an integral using antiderivatives because most antiderivatives have no known formulas; for example, there are no known formulas for the antiderivatives of rather familiar functions like

$$\sqrt{x^3 + 1}, \quad \sin(x^2), \quad e^{-x^2}$$

We know that these functions are indeed integrable because they are continuous (actually, differentiable) on their domains. Integrals of type

$$\int_a^b e^{-x^2} dx$$

arise in probability theory and their numerical values are of great practical interest. If we cannot find the precise value of such an integral then can we at least approximate that value?

To answer this question we abandon the search or antiderivatives and go back to the original definition of the integral. You may recall that by refining a partition and reducing its mesh we can estimate the value of an integral using a Riemann Sum.

We get different Riemann sums if we choose different values x_i^* in the subintervals $[x_{i-1}, x_i]$. But *where* in the subintervals do we pick these x_i^*?

A simple choice that often gives good approximations is when x_i^* is right in the middle of the subinterval.

The midpoint rule. The selection:

$$x_i^* = \frac{x_i + x_{i-1}}{2}$$

gives the approximation

$$\int_a^b f(x)dx \simeq \sum_{i=1}^N f\left(\frac{x_i + x_{i-1}}{2}\right) \Delta x_i$$

Integrals

For instance, suppose we want to estimate

$$\int_{-1}^{2} e^{-x^2} dx$$

A coarse (regular) partition $P_1 = \{-1, 0, 1, 2\}$ with $\Delta x = 1$ gives the approximation

$$\int_{-1}^{2} e^{-x^2} dx \simeq e^{-[(-1+0)/2]^2}(1) + e^{-[(0+1)/2]^2}(1) + e^{-[(1+2)/2]^2}(1)$$

$$= e^{-1/4} + e^{-1/4} + e^{-9/4}$$

$$\simeq 1.663$$

Not a bad approximation of the answer 1.62890552...![18]

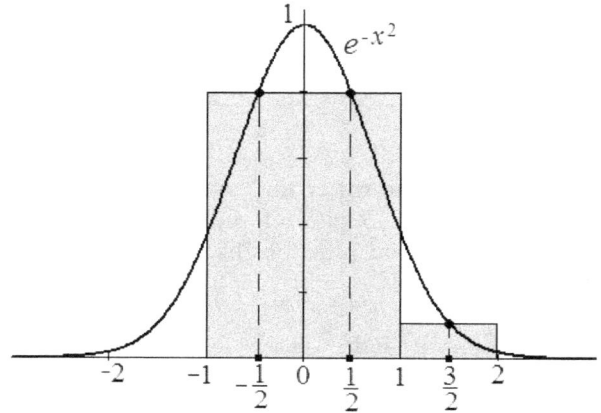

Figure 8.26: Estimating an area using the midpoint rule

Figure 8.26 illustrates the shaded strips whose total area approximates the area under the graph of the curve e^{-x^2} from $x = -1$ to $x = 2$. Although these strips do not closely match the region under the curve, the overestimates (shaded regions above the curve) and underestimates (unshaded regions below the curve and above the strips) tend to offset each other and reduce the approximation error. This offsetting does not happen universally for all curves but it does happen often enough to make the midpoint rule a useful method.

Exercise 119 *Repeat the above calculation but with the refined regular partition*

$$P_2 = \{-1, -\frac{1}{2}, 0, \frac{1}{2}, 1, \frac{3}{2}, 2\}$$

Thinking of integrals as areas, we can use other shapes than *rectangular* strips with which to estimate integrals. For instance, in Figure 8.27 we connect the points on the curve using straight lines to form *trapezoidal* strips.

---

[18] It is possible to obtain an "exact value" (irrational) for this integral in the form of an infinite series; we will discuss this further in Chapter 9, but even the infinite series has to be truncated to provide a numerical answer!

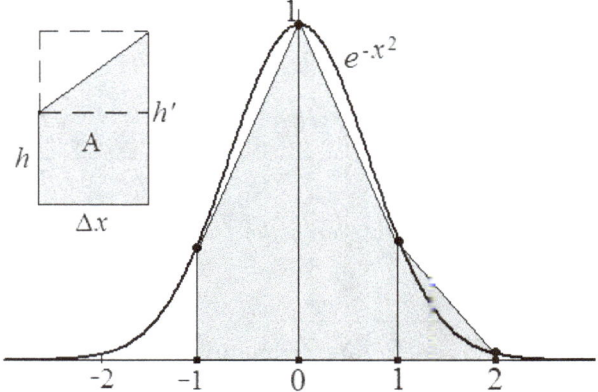

Figure 8.27: Estimating an area using the trapezoidal rule

The area of each trapezoidal strip such as the one shaded on the side in Figure 8.27 is

$$A = \frac{\Delta x}{2}(h + h')$$

As indicated, this shaded area is just the area of the rectangle of height h, which is $h\Delta x$, plus half the area of the rectangle on top of the first, which is $(h' - h)\Delta x/2$.

Under the graph of e^{-x^2} we have three such strips where the two heights of each strip are simply the values of e^{-x^2} at the end-points of the underlying subinterval. So we have, with $\Delta x = 1$:

$$\int_{-1}^{2} e^{-x^2} dx \simeq \frac{\Delta x}{2}[f(-1) + f(0)] + \frac{\Delta x}{2}[f(0) + f(1)] + \frac{\Delta x}{2}[f(1) + f(2)]$$
$$= \frac{\Delta x}{2}[f(-1) + 2f(0) + 2f(1) + f(2)] \qquad (8.50)$$
$$= \frac{1}{2}(e^{-(-1)^2} + 2e^0 + 2e^{-1^2} + e^{-2^2})$$
$$= \frac{1}{2}(e^{-1} + 2 + 2e^{-1} + e^{-4})$$
$$\simeq 1.561$$

Notice that the shaded trapezoids in Figure 8.27 match the area under the curve better than the shaded strips in Figure 8.26 corresponding to the midpoint rule.[19]

The second line of calculation in (8.50) extends to what is known as the *trapezoidal rule*.

---

[19]This improved geometrical matching is usually true but does not guarantee that the approximation to the integral is better. In fact, we see that the error for the trapezoidal case is greater than what we got from the midpoint rule because the overestimate-underestimate offset in the trapezoidal case is not as good as that in the midpoint rule.

Integrals

> **The trapezoidal rule.** Let $P = \{x_0, x_1, \ldots, x_N\}$ be any regular partition of the interval $[a, b]$ into N subintervals. Since $\Delta x = (b-a)/N$ we obtain:
>
> $$\int_a^b f(x)dx = \frac{b-a}{2N}[f(a) + 2f(x_1) + 2f(x_2) + \cdots + 2f(x_{N-1}) + f(b)] \qquad (8.51)$$

Exercise 120 *Estimate $\int_{-1}^{2} e^{-x^2} dx$ using the trapezoidal rule (8.51) with the regular partition of $[-1, 2]$ into six subintervals.*

Other numerical procedures for approximating integrals exist that give more accurate results than either the midpoint or the trapezoidal rule. The best known among such methods is *Simpson's rule*[20] which connects trios of adjacent points on the graph of $f(x)$ using local parabolic fittings to the curve.

For a further study of numerical integration you should check a standard text in *numerical analysis*. Our aim in this section was limited to providing simple, yet practical ways of integrating functions that we may run into but for which no antiderivative formulas are known.

Is that really a function?

The Fundamental Theorem of Calculus and numerical integration are more than practical tools for doing complex calculation. They also expand our intuition of what a function is. An integral like

$$\int_0^x e^{-u^2} du$$

defines a function $f(x)$ that is just as concrete as e^{-x^2}: if we can estimate the latter (and technically, that is the best we can do with e^{-x^2} for almost all values of x) then we can estimate $f(x)$ using numerical integration. We can make precise calculations with the resulting numbers and even plot an accurate graph of $f(x)$.

The Fundamental Theorem gives us information even without calculating numerical values. For instance, we know that $f(x)$ must be increasing because its derivative $f'(x) = e^{-x^2}$ is always positive.

In Chapter 9 we discover yet another powerful means of defining functions for which no simple formulas exist. Using infinite series of functions we can extend and refine our concept of "function" enormously. We find alternative ways of extending functions like $\int_0^x e^{-u^2} du$ or even of e^{-x^2} itself!

8.6 The improper Riemann integral: infinity made explicit

Earlier we discussed the function $f(x) = 1/\sqrt{x}$ which has an antiderivative $F(x) = 2\sqrt{x}$ that is defined for all $x \geq 0$. Since $f(x)$ is unbounded (has an infinite singularity) at 0 it is not Riemann integrable on any interval that contains 0, say, on $[0,1]$. Figure 8.17 illustrates the situation in terms of areas.

On the other hand, unlike $f(x)$, its antiderivative $F(x)$ is defined at 0; in fact, $F(1) - F(0) = 2$ so one side of (8.21) in FT1 is well-defined. Now let's take a closer look at the other side; if $0 < a < 1$

---
[20]Named after the English mathematician and inventor Thomas Simpson (1710-1761).

then $1/\sqrt{x}$ is continuous (hence integrable) on the interval $[a, 1]$ and the first part of the Fundamental Theorem FT1 gives

$$\int_a^1 \frac{1}{\sqrt{x}}dx = 2\sqrt{x}\Big|_a^1 = 2\sqrt{1} - 2\sqrt{a} = 2 - 2\sqrt{a}$$

If we take a sequence of values for a that converges to 0, say, $1/n$ then for each n

$$\int_{1/n}^1 \frac{1}{\sqrt{x}}dx = 2 - 2\sqrt{\frac{1}{n}}$$

As $n \to \infty$ we see that in the limit

$$\lim_{n \to \infty} \int_{1/n}^1 \frac{1}{\sqrt{x}}dx = \lim_{n \to \infty} \left(2 - 2\sqrt{\frac{1}{n}}\right) = 2 - \lim_{n \to \infty} 2\sqrt{\frac{1}{n}}$$

$$= 2 - 2\sqrt{\lim_{n \to \infty} \frac{1}{n}} = 2 - 2\sqrt{0} = 2$$

This argument is valid for *any* sequence x_n of positive real numbers that converges to 0 because

$$\lim_{n \to \infty} 2\sqrt{x_n} = 2\sqrt{\lim_{n \to \infty} x_n} = 2\sqrt{0} = 0$$

Therefore, we see that for every sequence x_n of positive real numbers that converges to 0

$$\lim_{n \to \infty} \int_{x_n}^1 \frac{1}{\sqrt{x}}dx = 2 = 2\sqrt{x}\Big|_0^1$$

This result highlights two significant issues, one philosophical and the other technical.

The first looks like a philosophically paradoxical issue; the unbounded region that is shaded in Figure 8.17 nevertheless appears to have a *finite* area of 2 square units. This is like a piece of land of area 2 (acres, say) that has an *infinite perimeter* since we never finish going around it. After our study of infinite series in Chapter 6, and especially the discussion of the infinite Tower of Boxes (and Gabriel's Horn) we recognize the situation as a familiar one, despite the philosophical mismatch between mathematical results and the material world. In the material world we can neither find nor create a plot of land having finite area but infinite perimeter such as the shaded region in Figure 8.17.

The second issue is mathematically profound and its resolution solves problems that arise from the occurrence of infinity in integrals. The Fundamental Theorem of Calculus (Part 1) *appears to hold* (in the limit) for a non-integrable function where we managed to get around the non-integrability at 0 by staying to the right and approaching 0 from a direction where the integral does in fact exist.

This observation may lead us to the conclusion that perhaps the FT1 holds whenever the function $f(x)$ happens to have an antiderivative, regardless of whether $f(x)$ is bounded or not.

But this conclusion is wrong and we can refute is by a simple example: the function $f(x) = 1/x$ has an antiderivative $F(x) = \ln x$ but

$$\lim_{n \to \infty} \int_{1/n}^1 \frac{1}{x}dx = \lim_{n \to \infty} \ln x\Big|_{1/n}^1 = \lim_{n \to \infty}\left(\ln 1 - \ln \frac{1}{n}\right) = -\lim_{n \to \infty}\left(\ln \frac{1}{n}\right)$$

Integrals 275

Since the logarithm function is continuous, we can move the last limit inside it to get:

$$\lim_{n\to\infty} \int_{1/n}^{1} \frac{1}{x} dx = -\ln\left(\lim_{n\to\infty} \frac{1}{n}\right) = -(-\infty) = \infty$$

To explain the difference between this case where $f(x) = 1/x$ and the previous case where $f(x) = 1/\sqrt{x}$ let's take a more careful look at the situation. There is indeed a subtle difference between the two cases: the antiderivative of $1/\sqrt{x}$ is $2\sqrt{x}$ which is defined at $x = 0$ but the antiderivative of $1/x$ which is $\ln x$ is not defined at 0. Therefore, the example of $1/x$ does not invalidate the notion that the FT1 may hold in the limit; we just need the antiderivative to be defined on the entire interval of integration. Then the antiderivative is automatically continuous by FT2 and gives a unique and consistent answer when we take the limit. Let's discuss this idea further.

The improper Riemann integral of unbounded functions.

Consider a function $f(x)$ that is continuous on a half-open interval $(a, b]$ but $f(x)$ is not bounded as the value of x in $(a, b]$ approaches a. An example is the function $1/\sqrt{x}$ on the interval $(0,1]$. We define the following.[21]

> **The improper integral, unbounded functions.** A continuous function $f(x)$ has an *improper Riemann integral* on the interval $(a, b]$ if the limit
>
> $$\lim_{n\to\infty} \int_{x_n}^{b} f(x) dx$$
>
> exists and has the *same value* for *every* sequence x_n in $(a, b]$ that converges to a. We usually write
>
> $$\int_{a}^{b} f(x) dx = \lim_{n\to\infty} \int_{x_n}^{b} f(x) dx$$
>
> although the left hand side does not mean that $f(x)$ is Riemann integrable on $[a, b]$. We commonly say the the *integral on the left converges to* the answer that we get on the right. If the limit does not exist for even one sequence x_n that converges to a then we say that $\int_{a}^{b} f(x) dx$ *diverges*.

A similar definition exists for the case where $f(x)$ is continuous on a half-open interval $[a, b)$ but not bounded as the value of x in $[a, b)$ approaches b. We write:

$$\int_{a}^{b} f(x) dx = \lim_{n\to\infty} \int_{a}^{x_n} f(x) dx$$

---

[21] In line with our stated plans, this definition uses sequences. But function limits provide a more succinct way of describing the limit in this and are therefore commonly used. For example,

$$\int_{a}^{b} f(x) dx = \lim_{u\to a^+} \int_{u}^{b} f(x) dx$$

where by $u \to a^+$ we mean that $u \to a$ while $u > a$. As we saw in Chapter 4 the two definitions are mathematically equivalent.

whenever the limit on the right exists and has the *same value* for *every* sequence x_n in $[a,b)$ that converges to b. Combinations of the above two cases are used to deal with other possible cases where $f(x)$ is unbounded at some point inside $[a,b]$; see below.

Now, suppose that $f(x)$ is continuous but unbounded on $(a,b]$ and it has a *bounded* antiderivative $F(x)$ on the closed interval $[a,b]$. Then FT2 says that $F(x)$ is continuous on $[a,b]$ so that if $x_n \to a$ as $n \to \infty$ then by FT1

$$\int_a^b f(x)dx = \lim_{n\to\infty} \int_{x_n}^b f(x)dx = \lim_{n\to\infty}[F(b) - F(x_n)] = F(b) - F(\lim_{n\to\infty} x_n) = F(b) - F(a)$$

We now see the precise manner in which the improper integral extends the Fundamental Theorem to unbounded functions using limits.

Exercise 121 *Explain why each of the following integrals is improper and find the value that each converges to, if the limit exists:*

$$(a) \int_0^1 \frac{1}{\sqrt[3]{x}} dx \qquad (b) \int_0^1 \frac{1}{x^2} dx \qquad (c) \int_1^5 \frac{1}{\sqrt{x-1}} dx$$

If both $f(x)$ and its antiderivative are unbounded (like the integral (b) in the above exercise) then the integral diverges.

Going a step further, the singularity of $f(x)$ may occur *between* a and b. The way to deal with this case is straightforward.

An infinite singularity in the interior of the integration interval. let's say that the $f(x)$ has an infinite singularity at $x = c$ where $a < c < b$. Then we split the integral as follows:

$$\int_a^b f(x)dx = \lim_{n\to\infty} \int_a^{x_n} f(x)dx + \lim_{n\to\infty} \int_{y_n}^b f(x)dx$$

If both of the two limits above exist for all sequences x_n in $[a,c)$ and all sequences y_n in $(c,b]$ that converge to c then the integral $\int_a^b f(x)dx$ converges to whatever number the sum of the two limits is. If either one of the two limits fails to exist then $\int_a^b f(x)dx$ diverges.

It is important to be aware of interior singularities, as they may be easy to overlook.
Here is why: consider the integral[22]

$$\int_{-1}^1 \frac{1}{x^2} dx$$

If the infinite singularity of the function $1/x^2$ at $x = 0$ is neglected then we may apply the FT1 (improperly, of course) to the antiderivative $-1/x$ to get:

$$\int_{-1}^1 \frac{1}{x^2} dx = -\frac{1}{x}\Big|_{-1}^1 = -\frac{1}{1} - \left(-\frac{1}{-1}\right) = -2$$

---

[22]The numbers ± 1 can be changed to any pair of real numbers a, b without changing the conclusion as long as $a < 0 < b$.

Integrals

Given that $1/x^2$ is never negative, this negative answer is quite absurd!

Infinite time horizon: a thought experiment.

Suppose that an object moves along a straight line, which we may take to be the x-axis; its velocity is given by the equation:
$$v(t) = \frac{12}{(t+1)^2}$$

Let's assume that the object's location is $x = 3$ (say, 3 meters to the right of our reference point $x = 0$) at time $t = 0$. Then the position of this object at a later time $t = T$ is given by

$$x(T) - x(0) = \int_0^T v(t)dt = \int_0^T \frac{12}{(t+1)^2}dt = 12\int_0^T (t+1)^{-2}dt$$

The difference on the left comes from the FT1 because $x(t)$ is an antiderivative of $v(t)$. We may use the power rule for integration to derive

$$x(T) - x(0) = 12[-(t+1)^{-1}]_0^T = -\frac{12}{t+1}\bigg|_0^T = \frac{-12}{T+1} - \frac{-12}{0+1} = 12 - \frac{12}{T+1}$$

Since $x(0) = 3$ the position of this object at time T is:

$$x(T) = 15 - \frac{12}{T+1}$$

The following table lists the position and velocity for several (integer) values of T:

T	0	1	2	3	4	5
$x(T)$	3	9	11	12	12.6	13
$v(T)$	12	3	1.33	0.75	0.48	0.33

The table shows that the object is moving further to the right but covering ever shorter distances and the object's velocity is always positive so it is always moving to the right, albeit ever more slowly.

Does this object move infinitely far away (always to the right) in infinite time?

To answer this question, let's take the limit of $T = n$ as $n \to \infty$:

$$\lim_{n \to \infty} x(n) = \lim_{n \to \infty}\left(15 - \frac{12}{n+1}\right) = 15 - \lim_{n \to \infty}\frac{12}{n+1} = 15 - 0 = 15$$

This means that we will never find the object more than 15 meters away from the origin, *no matter how long we wait*, even though the velocity is always positive! We have something similar to Zeno's paradox here in that we are adding distances an infinity of times without ever stopping but the object is not going infinitely far away.

Some insight is gained in this case by examining the velocity for large values of time T. Assuming that T is measured in seconds, we see that after only an hour ($T = 3600$ seconds) the velocity is:

$$v(3600) = \frac{12}{3601^2} \simeq 0.00000093 = 0.93 \times 10^{-6}$$

or less than a millionth of a meter per second,[23] and every 10 hours, the velocity drops by a factor of 100. Evidently, this slow rate of progress is not enough to move our object infinity far away.

But what if the rate of progress were a little faster? If $v(t) = 12/(t+1)$ (the denominator is not squared) then repeating our earlier calculation gives:

$$x(T) - x(0) = \int_0^T \frac{12}{t+1} dt = 12[\ln(t+1)]_0^T = 12\ln(T-1) - 12\ln(0+1) = 12\ln(T+1)$$

so the position at time T with $x(0) = 3$ as before is now:

$$x(T) = 3 + 12\ln(T+1)$$

For this function, taking the limit as $T = n \to \infty$:

$$\lim_{n\to\infty} x(n) = \lim_{n\to\infty} [3 + 12\ln(t+1)] = 3 + 12 \lim_{n\to\infty} \ln(t+1) = 3 + \infty = \infty$$

Here again the velocity goes to 0 over an infinite time horizon but the progress is much less slow; when $T = 3600$ the velocity is $v(3600) = 12/3601 \simeq 0.0033$ which is about 3600 times faster than in the previous case. Further, the velocity drops only by a factor of 10 every 10 hours. This object is slowing down but still moving fast enough to travel an infinite distance over an infinite time horizon.

The improper Riemann integral over unbounded intervals.

The above calculations remain valid if n is replaced with any sequence of real numbers x_n that diverges to infinity, that is, $x_n \to \infty$ as $n \to \infty$. We arrive at the following concept:

> **The improper integral, unbounded intervals.** A continuous function $f(x)$ has an *improper Riemann integral* on the interval $[a, \infty)$ where a is a fixed real number if the limit
>
> $$\lim_{n\to\infty} \int_a^{x_n} f(x)dx$$
>
> exists and has the *same value* for *every* sequence x_n in $[a, \infty)$ that diverges to ∞. If the above conditions are true then we write
>
> $$\int_a^\infty f(x)dx = \lim_{n\to\infty} \int_a^{x_n} f(x)dx$$
>
> and say that $\int_a^\infty f(x)dx$ *converges* to the stated limit when the limit exists as described above. If the limit does not exist for even one sequence x_n that diverges to infinity then we say that $\int_a^\infty f(x)dx$ *diverges*.

Improper integrals over unbounded intervals like $(-\infty, a]$ can be defined similarly ($x_n \to -\infty$ as $n \to \infty$).

Exercise 122 *Calculate each of the following improper integrals over unbounded intervals, if they converge:*

$$(a) \int_{0.01}^\infty \frac{1}{x^3} dx \qquad (b) \int_{10000}^\infty \frac{1}{\sqrt{x}} dx$$

---

[23]The width of a human hair has been found to range between 17 and 181 millionth of a meter, which averages to about 99 millionth of a meter (about 0.1 millimeter).

Integrals

The integral test for infinite series: a useful prospecting tool.

An interesting application of the concept of improper integral is developing a test that tells us whether an infinite series converges or diverges.

To set the stage, let's recall the p-series

$$\sum_{k=1}^{\infty} \frac{1}{k^2} = \frac{1}{1^2} + \frac{1}{2^2} + \frac{1}{3^2} + \cdots$$

whose n-th partial sum is

$$s_n = \sum_{k=1}^{n} \frac{1}{k^2} = \frac{1}{1^2} + \frac{1}{2^2} + \frac{1}{3^2} + \cdots + \frac{1}{n^2}$$

In Figure 8.28 we that s_n is also a Riemann sum corresponding to the regular partition $P_n = \{1, 2, 3, \ldots, n\}$ of the interval $[1, n]$ with subinterval points $x_i^* = i$, the *right* endpoint of each subinterval.

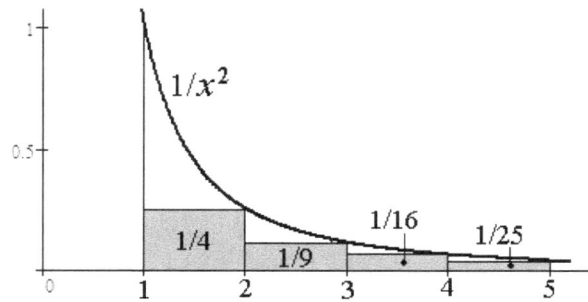

Figure 8.28: Illustrating convergence by the integral test

Interpreting the integral as area, from Figure 8.28 we deduce that

$$s_n < 1 + \int_1^n \frac{1}{x^2} dx$$

for every value of n since the shaded strips underestimate the area under the curve $1/x$. Now, let's take the limit as $n \to \infty$ to get:

$$\lim_{n \to \infty} s_n < 1 + \lim_{n \to \infty} \int_1^n \frac{1}{x^2} dx = 1 + \lim_{n \to \infty} \left(-\frac{1}{n} + \frac{1}{1} \right) = 1 - 0 + 1 = 2$$

This shows that

$$\sum_{k=1}^{\infty} \frac{1}{k^2} < 2$$

which means that the series on the left not only converges, but its value is less than 2.

The calculation of the integral $\int_1^n x^{-2} dx$ was easier than the series because of the Fundamental Theorem of Calculus and the existence of the antiderivative $-1/x$ for the function $1/x^2$.

Exercise 123 *Show that the improper integral $\int_1^\infty x^{-3/2} dx$ converges and use this fact to prove that the following p-series converges to a number less than 3:*

$$\sum_{k=1}^{\infty} \frac{1}{k^{3/2}} = \frac{1}{1^{3/2}} + \frac{1}{2^{3/2}} + \frac{1}{3^{3/2}} + \cdots = 1 - \frac{1}{2\sqrt{2}} + \frac{1}{3\sqrt{3}} + \cdots$$

The above calculations readily extend to any p-series *because the corresponding integral does not get any more difficult to evaluate.*

If $p > 1$ is any fixed real number then we have a convergent integral

$$\int_1^\infty \frac{1}{x^p} dx = \lim_{n\to\infty} \int_1^n x^{-p} dx = \lim_{n\to\infty} \left.\frac{x^{-p+1}}{-p+1}\right|_1^n$$

$$= \lim_{n\to\infty} \left(\frac{n^{-p+1}}{-p+1} - \frac{1^{-p-1}}{-p+1} \right)$$

$$= \lim_{n\to\infty} \left(-\frac{n^{p-1}}{p-1} + \frac{1}{p-1} \right)$$

$$= \frac{1}{p-1} - \lim_{n\to\infty} \frac{1}{(p-1)n^{p-1}} = \frac{1}{p-1}$$

If s_n is the n-th partial sum of the p-series

$$\sum_{k=1}^{\infty} \frac{1}{k^p} = \frac{1}{1^p} + \frac{1}{2^p} + \frac{1}{3^p} + \cdots \tag{8.52}$$

then s_n satisfies the inequality

$$\lim_{n\to\infty} s_n < 1 + \int_1^\infty \frac{1}{x^p} dx = 1 + \frac{1}{p-1} = \frac{p}{p-1}$$

Therefore, if $p > 1$ the p-series in (8.52) converges, as the p-series test also stated. But *now we can say a little bit more*: the sum of the entire series is less than $p/(p-1)$. For instance, if $p = 2$ then $\sum_{k=1}^{\infty} \frac{1}{k^2} < 2$ and if $p = 3$ then $\sum_{k=1}^{\infty} \frac{1}{k^3} < 3/2$.

What if $p \leq 1$?

Notice that $p/(p-1)$ can no longer be a meaningful upper bound. Further, if $0 < p \leq 1$ then we obtain a divergent integral: first, if $0 < p < 1$ then

$$\int_1^\infty \frac{1}{x^p} dx = \lim_{n\to\infty} \int_1^n x^{-p} dx = \lim_{n\to\infty} \left.\frac{x^{-p+1}}{-p+1}\right|_1^n$$

$$= \lim_{n\to\infty} \left(\frac{n^{1-p}}{1-p} - \frac{1^{1-p}}{1-p} \right) = \infty - \frac{1}{1-p} = \infty$$

Integrals

Also if $p = 1$ then

$$\int_1^\infty \frac{1}{x} dx = \lim_{n \to \infty} \ln x \Big|_1^n = \lim_{n \to \infty} (\ln n - \ln 1) = \infty$$

To be able to make use of these integrals, we must compare them to Riemann sums whose partial sums *exceed* the integrals on the corresponding intervals; see Figure 8.29.

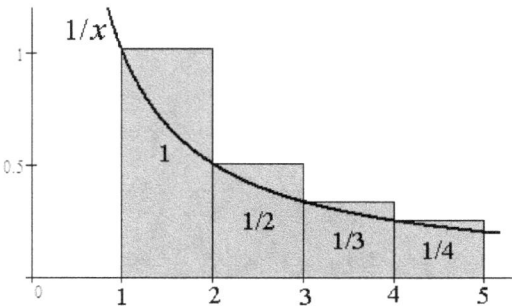

Figure 8.29: Illustrating divergence by the integral test

In Figure 8.29 we see that the n-th partial sum s_n of the series in (8.52) is a Riemann sum corresponding to the regular partition $P_n = \{1, 2, 3, \ldots, n\}$ of the interval $[1, n]$ with subinterval points $x_i^* = i - 1$, the *left* endpoint of each subinterval. For every n we see from Figure 8.29 that:

$$s_n > \int_1^n \frac{1}{x^p} dx$$

As $n \to \infty$ the right hand side diverges to ∞ which means that s_n does the same thing. We conclude that the p-series diverges when $0 < p \leq 1$.

The above calculations not only prove the p-series test that we mentioned earlier when discussing infinite series but in the convergent case, they also give us an upper bound for the value of the series.

If we take a closer look at the above discussion we notice that the specific functions $1/x^p$ are not really important; what matters is the convergence or divergence of the improper integrals. If we have an infinite series:

$$\sum_{k=1}^\infty f(k) \qquad (8.53)$$

whose terms are the values of a fixed function $f(x)$ at the integer values of $x = 1, 2, 3, \ldots$ then the aforementioned Riemann sums with either the left or the right endpoints can still be used for comparing the partial sums to integrals; see Figure 8.30.

In the left hand panel of Figure 8.30 we see that the area under the graph of $f(x)$ is greater than the value of the infinite series (shaded region) so if the improper integral converges then so does the infinite series. The right hand panel shows the reverse case: the area under the graph of $f(x)$ is less than the value of the infinite series (shaded region) so if the improper integral diverges then the

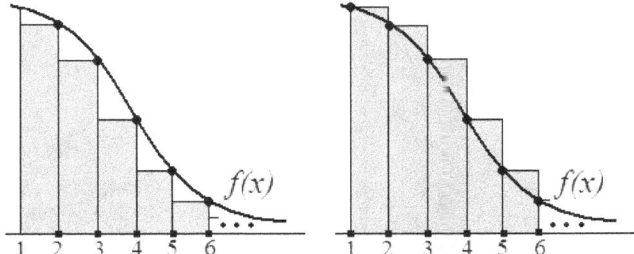

Figure 8.30: The two cases of the integral test

series does too. To make the comparisons that we just made $f(x)$ has to be a *positive function that decreases to zero*;[24] see Figure 8.30.

We summarize our discussion as the following test for finding the convergence or divergence of infinite series:[25]

> **The Integral Test:** Let $f(x)$ be a continuous, positive and decreasing function on the interval $[1, \infty)$. Then the infinite series (8.53) converges or diverges according to whether the improper integral below converges or diverges, respectively:
> $$\int_1^\infty f(x)dx$$

If its conditions hold then the integral test applies in a straightforward way by finding whether the improper integral converges or diverges.

To illustrate, consider the infinite series

$$\sum_{k=1}^\infty \frac{k}{e^k} = \frac{1}{e^1} + \frac{2}{e^2} + \frac{3}{e^3} + \cdots$$

Here $f(k) = k/e^k = ke^{-k}$ so we consider the function $f(x) = xe^{-x}$. If $x > 0$ then $xe^{-x} > 0$ so $f(x)$ is a positive function. Further, as a product of two differentiable functions, $f(x)$ is differentiable with derivative

$$f'(x) = (1)e^{-x} + x(-e^{-x}) = (1-x)e^{-x}$$

Since $f'(x) < 0$ when $x > 1$ it follows that $f(x)$ is decreasing on the interval $[1, \infty)$; see Figure 8.31.

Next, we check the improper integral:

$$\int_1^\infty xe^{-x}dx = \lim_{n \to \infty} \int_1^n xe^{-x}dx$$

---

[24] The italicized restrictions are necessary: the series in (8.53) diverges if the function values $f(k)$ do not converge to zero (recall the divergence test). Further, to make meaningful comparisons between the series and the integral, it is necessary that $f(x)$ be positive.

[25] The integral test also applies to infinite series that do not start with $k = 1$ like $\sum_{k=m}^\infty f(k)$ where m is any integer, as long as the improper integral $\int_m^\infty f(x)dx$ converges or diverges for a function $f(x)$ that is continuous, positive and decreasing on the interval $[m, \infty)$. The argument proving this general version is essentially the same as the one for $m = 1$.

Integrals

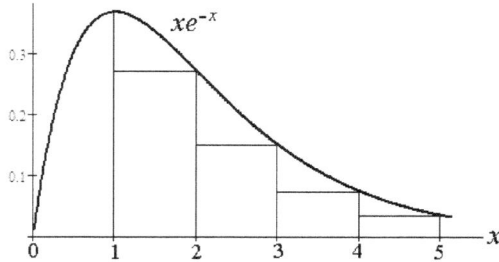

Figure 8.31: Using the integral test to show convergence

The *proper* integral on the right can be calculated using integration by parts:

$$\int_1^n xe^{-x}dx = -xe^{-x} - e^{-x}\Big|_1^n = -ne^{-n} - e^{-n} - (-e^{-1} - e^{-1}) = 2e^{-1} - ne^{-n} - e^{-n}$$

Now we take the limit as $n \to \infty$ to obtain

$$\int_1^\infty xe^{-x}dx = \lim_{n\to\infty}(2e^{-1} - ne^{-n} - e^{-n}) = 2e^{-1} - \lim_{n\to\infty} ne^{-n} - \lim_{n\to\infty} e^{-n} = 2e^{-1}$$

The limits are both zeros; the second is clear since $e^{-n} = 1/e^n$ and e^n becomes infinitely large, and the first can be inferred from Figure 8.31 since the curve xe^{-x} approaches 0 as $x \to \infty$ (a rigorous proof uses L'Hôpital's rule that we discuss later in Section ??.

Exercise 124 *(a) Use the integral test to prove that the following series diverges:*

$$\sum_{k=1}^\infty \frac{k}{k^2+1} = \frac{1}{2} + \frac{2}{5} + \frac{3}{10} + \cdots$$

(b) Prove that the following series converges:

$$\sum_{k=1}^\infty \frac{k}{k^3+1} = \frac{1}{2} + \frac{2}{9} + \frac{3}{28} + \cdots$$

In this case you may want to use the comparison test; note that

$$\frac{k}{k^3+1} < \frac{k}{k^3} < \frac{1}{k^2}$$

Would you want to use the comparison test for the series in (a) also?

Chapter 9

Infinite Series of Functions: *the wonders never cease!*

Earlier we discussed how to deal with irrational numbers using sequences of rational numbers. Although irrational numbers have infinite, non-periodic decimal expansions, we saw how to approximate them as accurately as needed. For instance, we can find as many digits of the decimal representation of $\sqrt{2}$ as is necessary to do any sort of practical calculation using the Newton-Raphson recursion, which has the nice feature of converging fast.

An important fact in this type of approximation was that the approximating numbers were *rational*. This allowed us to use simple, finite numbers to approximate each irrational number.

In this chapter we do something similar for functions. We show that transcendental functions like $\ln x$, $\sin x$, etc can be approximated using sequences of *polynomials*. The latter are simple algebraic combinations of the power functions x, x^2, x^3, ... like $2x - x^3$. Polynomials play a role in approximating transcendental functions (think of them as the irrational functions) that is analogous to rational numbers approximating irrational numbers. We may even liken the power functions to the positive integers!

The standard process of representing functions by converging sequences of polynomials involves infinite series of power functions that are appropriately called *power series*; they include the Taylor series of calculus. We discover that a large variety of transcendental functions can be expanded in the form of a power series; such functions will not be enigmatic any more.

We also discuss the infinite series of trigonometric functions that are known in applied mathematics as Fourier series. These are not power series but they succeed where power series fail: representing functions that are not infinitely differentiable. Extreme examples of continuous functions that fail to have a derivative even at a single point can be constructed using trigonometric series. Such functions are highly non-intuitive and their exact shapes are impossible to draw because they contain not the tiniest stretch of smoothness in them. We discuss extreme functions like that here because they illustrate the subtle, often overlooked complexities that infinity injects into our fuzzy, low-resolution intuition of what continuous means. A non-technical discussion of a function of this type, namely, the enigmatic "Weierstrass's function" concludes this chapter.

9.1 The geometric series as a function series: amazing power at a low cost

Let's begin with something familiar: the geometric series Recall that

$$1 + x + x^2 + x^3 + \cdots = \frac{1}{1-x} \quad \text{if } -1 < x < 1 \tag{9.1}$$

Here the fraction on the right hand side is a simple rational function that equals the sum of *all* power functions as long as x is a real number between -1 and 1. In our first encounter with infinite series x was a constant, namely, the common ratio; but we are now thinking of it as a variable that ranges in the open interval $(-1, 1)$. To distinguish this extended version from the series of constants, we refer to the series in (9.1) as the *geometric power series*.

An infinity of infinite series!

An interesting consequence of viewing x as a variable is the fact that (9.1) defines infinitely many infinite series of numbers together with their sums or values here is a partial list:

If $x = 1/2$ then

$$1 + \frac{1}{2} + \left(\frac{1}{2}\right)^2 + \left(\frac{1}{2}\right)^3 + \cdots = \frac{1}{1-(1/2)}$$

which can be written more succinctly as:

$$1 + \frac{1}{2} + \frac{1}{2^2} + \frac{1}{2^3} + \cdots = 2$$

Similarly, if $x = -1/2$ then

$$1 - \frac{1}{2} + \left(-\frac{1}{2}\right)^2 + \left(-\frac{1}{2}\right)^3 + \cdots = \frac{1}{1-(-1/2)}$$

or more succinctly,

$$1 - \frac{1}{2} + \frac{1}{2^2} - \frac{1}{2^3} + \cdots = \frac{2}{3}$$

If $x = 3/5$ then

$$1 + \frac{3}{5} + \left(\frac{3}{5}\right)^2 + \left(\frac{3}{5}\right)^3 + \cdots = \frac{1}{1-(3/5)}$$

and this reduces to

$$1 + \frac{3}{5} + \frac{9}{25} + \frac{27}{125} + \cdots = \frac{5}{2}$$

If $x = 1/\pi$ then

$$1 + \frac{1}{\pi} + \left(\frac{1}{\pi}\right)^2 + \left(\frac{1}{\pi}\right)^3 + \cdots = \frac{1}{1-(1/\pi)}$$

and this reduces to

$$1 + \frac{1}{\pi} + \frac{1}{\pi^2} + \frac{1}{\pi^3} + \cdots = \frac{\pi}{\pi - 1}$$

Series of Functions

And so on!

Many new power series born of just one!

Before studying the issue of convergence of infinite series of functions, let's consider a fabulous aspect of the geometric power series. The formula in (9.1) yields additional series of functions through substitution, differentiation and integration that I discuss here because they provide motivation for the rest of the chapter and well, because they are fun to derive!

As our first example, let's substitute $2u$ in (9.1) for x to get:

$$1 + 2u + (2u)^2 + (2u)^3 + \cdots = \frac{1}{1-2u}$$

This gives us a new infinite series formula:

$$1 + 2u + 4u^2 + 8u^3 + \cdots = \frac{1}{1-2u}$$

that is valid as long as $-1 < 2u < 1$ or equivalently, $-1/2 < u < 1/2$.

Next, let's substitute $x = u/2$ in (9.1) to obtain:

$$1 + \frac{u}{2} + \left(\frac{u}{2}\right)^2 + \left(\frac{u}{2}\right)^3 + \cdots = \frac{1}{1-u/2}$$

This simplifies to:

$$1 + \frac{u}{2} + \frac{u^2}{4} + \frac{u^3}{8} \cdots = \frac{2}{2-u}$$

As long as $-1 < u/2 < 1$ or equivalently, $-2 < u < 2$ the above formula is valid.

We can use a little algebra to find the infinite series for the function $1/(a - u)$ for any number $a > 0$ as follows:

$$\frac{1}{a-u} = \frac{1}{a(1-u/a)} = \frac{1}{a}\left[1 + \frac{u}{a} + \left(\frac{u}{a}\right)^2 + \left(\frac{u}{a}\right)^3 + \cdots\right]$$

where the substitution that works is $x = u/a$. Multiply the $1/a$ through and simplify to get:

$$\frac{1}{a-u} = \frac{1}{a} + \frac{u}{a^2} + \frac{u^2}{a^3} + \frac{u^3}{a^4} + \cdots$$

As long as $-1 < u/a < 1$ or equivalently, $-a < u < a$ the above series is valid.

Next, we substitute $x = -u$ in (9.1) to get

$$1 + (-u) + (-u)^2 + (-u)^3 + \cdots = \frac{1}{1-(-u)}$$

which leads to an alternating series:

$$1 - u + u^2 - u^3 + \cdots = \frac{1}{1+u} \tag{9.2}$$

This is valid as long as $-1 < -u < 1$ or equivalently, $-1 < u < 1$.

Exercise 125 *Find the infinite series for*
$$\frac{1}{2+u}$$
and the range of values of u for which your formula is valid.

Here is what $x = au^2$ gives where $a \neq 0$:

$$1 + au^2 + (au^2)^2 + (au^2)^3 + \cdots = \frac{1}{1 - au^2}$$
$$1 + au^2 + a^2u^4 + a^3u^6 + \cdots = \frac{1}{1 - au^2} \tag{9.3}$$

which is valid if $-1 < au^2 < 1$. To figure out the range for u more quickly consider writing these inequalities as a single absolute value inequality $|au^2| < 1$, or $|a|u^2 < 1$ since u^2 is never negative and can therefore be taken out of the absolute value sign without changing the inequality. Then:

$$u^2 < \frac{1}{|a|} \quad \text{or} \quad |u| < \frac{1}{\sqrt{|a|}}$$

Now the last inequality can be written without the absolute value on u as:

$$-\frac{1}{\sqrt{|a|}} < u < \frac{1}{\sqrt{|a|}}$$

For instance, if $a = -1$ in (9.3) then we get the series:

$$1 + (-1)u^2 + (-1)^2 u^4 + (-1)^3 u^6 + \cdots = \frac{1}{1 - (-1)u^2}$$
$$1 - u^2 + u^4 - u^6 + \cdots = \frac{1}{1 + u^2} \tag{9.4}$$

The range of validity of this formula is

$$-\frac{1}{\sqrt{|-1|}} < u < \frac{1}{\sqrt{|-1|}} \quad \text{or} \quad -1 < u < 1$$

Exercise 126 *Find the infinite series for the function*
$$\frac{1}{3+u^2}$$
and the range of values of u for which your formula is valid. It may help to combine two of the previous ideas by writing:

$$3 + u^2 = 3\left(1 + \frac{u^2}{3}\right) = 3\left(1 - \left(-\frac{1}{3}u^2\right)\right)$$

So how about $a = -1/3$?

Series of Functions 289

Some infinite series of functions do not seem to be of geometric type at first glance but in fact they are. For example, consider:

$$\sum_{n=1}^{\infty} e^{(n-1)u} = 1 + e^u + e^{2u} + \cdots \qquad (9.5)$$

This is not a series of power functions but the substitution $x = e^u$ turns it into (9.1) because $e^{(n-1)u} = (e^u)^{n-1} = x^{n-1}$. The range of values of u for which this series converges is calculated from that for (9.1):

$$-1 < e^u < 1$$

Since $e^u > 0$ for all u it follows that the inequality on the left hand side is true for all real values of u; as for the right hand side inequality, take the natural logarithm to free the u and obtain:

$$u = \ln e^u < \ln 1 = 0$$

Therefore, (9.5) converges for all negative values of u, that is, for all $u < 0$. Further, from (9.1) we also have the formula:

$$1 + e^u + e^{2u} + \cdots = \frac{1}{1 - e^u} \quad \text{for } u < 0$$

Exercise 127 *Prove each of the following equalities and find the ranges of values of u for which it is valid:*

$$1 + e^{-u} + e^{-2u} + e^{-3u} + \cdots = \frac{1}{1 - e^{-u}}$$

$$1 + \sin^2 u + \sin^4 u + \sin^6 u + \cdots = \frac{1}{\cos^2 u} = \sec^2 u$$

The above ideas using substitutions and algebra do not help us with the problem of finding the infinite series for a function like

$$\frac{1}{(1-x)^2}$$

It is true that this function is the same as $(1/(1-x))^2$ and (9.1) gives the infinite series for $1/(1-x)$; but then we must square an infinite series!

Not impossible, but the following observation makes our work much easier: suppose that we can take the derivative of the infinite series in (9.1) term by term, or one term at a time, and then add it all up to get the derivative of the series. We do not yet know if this is possible since the sum rule for derivatives applies only to a *finite* sum. But let's put the resolution of this problem off until a later section and take the derivative term by term to get:

$$\frac{d}{dx}(1 + x + x^2 + x^3 + \cdots + x^n + \cdots) = \frac{d}{dx}\left(\frac{1}{1-x}\right)$$

$$0 + 1 + 2x + 3x^2 + \cdots + nx^{n-1} + \cdots = \frac{-(-1)}{(1-x)^2}$$

where I used the quotient rule to take the derivative of the fraction on the right hand side.[1] Simplifying, we get:
$$1 + 2x + 3x^2 + \cdots + nx^{n-1} + \cdots = \frac{1}{(1-x)^2} \qquad (9.6)$$

Can you guess what series converges to the following?
$$\frac{1}{(1+x)^2}$$

Exercise 128 *By taking the derivative again term by term in (9.6) and doing a little algebra determine the function that the infinite series below converges to:*
$$1 + 3x + 6x^2 + 10x^3 + \cdots + \frac{n(n+1)}{2} x^{n-1} + \cdots$$

Having gotten some new series using the derivative, let's consider taking the integral next. Again, we have yet to determine if term by term integration is meaningfully possible but setting such details aside for now, we integrate the series in (9.2) to get:
$$\int (1 - u + u^2 - u^3 + \cdots) du = \int \frac{1}{1+u} du$$
$$u - \frac{u^2}{2} + \frac{u^3}{3} - \frac{u^4}{4} + \cdots = \ln(1+u) \qquad (9.7)$$

This is a nice, unexpected result that gives us the natural logarithm function as an infinite series! The theory of infinite series of functions is full of interesting and often unexpected results.

Setting[2] $u = 1$ in (9.7) answers a question about the alternating harmonic series that we came across earlier:
$$1 - \frac{1}{2} + \frac{1}{3} - \frac{1}{4} + \cdots = \ln 2$$

We already knew by the Alternating Series Test (AST) that the infinite series of numbers on the left hand side above converges; now we see that its limit or sum is actually $\ln 2$.

Exercise 129 *(a) Use the substitution $u = -1/2$ in (9.7), the properties of logarithm and a little algebra to obtain the following result:*
$$\sum_{n=1}^{\infty} \frac{1}{n 2^n} = \frac{1}{2} + \frac{1}{2(2^2)} + \frac{1}{3(2^3)} + \cdots = \ln 2$$

(b) The positive terms series on the left hand side above converges by the Ratio Test (RT). It converges to $\ln 2$ faster than the alternating series above, which converges only conditionally. Therefore, much fewer terms of the positive series are needed in order to approximate $\ln 2$ as compared with the alternating series. Consider adding the first 4 terms of each of the two series to estimate the value of $\ln 2$ to 4 decimal places and compare your answers with 0.6931 which is correct to 4 decimal places.

---

[1] Alternatively, the general power rule could be applied to $(1-x)^{-1}$.
[2] We will see later in this chapter that the equality (9.7) is valid for this value of u.

Series of Functions 291

9.2 Unexpected encounters with infinity

The geometric power series has many of the same basic features as the general power series that we discuss later in this chapter. In this section we take advantage of the simplicity of geometric series to identify and study an important occurrence of infinity which is not readily recognized.

Obviously the series is infinite, but this is not what I mean by the occurrence of infinity. Rather it is a subtle, even hidden occurrence that is responsible for a slow down in convergence as the value of x ranges over sets of numbers, known as the "intervals of convergence", ultimately switching from convergence to divergence at some point. If this type of infinity is not encountered then convergence may slow down but it does not switch to divergence.

The polynomial partial sums.

The equality in (9.1) certainly shows that adding infinitely many functions $x, x^2, x^3, \ldots$ is not mere fantasy. But the left hand side of (9.1) contains an occurrence of infinity that needs to be defined independently of the right hand side. *The equality in (9.1) does not define the left hand side;* rather it is saying that on the left hand side is a certain function of x that happens to be equal to the rational function on the right as long as $-1 < x < 1$.

This distinction is important because the right hand side of (9.1) is well defined as long as $x \neq 1$ whereas its left hand side is defined only for $-1 < x < 1$ because the geometric power series diverges if $x \geq 1$ or $x \leq -1$ (that is, $|x| \geq 1$).

Now how do we define the left hand side of (9.1)? If we go back to our first encounter with infinite series (of numbers) then we recall that we used the sequence of partial sums to define infinite series: the series was defined if that sequence converged to a real number. Consider the same for the series in (9.1):

$$s_0(x) = 1$$
$$s_1(x) = 1 + x$$
$$s_2(x) = 1 + x + x^2$$
$$s_3(x) = 1 + x + x^2 + x^3$$
$$\vdots$$

Notice that each partial sum $s_n(x)$ is a polynomial whose degree n is the index of the partial sum. Polynomials are defined on all of $\mathbb{R}$, so the partial sums $s_0(x), s_1(x), s_2(x), \ldots$ are all defined for all real numbers x. Figure 9.1 shows the partial sum polynomials $s_1(x)$ to $s_4(x)$ together with the graph of the rational function (dashed) over the interval $[-1, 1]$.

In Figure 9.1 we see that between -1 and 1 the partial sums line up with their rational function limit more closely, especially in the middle, as we add more terms (power functions of higher degrees). The partial sums drift away from the rational function near 1 or -1 and the departure is more noticeable near $x = 1$ where the rational function has a non-removable singularity.

Outside of the interval $(-1, 1)$ the more terms we add the greater the divergence becomes! Infinity's paw-prints are all over what we see in this figure, though they are not all easy to notice at first glance; let's take a closer look at what's going on here.

Partial sums converging near a singularity.

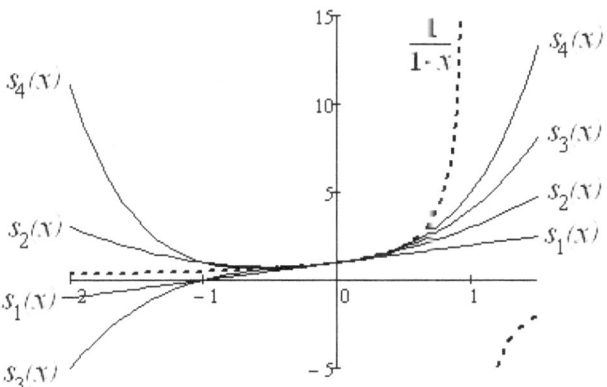

Figure 9.1: Several partial sums of the geometric power series

Recall that the common domain of all the partial sums is the set all real numbers, including $x = 1$. But the limit function $1/(1-x)$ has an infinite singularity at $x = 1$ and this raises the question as to how partial sums manage to converge to the limit function in the vicinity of $x = 1$. Figure 9.2 shows that convergence is not as rapid near this singularity as at other points.

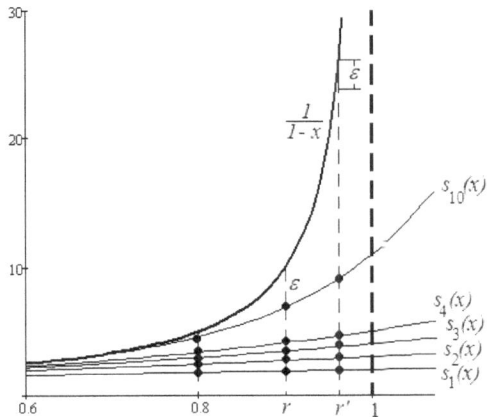

Figure 9.2: Convergence near the infinite singularity

If you look closely at Figure 9.2 you may notice that as the value of x approaches 1, the convergence seems to slow down (but not fully stop). To be more precise, let r be any number in $(-1, 1)$ and suppose that when $x = r$ the approximation $s_{10}(r)$, which is indicated by the dot on the partial sum curve $s_{10}(x)$ at $x = r$, just reaches within ε of the limit curve; that is,

$$\frac{1}{1-r} - s_{10}(r) = \varepsilon$$

Series of Functions

The exact value of ε is not important; just the usual small positive quantity that measures how closely $s_{10}(r)$ approximates $1/(1-r)$. The vertical sequence of dots on the various partial sum curves right above the value r approach a point on the curve $1/(1-r)$; in particular, the dot on the curve $s_{10}(x)$ at $x = r$ is within ε of the dot on the limit curve. But if we check the similar sequence of dots above $x = r'$ we see that these points fail to come within ε by a wide margin. We need $s_n(r')$ for a larger value of n in order to reach within ε of the limit. On the other hand, at $x = 0.8$ the value $s_4(0.8)$ already seems within ε of the limit curve and we do not actually need to go all the way to $n = 10$. Let's summarize this observation:

> **It matters where in the interval a point is!** Given a fixed $\varepsilon > 0$, the *index n of $s_n(x)$ needed to approximate the limit function $1/(1-x)$ within a distance ε depends on the value of x in $(-1, 1)$.*

Next, if we insert $x = r$ into the equality in (9.1) we get an infinite series of constants (vertically rising dots on partial sum curves) together with their limit:

$$1 + r + r^2 + r^3 + \cdots = \frac{1}{1-r} \tag{9.8}$$

Let's take a look at the convergence of the partial sums for the *numerical* series in (9.8). From our discussion of converging sequences of numbers, recall that for each (arbitrarily small) value of $\varepsilon > 0$ we must find an ε-index N_ε, a positive integer that is just big enough that for all indices $n \geq N_\varepsilon$

$$\left| \frac{1}{1-r} - (1 + r + r^2 + \cdots + r^n) \right| < \varepsilon \tag{9.9}$$

We can actually remove the absolute value sign since the sum of the infinite series is greater than the sum of any finite number of its terms. The calculation is further made simple by the fact that we earlier derived a formula for the finite geometric sum:

$$1 + r + r^2 + \cdots + r^n = \frac{1 - r^{n+1}}{1 - r}$$

So now (9.9) reduces to

$$\frac{1}{1-r} - \frac{1 - r^{n+1}}{1 - r} < \varepsilon$$

Simplify the left hand side to get

$$\frac{r^{n+1}}{1-r} < \varepsilon$$

Since we are looking for the index value N_ε we solve the above inequality for n as follows:

$$r^{n+1} < (1-r)\varepsilon$$

Taking the natural logarithm releases the exponent:

$$(n+1)\ln r < \ln((1-r)\varepsilon)$$

The direction of the inequality did not change because the natural logarithm is an increasing function. Now we divide by $\ln r$ and solve for n to get:

$$n > \frac{\ln((1-r)\varepsilon)}{\ln r} \qquad (9.10)$$

If you are wondering why the inequality changed from $<$ to $>$ then remember that $r<1$ so $\ln r < 0$ and division by a negative number changes the direction of the inequality.

The smallest integer that is greater than or equal to the number on the right hand side of the inequality in (9.10) is the ε-index for the numerical sequence $s_n(r)$:

$$N_\varepsilon = \left\lceil \frac{\ln((1-r)\varepsilon)}{\ln r} \right\rceil \qquad (9.11)$$

The most striking feature of the above number is that N_ε depends on the choice of r; there is no analog of this extra parameter for infinite sequences and series of numbers! Now N_ε is not just a number but a function $N_\varepsilon(r)$ that changes as r does even when ε is fixed.[3]

To illustrate the effect of this new parameter, let's fix the value of ε at say, $\varepsilon = 0.01$. The following table lists the values that we get for N_ε in (9.11) for each specified number r in the interval $(-1,1)$:

r	0.5	0.6	0.7	0.8	0.9	0.95	0.99
$N_\varepsilon(r)$	8	11	17	28	41	149	917

For example, if $r=0.5$ then

$$N_\varepsilon(0.5) = \left\lceil \frac{\ln((1-0.5)0.01)}{\ln 0.5} \right\rceil = \left\lceil \frac{\ln 0.005}{\ln 0.5} \right\rceil = \lceil 7.64 \rceil = 8$$

The table shows that in order for the partial sum $s_N(r)$ (and therefore, $s_n(r)$ for $n \geq N_\varepsilon$) to approximate the limit $1/(1-r)$ to within an error of $\varepsilon = 0.01$, *the value of N_ε must be higher the closer r is to 1.* For $r=0.5$ the approximation to within 0.01 is achieved by just 8 terms of the sum whereas for $r=0.95$ we require at least 149 terms to achieve the same accuracy. This result is consistent with, and sharpens what is shown in Figure 9.1, where the graphed partial sums $s_1(x)$ to $s_4(x)$ and even $s_{10}(x)$ are nowhere near the limit curve when the value of x is too close to 1.

The above table has another piece of information to offer: if the value of x, namely, r approaches 1 the value of N_ε becomes arbitrarily large; for $r=0.99$ the table shows that $N_\varepsilon(0.99) = 917$. For $r=0.999$ we can use (9.11) to find $N_\varepsilon(0.999) = 11,508$. Using a calculator or other computing device and (9.11) you can convince yourself that $N_\varepsilon(r) \to \infty$ as $r \to 1$. This is true regardless of the value of a fixed ε because the fraction

$$\frac{\ln((1-r)\varepsilon)}{\ln r} = \frac{\ln(1-r)}{\ln r} + \frac{\ln \varepsilon}{\ln r} \qquad (9.12)$$

blows up to infinity as $r \to 1$; see Figure 9.3 which illustrates the graph of the first fraction on the right hand side of (9.12). The second term only adds more infinity to the first because it has a denominator that converges to 0 while its numerator is a fixed number per fixed value of ε.[4]

---

[3]Technically, $N_\varepsilon(r)$ in (9.11) is a function (logarithmic) of two variables, r and ε. But when studying pointwise convergence we usually fix the value of ε to study the changes in convergence when r changes.

[4]You may note that $\ln r < 0$ since $r < 1$. We also generally assume that ε is arbitrarily small in our work on limits so we may assume that $\varepsilon < 1$. Then both $\ln \varepsilon$ and $\ln r$ are negative which makes the fraction $\ln \varepsilon / \ln r$ positive.

Series of Functions

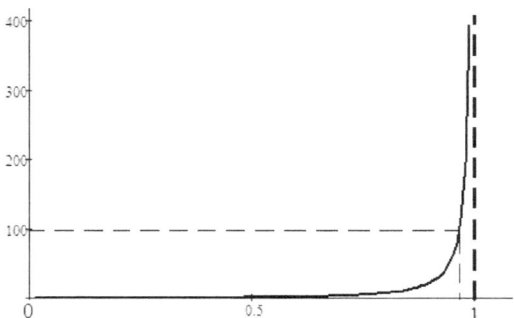

Figure 9.3: Exposing a hidden infinity in pointwise convergence

This occurrence of infinity hidden in the value of the ε-index N_ε because it does not appear explicitly in any of the functions and equations pertaining to the geometric power series or its partial sums. But its presence, if not its actual form was hinted at by the infinite singularity of the limit function $1/(1-x)$ at $x = 1$. The situation is more enigmatic near $x = -1$ where infinity hidden in the ε-index occurs without any traces in the graph.

Infinity hidden at the left endpoint.

Let's go back to Figure 9.1 and take a closer look at the other endpoint of $(-1, 1)$, namely, $x = -1$. The limit function $1/(1-x)$ is perfectly well-defined there: it equals $1/2$ so there are no singularities in the function. But from our study of geometric power series we know that the partial sums do not converge to anything if $x \le -1$ so we now consider what happens near $x = -1$ and just to the right of this number.

Let $-1 < r < 0$ and fix the value of $x = r$. By (9.1) the partial sums $s_n(x)$ at this specified value do converge to the number $1/(1-r)$. This is illustrated in Figure 9.4 where the partial sum polynomials $s_1(x)$ to $s_4(x)$ as well as $s_{19}(x)$ and $s_{20}(x)$ are plotted together with a portion of the limit function (the thick curve). Worth noticing in this diagram is the fact that all even-indexed partial sums pass through the point $(-1, 1)$ while all the odd-indexed ones pass through $(-1, 0)$. I leave it to you to explain why this is the case but stress the fact that we now see a type of oscillatory behavior that did not occur near $x = 1$.

In Figure 9.4 we see that the partial sums $s_n(-1)$ oscillate by jumping from -1 to 1 and back; so they cannot converge to any number, including $1/2$ (the value of the limit function $1/(1-x)$ at $x = -1$).

There are no singularities or jumps to infinity in Figure 9.4.

Does this mean that the divergence of partial sums at the left endpoint of -1 is not accompanied by any occurrences of the infinity?

In fact, infinity is there, *camouflaged*.

Let's see how it enters the picture here; as it happens, we have already done most of the calculation!

Going back to (9.9), since r is negative let's keep the absolute value and go on with our calculation

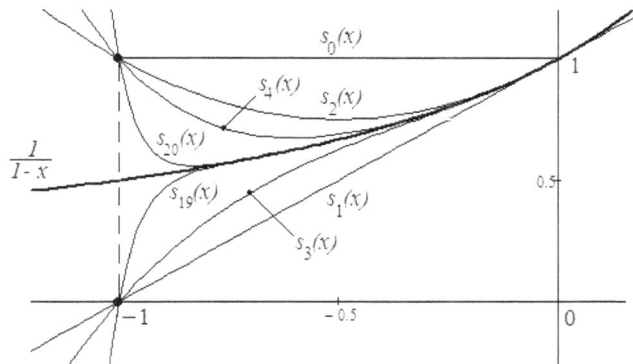

Figure 9.4: Several partial sums near the left hand endpoint

as before. We are led to the inequality:
$$\frac{|r|^{n+1}}{1-r} < \varepsilon$$
which can be solved for n just as before to yield the ε-index that once again depends on the value r:
$$N_\varepsilon(r) = \left\lceil \frac{\ln((1-r)\varepsilon)}{\ln|r|} \right\rceil \qquad (9.13)$$

The negativity of r plays a role in the numerator because if $r < 0$ then $1 - r > 1$ so $\ln(1-r) > 0$. But then $\ln|r| < 0$ because $|r| < 1$. Do we get a (nonsensical) negative value for N_ε then?

Not if $\varepsilon < 1/2$; because then since $r > -1$ we get
$$(1-r)\varepsilon < 2\varepsilon < 1$$
so $\ln((1-r)\varepsilon) > 0$ and as before, we get a positive value for N_ε. Taking $\varepsilon < 1/2$ is not a problem here since remember that in our work with limits we generally assume that ε is an arbitrarily small positive number.

Now, suppose that we set $\varepsilon = 0.01$ as in the case $x = 1$ but use (9.13) to create a table for N_ε corresponding to different values of r as follows:

r	-0.5	-0.6	-0.7	-0.8	-0.9	-0.95	-0.99
$N_\varepsilon(r)$	7	9	12	19	38	77	390

For example, if $r = -0.5$ then
$$N_\varepsilon(-0.5) = \left\lceil \frac{\ln((1+0.5)0.01)}{\ln|-0.5|} \right\rceil = \left\lceil \frac{\ln 0.015}{\ln 0.5} \right\rceil = \lceil 6.059 \rceil = 7$$

As we see from the above table, the value of N_ε increases substantially as r is chosen closer to -1. As in (9.12), we can prove that the value of N_ε blows up to infinity as follows:
$$\frac{\ln((1+r)\varepsilon)}{\ln|r|} = \frac{\ln((1+r)/2)}{\ln|r|} + \frac{\ln(2\varepsilon)}{\ln|r|}$$

Series of Functions   297

As long as $\varepsilon < 1/2$ both of the terms on the right hand side are positive and each one blows up to infinity.[5] It is remarkable that this happens even though the limit function $1/(1-x)$ and all of the partial sums $s_n(x)$ are continuous at $x = -1$ and all nearby points. Infinity is exhibited by the ε-index as a reaction to the divergence of the geometric power series for $x \leq -1$.

> **Convergence over sets of numbers.** *The occurrence of this infinity raises the issue that for a fixed $\varepsilon > 0$ there is no one N_ε large enough that works for all of the interval $(-1, 1)$ because we choose a value r for x that is arbitrarily close to 1 or to -1. Nevertheless, as long as the values $x = \pm 1$ are excluded, convergence does occur for each individual value of x. Thus, convergence at single points differs from convergence on entire sets, such as the interval $(-1, 1)$. This issue has profound consequences for infinite series of functions; we discuss some of these consequences later in this chapter by answering questions like: When does an infinite series of continuous functions (recall that all polynomials are continuous) converge to a continuous function? Can an infinite series of functions be integrated or differentiated term by term, one term at a time and then added up?*

9.3 Exploring the hidden infinity: a thought experiment

Before moving on to the general study of convergence for infinite sequences and series of functions, let's discuss a hypothetical "scientific study" where illustrate in a simple way how infinity may creep into the theory and cause divergence from experimentally measured values.

Suppose that a group of scientists devise a mathematical model of a phenomenon under observation. They derive a differential equation[6] whose solution quantifies the phenomenon being observed.

Initially, they cannot find a function that satisfies their differential equation, a commonplace occurrence with differential equations that model scientific phenomena. A possible course of action in this case is to derive the solution as an infinite series; we see how to do this in Section 9.6 below after introducing the concept of power series. Our hypothetical scientists obtain the following series:

$$f(x) = 1 - \frac{x^2}{2} + \frac{x^4}{4} - \frac{x^6}{8} + \cdots \qquad (9.14)$$

As we see later in this chapter, the function $f(x)$ that is expressed as an infinite series here is a solution of the differential equation and as such, it gives a mathematical description of the scientific phenomenon under observation.

The scientists use sophisticated instruments to experimentally measure the values of $f(x)$ to a high degree of accuracy. They successfully verify that their series agrees with their measurements for all values of x from about -1.4 to about 1.4. Within this range, they also find that using more terms of the series gives better agreements with the experimental measurements. The following table lists some of their findings:

---

[5] If you are computationally inclined then you may have noticed that the N_ε here is not as super-charged as in the case $x = 1$. In that case, in the fraction $\ln(1-r)(1/\ln r)$ both of the factors $\ln(1-r)$ and $1/\ln r$ reached infinitely large magnitudes while in the case of $x = -1$ the numerator $\ln(1-r)$ approaches $\ln 2$ as $r \to -1$ and only the factor $1/\ln |r|$ blows up. The infinite singularity at $x = 1$ asserts itself by kicking N_ε up twice as hard, so to speak.

[6] This differential equation is discussed in Section 9.6 below.

x	1.1	1.2	1.3	1.4	1.5
$s_{10}(x)$	0.62553	0.59707	0.62701	0.90946	2.18976
$s_{25}(x)$	0.62305	0.58128	0.53521	0.20637	-9.58961
$s_{50}(x)$	0.62305	0.58140	0.54211	0.68530	191.641
$f(x)$, measured	0.62305	0.58140	0.54201	0.50505	0.47059

They notice that the entries for the $x = 1.5$ column deviate significantly from the experimental results. Not only all values are way off from the correct number, things get worse, not better, by adding more terms of the series. The differential equation and the scientific hypotheses that led to it are examined and found to be sound. Our hypothetical scientists conclude that either their theory is incomplete (reluctantly) or the series solution is not valid beyond a certain range.

They set about to investigate. Not knowing how to solve the differential equation any other way, and not having a formula for $f(x)$, they plot the partial sums of their series to obtain the various curves that we see in Figure 9.5.

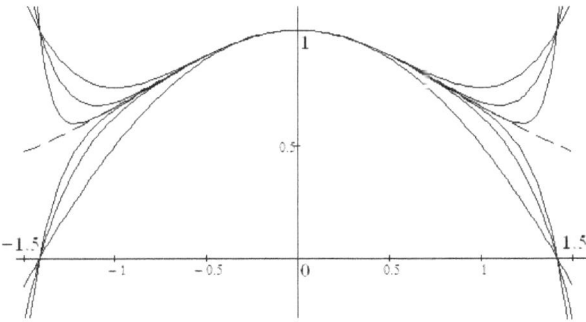

Figure 9.5: Divergence due to a hidden occurrence of infinity

The dashed curve represents the graph of $f(x)$ that is plotted using the "experimental data" obtained by our scientists. The figure shows that their series behaves well and agrees with experimental values for $-1.4 < x < 1.4$ (approximately).[7] But before the value of x reaches ± 1.5 the partial sums diverge from the experimental data and the more terms that are added the worse the results get. Evidently, the value of x at which all odd-indexed curves cross each other on the x-axis is a threshold number.

A breakthrough occurs when one of the scientists realizes that the infinite series in (9.14) is a geometric power series that can be written as:

$$\sum_{n=0}^{\infty} \left(\frac{-x^2}{2}\right)^n = \frac{1}{1-(-x^2/2)} \qquad (9.15)$$

Simplifying the fraction on the right hand side gives the sought-after formula for $f(x)$:

$$f(x) = \frac{2}{2+x^2} \qquad (9.16)$$

---

[7] The graphs of the first six partial sums of the series in (9.14) are shown as solid curves. Can you guess which partial sum each curve represents as we did earlier for Figure 9.4?

Series of Functions

Notice that this function is defined for all real values of x since the denominator is never less than 2 (when $x = 0$).

The scientists verify that this function gives correct results for all values of x obtained experimentally and its graph matches the dashed curve in Figure 9.5 perfectly. Corks are popped on champagne bottles and celebrations begin!

Following the discussion of (9.3) and Exercise 126 we can now see that the series in (9.14) converges to $f(x)$ over the interval $(-\sqrt{2}, \sqrt{2})$. Outside this interval, including at the values $x = \sqrt{2}$ and $x = -\sqrt{2}$ the series in (9.14) diverges and it is not meaningfully related to the function $f(x)$ in (9.16) which is well-defined for all real values of x.

Exercise 130 *Verify the above assertions about the convergence of the series in (9.14).*

It is worth emphasizing that *when we look at Figure 9.5 we see no traces of infinity*: there are no infinite values or even infinite slopes (vertical tangents) anywhere. But infinity is hidden in the ε-index and can be identified using an analysis entirely similar to that in the last section. The divergence of the series in (9.14) from the function $f(x)$ in (9.16) at the two endpoints $\pm\sqrt{2}$ may be rightfully attributed to this hidden occurrence of infinity.

This hidden infinity is also responsible for *the convergence slow-down* within the interval of convergence that we see in the last table of numbers above. Ignoring the entries in the right-most column, we see that in each of the other columns convergence to the value of $f(x)$ slows down as the value of x approaches $\sqrt{2}$, because more terms of the series are required to reach a given level of accuracy. This is a direct consequence of the increasing value of the ε-index: the larger this index is, the longer it takes for a sequence to come to within ε of the limit. This convergence slow-down is a clear sign of the occurrence of infinity in the ε-index.

In the next section, we discover that *even when a series converges to a function over some interval, this convergence may be subject to the occurrence of infinity in the ε-index* and the accompanying slow down.

9.4 Infinite sequences of functions: explore infinity's realm

For an infinite series of functions the partial sums form a sequence of functions. In the last section we saw that the partial sums of the geometric power series were in fact polynomial functions. In this section we define what we mean by a *convergent sequence of functions*, whether it is a partial sum sequence or not. The topic is quite interesting in its own right and its applications to the study of spaces of functions go far beyond the topic of infinite *series*.

So let's temporarily put the infinite series aside and examine the convergence of *sequences* of functions. Recall the ε and N definition of limits for number sequences in Chapter 4; when we apply the same idea to sequences of functions we discover that there are different ways in which sequences of functions may converge to limits, each way being important in its own context. We discuss two general types of convergence in this section that are of fundamental importance in the context of infinite series.

Pointwise convergence.

Consider a sequence of functions $f_n(x)$ where the variable x is in some nonempty (typically infinite) set S of real numbers. An example is the the sequence of partial sums $s_n(x) = 1+x+\cdots+x^n$ of the geometric power series with $S = (-1,1)$ that we discussed earlier. Each $s_n(x)$ is a polynomial function.

Let's consider each value of x separately; by setting $x = c$ where c is a fixed real number in the domain of every function $f_n(x)$ we get a sequence of constants $c_n = f_n(c)$. Now this is numerical sequence, not a sequence of functions, so we can use the ideas in Chapter 4. If it converges according to our earlier definition then its limit is a real number; let's call it $f(c)$ since it was calculated for the fixed value c. The equation

$$\lim_{n\to\infty} f_n(c) = f(c) \qquad (9.17)$$

states that our sequence of functions $f_n(x)$ *converges at the chosen point* $x = c$.

From Chapter 4 recall that the equality in (9.17) means this: for each $\varepsilon > 0$ there is an index N such that

$$|f_n(c) - f(c)| < \varepsilon \quad \text{for all } n \geq N$$

As before, N_ε denotes the least such value of N. If the equality in (9.17) holds for some value c of x then may or may not hold for other values of x. Consider the sequence $f_n(x) = nx$, each of which is a straight line that passes through the origin with slope n and let $S = (-\infty, \infty)$ be the set of all real numbers. At the value $x = 0$ we see that $f_n(0) = 0$ for every n so $\lim_{n\to\infty} f_n(0) = 0$. But if $c \neq 0$ then at $x = c$ the sequence $f_n(c)$ diverges to infinity because:

$$\lim_{n\to\infty} f_n(c) = \lim_{n\to\infty} nc = \begin{cases} \infty & \text{if } c > 0 \\ -\infty & \text{if } c < 0 \end{cases}$$

So this sequence of linear functions converges only for the single value $x = 0$.

It may happen that (9.17) holds for every fixed value of c (or point) in S; if so then for each x the limit $f(x)$ is a real number that depends on the choice of x. Putting all of these real numbers together defines a function $f: S \to \mathbb{R}$ and gives us the following definition.

Pointwise convergence and limit: A sequence of functions $f_n(x)$ *converges pointwise* to a function $f(x)$ on a set S if for all x in S the following is true:

$$\lim_{n\to\infty} f_n(x) = f(x)$$

The function $f(x)$ is the *pointwise limit* of the sequence of functions $f_n(x)$. We also write:

$$f_n \xrightarrow{p} f \quad \text{on } S$$

In the previous section we discussed the pointwise convergence of the sequence of partial sums of the geometric power series. Specifically, we showed that the sequence of partial sums $s_n(x)$ converges pointwise to the rational function $R(x) = 1/(1-x)$ on the set $S = (-1, 1)$. In symbols:

$$\lim_{n\to\infty} s_n(x) = \frac{1}{1-x} \quad \text{or} \quad s_n \xrightarrow{p} R \quad \text{on } (-1, 1)$$

We also showed that there were hidden occurrences of infinity in the ε- index; *these occurrences are common for pointwise converging sequences of functions.* Here is a another illustration that also happens to contain no singularities or explicit occurrences of infinity and makes an important point.

Consider the sequence of inverse tangent functions:

$$f_n(x) = \tan^{-1}(nx) \quad n = 1, 2, 3, \ldots \tag{9.18}$$

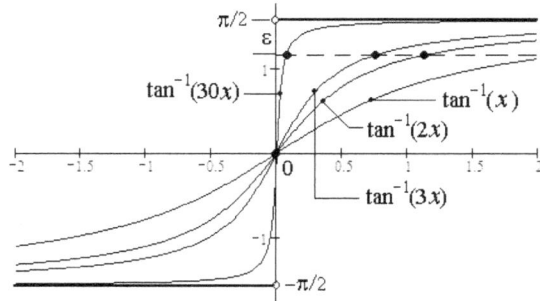

Figure 9.6: Continuous functions converging pointwise to a discontinuous function

Figure 9.6 shows the graphs of some of these functions. Although the inverse tangent function is defined on the set of all real numbers, in this example we arbitrarily restrict the set S to the interval $[-2, 2]$, so that there are no explicit occurrences of infinity in our discussion. All the interesting stuff happens around 0 anyway.

With the increasing value of n, we see in Figure 9.6 that the S-shaped curves are increasingly aligned with the three-piece, discontinuous function: [8]

$$f(x) = \begin{cases} \pi/2 & \text{if } x > 0 \\ 0 & \text{if } x = 0 \\ -\pi/2 & \text{if } x < 0 \end{cases} \tag{9.19}$$

This function is the pointwise limit of the sequence in (9.18). To prove this, fix the value of the variable x at a real number c in $[-2, 2]$. If $c > 0$ then $nc \to \infty$ as $n \to \infty$ so:

$$\lim_{n \to \infty} f_n(c) = \lim_{n \to \infty} \tan^{-1}(nc) = \frac{\pi}{2}$$

Similarly, if $c < 0$ then $nc \to -\infty$ as $n \to \infty$ so $\lim_{n \to \infty} f_n(c) = -\pi/2$. If $c = 0$ then $nc = 0$ for all n so $\tan^{-1}(nc) = \tan^{-1}(0) = 0$. Thus $f(x)$ in (9.19) is indeed the pointwise limit!

Now, consider the ε-index N_ε for a (small) positive ε. Figure 9.6 shows one such ε; we see that if we choose c greater than 1.5 then $f(c) - f_n(c) < \varepsilon$ for $n \geq 2$ (so $N_\varepsilon = 2$ for this ε) while if c is between 1 and 1.5 then $f(c) - f_n(c) < \varepsilon$ for $n \geq 3$ (so $N_\varepsilon = 3$). As c is chosen closer and closer to 0 a greater value of n is required to bring $f_n(c)$ within ε of $f(c)$; in Figure 9.6 we see that if c is 0.5 or 0.25 then $n \geq 30$ guarantees that $f(c) - f_n(c) < \varepsilon$ (though a smaller n might work too).

To actually calculate N_ε it is enough to consider either the positive half or the negative half since $f(x)$ and all $f_n(x)$ are symmetric with respect to the origin (they are odd functions). Looking at

---

[8] Remember that $\tan^{-1}(x)$ approaches $\pi/2$ as $x \to \infty$ and to $-\pi/2$ as $x \to -\infty$. Also worth a mention is the fact that the tangent line to the graph of $\tan^{-1}(nx)$ at the origin is just the line nx that we discussed earlier.

the positive half, since $f(c) = \pi/2$ is larger than $f_n(c)$ for each $c > 0$ we can drop the absolute value and write:

$$f(c) - f_n(c) < \varepsilon$$
$$\frac{\pi}{2} - \tan^{-1}(nc) < \varepsilon$$
$$\frac{\pi}{2} - \varepsilon < \tan^{-1}(nc)$$

The quantity $\pi/2 - \varepsilon$ on the left hand side is a fixed number per fixed ε; for instance, if $\varepsilon = 0.01$ then $\pi/2 - \varepsilon \simeq 1.56$. Now, let's take the tangent of the last inequality to cancel $\tan^{-1}$ and get:[9]

$$\tan\left(\frac{\pi}{2} - \varepsilon\right) < nc$$

Since $c > 0$ we can divide by c to obtain

$$\frac{1}{c}\tan\left(\frac{\pi}{2} - \varepsilon\right) < n$$

We now define N_ε as the smallest integer that is greater than the expression on the left hand side:

$$N_\varepsilon(c) = \left\lceil \frac{1}{c}\tan\left(\frac{\pi}{2} - \varepsilon\right) \right\rceil \qquad (9.20)$$

Notice that the factor $\tan(\pi/2 - \varepsilon)$ is a positive constant per fixed value of ε; for instance, if $\varepsilon = 0.01$ then $\tan(\pi/2 - \varepsilon) \simeq 92.62$. So as c gets closer to zero, $1/c$ blows up to infinity, taking $N_\varepsilon(c)$ along with it!

Exercise 131 *Let $\varepsilon = 0.01$ and use (9.20) to calculate the values of the ε-index N_ε corresponding to the following values of c:*

$$1,\ 0.5,\ 0.1$$

If you have a graphics device, consider sketching the graph of $\tan^{-1}(Nx)$ for N_ε corresponding to $c = 0.5$.

Pointwise convergence comes up short!

The hidden infinity in the ε-index of some pointwise converging sequences is responsible for significant shortcomings. For this reason, pointwise convergence is not strong enough to ensure the preservation of "nice" qualities. For instance, although every function $f_n(x) = \tan^{-1}(nx)$ above is differentiable everywhere, the pointwise limit function $f(c)$ is not even continuous, let alone be differentiable at $x = 0$. Similarly, when discussing the geometric power series, we discovered that even though every partial sum $s_n(x)$ is a polynomial, hence differentiable (so also continuous and integrable) at $x = 1$, the limit function $1/(1-x)$ with its infinite singularity is discontinuous and not integrable at $x = 1$. We conclude that:

---

[9] Since $\tan x$ is an increasing function when x is between $-\pi/2$ and $\pi/2$, the inequality direction remains unchanged after taking the tangent.

Series of Functions   303

The properties of being continuous or differentiable or integrable are not preserved by pointwise convergence.

In particular, knowing that $f_n \xrightarrow{p} f$ on an interval $[a, b]$ where f and every f_n is integrable, is not enough to allow the following operation:

$$\lim_{n \to \infty} \int_a^b f_n(x)dx = \int_a^b f(x)dx = \int_a^b \lim_{n \to \infty} f_n(x)dx \tag{9.21}$$

As we find later, being able to do this is crucial to proving that a given infinite series of functions is integrable term by term, a property that yields interesting results like the series expansion for the logarithm in (9.7).

A simple example where (9.21) fails is illustrated in Figure 9.7. Let's discuss this example because it also clearly illustrates the problem that occurs with the ε-index depending on the fixed value of x that is chosen; this dependence is often the reason why the ε-index blows up to infinity.

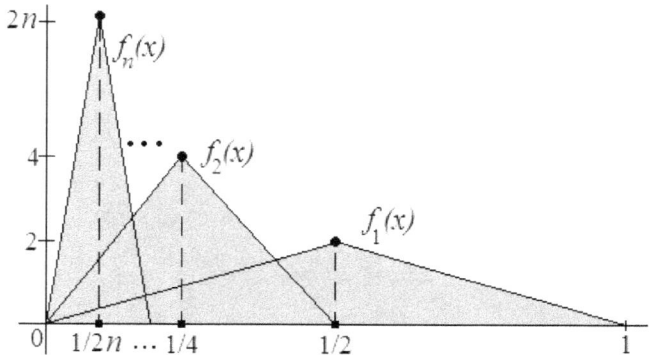

Figure 9.7: Showing that the integral and pointwise limit are not interchangeable

In Figure 9.7 we see a sequence of functions $f_1(x), f_2(x), \ldots, f_n(x), \ldots$ in the form of slanted and flat segments: the first function $f_1(x)$ is just the top sides of the shaded triangle having a height of 2 and the interval $[0,1]$ on the x-axis as its base. The second function $f_2(x)$ consists of the top sides of the shaded triangle of height 4 and having the interval $[0,1/2]$ as the base, together with the flat line segment from $1/2$ to 1; so $f_2(x) = 0$ on the interval $[1/2,1]$. Generally, the n-th function $f_n(x)$ consists of the top sides of the shaded triangle of height $2n$ and the interval $[0, 1/n]$ as base, plus a flat segment $f_n(x) = 0$ on $[1/n, 1]$. Notice that the lengths of the flat segments increase as n does, to eventually cover the interval $(0,1]$.

Each function $f_n(x)$ is continuous, so it is integrable. The integral of $f_n(x)$ is just the area of the n-th shaded triangle for every n; specifically:

$$\int_0^1 f_n(x)dx = \int_0^{1/n} f_n(x)dx + \int_{1/n}^1 f_n(x)dx = \frac{1}{2}(2n)\left(\frac{1}{n}\right) + 0 = 1$$

Since for every n the integral has a fixed value of 1, the sequence of integrals is a constant sequence of real numbers:[10]

$$\int_0^1 f_1(x)dx, \int_0^1 f_2(x)dx, \ldots, \int_0^1 f_n(x)dx, \ldots = 1, 1, \ldots, 1, \ldots$$

This sequence obviously converges to 1:

$$\lim_{n \to \infty} \int_0^1 f_n(x)dx = 1$$

On the other hand, the pointwise limit of the sequence f_n is the zero function $f(x) = 0$. To see why, consider any fixed value $x = c$ in the interval $[0,1]$. Now, either $c = 0$ or $c > 0$. If $c = 0$ then $f_n(0) = 0$ by the definition of the sequence (see Figure 9.7). If $c > 0$ then for all indices n large enough that $1/n < c$ (or $n > 1/c$) c falls in the interval $[1/n, 1]$ where $f_n(c) = 0 = f(c)$. This implies that $\lim_{n \to \infty} f_n(c) = 0$ and since c was an arbitrary value of x, we have actually proved that $\lim_{n \to \infty} f_n(x) = 0$ for every fixed value of x in $[0,1]$. In other words, $f_n \xrightarrow{p} 0$ on this interval.

But then the zero limit function $f(x) = 0$ is integrable as a constant function with integral

$$\int_0^1 f(x)dx = \int_0^1 0 dx = 0$$

We see that (9.21) is not true for this sequence of integrable functions!

It is worth mentioning that this example can be made more extreme. Suppose that the height of triangle n is n^2 instead of $2n$. Then its area is

$$\frac{1}{2}(n^2)\frac{1}{n} = \frac{n}{2}$$

which would lead to the sequence of integrals

$$\int_0^1 f_1(x)dx, \int_0^1 f_2(x)dx, \ldots, \int_0^1 f_n(x)dx, \ldots = \frac{1}{2}, 1, \frac{3}{2}, 2, \ldots, \frac{n}{2}, \ldots$$

This sequence diverges to infinity. Nevertheless, the pointwise limit $f(x)$ is still the zero function for the same reason as before, so its integral is still 0.

Given that the problem is due to the ε-index diverging to infinity, a pertinent question now is: Can we ensure that (9.21) is true by keeping the ε-index finite?

First, let's observe that the ε-index of the sequence $f_n(x)$ above is $N_\varepsilon = [1/c]$ where $[1/c]$ is the least integer that is greater than $1/c$. This is true because $f_n(c) = 0$ for every index n that exceeds $[1/c]$, so in particular $f_n(c) < \varepsilon$ for every positive value of ε (notice that N_ε in this case is independent of ε).

---

[10] The area of each shaded triangle is 1 in spite of the fact that its height $2n$ goes to infinity. This also means that the lengths of the two equal sides become infinitely large as $n \to \infty$ with the consequence that the perimeter goes to infinity (this is easy to prove since the lengths of the equal sides can be calculated using the Pythagorean theorem). The point $x = 1/2n$ on the x-axis at which the maximum height occurs goes to zero, where all functions have the value 0. So the triangles get infinitely tall but also stick-like narrow, as if disappearing into a black hole at the origin!

Next, since $N_\varepsilon = [1/c]$ we see that *N_ε will not go to infinity if and only if c is prevented from getting arbitrarily close to 0.* To accomplish this, we truncate the interval [0,1] to $[a, 1]$ where a can be as close to 0 as we like, but it is a *fixed* positive number. Now, if c is in $[a, 1]$ then $c \geq a$ and therefore, c cannot be chosen arbitrarily close to 0.

Now, let's see what happens if we restrict attention to $[a, 1]$...

Again, as $n \to \infty$ the functions $f_n(x)$ approach the zero function pointwise on $[a, 1]$ but now there is an ε-index that does not go to infinity. Since $c \geq a$ for every fixed value c of x in $[a, 1]$, it follows that $1/c \leq 1/a$ so we simply define $N_\varepsilon = [1/a]$ (for the interval $[a, 1]$). This gives $f_n(x) = 0$ for every $n \geq N_\varepsilon$ and all values of x in $[a, 1]$ because the triangular part of $f_n(x)$ has moved to the left of a. But since a does not change, N_ε is a constant so, in particular, it doesn't blow up to infinity when the value of x changes.

Now let's see how this solves the problem with the integrals!

To begin with, $\int_a^1 f(x)dx = 0$ since the limit function $f(x)$ is the zero function on $[a, 1]$; that is, $f(x) = 0$ for all x in $[a, 1]$. As for the integrals of $f_n(x)$ we just saw that $f_n(x) = 0$ for all x in $[a, 1]$ if $n \geq N_\varepsilon = [1/a]$ so $\int_a^1 f_n(x)dx = 0$ for all such large enough n. This means that the sequence of integrals is a sequence of constants that is eventually 0 after N_ε:

$$\int_a^1 f_1(x)dx, \ldots, \int_a^1 f_{N_\varepsilon - 1}(x)dx, \int_a^1 f_{N_\varepsilon}(x)dx, \ldots = I_1, \ldots, I_{N_\varepsilon - 1}, 0, 0, \ldots$$

where $I_1, \ldots, I_{N_\varepsilon - 1}$ are numbers less than 1 because the integrals do not represent the areas of *entire* shaded triangles (the tip of each triangle to the left of a is cut off). The above sequence ends in 0 so it converges to 0, which is the value of $\int_a^1 f(x)dx$. Therefore, (9.21) is true on $[a, 1]$!

If we interpret the integrals as measures of wealth then what happens in the above example is analogous to Elf Greedy's situation: If we stop at some large (but finite) step n then we may have an arbitrarily large fortune but if we wait infinitely long then we would end up with nothing!

Let's consider the elves' situation in the light of pointwise convergence. Figure 9.8 illustrates the outcomes for two days (Day 2 and Day 7).

The shaded bars indicate Greedy's fortune and the clear bars with thick boundary Generous's. Each bar shows the number of gold nuggets in the deposit-box compartment numbered on the horizontal axis.

In each day, the bars indicate the value of a function $f_n(x)$ that gives the distribution of gold nuggets in each elf's deposit box. For instance, if $n = 2$ (Day 2) then for Greedy $f_2(x)$ has the value 3 if $2 < x \leq 3$ and the value 4 if $3 < x \leq 4$ and $f_2(x) = 0$ for all other values of x. For Generous, $f_2(x)$ has the value 1 if $0 < x \leq 1$ and the value 3 if $2 < x \leq 3$ and $f_2(x) = 0$ for all other values of x.

Notice that while Generous's bars remain intact as days progress, Greedy's bars disappear even as they get more dense on the right. In the pointwise limit, Greedy's limit is the zero function while Generous's is the function whose graph consists of the sequence of clear bars in odd-numbered compartments. Note also that the areas of the bars are the integrals of the functions $f_n(x)$ described above.

More precisely, in Greedy's case the function sequence $f_n(x)$ converges pointwise to 0 on the interval $[0, \infty)$ because for every value r of x in this interval, by the Archimedean property there is an integer N that is greater, that is, $r < N$. It follows that $f_n(r) = 0$ for all $n \geq N$; in essence,

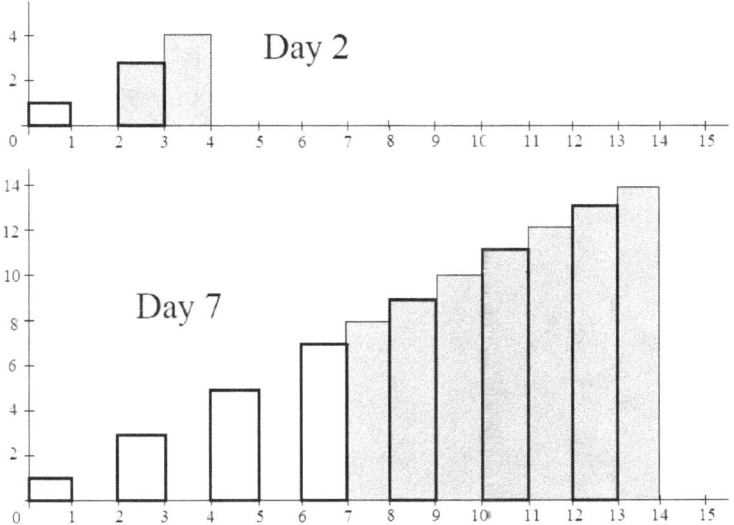

Figure 9.8: Snapshots of elves' fortune

where $f_n(x) \neq 0$ is past the number r for all $n \geq N$. Therefore,

$$\lim_{n \to \infty} f_n(r) = 0$$

that is, $f_n(x)$ converges to 0 at $x = r$. Now, r was an arbitrarily chosen value in the interval $[0, \infty)$ so we see that $f_n(x)$ converges to 0 pointwise on this interval and Greedy's gold nuggets disappear!

Exorcizing the hidden infinity: Uniform convergence.

In the case of triangle functions above, truncating the interval [0,1] blocked the value of N_ε from going to infinity and we saw that the problem with (9.21) being false was resolved. When thinking about this, it is important to keep in mind that *the truncation of the interval worked because it prevented N_ε from becoming infinitely large*. In other words, the only problem with the entire interval [0,1] was that it was possible to have $N_\varepsilon \to \infty$ in it.

On the truncated interval $[a, 1]$ we were able to define the ε-index N_ε as $[1/a]$, that is, independently of the value of x. This is something that can be generalized to what amounts to a strong type of convergence for sequences of functions.

> **Uniform convergence and limit.** A sequence of functions $f_n(x)$ *converges uniformly on a set S* of real numbers to a function $f(x)$ if for every $\varepsilon > 0$ there is an ε-index N_ε that does not depend on the value of x (it has the same value uniformly over all of S) and for all $n \geq N_\varepsilon$
> $$|f_n(x) - f(x)| < \varepsilon \quad \text{for all } x \text{ in } S \qquad (9.22)$$
> We also use the notation: $f_n \xrightarrow{u} f$ on S.

The above definition may seem abstract but it is rather easy to visualize. Let's write the absolute

Series of Functions

value inequality in (9.22) in the equivalent form as two inequalities:

$$-\varepsilon < f_n(x) - f(x) < \varepsilon$$

or:

$$f(x) - \varepsilon < f_n(x) < f(x) + \varepsilon$$

The last pair of inequalities say that for every $n \geq N_\varepsilon$ the graph of $f_n(x)$ over the entire set S is contained within a strip of width 2ε that is centered at the limit curve $f(x)$. Figure 9.9 illustrates this situation.

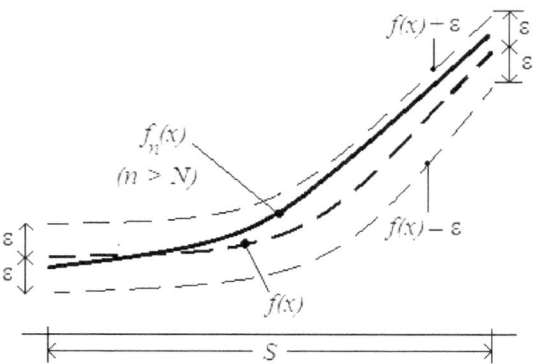

Figure 9.9: Illustrating uniform convergence

Let's call the strip of width 2ε the ε-strip about $f(x)$. Looking back at Figures 9.6 and 9.7 it is easy to see that if we draw ε-strips about the limit curves in each figure then the functions $f_n(x)$ can never be contained in the ε-strips *over the entire set S* no matter how large the index n is. The convergence in those examples is clearly not uniform.

Notice that every uniformly convergent sequence is automatically pointwise convergent which we may remember as:

$$f_n \xrightarrow{u} f \text{ on } S \text{ implies that } f_n \xrightarrow{p} f \text{ on } S.$$

The contrapositive of the above statement is also worth remembering:

If a sequence of functions f_n does not converge pointwise then it does not converge uniformly.

In the several examples that we discussed above we found that the *converse of the above statement is false*. In the case of triangle-top functions above, we can now say that that sequence converges to the zero function uniformly on $[a, 1]$ (no matter how close a is to 0) but only pointwise on $[0,1]$.

Exercise 132 *Show that the inverse tangent sequence $f_n(x) = \tan^{-1}(nx)$ converges uniformly on the interval $[a, \infty)$ if $a > 0$ no matter how close a may be to 0. All the pieces of the argument are already in place in our earlier calculations involving this sequence; we need only show that N_ε that we calculated will now be independent of the value c of x.*

You may have noticed that in our definition of uniform convergence we did not say: "N_ε must not go to infinity for fixed ε". We didn't have to, because the independence of N_ε from the value of x is equivalent to N_ε not going to infinity. If N_ε doesn't depend on the variable x then it doesn't change when ε is fixed; in particular, it won't go to infinity. Conversely, if N_ε is bounded and has a maximum value with changing x (ε fixed) then we simply re-define N_ε to be the maximum value and thus free it from dependence on x.

This is indeed what we did for the triangle functions on the interval $[a, 1]$; since $1/x \leq 1/a$ for every value of x in $[a, 1]$ we could define $N_\varepsilon = [1/a]$. We certainly can't do this on $[0,1]$ where the value of x can get arbitrarily close to 0.

Now, let's go back to the partial sums of the geometric power series and suppose that we truncate the interval $(-1, 1)$ to $[-a, a]$ where $a < 1$ but a can be as close to 1 as we like; it's value is of course constant or fixed. Then the value of r in (9.11) cannot approach 1 since $0 < r \leq a < 1$ and $1 - a \leq 1 - r$. Now, remembering that $\ln x$ is an increasing function, taking logarithms does not affect the inequalities, so we have:

$$\ln r \leq \ln a < 0 \quad \text{and} \quad \ln((1-a)\varepsilon) \leq \ln((1-r))\varepsilon < 0$$

Flipping the first pair now reverses the inequality:

$$\frac{1}{\ln r} \geq \frac{1}{\ln a}$$

But multiplying by a negative quantity also reverses the inequality:

$$\left(\frac{1}{\ln r}\right) \ln((1-a)\varepsilon) \leq \left(\frac{1}{\ln a}\right) \ln((1-a)\varepsilon) \quad \text{and:}$$

$$\left(\frac{1}{\ln r}\right) \ln((1-r)\varepsilon) \leq \left(\frac{1}{\ln r}\right) \ln((1-a)\varepsilon) \quad \text{since} \quad \frac{1}{\ln r} < 0$$

so we figure that:

$$\frac{\ln((1-r)\varepsilon)}{\ln r} = \left(\frac{1}{\ln r}\right) \ln((1-r)\varepsilon)$$

$$\leq \left(\frac{1}{\ln r}\right) \ln((1-a)\varepsilon)$$

$$\leq \left(\frac{1}{\ln a}\right) \ln((1-a)\varepsilon)$$

$$= \frac{\ln((1-a)\varepsilon)}{\ln a}$$

So if we define

$$N_\varepsilon = \left[\frac{\ln((1-a)\varepsilon)}{\ln a}\right]$$

then this N_ε is independent of any specific value of x and in particular, will not blow up to infinity. A similar calculation blocks the growth of N in (9.13). We conclude that:

Series of Functions 309

> *The sequence $s_n(x)$ of partial sums of the geometric power series converges uniformly to the function $1/(1-x)$ on the interval $[-a, a]$ if $0 < a < 1$ no matter how close a is to 1.*

The above statement is what makes good things possible for the geometric power series, such as the expansion of the natural logarithm in (9.7) in terms of power functions. We turn to this issue next.

On preserving limits, integrals and derivatives.

We have seen that if a sequence of continuous functions $f_n(x)$ converges pointwise to a function $f(x)$ then $f(x)$ is not necessarily continuous. However, things look much better if the convergence is uniform.

> **Uniform convergence preserves continuity.** *If $f_n \xrightarrow{u} f$ on S and f_n is continuous on S for each n then the uniform limit f is continuous on S.*

The proof of this theorem is a standard argument in real analysis; though it is short and simple we do not discuss it here so as to avoid technical details that are not pertinent to our focus on the role of infinity.

One interesting consequence of the preservation of continuity is the ability to *interchange or switch limits* which can be helpful in calculations.

> **Switching limits.** *If the functions $f_n(x)$ are continuous and converge uniformly (to some function) on a set S then*
>
> $$\lim_{n\to\infty} \lim_{k\to\infty} f_n(x_k) = \lim_{k\to\infty} \lim_{n\to\infty} f_n(x_k) \qquad (9.23)$$
>
> *for every sequence x_k in S that converges (to some point) in S.*

It is easy to see why this is true using the preservation of continuity theorem. By assumption we have $f_n \xrightarrow{u} f$ on S for some function f and $x_k \to x$ for some x in S. Since f_n is continuous on S for each n we have

$$\lim_{n\to\infty} \lim_{k\to\infty} f_n(x_k) = \lim_{n\to\infty} f_n(x) = f(x)$$

On the other hand, the limit function f is also continuous on S so

$$\lim_{k\to\infty} \lim_{n\to\infty} f_n(x_k) = \lim_{k\to\infty} f(x_k) = f(x)$$

So (9.23) is true!

If the convergence of $f_n(x)$ to $f(x)$ is only pointwise but *not uniform* then (9.23) can fail quite spectacularly!

Consider the sequence of inverse tangent functions $f_n(x) = \tan^{-1}(nx)$. If x_k is any sequence that converges to 0 as $k \to \infty$ then $nx_k \to 0$ for each fixed n so we have:

$$\lim_{n\to\infty} \lim_{k\to\infty} f_n(x_k) = \lim_{n\to\infty} \lim_{k\to\infty} \tan^{-1}(nx_k) = \lim_{n\to\infty} \tan^{-1}(0) = \lim_{n\to\infty} 0 = 0$$

On the other hand, $nx_k \to \infty$ for each fixed k as $n \to \infty$ and since $\tan^{-1} x \to \pi/2$ as $x \to \infty$ we have:

$$\lim_{k\to\infty} \lim_{n\to\infty} f_n(x_k) = \lim_{k\to\infty} \lim_{n\to\infty} \tan^{-1}(nx_k) = \lim_{k\to\infty} \frac{\pi}{2} = \frac{\pi}{2}$$

So (9.23) is false for *every* sequence x_k of real numbers that converges to 0. For instance, $1/k \to 0$ as $k \to \infty$ so

$$\lim_{n \to \infty} \lim_{k \to \infty} \tan^{-1}\left(\frac{n}{k}\right) \neq \lim_{k \to \infty} \lim_{n \to \infty} \tan^{-1}\left(\frac{n}{k}\right)$$

Exercise 133 *Use the fact that $f_n(x) = \tan^{-1}(nx)$ converges uniformly on $[a, \infty)$ for each fixed $a > 0$ (see Exercise 132) to prove that*

$$\lim_{n \to \infty} \lim_{k \to \infty} \tan^{-1}(nx_k) = \lim_{k \to \infty} \lim_{n \to \infty} \tan^{-1}(nx_k) = \frac{\pi}{2}$$

for every convergent sequence x_k in $[a, \infty)$. You may start by assuming that $x_k \to b$ for some unspecified number b where of course, $b \geq a$.

Next, let's consider the preservation of integrals. We take S to be an interval because we defined the (Riemann) integral on intervals.

> **Uniform convergence preserves the integral.** If $f_n \xrightarrow{u} f$ on an interval $[a, b]$ and f_n is integrable on $[a, b]$ for each n then the uniform limit f is integrable on $[a, b]$ and
>
> $$\lim_{n \to \infty} \int_a^b f_n(x)\,dx = \int_a^b f(x)\,dx \tag{9.24}$$

I leave out the proof; you can find it in typical real analysis textbooks. Notice that because $f(x) = \lim_{n \to \infty} f_n(x)$, (9.24) essentially says that the limit and the integral can be switched:

$$\lim_{n \to \infty} \int_a^b f_n(x)\,dx = \int_a^b \lim_{n \to \infty} f_n(x)\,dx$$

The limit inside the integral above is a uniform limit not pointwise, although the notation is vague (harmlessly, given the context). The limit *outside* the integral is a simple limit of a *numerical* sequence, namely, the sequence of numbers: $\int_a^b f_n(x)\,dx = I_n$

We have already seen that if the limit is pointwise then (9.24) is false; see our earlier discussion of the sequence of triangle functions. We discuss the usefulness of (9.24) in the next section when studying the term by term integration of infinite series.

Finally, we come to derivatives! Does uniform convergence preserve differentiation too? Specifically, if $f_n \xrightarrow{u} f$ on a set S and each f_n is differentiable at every point of S can we say that f is differentiable too, and furthermore, $\lim_{n \to \infty} f_n'(x) = f'(x)$ for every x in S?

If you have been meticulous in your reading so far then you may have noticed that we are requiring something more for derivatives here; namely, that $f_n \xrightarrow{u} f$ imply also that $f_n' \xrightarrow{u} f'$ on S. We didn't get this much mileage out of either continuity or integrability; the issue doesn't even arise in the former case and it is moot in the latter case because after integrating $f_n(x)$ we get a numerical sequence $\int_a^b f_n(x)\,dx$ that doesn't require special consideration.

Expecting too much often leads to disappointment and that happens here too.[11]

---

[11] In math as in life generally!

Consider the sequence $f_n(x) = (1/n)\sin(nx)$ of differentiable functions on the set of all real numbers $(-\infty, \infty)$. For each fixed value of $x = c$ we have:

$$\lim_{n \to \infty} f_n(c) = \lim_{n \to \infty} \frac{\sin(nc)}{n} = 0$$

because $-1 \leq \sin(nc) \leq 1$ for all n so the fraction goes to 0 as its denominator becomes infinitely large. So this sequence converges pointwise to the zero function $f(x) = 0$ on $(-\infty, \infty)$. It is easy to see that the convergence is actually uniform:

$$|f_n(x) - f(x)| = \left| \frac{\sin(nx)}{n} - 0 \right| = \frac{|\sin(nx)|}{n} \leq \frac{1}{n}$$

so for each $\varepsilon > 0$ we ensure $|f_n(x) - f(x)| < \varepsilon$ by choosing n large enough that $1/n < \varepsilon$; that is, $n > 1/\varepsilon$ which gives an ε-index $N_\varepsilon = \lceil 1/\varepsilon \rceil$ that does not depend on the value of x. So we see that $f_n \xrightarrow{u} 0$ on $(-\infty, \infty)$. Note that the zero function is differentiable since it is a constant function.

On the other hand, looking at the derivatives:

$$f'_n(x) = \frac{1}{n} \cos(nx) \frac{d}{dx}(nx) = \frac{1}{n} \cos(nx)(n) = \cos(nx)$$

we see that for a fixed value $x = c$ the sequence of derivatives $f'_n(c) = \cos(nc)$ does not converge for all choices of c. For instance, if $x = \pi$ then

$$f'_n(\pi) = \cos(n\pi) = \begin{cases} -1 & \text{if } n \text{ is odd} \\ 1 & \text{if } n \text{ is even} \end{cases}$$

So the sequence $f'_n(\pi)$ does not converge. We conclude that the sequence of derivatives $f'_n(x)$ doesn't converge pointwise, and therefore, it can't converge uniformly.

Fortunately, *there is a fix for this problem*, which is important because we would *very much like* to be able to take the derivative of a series of functions term by term!

To motivate the result on derivatives, let's go back to our series of functions above, namely, $f_n(x) = (1/n)\sin(nx)$ but this time consider their *antiderivatives*:

$$F_n(x) = -\frac{\cos(nx)}{n^2}$$

You can easily verify that $F'_n(x) = (1/n)\sin(nx)$ so $F_n(x)$ is a valid antiderivative of $f_n(x)$. Now, because:

$$|F_n(x) - F(x)| = \left| -\frac{\cos(nx)}{n^2} - 0 \right| = \frac{|\cos(nx)|}{n^2} \leq \frac{1}{n^2}$$

by solving the inequality $1/n^2 < \varepsilon$ we obtain an ε-index $N_\varepsilon = \lceil 1/\sqrt{\varepsilon} \rceil$ that is independent of the value of x. We see therefore that the sequence of antiderivatives converges uniformly to the zero function $F(x) = 0$.

This suggests that we may want to *assume* that the sequence of *derivatives* converges uniformly, then *from that assumption* conclude that the original sequence of functions converges uniformly (as antiderivatives of their own derivatives!). This and one other technical hypothesis give us the following odd-looking but useful result:

> **Derivatives and uniform convergence.** *Let $f_n(x)$ be a sequence of functions with continuous derivatives on an interval $[a,b]$. If (i) the sequence of derivatives $f_n'(x)$ converges uniformly on $[a,b]$ to a function $g(x)$ and (ii) there is x_0 in $[a,b]$ such that the sequence of numbers $f_n(x_0)$ has a limit then $f_n \xrightarrow{u} f$ on $[a,b]$ where $f(x)$ is differentiable and $f'(x) = g(x)$.*

The proof when the derivatives $f_n'(x)$ are continuous utilizes the Fundamental Theorem of Calculus, which is not surprising following our discussion of antiderivatives earlier. This proof is found in typical real analysis textbooks.

If the aforementioned x_0 in (b) above does not exist then the derivative theorem above may be false. Consider $f_n(x) = \ln(nx)$ on, say, $[1,2]$ and note that the sequence of derivatives:

$$f_n'(x) = \frac{1}{nx} n = \frac{1}{x}$$

which is a constant sequence (only one term, $1/x$, repeats) does converge (to $1/x$) as $n \to \infty$ and this convergence is (trivially) uniform. However, for *every* fixed value c of x in $[1,2]$ $\ln(nc) = \ln n + \ln c \to \infty$ as $n \to \infty$. So this sequence can't converge (uniformly or pointwise) on $[1,2]$.

When uniform convergence comes up short: length not preserved.

We have seen so far in this section that uniform convergence works differently for derivatives than it does for continuity or for integrals. This is a good place to show that uniform convergence does not generally preserve lengths and to do so, I will go back to a topic we discussed earlier: the anomalous convergence of staircase curves. We saw in the introductory chapter that these converge uniformly to the diagonal of a square but their length does not converge to the length of the diagonal.

We have seen that uniform convergence preserves integrals and we also know that areas may be calculated as integrals. This explains why Riemann sums work fine as *areas* under the staircase paths that converge to the area under the diagonal of the square.

Now, the staircase paths converge to the diagonal of the square uniformly. This is easy to see from the fact that the difference between any point (x,y) on a staircase path and the point (x,x) on the diagonal is at most the length $1/n$ of each riser, so the curves approaches 0 in our construction with the same value of ε for all values of x (that is, the ε-index is finite). But why don't the lengths of these paths converge to the length of the diagonal?

The main cause of anomalous convergence is a case of *competing sequences*: when one sequence dominates, the lengths converge; otherwise they don't. It is possible to explain this without discussing technical formulas that give the lengths of arcs of curves in the form of integrals. The calculations are straightforward and don't require advanced concepts. To simplify our work, let's rotate the square so that the corners of each step point upward as in Figure 9.10 which shows four stages of the process, with one, two, three and four steps, all plotted together.

I have also normalized, or re-scaled the diagonal's length to 1 rather than $\sqrt{2}$ to simplify the numbers (equivalent to considering a square with sides of length $1/\sqrt{2}$ instead of 1).

The largest triangle in Figure 9.10 is just half of the original square so its height is $c_1 = 1/2$ (half of the diagonal length); however, the only general assumption on c_1 in our discussion is that it be positive. Next, we split the interval $[0,1]$ (the current diagonal) in half and draw the two isoceles

Series of Functions   313

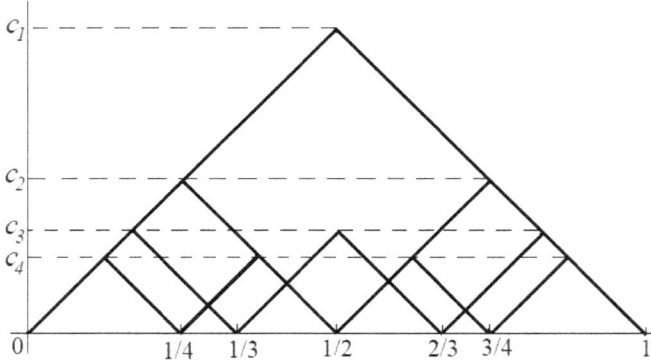

Figure 9.10: Illustrating staircase curves

triangles shown: the one on the left has the interval $[0, 1/2]$ as its bottom side and the one on the right has $[1/2, 1]$ as its bottom side. Each of these triangles has a height c_2; in general, we only assume that $0 < c_2 < c_1$ but in Figure 9.10 $c_2 = 1/4$ so the pairs of equal sides fall on the equal sides of the large triangle, consistent with the original staircase construction in the first chapter of this book.

The third stage splits the interval $[0,1]$ into three equal pieces and we get three triangles each of height c_3 (in Figure 9.10 $c_3 = 1/6$). The fourth stage creates four triangles each of height c_4 and so on. The general assumptions concerning the heights c_1, c_2, etc are as follows:

$$c_1 > c_2 > c_3 > \cdots > 0 \quad \text{and} \quad c_n \to 0 \text{ as } n \to \infty \tag{9.25}$$

Figure 9.10 illustrates the special case $c_n = 1/(2n)$. Let's consider the generic stage n. As we see in Figure 9.11, each of the n triangles has a side of length $1/n$ on the x-axis.

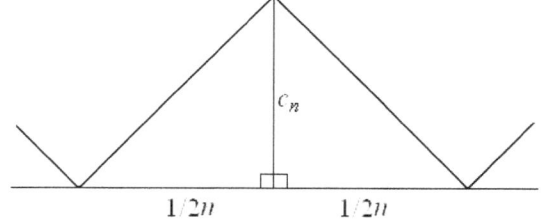

Figure 9.11: A generic triangle

The other two sides are equal and by the Pythagorean theorem, each of them has length:

$$\sqrt{c_n^2 + \left(\frac{1}{2n}\right)^2} = \frac{1}{2n}\sqrt{(2nc_n)^2 + 1}$$

There are $2n$ of these sides corresponding to the n triangles so the total length of all of these sides (the length of the entire staircase) is

$$L_n = 2n \left[\frac{1}{2n} \sqrt{(2nc_n)^2 + 1} \right] = \sqrt{(2nc_n)^2 + 1} \qquad (9.26)$$

We are now in a position to take the limit as $n \to \infty$. We distinguish three possible cases:

Case 1: *There is a number $c > 0$ such that $c_n = c/(2n)$ for every $n = 1, 2, 3, \ldots$ then*

$$\lim_{n \to \infty} L_n = \sqrt{c^2 + 1} \qquad (9.27)$$

This is easy to explain: we have $2nc_n = c$ so

$$\lim_{n \to \infty} L_n = \lim_{n \to \infty} \sqrt{c^2 + 1} = \sqrt{c^2 + 1}$$

In particular, in the case shown in Figure 9.10 where $c = 1$ we see that

$$\lim_{n \to \infty} L_n = \sqrt{2} \simeq 1.4142$$

This is greater than 1, namely, the length of the interval $0,1]$ which all staircase paths converge to, similarly to what we discussed in the first chapter. But we can say even more: by adjusting the value of c we can make the length of our "staircase" to be any (positive) real number that we like! For instance, to get staircases of length $5/4 = 1.25$ at each stage, we set

$$\sqrt{c^2 + 1} = \frac{5}{4}$$

Solving this equation gives the value of c that gives the desired length:

$$c^2 + 1 = \frac{25}{16}$$

$$c = \sqrt{\frac{9}{16}} = \frac{3}{4} = 0.75$$

This is as if each time that we partition, we press down on or shorten the height of every triangle in the partition by 25%; see Figure 9.12 which shows a generic triangular piece of the staircase curve at stage n for different values of c.

This reduction in height reduces the lengths of the two equal sides enough for each triangle to reduce the total length from $\sqrt{2}$ to 1.25. If we choose $c > 1$ then with each partition we pull up, or extend the height of each triangle so the total length gets larger too.

On the other hand, with $c > 0$ the length $\sqrt{c^2 + 1}$ of every staircase curve is always greater than 1 no matter how small the value of c. The discrepancy in the convergence of lengths cannot be removed in this case!

Let's consider a more dynamic case next.

Case 2: *If the sequence c_n converges to 0 fast enough that $nc_n \to 0$ as $n \to \infty$ then*

$$\lim_{n \to \infty} L_n = 1$$

Series of Functions   315

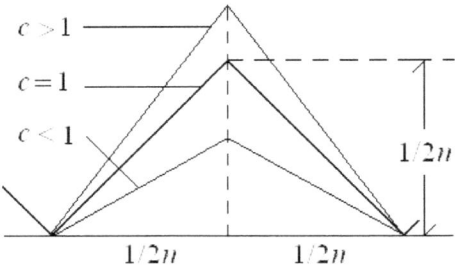

Figure 9.12: Adjusting heights using c_n

This is clear from (9.26). *In this case, what is expected actually happens*: as the staircase paths flatten down to the interval [0,1] their lengths approach 1, the length of the interval [0,1]. A simple example of this case is $c_n = 1/n^2$; with this choice we see that

$$nc_n = n\frac{1}{n^2} = \frac{1}{n} \to 0 \quad \text{as } n \to \infty$$

In this case, each time that we partition [0,1] we compress the heights of each triangle in Figure 9.12 by a factor that is greater than at the immediately preceding step. At stage $n = 2$ the compression is by a factor of 2, at stage $n = 3$ the compression is by a factor of 3 and so on. This increasing compression of heights with each partition reduces the total length of the curve enough to cause the length sequence to converge to the expected value 1.

Case 3: *If the sequence c_n converges to 0 so slowly that $nc_n \to \infty$ as $n \to \infty$ then*

$$\lim_{n \to \infty} L_n = \infty$$

Again, this is clear from (9.26). A simple example of this case is $c_n = 1/\sqrt{n}$ which converges to 0 but

$$nc_n = n\frac{1}{\sqrt{n}} = \sqrt{n} \to \infty \quad \text{as } n \to \infty$$

Case 3 is especially startling because it implies that a (roughly) triangular region with a finite area[12] may have an arbitrarily long perimeter. If you have had exposure to fractals then this may remind you of a fractal like "Koch's snowflake" that has a finite area but an infinite perimeter; see the section on the snowflake blow. But there is an important difference here: the limiting curve isn't a fractal but a straight line, namely the unit interval on the x-axis. So the limiting perimeter, being the perimeter of a triangle, is also finite, unlike the fractal. We can *liken* the curve to a fractal for large n because of the many bumps that it develops; but since this comparison works only for finite values of n we don't have a true fractal for comparison.

Figure 9.13 highlights the difference between cases 1 and 3 above.

The tall curve, resembling a comb shows the left half of the curve in Case 3 (from $x = 0$ to $x = 0.5$) for $n = 100$ and $c_n = 1\sqrt{n}$. The little bumps at the bottom correspond to $c_n = 1/(2n)$.[13]

---

[12]The area of the largest triangle in Figure 9.10 is 1/4, or half the area of the square of side $1/\sqrt{2}$.

[13]The curves in Figure 9.13 are actually the graphs of smooth functions $c_n \sin^2(n\pi x)$ for the two stated choices of the sequence c_n. For large values of n the smooth and non-smooth curves (like the staircase) can be told apart only by zooming in sufficiently and for our purpose here, as we will soon see both types of curves imply the same thing.

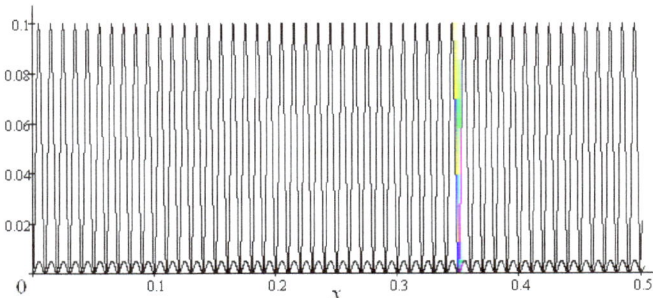

Figure 9.13: The length of a bounded oscillating curve may grow without bound

The right halves of both curves (from $x = 0.5$ to $x = 1$) are identical to what is shown in Figure 9.13.

Evidently the tall, comb-shaped curve in Figure 9.13 has a much longer total length than the much smaller curve at the bottom. The difference between these lengths grows even larger as n increases, but in the end both processes lead to the interval $[0,1]$ on the x-axis which has length 1.

I summarize the discussion so far as follows.

> The lengths of staircase curves converge to 1 only if nc_n converges to 0, that is, if c_n converges to 0 fast enough so as to suppress the growth in the size of n. This may be intuitively interpreted as pressing the staircase curves down more and more while partitioning the interval $[0,1]$. Otherwise, the lengths of the staircase curves do not converge to 1; they may converge to any positive number greater than 1 if we don't press down on them or they may even diverge to infinity if nc_n diverges to infinity even as c_n converges to 0.

A point worth mentioning before we wrap up this subsection is that the non-smooth, jagged shape of the staircase curves is not responsible for the anomaly. We obtain the same results with very smooth (infinitely differentiable) curves too. As an example, consider the sequence of smooth functions

$$f_n(x) = c_n \sin^2(n\pi x), \quad 0 \leq x \leq 1$$

where c_n is a decreasing sequence of positive numbers as in (9.25). Figure 9.14 illustrates four of these functions with $c_n = 1/(2n)$.

There is a standard formula in calculus for the lengths of these curves:

$$L_n = \int_0^1 \sqrt{1 + [f_n'(x)]^2}\,dx$$

The derivative $f_n'(x)$ is found using the chain rule:

$$f_n'(x) = 2n\pi c_n \sin(n\pi x) \cos(n\pi x)$$
$$= n\pi c_n \sin(2n\pi x)$$

The last step is justified by the trigonometric double-angle identity $2\sin\theta\cos\theta = \sin(2\theta)$. Inserting this into the formula gives

$$L_n = \int_0^1 \sqrt{1 + \pi^2 n^2 c_n^2 \sin^2(2n\pi x)}\,dx, \quad n = 1, 2, 3, \ldots$$

Series of Functions 317

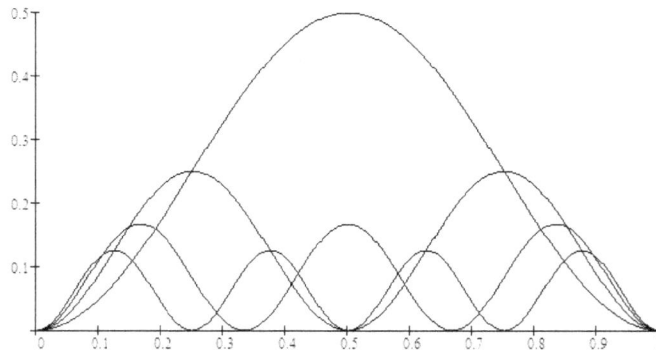

Figure 9.14: A sequence of smooth curves that converges uniformly

This formula shows that if $nc_n \to 0$ as $n \to \infty$ then using the theory that we have already discussed in this section and earlier in the book we obtain:

$$\lim_{n \to \infty} L_n = \int_0^1 \sqrt{1 + \pi^2 \lim_{n \to \infty} n^2 c_n^2 \sin^2(2n\pi x)} dx = \int_0^1 \sqrt{1 + 0} dx = 1$$

This result is similar to what we got for Case 2 of the staircase curves.

If nc_n does not converge to 0 then it is more difficult to deal with the integral directly due to a variety of technical issues. But we can talk about the other cases using a graphical argument and what have already learned about the staircase curves. If we superimposes the staircase graph in Figure 9.10 on the curves in Figure 9.14 we get what is shown in Figure 9.15.

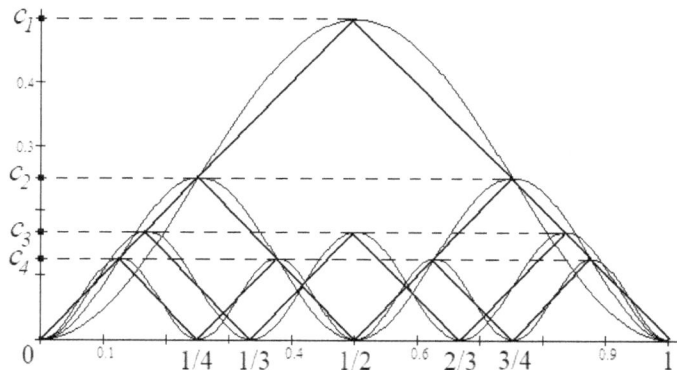

Figure 9.15: Superimposing staircase curves on smooth ones

Each straight line segment in Figure 9.15 is shorter than the piece of a smooth sine curve that joins its two ends. Therefore, the total length of the sine curve at each stage is larger than the total length the corresponding staircase curve. It follows that the lengths of the sine curves do not approach 1 as $n \to \infty$ if nc_n does not converge to 0; further, if $nc_n \to \infty$ then the lengths of the sine curves actually diverge to infinity, nowhere near the length of the curve that they converge (uniformly) to, namely, the segment from 0 to 1 on the x-axis.

9.5 Infinite series of functions: the magical infinity show!

The functions $f_n(x)$ in a sequence of functions can be added up to generate create an infinite series of functions just like we defined an infinite series of numbers earlier by adding up a given sequence of constants a_n. In fact, since constants are actually constant functions $f_n(x) = a_n$ (independent of x) we see that our earlier study is a very special case of what we discuss in this chapter!

Given an infinite sequence of functions over a given nonempty set S:

$$f_1(x), f_2(x), \ldots, f_n(x), \ldots \qquad x \in S \tag{9.28}$$

we derive the corresponding sequence of partial sums:

$$s_1(x) = f_1(x)$$
$$s_2(x) = f_1(x) + f_2(x)$$
$$s_3(x) = f_1(x) + f_2(x) + f_3(x)$$
$$\vdots$$
$$s_n(x) = f_1(x) + f_2(x) + \cdots + f_n(x)$$
$$\vdots$$

It worth noticing here that if $f_n(x)$ is continuous, integrable or differentiable for every n then $s_n(x)$ also has the same properties because of the various sum rules that we previously discussed.

Infinite series of functions. Suppose that the sequence $s_n(x)$ of partial sums of the sequence in (9.28) converges pointwise (or uniformly) on S to a function $f(x)$. Then we write

$$\sum_{n=1}^{\infty} f_n(x) = \lim_{n \to \infty} s_n(x) = f(x) \qquad x \in S \tag{9.29}$$

We say that *the infinite series on the left hand side converges pointwise to $f(x)$ on S if $s_n \xrightarrow{p} f$* and likewise, *the infinite series converges uniformly to $f(x)$ on S if $s_n \xrightarrow{u} f$*.

In our earlier discussion of the geometric power series we proved the following:

$$\sum_{n=1}^{\infty} x^{n-1} = 1 + x + x^2 + \cdots = \frac{1}{1-x} \quad \text{pointwise on } (-1, 1)$$

$$\sum_{n=1}^{\infty} x^{n-1} = 1 + x + x^2 + \cdots = \frac{1}{1-x} \quad \text{uniformly on } [-r, r],\ 0 < r < 1$$

For the geometric power series, the functions being added are the power functions $f_n(x) = x^{n-1}$, with the limit function (or sum) $f(x) = 1/(1-x)$. We saw how the truncation of the interval $(-1, 1)$ to $[-r, r]$ led to uniform convergence by blocking the ε-index from blowing up to infinity.

Series of Functions 319

The Cauchy criterion and the Weierstrass test.

For most infinite series of functions a simple formula for $f(x)$ does not exist; the geometric power series is a rare example of a "nice" series in this sense. Convergence tests exist based on the *Cauchy convergence criterion* which, as in the case of sequences and series of numbers, does not require a knowledge of the limit.

> **The Cauchy criterion for uniform convergence (CC).** *The infinite series $\sum_{n=1}^{\infty} f_n(x)$ converges uniformly on a nonempty set S if and only if for each $\varepsilon > 0$ the ε-index N_ε is independent x in S and*
> $$|s_m(x) - s_k(x)| < \varepsilon \quad \text{for all } x \text{ in } S \text{ and all } m, k \geq N_\varepsilon \qquad (9.30)$$

Suppose that $m > k$ and notice that the difference above is in fact a sum:
$$s_m(x) - s_k(x) = f_{k+1}(x) + \cdots + f_m(x)$$
so the Cauchy condition (9.30) can be written as
$$|f_{k+1}(x) + \cdots + f_m(x)| < \varepsilon \quad \text{for all } x \text{ in } S \text{ and all } m > k \geq N_\varepsilon \qquad (9.31)$$

The fact that in (CC) we didn't need to specify a limit function $f(x)$ makes the Cauchy criterion very useful. Also significant is the fact that the statement of (CC) is an *if and only if*, or a *necessary and sufficient* condition. We use both directions to prove the following corollary that gives us a basic but useful convergence test for infinite series of functions.

> **The Weierstrass Test for Uniform Convergence (WT).** *Suppose that each of the functions $f_n(x)$ is bounded by a positive real number M_n:*
> $$|f_n(x)| \leq M_n \quad \text{for all } x \in S \qquad (9.32)$$
> *If the infinite series of constants $\sum_{n=1}^{\infty} M_n$ converges (that is, $\sum_{n=1}^{\infty} M_n < \infty$) then the series $\sum_{n=1}^{\infty} f_n(x)$ converges uniformly on S.*

The proof of (WT) is simple. First, let's observe that if the sequence $\sum_{n=1}^{\infty} M_n$ converges then, thinking of each number M_n as a constant function, (CC) implies that for each $\varepsilon > 0$ there is a finite N_ε such that
$$M_{m+1} + \cdots + M_n < \varepsilon \quad \text{for all } m, n \geq N_\varepsilon \qquad (9.33)$$

We didn't need the absolute value here because the numbers M_n are positive for every n. We shall use (9.33) to show that (9.31) holds. By the triangle inequality we may write the left hand side in (9.31) as:
$$|f_{m+1}(x) + \cdots + f_n(x)| \leq |f_{m+1}(x)| + \cdots + |f_n(x)| \qquad (9.34)$$

By (9.33) and our hypothesis in (9.32):
$$|f_{m+1}(x)| + \cdots + |f_n(x)| \leq M_{m+1} + \cdots + M_n < \varepsilon$$

so from (9.34) it follows that (9.31) holds. Now, applying (CC) once more we may conclude that $\sum_{n=1}^{\infty} f_n(x)$ converges uniformly on S.

To illustrate the utility of (WT) (and indirectly, also of (CC)) consider the following infinite series of functions:
$$\sum_{n=1}^{\infty} \frac{x^3+n}{x+n^3} \qquad 0 \leq x \leq 10 \qquad (9.35)$$

Does this series converge uniformly on the interval [0,10]? We check each term of the sum:
$$f_n(x) = \frac{x^3+n}{x+n^3}$$
to see if it is bounded by some number M_n. Notice that:
$$x^3 + n \leq 10^3 + n = 1000 + n \quad \text{since } x \leq 10 \qquad (9.36)$$
$$x + n^3 \geq 0 + n^3 = n^3 \quad \text{since } x \geq 0$$

Therefore, since taking reciprocals reverses an inequality:
$$f_n(x) = \frac{1}{x+n^3}(x^3+n) \leq \frac{1}{n^3}(n+1000) = \frac{1}{n^2} + \frac{1000}{n^3}$$

The last expression on the right hand side is a good candidate for our M_n (it depends only on n). In fact, if we define it to be M_n for each n then:
$$\sum_{n=1}^{\infty} M_n = \sum_{n=1}^{\infty}\left(\frac{1}{n^2}+\frac{1000}{n^3}\right) = \sum_{n=1}^{\infty}\frac{1}{n^2} + \sum_{n=1}^{\infty}\frac{1000}{n^3}$$

The sum $\sum_{n=1}^{\infty} 1/n^2$ converges as a p-series with power $p = 2 > 1$ and the sum $\sum_{n=1}^{\infty} 1/n^3$ is likewise a convergent p-series. Since $\sum_{n=1}^{\infty} 1/n^3 < \infty$ we conclude that $\sum_{n=1}^{\infty} 1000/n^3 < \infty$ too so as a whole, $\sum_{n=1}^{\infty} M_n < \infty$. Now, we invoke the (WT) to conclude that the series in (9.35) converges uniformly on [0,10].

In fact, you may have noticed that we used the number 10 only marginally, in (9.36). Instead of 10, if we used an arbitrary positive real number b then we would end up with
$$x^3 + n \leq b^3 + n \quad \text{since } x \leq b$$

If in the rest of the argument we use b^3 instead of 1000 then the argument works just as well and proves that the series in (9.35) converges uniformly on the interval $[0, b]$.

The next exercise uses essentially the same argument.

Exercise 134 *Prove that the following series converges uniformly on the interval $[0, b]$ where b is any fixed, positive number:*
$$\sum_{n=1}^{\infty} \frac{2x+1}{x+n^2}$$

For another example, consider the following series:
$$\sum_{n=1}^{\infty} \frac{\sin(nx)}{n\sqrt{n}} \qquad (9.37)$$

Series of Functions

whose terms are not always positive. Let's see why this series converges uniformly on the set of all real numbers $(-\infty, \infty)$. The absolute value of each term is bounded:

$$\left|\frac{\sin(nx)}{n\sqrt{n}}\right| = \frac{|\sin(nx)|}{n\sqrt{n}} = \frac{|\sin(nx)|}{n^{3/2}} \leq \frac{1}{n^{3/2}}$$

because $|\sin x| \leq 1$ for all real numbers x. If we define $M_n = 1/n^{3/2}$ then

$$\sum_{n=1}^{\infty} M_n = \sum_{n=1}^{\infty} \frac{1}{n^{3/2}} < \infty$$

since a p-series with power $p = 3/2 > 1$ converges. So we may invoke the (WT) to conclude that the series in (9.37) converges uniformly on the set $(-\infty, \infty)$.

Exercise 135 *Prove that the following series converges uniformly on the interval $[-10, 10]$:*

$$\sum_{n=1}^{\infty} \frac{x^n}{n!} \qquad (9.38)$$

Once you identify the appropriate M_n you may use the ratio test to show that $\sum_{n=1}^{\infty} M_n < \infty$. Note that the number 10 is not important here; consider extending your proof to the interval $[-b, b]$ for any fixed, positive real number b.

A word of caution about the (WT) is in order here; unlike (CC), (WT) is only a sufficient condition, not necessary and sufficient (if and only if). So if the series $\sum_{n=1}^{\infty} M_n$ diverges then we cannot say that the function series doesn't converge uniformly; it may. In fact, if we drop $\sqrt{n}$ from the series in (9.37) then the new series still converges uniformly but not by the (WT), which doesn't work with the obvious choice $M_n = 1/n$. In a test that you can find in most typical real analysis textbooks, Dirichlet used a result of the Norwegian mathematician Niels Henrik Abel (1802-1829) to prove the convergence of the series in (9.37) without the factor $\sqrt{n}$. Abel is better known for his work in algebra.

Integration and differentiation term by term of a function series.

Looking back at the definition of infinite series of functions in (9.29) we see that the idea of taking limits, derivatives or integrals is a familiar one: we just use the results that we already found for sequences. The corresponding results for series are low-hanging fruits for easy picking!

The underlying theme is simple: *what is true for finite sums is true for infinite series if the convergence is uniform.*

Continuity. *If the series $\sum_{n=1}^{\infty} f_n(x)$ converges uniformly to a function $f(x)$ on a nonempty set S and every function $f_n(x)$ is continuous on S then $f(x)$ is continuous on S too.*

This theorem is easy to prove using the corresponding result for sequences. One of our assumptions is that $f_n(x)$ is continuous for every n; so each partial sum $s_n(x)$ is continuous because each is a finite sum of continuous functions. Our other assumption is that $s_n \xrightarrow{u} f$ on S and since the uniform limit of a sequence of continuous functions is continuous, it follows that $f(x)$ is continuous on S. Done!

This is useful information that is not obvious; for example, we don't have formulas for functions that the series in (9.35) or in (9.37) converge to but we know that those functions must be continuous because every term of each of their series is a continuous function and each of the series converges uniformly on the stated sets.

We can get a little bit more mileage out of the continuity theorem above because $f(x)$ is continuous on S if and only if it is continuous at every x in S. To illustrate, consider the fact that the geometric power series converges only pointwise on the interval $(-1,1)$ but its limit $1/(1-x)$ is continuous on *all of this interval*. We recall that the series converges uniformly on intervals of type $[-r, r]$ if $0 < r < 1$. Now, pick any value of x in $(-1,1)$, say $x = c$. We can then choose r so that $-r < c < r$ and still have $0 < r < 1$. Since this is possible for every real number x in $(-1,1)$, it follows that the limit is continuous on all of $(-1,1)$. This observation is not limited to the geometric power series; it extends to other series and their uniform limits since it depends only incidentally on that specific series or its limit.

Exercise 136 *Call the limit function of the series in (9.38) $f(x)$. Explain why $f(x)$ is continuous on $(-\infty, \infty)$.*[14]

The next result is another useful consequence of uniform convergence.

> **Integration term by term.** *If the series $\sum_{n=1}^{\infty} f_n(x)$ converges uniformly to a function $f(x)$ on an interval $[a, b]$ and every function $f_n(x)$ is integrable on $[a, b]$ then $f(x)$ is integrable on $[a, b]$ and*
> $$\int_a^b f(x)dx = \sum_{n=1}^{\infty} \int_a^b f_n(x)dx \qquad (9.39)$$

Again, we use the result for sequences. Since the sum of a (finite) number of integrable functions is integrable by the sum rule, we conclude that the partial sum $s_n(x)$ is integrable on $[a, b]$ for every n and:

$$\int_a^b s_n(x)dx = \int_a^b [f_1(x) + \cdots + f_n(x)]dx$$
$$= \int_a^b f_1(x)dx + \cdots + \int_a^b f_n(x)dx$$

---

[14] Using power series, we will soon find that in this case, $f(x) = e^x - 1$.

Series of Functions

Taking the limit as $n \to \infty$ gives the infinite series (of numbers):

$$\lim_{n\to\infty} \int_a^b s_n(x)dx = \lim_{n\to\infty} \left(\int_a^b f_1(x)dx + \cdots + \int_a^b f_n(x)dx \right)$$

$$= \int_a^b f_1(x)dx + \cdots + \int_a^b f_n(x)dx + \cdots$$

$$= \sum_{n=1}^{\infty} \int_a^b f_n(x)dx$$

On the other hand, since the sequence $s_n(x)$ converges uniformly, $\lim_{n\to\infty} s_n(x) = f(x)$ is integrable and we can move the limit outside the integral to get:

$$\int_a^b f(x)dx = \int_a^b \lim_{n\to\infty} s_n(x)dx = \lim_{n\to\infty} \int_a^b s_n(x)dx$$

This shows that (9.39) is true.

Let us now take a fresh look at the series (9.7). The geometric power series with alternating signs converges uniformly on any interval $[-r, r]$ if $0 < r < 1$:

$$1 - x + x^2 - x^3 + \cdots = \frac{1}{1+x}$$

Let u be any number[15] in the interval $(-1, 1)$ and integrate the above series term by term from 0 to u to get:

$$\int_0^u 1 dx - \int_0^u x dx + \int_0^u x^2 dx - \int_0^u x^3 dx + \cdots = \int_0^u \frac{1}{1+x} dx$$

After you finish the integration you get precisely the equality in (9.7)!

As you might have guessed, there is also a result that lays out the conditions under which we can take the derivative of a series term by term.

> **Differentiation term by term.** Let $f_n(x)$ be a sequence of functions with continuous derivatives $f_n'(x)$ on an interval $[a, b]$. If the series $\sum_{n=1}^{\infty} f_n'(x)$ of derivatives converges uniformly to a function $g(x)$ and if for some x_0 in $[a, b]$ the series of constants $\sum_{n=1}^{\infty} f_n(x_0)$ converges then the series $\sum_{n=1}^{\infty} f_n(x)$ converges uniformly to a differentiable function $f(x)$ on $[a, b]$ and $f'(x) = g(x)$.

Again, the proof is straightforward by going back to the partial sums. Since $s_n(x)$ is a finite sum of differentiable functions and the derivative sum rule gives us

$$s_n'(x) = \sum_{k=1}^{n} f_k'(x)$$

for every n we see that the sequence of functions $s_n(x)$ satisfies the conditions that we laid out earlier for taking derivatives of sequences of functions and their uniform limit.

---

[15] As long as $u \neq \pm 1$ we can find r between 0 and 1 such that the interval $[0, u]$ (or $[u, 0]$ if u is negative) is contained in $[-r, r]$.

9.6 Power series and Taylor expansions: transcendental functions explained

There are two special types of infinite series of functions that appear frequently in applied mathematics. We discuss the first of these in this section and leave the second, called Fourier series for later in this chapter.

> A **power series** is an infinite series of (translated or shifted) power functions:
> $$\sum_{n=0}^{\infty} c_n(x-c)^n = c_0 + c_1(x-c) + c_2(x-c)^2 + \cdots \qquad (9.40)$$
> where c and c_n are real numbers and x is a variable in a set S of real numbers. The number c is the *center* of the power series and the numbers c_n are its *coefficients*.

Notice that the standard notation for power series starts it at the *zeroth term* $a_0(x-c)^0$; defining $(x-c)^0 = 1$ we see that the zeroth term of every power series is always a number c_0. The *n-th term* is the power function $f_n(x) = c_n(x-c)^n$ so the theory of the previous sections applies to this type of series too.

If $c = 0$ then the power series in (9.40) takes a simpler-looking form:

$$\sum_{n=0}^{\infty} c_n x^n = c_0 + c_1 x + c_2 x^2 + \cdots \qquad (9.41)$$

which is a power series centered at zero. You might wonder, given the simpler form of the power series (9.41) why bother with any other center? The answer: for practical reasons; we need the greater flexibility for the power series expansions of functions like the natural logarithm $\ln x$ that are not defined at $x = 0$.

The geometric power series is the simplest example of a power series centered at zero because all of its coefficients are equal: $c_n = a$ for all $n \geq 0$ (we set $a = 1$ earlier to simplify the notation easier without missing anything of significance). The most general version of the geometric power series is:

$$\sum_{n=0}^{\infty} a(x-c)^n = a + a(x-c) + a(x-c)^2 + \cdots$$

This has a nonzero center c and (common) coefficient $c_n = a$. Recall from our earlier discussion that this series converges if:

$$-1 < x - c < 1$$
$$c - 1 < x < c + 1$$

and for this range of values of x, the following formula is valid:

$$a + a(x-c) + a(x-c)^2 + \cdots = \frac{a}{1-(x-c)} = \frac{a}{1+c-x}$$

Finding the range of values of x for which the general power series in (9.40) converges requires a different, more general test. Of course, every power series converges at its center $x = c$ because all

Series of Functions

terms of the series (except maybe the first) drop out. Beyond the center, let's recall that the limit ratio test (LRT) extended the geometric series test; we may use the LRT here too to calculate the range of values of x for which the series in (9.40) converges.

For each fixed value of x we must have:[16]

$$\lim_{n\to\infty}\left|\frac{(n+1)\text{-st term}}{n\text{-th term}}\right| = \lim_{n\to\infty}\left|\frac{c_{n+1}(x-c)^{n+1}}{c_n(x-c)^n}\right| = \lim_{n\to\infty}\left|\frac{c_{n+1}}{c_n}\right||x-c| < 1 \qquad (9.42)$$

We assume for simplicity that the limit above exists, and call it ρ (Greek *rho*):

$$\rho = \lim_{n\to\infty}\left|\frac{c_{n+1}}{c_n}\right|$$

Then either $\rho = 0$ in which case (9.42) is true for all real values of x, or $\rho \neq 0$ in which case:

$$|x - c| < \frac{1}{\rho} \qquad (9.43)$$

guarantees that the power series converges. The above inequality may be written as:

$$-\frac{1}{\rho} < x - c < \frac{1}{\rho}$$
$$c - \frac{1}{\rho} < x < c + \frac{1}{\rho}$$

The number $1/\rho$ defines the range of values of x before and after c for which the power series converges. If we define $R = 1/\rho$ then we see that the power series converges if:

$$c - R < x < c + R$$

Also notice that if $x < c - R$ or $x > c + R$ then the (LRT) implies that the series diverges.

In our future calculations we are interested in finding R rather than $\rho = 1/R$. So we modify the calculation to yield the value of R directly by defining:

$$R = \lim_{n\to\infty}\left|\frac{c_n}{c_{n+1}}\right| \qquad (9.44)$$

These facts explain the following definition:

> The **radius of convergence** of the power series in (9.40) is the real number R. If $R = \infty$ (the limit in (9.44) diverges to infinity) then the power series converges for all real numbers. If $R = 0$ then the power series converges only at its center $x = c$.
>
> The **interval of convergence** of the power series in (9.40) is the open interval $(c-R, c+R)$ if R is a real number, and it is $(-\infty, \infty)$ if $R = \infty$. Note that the interval of convergence refers to the *pointwise convergence*. In this interval the power series *converges absolutely* (recall that this means regardless of sign changes) because we used the absolute value in finding the radius of convergence. If $R < \infty$ and either $x < c - R$ or $x > c + R$ then the power series *diverges*.

---

[16]These operations assume that $a_n \neq 0$ for all n. For most power series of interest to us here, a relabeling of indices takes care of the matter, as in our analysis of the series in (9.47) below. But the easiest way of avoiding this problem is to use the *root test*, a convergence test that we don't discuss in this book but is technically more versatile; see a standard real analysis textbook for details.

Let's consider the following power series as an illustration:

$$\sum_{n=0}^{\infty} n(x-c)^n = (x-c) + 2(x-c)^2 + 3(x-c)^3 + \cdots$$

Here we have $c_n = n$ so that:

$$R = \lim_{n\to\infty} \left|\frac{c_n}{c_{n+1}}\right| = \lim_{n\to\infty} \frac{n}{n+1} = 1$$

Therefore, the radius of convergence is 1 and the interval of absolute convergence is $(c-1, c+1)$. In particular, if the center is at 0 then $c = 0$ and the interval of convergence is $(-1, 1)$.

For another illustration, consider the series:

$$\sum_{n=0}^{\infty} \frac{(x-c)^n}{n!} = 1 + (x-c) + \frac{(x-c)^2}{2} + \frac{(x-c)^3}{6} + \cdots \tag{9.45}$$

Here $c_n = 1/n!$ so that:

$$R = \lim_{n\to\infty} \left|\frac{c_n}{c_{n+1}}\right| = \lim_{n\to\infty} \frac{1/n!}{1/(n+1)!} = \lim_{n\to\infty} \frac{(n+1)n!}{n!} = \lim_{n\to\infty} (n+1) = \infty$$

Since $R = \infty$ the interval of (absolute) convergence of (9.45) is $(-\infty, \infty)$, regardless of the value of c.

Here is a power series that converges only at its center:

$$\sum_{n=0}^{\infty} n!(x-c)^n = 1 + (x-c) + 2(x-c)^2 + 6(x-c)^3 + \cdots$$

In this case, $c_n = n!$ and we find that:

$$R = \lim_{n\to\infty} \left|\frac{c_n}{c_{n+1}}\right| = \lim_{n\to\infty} \frac{n!}{(n+1)!} = \lim_{n\to\infty} \frac{1}{n+1} = 0$$

This means that the series converges only when $x = c$.

Exercise 137 *Find the radius and the interval of (absolute) convergence for each of the following power series:*

$$\sum_{n=0}^{\infty} \frac{n}{2^n}(x-1)^n = \frac{1}{2}(x-1) + \frac{1}{2}(x-1)^2 + \frac{3}{8}(x-1)^3 + \cdots$$

$$\sum_{n=0}^{\infty} 2^n(x-1)^n = 1 + 2(x-1) + 4(x-1)^2 + 8(x-1)^3 + \cdots \tag{9.46}$$

$$\sum_{n=0}^{\infty} 2^{n^2}(x-1)^n = 1 + 2(x-1) + 16(x-1)^2 + 512(x-1)^3 + \cdots$$

Series of Functions

Zero coefficients. Many power series contain zero coefficients, which presents a problem when applying the (LRT). Let's see how to find the radius and interval of convergence for the following power series:

$$\sum_{n=0}^{\infty} \frac{x^{2n}}{3^n} = 1 + \frac{x^2}{3} + \frac{x^4}{3^2} + \frac{x^6}{3^3} + \cdots \qquad (9.47)$$

If we use the notation in (9.41) in a strict sense then $c_n = 0$ for all odd values of n and we cannot carry out the calculation in (9.42). However, notice that we may substitute an alternative variable u for x^2 and find that $x^4 = u^2$, $x^6 = u^3$, etc. Using this relabeling, our series may be written as:

$$\sum_{n=0}^{\infty} \frac{u^n}{3^n} = 1 + \frac{u}{3} + \frac{u^2}{3^2} + \frac{u^3}{3^3} + \cdots \qquad (9.48)$$

In this form, we have $c_n = 1/3^n$ so that:

$$R = \lim_{n \to \infty} \left| \frac{c_n}{c_{n+1}} \right| = \lim_{n \to \infty} \left| \frac{3^{n+1}}{3^n} \right| = \lim_{n \to \infty} 3 = 3$$

This yields a radius of convergence $R = 3$ and (with $c = 0$) an interval of convergence $(-3, 3)$ for the series (9.48). What does this say about the original series (9.47)?

Let's go back to (9.43) and rewrite it with R and u as:

$$|u - c| < R$$

With $c = 0$, $R = 3$ and $u = x^2$ we now have:

$$|x^2| < 3 \quad \text{or} \quad |x|^2 < 3$$

Taking square root gives:

$$|x| < \sqrt{3}$$

and this inequality can be written as

$$-\sqrt{3} < x < \sqrt{3}$$

This means that the radius of convergence of (9.47) is $R = \sqrt{3}$ and its interval of convergence is $(-\sqrt{3}, \sqrt{3})$. The radius and the interval of convergence of (9.47) are smaller than those for (9.48). Indeed, if we pick a number that is in the interval of convergence of (9.48) but not in that of (9.47), say, $x = 2$ and insert this in (9.47) for x then we obtain:

$$1 + \frac{2^2}{3} + \frac{2^4}{3^2} + \frac{2^6}{3^3} + \cdots = 1 + \frac{4}{3} + \frac{4^2}{3^2} + \frac{4^3}{3^3} + \cdots$$

The last series is geometric with common ratio $4/3 > 1$ so it diverges and shows that 2 cannot be in the interval of convergence of (9.47).

If the occurrences of the zero coefficients are more complicated then the above approach will not work and we need a more versatile test, such as as the *root test* that you can find in a standard analysis textbook.

It is worth mentioning that the series in (9.48) is actually a geometric power series whose interval of convergence can be determined using earlier methods; further, we know what actual function it converges to:

$$1 + \frac{u}{3} + \frac{u^2}{3^2} + \frac{u^3}{3^3} + \cdots = \frac{1}{1 - u/3} = \frac{3}{3 - u}$$

Now replacing u with x^2 gives:

$$1 + \frac{x^2}{3} + \frac{x^4}{3^2} + \frac{x^6}{3^3} + \cdots = \frac{3}{3 - x^2}$$

It is well worth checking to see if a given series is a geometric power series in disguise before plunging into ratio test or other calculations because that observation will not only save time and effort, it also gives us the limit function–a rare prize indeed!

Exercise 138 *Use a suitable substitution to show that the power series in (9.46) above is geometric. Then use this fact to find its interval of convergence, as well as the limit function that it converges to on that interval.*

The end-points of the interval of convergence. The interval of convergence of the power series in (9.40) is $(c - R, c + R)$. If R is a positive real number then we know that the series converges absolutely in this interval and, by the (LRT), the series diverges outside this interval. But *what happens at the end-points $c \pm R$?*

At the end-points the (LRT) fails so the series may converge or diverge. If the series converges at one of these points then it may converge conditionally or absolutely.

To illustrate, consider the series:

$$\sum_{n=0}^{\infty} \frac{x^n}{(n+1)^2} = \frac{1}{1^2} + \frac{x}{2^2} + \frac{x^2}{3^2} + \frac{x^3}{4^2} + \cdots \qquad (9.49)$$

Here we have $c_n = 1/(n+1)^2$ and $c_{n+1} = 1/(n+2)^2$ so:

$$R = \lim_{n \to \infty} \left| \frac{c_n}{c_{n+1}} \right| = \lim_{n \to \infty} \frac{(n+2)^2}{(n+1)^2} = \lim_{n \to \infty} \left(\frac{n+2}{n+1} \right)^2 = \left(\lim_{n \to \infty} \frac{n+2}{n+1} \right)^2 = 1^2 = 1$$

So the interval of absolute convergence is $(-1, 1)$. The end-points in this case are ± 1 and we check each separately:

$$x = 1: \qquad \sum_{n=0}^{\infty} \frac{1^n}{(n+1)^2} = \frac{1}{1^2} + \frac{1}{2^2} - \frac{1}{3^2} + \frac{1}{4^2} + \cdots$$

You may recognize the last series above as a p-series with power $p = 2 > 1$. Since such a series converges we conclude that the power series (9.49) converges when $x = 1$. Next, let's see what happens when $x = -1$:

$$x = -1: \qquad \sum_{n=0}^{\infty} \frac{(-1)^n}{(n+1)^2} = \frac{1}{1^2} - \frac{1}{2^2} + \frac{1}{3^2} - \frac{1}{4^2} + \cdots$$

Series of Functions

This is an alternating series which converges absolutely, as we saw with $x = 1$. So the power series (9.49) converges when $x = -1$ too, and the interval of convergence is $[-1, 1]$ with convergence being absolute at all points, including both of the endpoints.

Now, consider the similar power series:

$$\sum_{n=0}^{\infty} \frac{x^n}{n+1} = 1 + \frac{x}{2} + \frac{x^2}{3} + \frac{x^3}{4} + \cdots$$

Here we have $c_n = 1/(n+1)$ and $c_{n+1} = 1/(n+2)$ so:

$$R = \lim_{n \to \infty} \left| \frac{c_n}{c_{n+1}} \right| = \lim_{n \to \infty} \frac{(n+2)}{(n+1)} = 1$$

So the interval of absolute convergence is $(-1, 1)$. The end-points in this case are again ± 1 and we check each separately:

$$x = 1: \quad \sum_{n=0}^{\infty} \frac{1^n}{n+1} = \frac{1}{1} + \frac{1}{2} + \frac{1}{3} + \frac{1}{4} + \cdots$$

You may recognize the last series above as the harmonic series (a p-series with power $p = 1$) which diverges. We conclude that the power series diverges when $x = 1$. Next, let's see what happens when $x = -1$:

$$x = -1: \quad \sum_{n=0}^{\infty} \frac{(-1)^n}{n+1} = \frac{1}{1} - \frac{1}{2} + \frac{1}{3} - \frac{1}{4} + \cdots$$

This is an alternating series that converges conditionally. So the power series converges when $x = -1$. We conclude that the interval of convergence is $[-1, 1)$ with convergence being conditiional at $x = -1$.

Uniform convergence and the integration of power series.

We have seen how a power series converges pointwise (absolutely or conditionally) in its interval of convergence and possibly at the end-points of that interval. But what about *uniform convergence*? This is quite important if we want to take derivative of, or integrate a power series term by term. We have already discussed uniform convergence for series of functions generally so we only need to translate what the earlier theorems say about power series in particular, where $f_n(x) = c_n x^{n-1}$ for all n.

We can apply Weierstrass's test for uniform convergence (WT) that we discussed earlier to power series too. Let $0 < r < R$ where $(c - R, c + R)$ is the interval of (pointwise, absolute) convergence of the power series.

The following numerical series converges by the limit ratio test (LRT):

$$\sum_{n=0}^{\infty} |c_n| r^n < \infty$$

because

$$\lim_{n \to \infty} \left(\left| \frac{c_{n+1}}{c_n} \right| \frac{r^{n+1}}{r^n} \right) = \frac{1}{R} \frac{r}{1} = \frac{r}{R} < 1$$

Next, for each value of x in the closed interval $[c-r, c+r]$,

$$c - r \leq x \leq c + r$$
$$-r \leq x - c \leq r$$

so that $|x - c| \leq r$. This observation implies that

$$|c_n||x - c|^n \leq |c_n|r^n$$

So if we choose $M_n = |c_n|r^n$ then $|c_n||x-c|^n \leq M_n$ for every x in $[c-r, c+r]$ and $\sum_{n=0}^{\infty} M_n < \infty$; under these circumstances, (WT) implies that the power series converges uniformly (and absolutely) in the closed and bounded interval $[c - r, c + r]$.

Note that r can be arbitrarily close to R if $R < \infty$ and r can be any positive real number otherwise.

> **Uniform convergence of power series.** *The power series $\sum_{n=0}^{\infty} c_n(x - c)^n$ with radius of convergence $R > 0$ converges uniformly (and absolutely) on every closed and bounded interval $[c - r, c + r]$ with $0 < r < R$. In particular, if $R = \infty$ then the power series converges uniformly on every bounded set of real numbers.*

Evidently, a key word for uniform convergence of a power series is *bounded* set. Can a power series converge uniformly on an unbounded set such as $[0, \infty)$ or $(-\infty, \infty)$? We discuss this issue later after studying the Taylor series where we have information about the limit functions to which power series converge. As you may be guessing, if the interval of convergence is unbounded then the ε-index for the converging partial sum sequence $s_n(x)$ cannot be blocked from blowing up to infinity if we do not stop at some convenient, but finite number.

Now we turn to the important issues of term by term integration and term by term differentiation of power series.

> **Term by term integration of power series.** *If a power series $\sum_{n=0}^{\infty} c_n(x - c)^n$ converges with radius $R > 0$ to a function $f(x)$ and a, b are any two numbers in the interval of convergence with $a < b$ then:*
>
> $$\int_a^b f(x)dx = \sum_{n=0}^{\infty} \int_a^b c_n(x-c)^n dx = \sum_{n=0}^{\infty} \frac{c_n}{n+1}[(b-c)^{n+1} - (a-c)^{n+1}]$$

You may be wondering about what happened to bounded intervals. This theorem implicitly uses the fact that integration is being done from a to b, so there is a finite value $r < R$ such that

$$c - R < c - r \leq a < b \leq c + r < c + R$$

Therefore, all the action is confined to the closed and bounded interval $[c-r, c+r]$; this interval contains $[a, b]$ and the power series converges uniformly in it. So our earlier result on the term by term integration of series of functions applies.

The power series expansion of the natural logarithm.

The above theorem fully justifies the series expansion of the natural logarithm in (9.7) that we found earlier; we can be more precise about it now and obtain other versions of this useful formula.

Series of Functions

The geometric power series $\sum_{n=0}^{\infty}(x-c)^n$ has an interval of convergence $(c-1, c+1)$ which of course, always contains the center c. Now, if u is any number in the interval of convergence then:

$$\int_c^u \frac{1}{1-(x-c)}dx = \sum_{n=0}^{\infty}\int_c^u (x-c)^n dx = \sum_{n=0}^{\infty}\frac{(u-c)^{n+1}}{n+1}$$

$$\ln 1 - \ln(c+1-u) = \sum_{n=0}^{\infty}\frac{1}{n+1}[(u-c)^{n+1} - 0^{n+1}]$$

$$\ln(c+1-u) = -\sum_{n=0}^{\infty}\frac{(u-c)^{n+1}}{n+1} \qquad c-1 < u < c+1$$

In particular, if $c = 0$ then:

$$\ln(1-u) = -\sum_{n=0}^{\infty}\frac{u^{n+1}}{n+1} = -u - \frac{u^2}{2} - \frac{u^3}{3} - \cdots \qquad -1 < u < 1 \qquad (9.50)$$

This is a slightly different-looking (but equivalent) expansion of the logarithm than the one in (9.7).

To obtain a power series for $\ln x$ itself, set $x = 1 - u$ in (9.50) to get:

$$\ln x = \sum_{n=0}^{\infty}\frac{-(1-x)^{n+1}}{n+1} \qquad -1 < 1-x < 1$$

The inequalities simplify by first adding -1 to get: $-2 < -x < 0$ and then multiplying by -1 to obtain: $0 < x < 2$. Also we may write the numerator inside the series as:

$$-(1-x)^{n+1} = -(-1)^{n+1}(x-1)^{n+1} = (-1)^n(x-1)^{n+1}$$

Putting these modifications together gives us the following power series that is in the form that we defined earlier:

$$\ln x = \sum_{n=0}^{\infty}\frac{(-1)^n(x-1)^{n+1}}{n+1} = \frac{x-1}{1} - \frac{(x-1)^2}{2} + \frac{(x-1)^3}{3} - \cdots \qquad 0 < x \leq 2 \qquad (9.51)$$

This series is the *standard expansion of* $\ln x$ *as a power series centered at* $c = 1$, the midpoint of the interval of convergence $(0, 2]$.

Figure 9.16 shows the graph of $\ln x$ (thick curve) and its approximation using the first four partial sums $s_0(x), s_1(x), s_2(x), s_3(x)$ of the series in (9.51).

We know that the series in (9.51) converges absolutely on the interval $(0, 2)$ and uniformly on any closed interval $[a, b]$ that is contained in $(0, 2)$. We also see that $x \leq 2$ in (9.51); let's check what happens at $x = 2$. Inserting this number for x in (9.51) gives:

$$\ln 2 = 1 - \frac{1}{2} + \frac{1}{3} - \frac{1}{4} + \cdots \qquad (9.52)$$

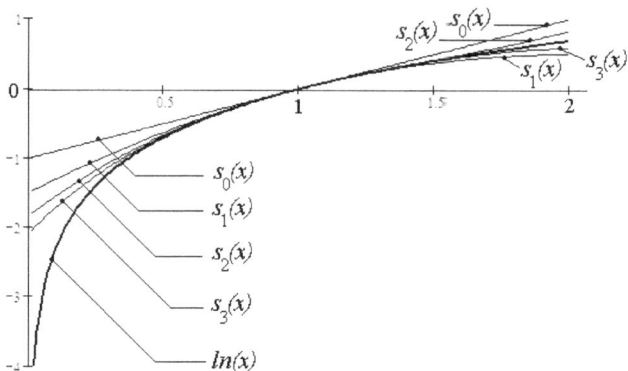

Figure 9.16: Partial sums of the power series converging to $\ln x$

This is a valid equality since the infinite series of numbers on the right hand side is a convergent alternating (harmonic) series. We conclude that the series in (9.51) converges on the interval $(0, 2]$ with conditional convergence at 2. If $x = 0$ then the series:

$$-1 - \frac{1}{2} - \frac{1}{3} - \cdots$$

on the right hand side of (9.51) diverges to infinity ($-\infty$ to be more precise, as the negative of the harmonic series) which is consistent with the behavior of $\ln x$ as x approaches 0.

It is also very important to remember that *the power series in (9.51) does not converge to anything outside the interval $(0, 2]$ and therefore, it is not equal to $\ln x$ even though $\ln x$ is well-defined for $x > 2$.*

Do you see why we cannot get a power series for $\ln x$ that is centered at zero, which would certainly be convenient?

Power series like (9.7) or (9.50) that are centered at zero are sometimes used for translated or shifted versions of $\ln x$ when it simplifies calculations.

Exercise 139 *When paired with the properties of the logarithm, the series (9.51) or its equivalents centered at 0 can be used to estimate the numerical values of the logarithms of (positive) real numbers. If $x > 1$ the property $\ln(1/x) = -\ln x$ extends the usefulness of (9.51) to all positive real numbers. For instance,*

$$\ln 2 = -\ln \frac{1}{2}$$

and the series for $\ln(1/2)$ in (9.51) converges faster than the alternating harmonic series above because $1/2$ is in the interval of absolute convergence, and also it is closer to the center than 2 is.

(a) Estimate the following logarithms using the first 4 terms of the series in (9.51):

$$\ln 3 \qquad \ln 4.5$$

Series of Functions

(b) If you were to use a few terms of the series in (9.51) to estimate $\ln 15$ *which of the following methods do you think gives better values, using the same number of terms of the series in each case?*

$$\ln 15 = -\ln \frac{1}{15}$$

$$\ln 15 = \ln 3 + \ln 5 = -\ln \frac{1}{3} - \ln \frac{1}{5}$$

You can check by adding the first four terms (or more for better accuracy, as long as you use the same number of terms in each of the two cases) of (9.51) and compare the answers to what a calculator gives for $\ln 15$. *Can you explain why the second way gives consistently better results no matter how many terms of the series you use? Figure 9.16 offers some insights!*

Inverse tangent power series and the Madhava-Gregory-Leibniz series for π.

We have met the inverse tangent function $\tan^{-1} x$ before in different contexts. Now we discuss its power series and the historical significance that is attached to the infinite series that is obtained when $x = 1$. In the process, we discover another interesting application of term by term integration of power series.

Let's start with the series in (9.4):

$$1 - u^2 + u^4 - u^6 + \cdots = \frac{1}{1 + u^2}$$

that we obtained earlier with the help of the geometric power series. This power series converges in the interval $(-1, 1)$. If x is any number in this interval then integrating the above series gives us:

$$\int_0^x (1 - u^2 + u^4 - u^6 + \cdots) du = \int_0^x \frac{1}{1 + u^2} du$$

$$\left[u - \frac{u^3}{3} + \frac{u^5}{5} - \frac{u^7}{7} + \cdots \right]_{u=0}^{u=x} = \int_0^x \frac{1}{1 + u^2} du$$

$$x - \frac{x^3}{3} + \frac{x^5}{5} - \frac{x^7}{7} + \cdots = \int_0^x \frac{1}{1 + u^2} du \qquad (9.53)$$

But what is the integral on the right hand side? This integral is found using a trigonometric substitution and the process is simple enough that is discussed in most freshman calculus courses; however, I will do it backwards by showing that the derivative of $\tan^{-1} u$ is $1/(1 + u^2)$ using the chain rule.

First, if we set $\theta = \tan^{-1} u$ (we assume here that $-\pi/2 < \theta < \pi/2$) then $\tan \theta = u$ so taking the derivative with respect to u we get:

$$\frac{d}{du} \tan \theta = \frac{du}{du}$$

$$\left(\frac{d}{d\theta} \tan \theta \right) \frac{d\theta}{du} = 1$$

$$\sec^2 \theta \frac{d\theta}{du} = 1$$

$$\frac{d}{du} \tan^{-1} u = \frac{1}{\sec^2 \theta} = \cos^2 \theta \qquad (9.54)$$

To complete the picture we must insert $\tan^{-1} u$ for θ on the right hand side too, but the resulting expression doesn't help us deal with the integral in (9.53). So let's observe instead that:

$$\tan\theta = u \Rightarrow \sin\theta = u\cos\theta \Rightarrow \sin^2\theta = u^2\cos^2\theta$$

Next, we use the trigonometric Pythagorean identity $\sin^2\theta + \cos^2\theta = 1$ to eliminate $\sin^2\theta$:

$$1 - \cos^2\theta = u^2\cos^2\theta$$
$$1 = (1+u^2)\cos^2\theta$$
$$\frac{1}{1+u^2} = \cos^2\theta$$

We may now go back to (9.54) and eliminate $\cos^2\theta$ to get:

$$\frac{d}{du}\tan^{-1} u = \frac{1}{1+u^2}$$

We're almost there; using this and the Fundamental Theorem of Calculus in (9.53) gives us:

$$x - \frac{x^3}{3} + \frac{x^5}{5} - \frac{x^7}{7} + \cdots = \tan^{-1} x, \qquad -1 \leq x \leq 1 \tag{9.55}$$

since $\tan^{-1} 0 = 0$.

The interval of convergence of the above power series includes $x = 1$ where the series:

$$1 - \frac{1}{3} + \frac{1}{5} - \frac{1}{7} + \cdots$$

converges by the alternating series test (AST).[17] On the other side of (9.55) $\tan^{-1} 1 = \pi/4$ because $\tan(\pi/4) = 1$; therefore,

$$1 - \frac{1}{3} + \frac{1}{5} - \frac{1}{7} + \cdots = \frac{\pi}{4} \tag{9.56}$$

The above discussion is similar to the case of $\ln 2$ in (9.52) above in more ways than one, including the fact that they are really simple alternating series that converge to irrational numbers.

The series in (9.56) attracted special attention over the course of history, by being a simple way of accessing π mentally; it appeared in one of Leibniz's publications around 1676 and became known as "Leibniz's series". But it had already appeared in a 1668 publication of the Scottish mathematician James Gregory (1638-1675) so the name Gregory-Leibniz series has also been used. Later, the series was discovered in works of the old Kerala school in India, founded by Madhava of Sangamagrama (1340-1425). In any case, the series in (9.56) was known well over 200 years before Gregory's publication, hence the name Madhava-Gregory-Leibniz series.

Like the series in (9.52) for $\ln 2$, the series in (9.56) converges too slowly to be practical for the purpose of calculating π. We can certainly improve on the convergence rate by using another angle;

---

[17]The same is true for $x = -1$ where the series converges to $-\pi/4$.

Series of Functions

for instance, if we use $\theta = \pi/6$ then $\tan\theta = 1/\sqrt{3}$ is in the interior of the interval of convergence and the series that we obtain is:

$$\frac{1}{\sqrt{3}} - \frac{(1/\sqrt{3})^3}{3} + \frac{(1/\sqrt{3})^5}{5} - \frac{(1/\sqrt{3})^7}{7} + \cdots = \tan^{-1}\frac{1}{\sqrt{3}}$$

$$\frac{1}{\sqrt{3}} - \frac{1}{\sqrt{3}}\frac{1}{3(3)} + \frac{1}{\sqrt{3}}\frac{1}{5(3)^2} - \frac{1}{\sqrt{3}}\frac{1}{7(3)^3} + \cdots = \frac{\pi}{6}$$

If we pull out $1/\sqrt{3}$ and multiply by 6 we get the following series for π

$$\pi = \frac{6}{\sqrt{3}}\left(1 - \frac{1}{3(3)} + \frac{1}{5(3)^2} - \frac{1}{7(3)^3} + \cdots\right)$$

Madhava actually used 21 terms of the above series to obtain a value of π that was accurate to 11 decimal places:

$$\pi = 3.1415926535922\ldots$$

Other series that involve π are known that not only converge fast but they also do not involve irrational numbers like $\sqrt{3}$; we discuss one such series below that involves the inverse of the sine function.

Taking derivatives of power series.

Being able to take the derivative of a power series term by term is possibly the most important property of such series because it leads to the important concept of Taylor series that we will discuss shortly.

> **Taking the derivative of a power series term by term.** *A power series $\sum_{n=0}^{\infty} c_n(x-c)^n$ with radius of convergence $R > 0$ converges to a differentiable (thus continuous) function $f(x)$ on its interval of convergence $(c-R, c+R)$ and the derivative of $f(x)$ is obtained by taking the derivative of the power series term by term to obtain:*
>
> $$f'(x) = \sum_{n=1}^{\infty} nc_n(x-c)^{n-1} \qquad (9.57)$$
>
> *The interval of convergence of the series in (9.57) is also $(c-R, c+R)$.*

We derive the identity in (9.57) below. To prove the above theorem, recall that the series $\sum_{n=0}^{\infty} c_n(x-c)^n$ converges (trivially to c_0) when $x = c$. To use our earlier results, we must also show that the series in (9.57) converges on $(c-R, c+R)$. Using the limit comparison test (LRT) we find that for each fixed value of x (other than c) in $(c-R, c+R)$:

$$\lim_{n\to\infty}\left|\frac{(n+1)c_{n+1}(x-c)^n}{nc_n(x-c)^{n-1}}\right| = \lim_{n\to\infty}\left(\frac{n+1}{n}\right)\lim_{n\to\infty}\left|\frac{c_{n+1}}{c_n}\right||x-c| = (1)\left(\frac{1}{R}\right)|x-c| = \frac{|x-c|}{R} < 1$$

The last inequality is true because by assumption, $|x-c| < R$. So the series in (9.57) converges for every value of x in $(c-R, c+R)$. If $|x-c| > R$ then the last inequality above gets reversed and the (LRT) implies that the series in (9.57) diverges. Therefore, the series in (9.57) has the same interval and radius of convergence as the original series. Further, the series in (9.57) is also a power series (with coefficients nc_n) so by our earlier results, it converges uniformly in every interval of type $[c-r, c+r]$ if $0 < r < R$. So we may use the theorem on the term by term differentiation of function series to conclude that the original series also converges uniformly on $[c-r, c+r]$ to a function $f(x)$ that is differentiable and its derivative $f'(x)$ is given by the series in (9.57).

Exercise 140 *Consider the geometric power series:*

$$\sum_{n=0}^{\infty} \frac{x^n}{2^n} = 1 + \frac{x}{2} + \frac{x^2}{4} + \frac{x^3}{8} + \cdots$$

(a) What function $f(x)$ does this series converge to? What is the interval of convergence?
(b) Find the formula for the derivative $f'(x)$ and give its power series.

A useful and very important aspect of the above theorem is that it can be applied repeatedly to obtain power series for derivatives of higher order.

By repeatedly applying the above theorem and taking derivatives, we see that if $f(x) = \sum_{n=0}^{\infty} c_n (x-c)^n$ where the power series has a positive radius of convergence R then all higher order derivatives of $f(x)$ exist on the same interval of convergence $(c-R, c+R)$ and they are found by taking the derivative of the original power series term by term.

$$\begin{aligned}
f'(x) &= \frac{d}{dx} \sum_{n=0}^{\infty} c_n(x-c)^n \\
&= \frac{d}{dx}\left(c_0 + c_1(x-1)^1 + c_2(x-1)^2 + c_3(x-1)^3 + \cdots\right) \\
&= 0 + c_1 + 2c_2(x-1)^1 + 3c_3(x-1)^2 + 4c_4(x-1)^3 + \cdots \\
&= \sum_{n=1}^{\infty} nc_n(x-c)^{n-1}
\end{aligned}$$

This explains where the formula in (9.57) comes from. Similarly, for the second derivative:

$$\begin{aligned}
f''(x) &= \frac{d^2}{dx^2} \sum_{n=0}^{\infty} c_n(x-c)^n \\
&= \frac{d}{dx} \sum_{n=1}^{\infty} nc_n(x-c)^{n-1} \\
&= \frac{d}{dx}\left(c_1 + 2c_2(x-1)^1 + 3c_3(x-1)^2 + 4c_4(x-1)^3 + \cdots\right) \\
&= 0 + (1)(2)c_2 + (2)(3)(x-1)^1 + (3)(4)(x-1)^2 + \cdots \\
&= \sum_{n=2}^{\infty} n(n-1)c_n(x-c)^{n-2}
\end{aligned}$$

Series of Functions   337

Continuing this procedure k times leads to the following general formula:[18]

$$f^{(k)}(x) = \frac{d^k}{dx^k} \sum_{n=0}^{\infty} c_n(x-c)^n = \sum_{n=k}^{\infty} n(n-1)(n-2)\cdots(n-k+1)c_n(x-c)^{n-k} \quad (9.58)$$

Exercise 141 *Use the above procedure to show that:*

$$f'''(x) = \sum_{n=3}^{\infty} n(n-1)(n-2)c_n(x-c)^{n-3}$$

It is possible to write the derivative formulas in the standard power series notation, which starts from $n = 0$. For the first derivative, we shift the index of summation $n \to n+1$ as follows: define $m = n - 1$ so that if n starts from 1 then m starts from 0:

$$f'(x) = \sum_{m=0}^{\infty} (m+1)c_{m+1}(x-c)^m$$

Since the summation index is a dummy variable (it is only a place-holder) we can use any letter so using n again to keep formulas looking uniform, we get:

$$f'(x) = \sum_{n=0}^{\infty} (n+1)c_{n+1}(x-c)^n \quad (9.59)$$

Similarly, for the second derivative define $m = n - 2$ to get:

$$f''(x) = \sum_{m=0}^{\infty} (m+2)(m+1)c_{m+2}(x-c)^m$$

which may be rewritten with n again instead of m to look like:

$$f''(x) = \sum_{n=0}^{\infty} (n+2)(n+1)c_{n+2}(x-c)^n$$

This process when continued k times gives:

$$f^{(k)}(x) = \sum_{n=0}^{\infty} (n+k)(n+k-1)\cdots(n+1)c_{n+k}(x-c)^n \quad (9.60)$$

In the form (??) we see more easily that all higher derivatives are also ordinary power series; for example, we may read (??) as:

$$f^{(k)}(x) = \sum_{n=0}^{\infty} c'_n(x-c)^n \quad \text{with } c'_n = (n+k)(n+k-1)\cdots(n+1)c_{n+k}$$

---

[18]The logical validity of (9.58) is proved using a basic proof method called *mathematical induction*, a detail that we do not get into here but is often encountered in college courses for math and science majors in their first two years.

Exercise 142 *Show that the following power series has interval of convergence $(-\infty, \infty)$:*

$$\sum_{n=0}^{\infty} \frac{x^n}{n!} = 1 + \frac{x}{1!} + \frac{x^2}{2!} + \frac{x^3}{3!} + \cdots$$

Let $E(x)$ be the function that the above series converges to on $(-\infty, \infty)$. Find $E'(x)$ and show that $E'(x) = E(x)$. We will soon see that $E(x) = e^x$, the natural exponential function.

Infinitely differentiable functions and the Taylor series.

An important benefit of being able to take the derivative of a power series repeatedly is the following:[19]

Infinitely differentiable functions. If a power series converges to a function $f(x)$ then $f(x)$ has derivatives of all orders defined on its interval of convergence. We say that such a function is *infinitely differentiable*.

For example, the function $1/(1-x)$ is infinitely differentiable on the interval $(-1, 1)$ where the geometric power series converges to it. Similarly, $\ln x$ is infinitely differentiable on $(0,2)$, the interval of convergence of its power series in (9.51).[20] On the other hand, $f(x) = x^{4/3}$ is defined on $(-\infty, \infty)$ and has a derivative $f'(x) = (4/3)x^{1/3}$ that is also defined on $(-\infty, \infty)$; but $f''(x) = (4/9)x^{-2/3}$ is not defined on $(-\infty, \infty)$ because $f'(0)$ cannot be defined; so $x^{4/3}$ is differentiable on $(-\infty, \infty)$, but only once.

More generally, the power function:

$$f(x) = x^{n+1/3}$$

where n is a positive integer is differentiable n times on $(-\infty, \infty)$ but no more; each time that we take the derivative, we reduce n by one until after n times only the power $1/3$ is left. Taking the derivative again leads to a negative power $-2/3$ and division by 0 at $x = 0$.

A basic type of function that is infinitely differentiable is a polynomial function that we met earlier when discussing algebraic numbers; for instance, the polynomial $f(x) = x^3 - 4x + 2$ has derivatives:

$$f'(x) = 3x^2 - 4 + 0 = 3x^2 - 4$$
$$f''(x) = 6x - 0 = 6x$$
$$f'''(x) = 6$$
$$f^{(4)}(x) = 0$$

Since the derivative of 0 is again 0 it follows that $f^{(k)}(x) = 0$ for all $k > 3$. In general, by taking the derivatives repeatedly, we see that a polynomial of degree n such as:

$$P(x) = x^n + a_1 x^{n-1} + a_2 x^{n-2} + \cdots + a_{n-1} x + a_n$$

---

[19] In analogy with complex analysis, these functions are sometimes called "real analytic functions".
[20] The power series for $\ln x$ does not converge uniformly at $x = 2$ so we disregard this value of x.

Series of Functions 339

is infinitely differentiable on $(-\infty, \infty)$ with $P^{(k)}(x) = 0$ for all $k > n$.

Infinitely differentiable functions may seem (and are) very special; yet, remarkably, we will soon discover that most of the functions that are familiar to us, including e^x, $\sin x$, $\cos x$ etc are all of this type.

The following is another important piece of the series expansion puzzle.

> **The coefficients.** *If a function $f(x) = \sum_{n=0}^{\infty} c_n(x-c)^n$ is the sum of a power series on a nonempty interval $(c-R, c+R)$ then the coefficients of the power series are given by:*
> $$c_n = \frac{f^{(n)}(c)}{n!} \tag{9.61}$$

With regard to the notation, recall that $0! = 1$ by definition; we also define $f^{(0)}(x) = f(x)$, that is, by the zero-th derivative of $f(x)$ we simply mean $f(x)$ itself.

This is a simple consequence of the formula (9.60). If we set $x = c$ in that formula then we obtain:
$$f^{(k)}(c) = (0+k)(0+k-1)\cdots(0+1)c_{0+k} = k!c_k$$
because all higher degree terms contain the difference $x - c$ and they drop out. Solving for c_k gives $c_k = f^{(k)}(c)/k!$ which is the same as (9.61). Talk about a low-hanging fruit!

The significance of the formula in (9.61) is that using it we can go backward, *starting from a function and building its power series by calculating the coefficients*. One more piece of information ensures that the power series that we get using this process is *the* power series for the function.

> **Uniqueness of the power series.** *If $\sum_{n=0}^{\infty} d_n(x-c)^n = f(x)$ is another power series for $f(x)$ on the same interval of convergence $(c-R, c+R)$ then $c_n = d_n$ for all $n \geq 0$.*

The proof of this statement is also easy. Using (9.61) we see that
$$d_n = \frac{f^{(n)}(c)}{n!} = c_n$$
which is exactly what we wanted to prove!

Let's put this new information to good use by finding the power series (centered at $c = 0$) for the natural exponential function $f(x) = e^x$. We know that:
$$f'(x) = e^x, \quad f''(x) = e^x, \quad f'''(x) = e^x, \quad \text{and so on}$$

In short, $f^{(n)}(x) = e^x$. Therefore, $f^{(n)}(0) = e^0 = 1$ for all $n \geq 0$ and (9.61) gives $c_n = 1/n!$ These are the coefficients of the power series:
$$e^x = \sum_{n=0}^{\infty} \frac{x^n}{n!} = 1 + \frac{x}{1!} + \frac{x^2}{2!} + \frac{x^3}{3!} + \cdots \tag{9.62}$$

whose interval of convergence you found to be $(-\infty, \infty)$ in Exercise 142. Since the power series for e^x is unique on $(-\infty, \infty)$ the formula in (9.62) gives the power series centered at 0 for e^x. As a by-product, setting $x = 1$ in (9.62) gives the following series for the irrational number e:
$$e = 1 + \frac{1}{1!} + \frac{1}{2!} + \frac{1}{3!} + \cdots \tag{9.63}$$

This series converges to the irrational value of e rather quickly; the first 7 terms give $e \simeq 2.71806$ which is correct to three decimal places. Just harvesting ripe, low-hanging fruits now!

Let's summarize the preceding discussion as follows:

The power series expansion of a function (Taylor-Maclaurin series). If a function $f(x)$ has a power series centered at a number c then $f(x)$ is uniquely given by:

$$f(x) = \sum_{n=0}^{\infty} \frac{f^{(n)}(c)}{n!} (x-c)^n \qquad (9.64)$$

on the interval of convergence of the power series.

The power series on the right hand side of (9.64) is called the *Taylor series*[21] for $f(x)$ about $x = c$ (or the *Maclaurin series*,[22] if $c = 0$).

It is worth mentioning here that the Maclaurin series of a polynomial $P(x)$ is $P(x)$; for example, if $P(x) = x^2 - x + 3$ then

$$P'(x) = 2x - 1, \quad P''(x) = 2, \quad P^{(n)}(x) = 0 \text{ for } n \geq 3$$

so that

$$P(0) = 3, \quad P'(0) = -1, \quad P''(0) = 2, \quad P^{(n)}(0) = 0 \text{ for } n \geq 3$$

Therefore,

$$\sum_{n=0}^{\infty} \frac{P^{(n)}(0)}{n!} x^n = 3 + \frac{(-1)x}{1!} + \frac{(2)x^2}{2!} + \frac{(0)x^3}{3!} + \frac{(0)x^4}{4!} + \cdots$$
$$= 3 - x + x^2$$
$$= P(x)$$

So we may always think of a polynomial as its own Taylor series about 0. Of course, the Taylor series of a polynomial is a different polynomial (but still of the same degree) if the center is not zero.

Exercise 143 *By calculating the coefficients c_n show that the Taylor series for the polynomial $P(x) = x^2 - x + 3$ about $c = -1$ is:*

$$5 - 3(x+1) + (x+1)^2$$

Multiplying this expression out and simplifying it gives the original $P(x)$ above.

Now let's calculate the Maclaurin series for $f(x) = \sin x$. We find the function value and its higher derivatives at 0:

$$f(0) = \sin 0, \quad f'(x) = \cos x, \text{ so } f'(0) = \cos 0 = 1$$
$$f''(x) = -\sin x, \text{ so } f''(0) = -\sin 0 = 0$$
$$f'''(x) = -\cos x, \text{ so } f'''(0) = -\cos 0 = -1$$
$$f^{(4)}(x) = \sin x, \text{ so } f^{(4)}(0) = \sin 0 = 0$$

---

[21] Named after the English mathematician Brook Taylor (1685-1731).
[22] Named after the Scottich mathematician Colin Maclaurin (1698-1746).

Series of Functions 341

Since $f^{(4)}(x) = f(x)$ the set of values: $0, 1, 0, -1$ repeats every 4 steps. From (9.61) we obtain the coefficients of the power series:

$$c_0 = \frac{f^{(0)}(0)}{0!} = \frac{f(0)}{1} = 0, \qquad c_1 = \frac{f'(0)}{1!} = \frac{1}{1} = 1$$

$$c_2 = \frac{f''(0)}{2!} = \frac{0}{2} = 0, \qquad c_3 = \frac{f'''(0)}{3!} = -\frac{1}{3!}$$

The next four coefficients cycle the same four numerators but the denominators simply contain larger factorials:

$$c_4 = \frac{f^{(0)}(0)}{4!} = \frac{f(0)}{4!} = 0, \qquad c_5 = \frac{f'(0)}{5!} = \frac{1}{5!}$$

$$c_6 = \frac{f''(0)}{6!} = \frac{0}{6!} = 0, \qquad c_7 = \frac{f'''(0)}{7!} = -\frac{1}{7!}$$

The pattern is evident: all even-indexed coefficients are 0 and they do not appear in the power series, while the odd coefficients alternate in sign. A little reflection shows that the power series is the following:

$$\sin x = x - \frac{x^3}{3!} + \frac{x^5}{5!} - \frac{x^7}{7!} + \cdots = \sum_{n=0}^{\infty} \frac{(-1)^n x^{2n+1}}{(2n+1)!} \qquad (9.65)$$

I emphasize that *this formula is valid if x is in radians*. The radius of convergence is found using the same idea that we used earlier for the series in (9.47). In this case, we factor out an x and substitute $u = x^2$ to get the relabeled series:

$$x \sum_{n=0}^{\infty} \frac{(-1)^n x^{2n}}{(2n+1)!} = x \sum_{n=0}^{\infty} \frac{(-1)^n u^n}{(2n+1)!}$$

The coefficients of the last series are: $c_n = (-1)^n/(2n+1)!$ so:

$$R = \lim_{n \to \infty} \left| \frac{c_n}{c_{n+1}} \right| = \lim_{n \to \infty} \left| \frac{(-1)^n (2(n+1)+1)!}{(-1)^{n+1}(2n+1)!} \right| = \lim_{n \to \infty} \frac{(2n+3)!}{(2n+1)!} = \infty$$

To see how the limit diverges to infinity, notice that $(2n+3)! = (2n+3)(2n+2)(2n+1)!$ so that

$$\frac{(2n+3)!}{(2n+1)!} = \frac{(2n+3)(2n+2)(2n+1)!}{(2n+1)!} = (2n+3)(2n+2)$$

We see that the product on the far end on the right diverges to infinity as $n \to \infty$.

Therefore, the interval of convergence of the power series with u is $(-\infty, \infty)$. Of course, $u = x^2$ is never negative but it can be infinitely large so x can be any real number, that is, with values ranging in $(-\infty, \infty)$; conveniently, this also happens to be the domain of $\sin x$.

Exercise 144 *(a) By finding the coefficients as we did above for $\sin x$, show that the Maclaurin series for $\cos x$ is the following, with interval of convergence $(-\infty, \infty)$:*

$$\cos x = 1 - \frac{x^2}{2!} + \frac{x^4}{4!} - \frac{x^6}{6!} + \cdots = \sum_{n=0}^{\infty} \frac{(-1)^n x^{2n}}{(2n)!} \qquad (9.66)$$

(b) Alternatively, derive the above series expansion for $\cos x$ the quick way by taking the derivative of the series in (9.65) term by term.

Even and odd functions expand in even and odd powers. Notice that all of the terms in (9.65) are odd powers of x and all the terms in (9.66) are even powers of x. Recall that $\sin x$ is an odd function while $\cos x$ is even. In their power series expansions we find a direct link between odd functions and odd-degree power functions, and similarly, even functions and even-degree power functions. Based on this observation, you can tell that the following Maclaurin series that contains only odd power of x can converge only to an odd function:

$$\sum_{n=1}^{\infty} a_n x^{2n-1} = a_1 x + a_2 x^3 + a_3 x^5 + \cdots$$

regardless of the values of the coefficients a_1, a_2, a_3, etc (as long as at least one of these is nonzero). For instance, the power series in (9.55) shows that the inverse tangent $\tan^{-1} x$ is an odd function. On the other hand, the Maclaurin series:

$$\sum_{n=0}^{\infty} b_n x^{2n} = b_0 + b_1 x^2 + b_2 x^4 + b_3 x^6 + \cdots$$

can converge only to an even function regardless of the values of the coefficients b_0, b_1, b_2, etc.

We encounter additional series of both types later on. I end this discussion with an amazing little identity discovered by Euler that you may have already seen before. We need a brief use of the imaginary numbers that will not significantly take us off track.

Euler's remarkable identity.

We write $\sqrt{-1} = i$; this is the *imaginary unit* in the complex numbers and is algebraically defined as one of the two non-real solutions of the equation $x^2 + 1 = 0$ (the other being $-i$). Note that the positive-integer powers of i follow the following pattern:

$$i^2 = -1, \quad i^3 = i^2 i = -i, \quad i^4 = i^2 i^2 = 1, \quad i^5 = i, \quad i^6 = -1, \quad i^7 = -i, \quad i^4 = 1, \ldots$$

The pattern $i, -1, -i, 1$ repeats and produces a sequence of period 4. Now, without worrying about technical issues, let's take a look at the Maclaurin series for e^{ix} in which we replace x with ix to get:

$$e^{ix} = \sum_{n=0}^{\infty} \frac{(ix)^n}{n!} = 1 + \frac{ix}{1!} - \frac{x^2}{2!} - \frac{ix^3}{3!} + \frac{x^4}{4!} + \frac{ix^5}{5!} - \frac{x^6}{6!} - \frac{ix^7}{7!} + \cdots$$

If we separate the terms without the i (the *real part* of the series) from the terms with the i (the *imaginary part* of the series) and factor out the i then we get:

$$e^{ix} = \left(1 - \frac{x^2}{2!} + \frac{x^4}{4!} - \frac{x^6}{6!} + \cdots\right) + i\left(\frac{x}{1!} - \frac{x^3}{3!} + \frac{x^5}{5!} - \frac{x^7}{7!} + \cdots\right)$$

Series of Functions   343

Now it is easy to recognize the series for $\cos x$ on the left (the real part of e^{ix}) and the series for $\sin x$ on the right (the imaginary part of e^{ix}). This shows that:

$$e^{ix} = \cos x + i \sin x \qquad (9.67)$$

The above identity is commonly known as *Euler's formula* but it was known earlier to the English mathematician Roger Cotes (1682-1716) who was a colleague of Newton and is best known nowadays for the numerical integration method named the "Newton-Cotes method".[23]

The identity in (9.67) is arguably one of the most important and useful formulas in all of mathematics. In particular, if we set $x = \pi$ in (9.67) then we obtain the intriguing equality:

$$e^{i\pi} = -1$$

that is known as *Euler's identity*.

Further, if n is a positive integer then:

$$(\cos x + i \sin x)^n = (e^{ix})^n = e^{inx} = \cos(nx) + i \sin(nx)$$

which is known as *de Moivre's identity*.[24] This is the fastest derivation of this identity!

Euler's formula shows that the complex exponential function e^{ix} is periodic with period 2π which is in sharp contrast to the real exponential function e^x which is strictly increasing for real values of x. Further, while e^x is always positive, e^{ix} may be positive or negative as $e^{i\pi}$ shows. The general exponential function e^z where $z = x + iy$ is an extension of e^{ix} to the entire complex plane and is discussed fully in standard textbooks on complex analysis.

An interesting by-product of (9.67) is that the real functions $\sin x$ and $\cos x$ can be written in terms of the complex exponential e^{ix} as follows:

$$e^{ix} + e^{-ix} = \cos x + i \sin x + \cos(-x) + i \sin(-x)$$

Since $\cos x$ is even and $\sin x$ is odd, this can be written as:

$$e^{ix} + e^{-ix} = \cos x + i \sin x + \cos x - i \sin x = 2 \cos x$$

Dividing by 2 gives the identity:

$$\cos x = \frac{e^{ix} + e^{-ix}}{2}$$

Similarly,

$$e^{ix} - e^{-ix} = \cos x + i \sin x - (\cos x - i \sin x) = 2i \sin x$$

so dividing by $2i$ gives:

$$\sin x = \frac{e^{ix} - e^{-ix}}{2i}$$

The important thing about these relations is that we can use the complex exponential to simplify some calculations involving the basic real trigonometric functions. For example, we can quickly prove the following angle sum identities:

$$\cos(x+y) = \cos x \cos y - \sin x \sin y \qquad (9.68)$$

$$\sin(x+y) = \sin x \cos y + \cos x \sin y \qquad (9.69)$$

---

[23] Cotes had come across the identity $ix = \ln(\cos x + i \sin x)$ sometime before 1714. This identity is clearly the same as Euler's but in logarithmic form.

[24] Named after the French mathematician Abraham de Moivre (167-1754).

Derivation of these identities using elementary geometry and right triangles is not difficult but by no means transparent. On the other hand, from (9.67) we derive:

$$e^{i(x+y)} = \cos(x+y) + i\sin(x+y) \qquad (9.70)$$

The left hand side can be written as:

$$e^{i(x+y)} = e^{ix}e^{iy} = (\cos x + i\sin x)(\cos y + i\sin y)$$

Multiplying out the product on the right above and collecting real and imaginary terms we get:

$$e^{i(x+y)} = \cos x \cos y - \sin x \sin y + i(\sin x \cos y + \cos x \sin y)$$

Since the left hand side above matches the left hand side of (9.70) we can set the real parts of the right hand sides equal to get (9.68). Similarly, setting the imaginary parts equal gives (9.69). We can use (9.68) and (9.69) to derive some rather important formulas that are used in the section on Fourier series below; see Appendix 11.2.

The amazing binomial series, the inverse sine... and π again.

From elementary algebra you may recall the binomial formulas:

$$(a+b)^2 = a^2 + 2ab + b^2$$
$$(a+b)^3 = a^3 + 3a^2b + 3ab^2 + b^3$$

and so on, where the coefficients are easy to calculate for higher powers using the Pascal triangle. As long as the power is a positive integer, the expansion has a finite number of terms; in fact, the general binomial formula *when n is a positive integer* is:

$$(a+b)^n = a^n + na^{n-1}b + \frac{n(n-1)}{2}a^{n-2}b^2 + \frac{n(n-1)(n-2)}{6}a^{n-3}b^3 + \cdots + b^n \qquad (9.71)$$
$$= \sum_{k=0}^{n} \frac{n(n-1)(n-2)\cdots(n-k+1)}{k!} a^{n-k}b^k$$

If n is not a positive integer or 0 then the binomial expression is no longer a finite sum. You may be wondering what possible meaning $(a+b)^p$ can have if p is a negative integer, a rational number or worst of all, an irrational number. It turns out that for non-integer values of p calculus has found a meaning for $(a+b)^p$ as an infinite series! We now describe this series and show one of its surprising uses.

To explain this meaning, consider the *binomial function*

$$f(x) = (1+x)^p$$

where the constant p is a real number and let's find the Maclaurin series. We start by calculating

Series of Functions

the derivatives:

$$f'(x) = p(1+x)^{p-1},$$
$$f''(x) = p(p-1)(1+x)^{p-2},$$
$$f'''(x) = p(p-1)(p-2)(1+x)^{p-3},$$
$$\vdots$$
$$f^{(n)}(x) = p(p-1)(p-2)\cdots(p-n+1)(1+x)^{p-n}$$
$$\vdots$$

Notice that if p is not a positive integer then the above list of derivatives is infinite. Next, we obtain the values of the derivatives at $x = 0$

$$f'(0) = p,$$
$$f''(0) = p(p-1),$$
$$f'''(0) = p(p-1)(p-2),$$
$$\vdots$$
$$f^{(n)}(0) = p(p-1)(p-2)\cdots(p-n+1)$$
$$\vdots$$

Inserting these numbers in the series in (9.64) and noting that $f(0) = 1$ we get:

$$\begin{aligned}(1+x)^p &= 1 + \sum_{n=1}^{\infty} \frac{f^{(n)}(0)}{n!} x^n \\ &= 1 + \sum_{n=1}^{\infty} \frac{p(p-1)(p-2)\cdots(p-n+1)}{n!} x^n \\ &= 1 + \frac{p}{1!}x + \frac{p(p-1)}{2!}x^2 + \frac{p(p-1)(p-2)}{3!}x^3 + \cdots\end{aligned} \quad (9.72)$$

The series in (9.72) is called the *binomial series*.

To find the interval of convergence for the binomial series, we note that the coefficients are

$c_n = p(p-1)\cdots(p-n+1)/n!$ so the radius of convergence is:

$$\begin{aligned} R &= \lim_{n\to\infty} \left|\frac{c_n}{c_{n+1}}\right| \\ &= \lim_{n\to\infty} \left|\frac{p(p-1)\cdots(p-n+1)(n+1)!}{p(p-1)\cdots(p-n+1)\,(p-n)n!}\right| \\ &= \lim_{n\to\infty} \left|\frac{(n+1)n!}{(p-n)n!}\right| \\ &= \left|\lim_{n\to\infty} \frac{n+1}{p-n}\right| \\ &= \left|\lim_{n\to\infty} \frac{1+1/n}{-1+p/n}\right| \\ &= 1 \end{aligned}$$

Therefore, *the interval of convergence of the binomial series in (9.72) is $(-1,1)$*. Convergence at the end-points is not guaranteed and depends on the value of p; see the examples below. In particular, if $p < 0$ then $x = -1$ is not even in the domain of the binomial function $(1+x)^p$.

Let's check how this binomial series relates to some of the series that we discussed previously. If $p = n$, a positive integer, then

$$(1+x)^n = 1 + \frac{n}{1!}x + \frac{n(n-1)}{2!}x^2 + \cdots + \frac{n(n-1)\cdots(1)}{n!}x^n \tag{9.73}$$

where all terms after the last one listed above drop out since the coefficients of all the remaining terms of the series contain a factor $(n-n)$ and are therefore zeros. The above series is indeed what (9.71) gives with $a = 1$ and $b = x$. Actually, we may obtain (9.71) as a special case of (9.72); here is how: Since one of a or b is the larger of the two[25] suppose that $a > b$ and notice that

$$(a+b)^n = a^n\left(1 + \frac{b}{a}\right)^n$$

So if we set $x = b/a$ in (9.73), which is fine since b/a is in the interval of convergence $(-1,1)$, and then multiply the sum by a^n we obtain (9.71). Incidentally, this also proves (9.71)!

Next, set $p = -1$ in (9.72) to get:

$$(1+x)^{-1} = 1 + \frac{(-1)}{1!}x + \frac{(-1)(-2)}{2!}x^2 + \frac{(-1)(-2)(-3)}{3!}x^3 + \cdots$$
$$\frac{1}{1+x} = 1 - x + x^2 - x^3 + \cdots$$

You may recognize the last series as the geometric series in (9.2). In fact, by changing x to $-x$ above we obtain the original geometric power series as a special case of the binomial series (9.72). This also means that all other series that we obtained from the geometric power series, including the

---

[25] I am assuming that a and b are positive. If one or both happens to be negative then a simple modification of this argument takes care of the problem.

Series of Functions   347

series for $\ln x$ and $\tan^{-1} x$ are special cases of the binomial series, if we include the latter's integrals and derivatives!

Next, let's consider a non-integer value for p. I discuss the case $p = -1/2$ and leave $p = 1/2$ as an exercise. If we set $p = -1/2$ in (9.72) then we get:

$$(1+x)^{-1/2} = 1 + \frac{-1/2}{1!}x + \frac{(-1/2)(-3/2)}{2!}x^2 + \frac{(-1/2)(-3/2)(-5/2)}{3!}x^3 + \cdots$$

$$\frac{1}{\sqrt{1+x}} = 1 - \frac{1}{1!2}x + \frac{(1)(3)}{2!2^2}x^2 - \frac{(1)(3)(5)}{3!2^3}x^3 + \cdots \qquad (9.74)$$

or more compactly,

$$\frac{1}{\sqrt{1+x}} = \sum_{n=0}^{\infty} \frac{(-1)^n (3) \cdots (2n-1)}{n! 2^n} x^n$$

Exercise 145 *Find the binomial series for $\sqrt{1+x}$.*

An obvious application of the binomial series is estimating roots of rational numbers; a few centuries ago when there were no computers or calculators, being able to estimate roots quickly was a major issue.[26] Newton was familiar with the binomial series before it became well-known. Speedy publication or wide communication of ideas and results was not possible those days, so he managed to impress many smart people of his time by doing seemingly impossible calculations!

To illustrate this application of the binomial series let's use it to estimate $\sqrt{2.5}$. The "obvious" substitution $x = 1.5$ in $(1+x)^p$ with $p = 1/2$ doesn't work because 1.5 is not in the interval of convergence $(-1, 1)$. But

$$\sqrt{2.5} = \sqrt{\frac{5}{2}} = \left(\frac{5}{2}\right)^{1/2} = \left(\frac{2}{5}\right)^{-1/2} = \left(1 - \frac{3}{5}\right)^{-1/2}$$

and $3/5 = 0.6$ is well within the interval of convergence of the binomial series for $(1+x)^{-1/2}$. We use the series in (9.74) with $x = -0.6$. If we use just the first four terms then we obtain the approximation:

$$\sqrt{2.5} \simeq 1 - \frac{1}{1!2}(-0.6) + \frac{(1)(3)}{2!2^2}(-0.6)^2 - \frac{(1)(3)(5)}{3!2^3}(-0.6)^3$$

$$= 1 + 0.3 + \frac{3(0.36)}{8} + \frac{1.5(0.36)}{8} = 1.5025$$

which is not a bad estimate of $\sqrt{2.5}$ whose value rounded to four decimal places is 1.5811.

Exercise 146 *Use the first four terms of your series in Exercise 145 to estimate $\sqrt{1.2}$.*

---
[26] Recall that the Newton-Raphson method, when properly used, is the fastest known way of estimating any root in a simple and quick way.

As I mentioned earlier, the binomial series is one of the most versatile power series known. Let me illustrate this by substituting $-u^2$ for x in (9.74). This substitution doesn't change the interval of convergence since $-u^2$ is in the interval $(-1, 1)$ if and only if x is. We obtain the equality:

$$\frac{1}{\sqrt{1-u^2}} = 1 - \frac{1}{1!2}(-u^2) + \frac{(1)(3)}{2!2^2}(-u^2)^2 - \frac{(1)(3)(5)}{3!2^3}(-u^2)^3 + \cdots$$

$$\frac{1}{\sqrt{1-u^2}} = 1 + \frac{1}{1!2}u^2 + \frac{(1)(3)}{2!2^2}u^4 + \frac{(1)(3)(5)}{3!2^3}u^6 + \cdots \quad (9.75)$$

The left hand side of (9.75) happens to be the derivative of the *inverse sine function* $\sin^{-1} u$; to verify this we use the chain rule as we did in the inverse tangent case earlier: if $\theta = \sin^{-1} u$ then $\sin \theta = u$ so

$$\frac{d}{du}\sin\theta = \frac{du}{du}$$

$$\cos\theta \frac{d\theta}{du} = 1$$

$$\frac{d}{du}\sin^{-1} u = \frac{1}{\cos\theta}$$

Using the trigonometric Pythagorean identity and a little algebra we find that

$$\cos^2\theta = 1 - \sin^2\theta = 1 - u^2$$

$$\cos\theta = \sqrt{1-u^2}$$

so putting it all together we get:

$$\frac{d}{du}\sin^{-1} u = \frac{1}{\sqrt{1-u^2}}$$

as claimed. Now, if x is in $(-1,1)$ then we may integrate (9.75) term by term to obtain:

$$\int_0^x \frac{1}{\sqrt{1-u^2}} du = \left[u + \frac{u^3}{1!2(3)} + \frac{(1)(3)u^5}{2!2^2(5)} + \frac{(1)(3)(5)u^7}{3!2^3(7)} + \cdots\right]_{u=0}^{u=x}$$

$$\sin^{-1} x = x + \frac{x^3}{1!2(3)} + \frac{(1)(3)x^5}{2!2^2(5)} + \frac{(1)(3)(5)x^7}{3!2^3(7)} + \cdots \quad (9.76)$$

This is the standard inverse-sine Maclaurin series that we can write more compactly as:

$$\sin^{-1} x = \sum_{n=0}^{\infty} \frac{(1)(3)\cdots(2n-1)x^{2n+1}}{n!2^n(2n+1)}, \quad -1 < x < 1$$

I note in passing that this series shows that $\sin^{-1} x$ is an odd function. Let's use (9.76) to approximate π one more time! Note that $\sin(\pi/6) = 1/2$ so that $\sin^{-1}(1/2) = \pi/6$. Now we insert $x = 1/2$ in (9.76) and use four of its terms to obtain:

$$\frac{\pi}{6} \simeq \frac{1}{2} + \frac{(1/2)^3}{1!2(3)} + \frac{(1)(3)(1/2)^5}{2!2^2(5)} + \frac{(1)(3)(5)(1/2)^7}{3!2^3(7)}$$

$$= \frac{1}{2} + \frac{1}{2^4(3)} + \frac{3}{2^8(5)} + \frac{5}{2^{11}(7)}$$

Multiplying by 6 and doing the arithmetic gives the following approximation of π rounded to four decimal places:
$$\pi \simeq 3.1412$$

The difference between this number and the actual value of π is less than 0.0004. Using just 18 terms of (9.76) with $x = 1/2$ gives the estimate:

$$\pi \simeq 3.14159265358959$$

which is already better than the one using 21 terms of the inverse tangent series earlier, and as an extra bonus, this new approximation is *rational*, i.e., it does not include irrational factors like $\sqrt{3}$. It is worth remembering that *both the inverse sine and the inverse tangent series are integrals of power series that we can easily obtain from the binomial series.*

Approximating functions using Taylor polynomials.

We previously saw that a finite number of terms of an infinite series may be used to approximate a real number, like truncating the series in (9.63) to approximate e or the truncated inverse sine and inverse tangent series that we used to estimate π. The same idea holds for functions; by truncating the Taylor series for a function $f(x)$, the resulting polynomial (a partial sum of the Taylor series) approximates $f(x)$ on the interval of convergence. This is illustrated in Figures 9.1, 9.2 and 9.4 for the rational function $1/(1-x)$ on the interval $(-1,1)$. This is the interval of convergence of the geometric power series, which is none other than the Maclaurin series for $1/(1-x)$. Also Figure 9.16 shows the approximating polynomials, or partial sums, for $\ln x$ on the interval of convergence $(0,2)$.

These figures also indicate another important fact; *the Taylor series converges exactly at its center $x = c$ (where every term of the series except the first term $f(c)$ drops out) but the convergence generally slows down as we move away from the center*, requiring more terms of the series to achieve the same level of accuracy.

Let's formalize these approximating polynomials now.

Taylor polynomials. The truncation of the series in (9.64) leaves a polynomial:

$$f(c) + \frac{f'(c)}{1!}(x-c) + \frac{f''(c)}{2!}(x-c)^2 + \cdots + \frac{f^{(n)}(c)}{n!}(x-c)^n = \sum_{j=0}^{n} \frac{f^{(j)}(c)}{j!}(x-c)^j$$

This finite sum, which is actually the n-th partial sum of the Taylor series, is commonly known as the *Taylor polynomial of degree n*, assuming that the number $f^{(n)}(c) \neq 0$.

A note on terminology is in order: the n-th partial sum of a Taylor series actually consists of $n+1$ terms since we start counting partial sums from $s_0(x)$ for power series. Further, the n-th partial sum is not always a polynomial of degree n because $f^{(n)}(c)$ may be 0. For example, the 4th partial sum of the series in (9.64) for $\sin x$ is:

$$0 + \frac{(1)x}{1!} + \frac{(0)x^2}{2!} + \frac{(-1)x^3}{3!} + \frac{(0)x^4}{4!} = x - \frac{x^3}{3!}$$

which is a polynomial of degree 3, not 4. In fact, since every other term is zero, the third and 4th partial sums are identical; the same is true of the partial sums $s_n(x)$ and $s_{n+1}(x)$ for every n. The series in (9.65) shows only the distinct, nonzero terms of the Taylor series for $\sin x$. Also note that *the distinct Taylor polynomials for $\sin x$ all have odd degrees*, as indicated by the powers of x in the series.

Figure 9.17 illustrates the graph of $\sin x$ (the thick curve) and the first four (distinct) Taylor polynomials (the indicated partial sums).

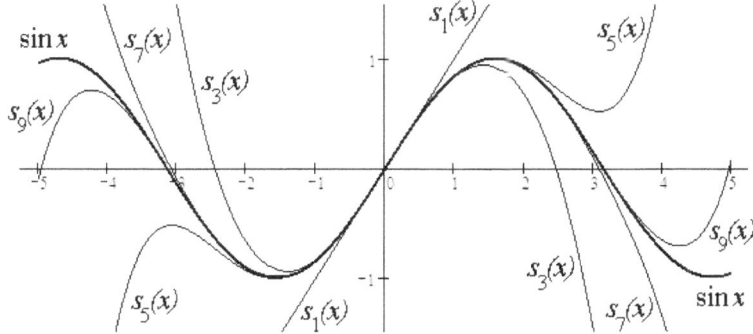

Figure 9.17: Several Taylor Polynomials for $\sin x$

We clearly see in Figure 9.17 that the approximations improve over larger stretches of the x-axis with the increasing degree of the polynomial. The partial sum $s_9(x)$, namely, the Taylor polynomial of degree 9, is practically indistinguishable from $\sin x$ on the interval $[-\pi, \pi]$ a full period of $\sin x$ so we can get accurate estimates of $\sin x$ for every value of x using just $s_9(x)$. If you have a device capable of graphing multiple curves, consider creating similar figures for the functions $\cos x$ and e^x. It is not without fun to see it all together!

Exercise 147 *Calculate $\sin x$ as well as $s_1(x)$, $s_3(x)$, $s_5(x)$, $s_7(x)$, $s_9(x)$ for $x = \pi/4, \pi/2, \pi$ to see how the different approximations compare.*

The integration of the Taylor polynomial for a function $f(x)$ gives us a way of approximating the integral of $f(x)$ within the interval of convergence of its Taylor series. To illustrate, consider $f(x) = \sin(x^2)$. There is no known antiderivative for this function but we may easily obtain the Taylor series for it by simply replacing x by x^2 in (9.65):

$$\sin(x^2) = x^2 - \frac{(x^2)^3}{3!} + \frac{(x^2)^5}{5!} - \frac{(x^2)^7}{7!} + \cdots$$
$$= x^2 - \frac{x^6}{3!} + \frac{x^{10}}{5!} - \frac{x^{14}}{7!} + \cdots \qquad (9.77)$$

Suppose we retain the first three nonzero terms to get the Taylor polynomial (of degree 10) as an approximation for $\sin(x^2)$:

$$\sin(x^2) \simeq x^2 - \frac{x^6}{3!} + \frac{x^{10}}{5!}$$

Series of Functions 351

Since this is the Taylor polynomial about 0 we get better estimates for integrals taken over intervals that surround 0. For instance, integrating from 0 to 1 (remember, in radians!) we have:

$$\int_0^1 \sin(x^2)dx \simeq \int_0^1 \left(x^2 - \frac{x^6}{3!} + \frac{x^{10}}{5!}\right)dx$$

$$= \left[\frac{x^3}{3} - \frac{x^7}{3!(7)} + \frac{x^{11}}{5!(11)}\right]_{x=0}^{x=1}$$

$$= \frac{1}{3} - \frac{1}{6}\left(\frac{1}{7}\right) + \frac{1}{120}\left(\frac{1}{11}\right)$$

$$\simeq 0.310281$$

where I rounded the last number to 6 decimal places.

But how good an approximation is this number?

It is tempting to use the almighty calculator (or computer) that, magically, has answers for everything numerical.[27]

But we don't actually need a calculator, or even the exact value, to get some idea as to how good an answer we have!

The series (9.77) is an alternating series for which you may recall we found an error bound earlier. For the particular problem here, the approximation error in using the first three terms of the series is less than the magnitude of the 4th term, that is:

$$\left|-\int_0^1 \frac{x^{14}}{7!}dx\right| = \left|\frac{x^{15}}{7!(15)}\right|_{x=0}^{x=1} = \left|\frac{1}{7!(15)}\right| < 0.000093$$

So although we don't know the exact value of the integral $\int_0^1 \sin(x^2)dx$ we can confidently declare that our answer 0.310281 is off from the exact value by less than 0.000093. If this is not small enough for some purpose then we simply add more terms of the series until the error is within acceptable bounds.

Exercise 148 *(a) Find the Maclaurin series for e^{-x^2} by replacing x with $-x^2$ in the power series for e^x. Explain why this series converges on $(-\infty, \infty)$.*

(b) Use the first 4 terms of your answer in (a) to estimate the integral:

$$\int_0^1 e^{-x^2}dx$$

Without using a calculator, how good an estimate of the actual integral is your answer with just these 4 terms?

---

[27]While keeping in mind that calculators and computers don't really do anything that has not already been programmed or built into them (in particular, no machine knows the exact value of this integral), let's try it. With a scientific calculator's integration button I effortlessly get 0.310268, rounded to 6 decimal places. The relative percentage error of the number that we calculated is:

$$\frac{0.310281 - 0.310268}{0.310268}(100) = 0.0042\%$$

which is less than half of one-hundredth of one percent; not bad with just three terms of the infinite series!

But what if the power series is not alternating?

There is a general formula for getting the error bound when using a Taylor polynomial to estimate a function $f(x)$. This formula is based on the so-called *remainder term* of a Taylor series and it applies *whether the series is alternating or not*. We do not discuss this matter any further in this book but you can find some discussion of it in most standard calculus textbooks.

9.7 Fine tuning the infinite: l'Hôpital's rule

In this section we take another look at the expression 0/0. We saw earlier that this expression does not represent a fixed number; you may recall that *the derivative of every function has the form 0/0 since the derivative exists only when the numerator and the denominator of the difference ratio approach 0 simultaneously*. Clearly then 0/0 is not a fixed quantity like 0/1 or 2/3 but a non-specific or indeterminate quantity that can have any value.

The derivative is not the only context where 0/0 shows up; we come across it generally when calculating limits at points of discontinuity of a function: *if the discontinuity is removable then the limit has a numerical value*. Let's explore this aspect now.

Suppose that x_n is an arbitrary sequence of real numbers that converges to 0 and consider the following limits:

$$\lim_{n\to\infty} \frac{\sin x_n}{x_n^2} \tag{9.78}$$

$$\lim_{n\to\infty} \frac{\sin(x_n^2)}{x_n} \tag{9.79}$$

$$\lim_{n\to\infty} \frac{\sin x_n}{2x_n} \tag{9.80}$$

$$\lim_{n\to\infty} \frac{\sin(2x_n)}{x_n} \tag{9.81}$$

We have already seen that $2x$, x^2, $\sin x$ and $\sin(2x)$ are continuous functions so all of the sequences $2x_n$, x_n^2, $\sin x_n$ and $\sin(2x_n)$ converge to 0 as $n \to \infty$ (remember that $\sin 0 = 0$). Since all four limits end up with a 0 in the denominator, as we saw earlier we cannot find these limits by simply distributing the limit over the fractions.

But does this mean that these limits don't exist? Or, because they all lead to the same expression 0/0, can we conclude that if one exists then so do the others and they are all equal?

>**A numerical experiment.** *We can get a better sense of what's going on with the limits in (9.78)-(9.81) by conducting a numerical experiment using a calculator; the following table lists some of the values of these limits (rounded to 3 decimal places) for the sequence $x_n = 1/n$ which converges to 0 as $n \to \infty$.*[28]

---

[28] If we use other sequences that converge to 0, like $1/n^2$ or $-1/\sqrt{n}$, we get a table with different numbers but all the numbers follow the same trends as shown in the table here.

Series of Functions   353

n	1	20	40	60	80	100
$(\sin x_n)/x_n^2$	0.841	19.992	39.996	59.997	79.998	99.998
$(\sin x_n^2)/x_n$	0.841	0.05	0.025	0.017	0.012	0.001
$(\sin x_n)/(2x_n)$	0.421	0.5	0.5	0.5	0.5	0.5
$(\sin 2x_n)/x_n$	0.909	1.997	1.999	2	2	2

This table of calculated values shows four totally different outcomes for (9.78)-(9.81). The limit in (9.78) appears to be diverging to infinity, the one in (9.79) seems to converge to zero, for (9.80) we're getting 1/2 and the limit in (9.81) is evidently 2.

This numerical experiment confirms that the expression 0/0 cannot have a well-defined value; not surprisingly, it is called an *indeterminate form*.

Calculating limits using Taylor polynomials.

We can take the mystery out of why the three limits turn out as they do in above table by looking at the power series expansions of the fractions.

A dose of power series to clear our heads. We have already found the power series expansions of $\sin x$ and $\sin(x^2)$ in (9.65) and (9.77), respectively. Using these we find that:

$$\frac{\sin x_n}{x_n^2} = \frac{1}{x_n^2}\left(x_n - \frac{x_n^3}{3!} + \frac{x_n^5}{5!} - \frac{x_n^7}{7!} + \cdots\right) = \frac{1}{x_n} - \frac{x_n}{3!} + \frac{x_n^3}{5!} - \frac{x_n^5}{7!} + \cdots$$

As $x_n \to 0$ all terms after the first one drop out while the first term diverges to infinity. Therefore, there is no limit in (9.78). Next,

$$\frac{\sin(x_n^2)}{x_n} = \frac{1}{x_n}\left(x_n^2 - \frac{x_n^6}{3!} + \frac{x_n^{10}}{5!} - \frac{x_n^{14}}{7!} + \cdots\right) = x_n - \frac{x_n^5}{3!} + \frac{x_n^9}{5!} - \frac{x_n^{13}}{7!} + \cdots$$

As $x_n \to 0$ all of the terms of the above series go to zero, leaving us with just 0 as the limit. Next,

$$\frac{\sin x_n}{2x_n} = \frac{1}{2x_n}\left(x_n - \frac{x_n^3}{3!} + \frac{x_n^5}{5!} - \frac{x_n^7}{7!} + \cdots\right) = \frac{1}{2} - \frac{x_n^2}{3!2} + \frac{x_n^4}{5!2} - \frac{x_n^6}{7!2} + \cdots$$

As $x_n \to 0$ all terms after the first one drop out, leaving us with a limit of 1/2, which is the limit in (9.80). The fourth limit is just as easily explained:

$$\frac{\sin(2x_n)}{x_n} = \frac{1}{x_n}\left(2x_n - \frac{(2x_n)^3}{3!} + \frac{(2x_n)^5}{5!} - \cdots\right) = 2 - \frac{8x_n^2}{3!} + \frac{32x_n^4}{5!} - \cdots$$

As $x_n \to 0$ all terms after the first one drop out, leaving a limit of 2 in (9.81).

We see from the above series that when x_n is very close to zero, each of the functions on the left hand sides of the above four equalities is closely represented by the first term of the corresponding series on the right hand side.

It is worth emphasizing that the higher degree terms drop out when we take the limit. So to calculate a 0/0 indeterminate form we only need to find the first few terms of the series, or more precisely, a Taylor polynomial of high enough degree, but not the entire infinite series.

Here's an example where x_n does not approaches a number other than 0:

$$\lim_{n\to\infty} \frac{\ln x_n}{x_n - 1} \qquad (9.82)$$

where $x_n \to 1$ as $n \to \infty$. Since $\ln x_n$ and $x_n - 1$ both go to 0 we have a 0/0 form. We list just a few terms of the Taylor series for $\ln x$ in (9.51) where we may assume that x_n is in the interval (0,2]:

$$\ln x_n = \frac{x_n - 1}{1} - \frac{(x_n - 1)^2}{2} + \frac{(x_n - 1)^3}{3} - \cdots$$

This implies that:

$$\frac{\ln x_n}{x_n - 1} = 1 - \frac{(x_n - 1)}{2} + \frac{(x_n - 1)^2}{3} - \cdots$$

Since all the terms after the first contain some positive integer power of $(x_n - 1)$ they converge to 0 so we get:

$$\lim_{n\to\infty} \frac{\ln x_n}{x_n - 1} = \lim_{n\to\infty}\left(1 - \frac{(x_n - 1)}{2} + \cdots\right) = 1 - 0 + 0 - \cdots = 1$$

Exercise 149 *Show that each of the following limits has the 0/0 indeterminate form when x_n is any sequence that converges to 0 as $n \to \infty$. Then find the limit in each case:*

(a) $\displaystyle\lim_{n\to\infty} \frac{\cos x_n - 1}{x_n}$
(b) $\displaystyle\lim_{n\to\infty} \frac{\cos x_n - 1}{x_n^2}$
(c) $\displaystyle\lim_{n\to\infty} \frac{e^{x_n} - 1}{x_n}$
(d) $\displaystyle\lim_{n\to\infty} \frac{\sin(4x_n)}{3x_n}$

An important technical issue needs to be addressed in using power series as we did above: can we multiply a power series by a function, or take its limit term by term?

The answer is yes, typically; some of the tools that we developed earlier are needed, like the Weierstrass test for uniform convergence (WT), the ratio test and the like.[29]

L'Hôpital's rule for the $\frac{0}{0}$ indeterminate form.

Let's ignore the technical issues here and explore the general problem. Suppose that there is a pair of functions $f(x)$ and $g(x)$ such that

$$f(a) = g(a) = 0 \qquad (9.83)$$

for a real number a. Then the fraction $f(a)/g(a)$ is undefined but as we saw above, the ratio $f(x)/g(x)$ of the two functions may nevertheless approach a specific real number as x approaches a.

---

[29]There are several other indeterminate forms that are written as ∞/∞, $0 \times \infty$, $\infty - \infty$, etc. We discuss the first of these later in this section and you can find some discussion of the rest in most standard calculus textbooks.

Series of Functions 355

The behaviors of the two functions are not summarized in their values at $x = a$; what the functions are doing when x is in the vicinity of a involves much more detail than that.

To be specific, suppose that both $f(x)$ and $g(x)$ have Taylor series expansions centered at $x = a$. Then the function values $f(a)$ and $g(a)$ are just the first terms in each series; there are infinitely many more terms that are defined by the higher derivatives of $f(x)$ and $g(x)$ at $x = a$. If 9.83 is true then the Taylor series look like:

$$f(x) = \frac{f'(a)}{1!}(x-a) + \frac{f''(a)}{2!}(x-a)^2 + \frac{f'''(a)}{3!}(x-a)^3 + \cdots$$
$$g(x) = \frac{g'(a)}{1!}(x-a) + \frac{g''(a)}{2!}(x-a)^2 + \frac{g'''(a)}{3!}(x-a)^3 + \cdots$$

If x_n is an arbitrary sequence that converges to a (but $x_n \neq a$ for all n) then using the above Taylor series and a little bit of algebra we find that:

$$\frac{f(x_n)}{g(x_n)} = \frac{f'(a)(x_n - a) + f''(a)(x_n - a)^2/2! + f'''(a)(x_n - a)^3/3! + \cdots}{g'(a)(x_n - a) + g''(a)(x_n - a)^2/2! + g'''(a)(x_n - a)^3/3! + \cdots}$$
$$= \frac{(x_n - a)[f'(a) + f''(a)(x_n - a)/2! + f'''(a)(x_n - a)^2/3! + \cdots]}{(x_n - a)[g'(a) + g''(a)(x_n - a)/2! + g'''(a)(x_n - a)^2/3! + \cdots]}$$
$$= \frac{f'(a) + f''(a)(x_n - a)/2! + f'''(a)(x_n - a)^2/3! + \cdots}{g'(a) + g''(a)(x_n - a)/2! + g'''(a)(x_n - a)^2/3! + \cdots}$$

Now, if $g'(a) \neq 0$ then as $x_n \to a$ with $n \to \infty$ all the terms containing $x_n - a$ drop out and we get the limit:

$$\lim_{n \to \infty} \frac{f(x_n)}{g(x_n)} = \lim_{n \to \infty} \frac{f'(a) + f''(a)(x_n - a)/2! + \cdots}{g'(a) + g''(a)(x_n - a)/2! + \cdots} = \frac{f'(a)}{g'(a)}$$

Note that the last fraction on the right does not involve division by 0 so it is a well-defined quantity. Now, since both $f(x)$ and $g(x)$ have Taylor series, they are infinitely differentiable. In particular, this implies that the derivatives $f'(x)$ and $g'(x)$ are continuous functions and we can write: $\lim_{n \to \infty} f'(x_n) = f'(a)$ and $\lim_{n \to \infty} g'(x_n) = g'(a)$. Therefore,

$$\lim_{n \to \infty} \frac{f(x_n)}{g(x_n)} = \lim_{n \to \infty} \frac{f'(x_n)}{g'(x_n)} \quad \text{if } g'(a) \neq 0$$

This equality has come to be known as l'Hôpital's rule since it first appeared in the 1696 book by the French mathematician Guillaume de l'Hôpital (1661-1704).[30] It is now the predominant tool for resolving indeterminate forms; its precise formulation is the following, for which it is helpful to review the concept of "function limits" that we discussed earlier:

---
[30]It is believed that the method was first introduced to l'Hôpital in 1694 by the Swiss mathematician Johann Bernoulli (1667-1748) who was hired as a tutor by l'Hôpital. Bernoulli signed a contract that gave l'Hôpital the right to use his discoveries without reference. This practice was not uncommon at the time.

L'Hôpital's rule transforms the limit of a 0/0 indeterminate form to the limit of the ratio of the derivatives: *Let $f(x)$ and $g(x)$ be differentiable functions with continuous derivatives $f'(x)$ and $g'(x)$ at $x = a$. If $f(a) = g(a) = 0$ then*

$$\lim_{x \to a} \frac{f(x)}{g(x)} = \lim_{x \to a} \frac{f'(x)}{g'(x)} \qquad (9.84)$$

provided that the limit on the right hand side exists.

For example, if $x_n \to 0$ as $n \to \infty$ then we quickly find that:

$$\lim_{n \to \infty} \frac{\sin x_n}{2x_n} = \lim_{n \to \infty} \frac{\cos x_n}{2} = \frac{\lim_{n \to \infty} \cos x_n}{2} = \frac{\cos 0}{2} = \frac{1}{2}$$

A similar calculation can be used for the limits in (9.79) and (9.81).

Exercise 150 *Use l'Hôpital's rule to find each of the following limits, assuming that x_n is any unspecified sequence that converges to 0 as $n \to \infty$:*

$$(a) \quad \lim_{n \to \infty} \frac{\sin(4x_n)}{\sin(3x_n)} \qquad (b) \quad \lim_{n \to \infty} \frac{\cos x_n - 1}{\sin x_n}$$

Notice that (9.84) is valid only if $g'(a) \neq 0$.
What if $g'(a) = 0$?
In that case, there are two possibilities: either $f'(a) = 0$ or $f'(a) \neq 0$. In the latter case, the ratio $f(x)/g(x)$ does not have a limit; we saw an example of this case in (9.78) where $g'(x) = 2x$ is zero at $x = 0$.

On the other hand, if $f'(a) = 0$ then the limit may still exist. In this case, the Taylor series for $f(x)$ and $g(x)$ both lose their second terms also and reduce to:

$$f(x) = \frac{f''(a)}{2!}(x-a)^2 + \frac{f'''(a)}{3!}(x-a)^3 + \cdots$$

$$g(x) = \frac{g''(a)}{2!}(x-a)^2 + \frac{g'''(a)}{3!}(x-a)^3 + \cdots$$

Notice that our earlier argument can be repeated: If $g''(c) \neq 0$ then

$$\lim_{x \to a} \frac{f(x)}{g(x)} = \lim_{x \to a} \frac{f''(a)/2!}{g''(a)/2!} = \lim_{x \to a} \frac{f''(a)}{g''(a)} \qquad (9.85)$$

Of course, if $g''(a) = 0$ then we have two cases for $f''(a)$ that we need to consider, so we proceed the way that we did earlier when discussing $g'(a)$.

What (9.85) shows is that when the derivative of the denominator of the fraction is also zero then we may *apply l'Hôpital's rule again, this time to the ratio of derivatives*, by proceeding to the second derivative. To illustrate this possibility, consider the following limit where $x_n \to 0$ as $n \to \infty$:

$$\lim_{n \to \infty} \frac{\cos x_n - 1}{x_n^2} \qquad (9.86)$$

Series of Functions

Because $\cos x_n \to 1$ as $x_n \to 0$ we have an indeterminate form 0/0. Applying l'Hôpital's rule gives:
$$\lim_{n\to\infty} \frac{\cos x_n - 1}{x_n^2} = \lim_{n\to\infty} \frac{-\sin x_n}{2x_n}$$

The last limit is again the indeterminate form 0/0 because $\sin 0 = 0$. So let's apply l'Hôpital's rule to this fraction:
$$\lim_{n\to\infty} \frac{-\sin x_n}{2x_n} = \lim_{n\to\infty} \frac{-\cos x_n}{2} = -\frac{\cos 0}{2} = -\frac{1}{2}$$

This is the same answer that we get by doing Exercise 149(b) using the Maclaurin series method!

Exercise 151 *Suppose that x_n is any sequence that converges to 0 as $n \to \infty$. Find the following limits by applying l'Hôpital's rule more than once:*

$$(a) \quad \lim_{n\to\infty} \frac{x_n - \sin x_n}{1 - \cos x_n} \qquad (b) \quad \lim_{n\to\infty} \frac{x_n - \sin x_n}{x_n^3}$$

Using power series to find the limit of a 0/0 form as we did above comes with the assumption that the functions $f(x)$ and $g(x)$ are infinitely differentiable. But the limits in (9.84) involve only the first derivatives of these functions so our argument above is based on assumptions that are too strong.

A proper justification of l'Hôpital's rule requires technical details that can be found in many introductory books on real analysis or advanced calculus. But to give you a sense of how the argument goes, here's the basic idea:

Let x_n be an arbitrary sequence that converges to a but $x_n \neq a$ for all n. Doing a little algebra and using the definition of the derivative (but ignoring, for the sake of this argument, the possibility that we may be dividing by 0 where $g(x_n) = g(a)$):

$$\lim_{x\to a} \frac{f(x)}{g(x)} = \lim_{n\to\infty} \frac{f(x_n) - f(a)}{g(x_n) - g(a)}$$
$$= \lim_{n\to\infty} \frac{[f(x_n) - f(a)]/(x_n - a)}{[g(x_n) - g(a)]/(x_n - a)}$$
$$= \frac{f'(a)}{g'(a)}$$

Since we also assumed that the derivatives $f'(x)$ and $g'(x)$ are continuous at $x = a$ and $g'(a) \neq 0$ the last quantity can be written as a limit again:

$$\frac{f'(a)}{g'(a)} = \lim_{n\to\infty} \frac{f'(x_n)}{g'(x_n)} = \lim_{x\to a} \frac{f'(x)}{g'(x)}$$

which gives us the right hand side of (9.84).

For further insights about l'Hôpital's rule that don't require technical material, consider graphing the functions next to the limit symbols in Exercise 149 and in (9.78)-(9.82) and examine how each curve behaves near the limit value $x = a$ in each case.

L'Hôpital's rule for the $\frac{\infty}{\infty}$ indeterminate form. There is a reciprocal and often useful version of l'Hôpital's rule that merits a mention here in part because it involves explicit occurrences of infinity.

The $\frac{\infty}{\infty}$ indeterminate form. Let x_n be a sequence that converges to a real number a (or to ∞ or $-\infty$) as $n \to \infty$. If $f(x)$ and $g(x)$ are functions such that each of $f(x_n)$ and $g(x_n)$ diverges to either ∞ or $-\infty$ as $n \to \infty$ then the fraction

$$\frac{f(x_n)}{g(x_n)}$$

produces an instance of the ∞/∞ *indeterminate form*. L'Hôpital's rule can be extended to this type of indeterminate form as follows:

$$\lim_{n \to \infty} \frac{f(x_n)}{g(x_n)} = \lim_{n \to \infty} \frac{f'(x_n)}{g'(x_n)} \qquad (9.87)$$

provided that all the derivatives and the limit on the right hand side exists.

Again the proof of this equality is too technical for our discussion but we can check that it is true using an example. Consider the following limit:

$$\lim_{n \to \infty} \frac{2n}{e^n} \qquad (9.88)$$

Note that both $2n$ and e^n diverge to infinity as $n \to \infty$. Let's identify $f(x) = 2x$ and $g(x) = e^x$. Since $f'(x) = 2$ and $g'(x) = e^x$ using (9.87) gives:

$$\lim_{n \to \infty} \frac{2n}{e^n} = \lim_{n \to \infty} \frac{2}{e^n} = 0$$

because $e^n \to \infty$ as $n \to \infty$. The following table shows what happens to the fraction $2n/e^n$ as n gets larger and larger:

n	1	5	10	15	20
$2n/e^n$	0.736	0.067	0.00091	0.0000092	0.000000082

These numbers confirm the validity of our conclusion that the limit is 0.

As with the earlier version of l'Hôpital's rule, this version can also be applied repeatedly. For example, let's consider the following limit:

$$\lim_{n \to \infty} \frac{n^2}{e^n}$$

which is of the ∞/∞ type. Applying (9.87) once gives:

$$\lim_{n \to \infty} \frac{n^2}{e^n} = \lim_{n \to \infty} \frac{2n}{e^n}$$

Series of Functions   359

The new limit is again an $\infty infty$ form and as we have just seen, applying l'Hôpital's rule again gives the answer 0.

Exercise 152 *Calculate each of the following limits:*

$$(a) \quad \lim_{n \to \infty} \frac{n^3 + n^2}{e^n + 2n} \qquad (b) \quad \lim_{n \to \infty} \frac{n^3 + n^2}{n^3 + 2n}$$

The sequence x_n is usually arbitrary in (9.87). For example, if x_n is any sequence of real numbers that diverges to ∞ then

$$\lim_{n \to \infty} \frac{2x_n}{e^{x_n}} = \lim_{n \to \infty} \frac{2}{e^{x_n}} = 0$$

As an illustration, consider the (function) limit

$$\lim_{x \to 0^+} \frac{\ln x}{x - \ln x} \tag{9.89}$$

Here we identify $f(x) = \ln x$ and $g(x) = x - \ln x$ and note that $\ln x \to -\infty$ as $x \to 0$ *from the right* (because $\ln x$ is not defined to the left of 0 the above limit requires the values of x to approach 0 from the right). Therefore, the numerator diverges to $-\infty$ and the denominator to ∞ resulting in an ∞/∞ form.

To show that the limit in (9.89) exists and find its value, we must show that for every sequence x_n that converges to 0 (and $x_n > 0$ for all n) the limit

$$\lim_{n \to \infty} \frac{\ln x_n}{x_n - \ln x_n} \tag{9.90}$$

exists *and has the same value* regardless of the choice of the sequence x_n. Applying (9.87) we get:

$$\lim_{n \to \infty} \frac{\ln x_n}{x_n - \ln x_n} = \lim_{n \to \infty} \frac{1/x_n}{1 - 1/x_n}$$

The last limit is again of the ∞/∞ variety so applying l'Hôpital's rule to it gives:

$$\lim_{n \to \infty} \frac{1/x_n}{1 - 1/x_n} = \lim_{n \to \infty} \frac{-1/x_n^2}{1/x_n^2} = \lim_{n \to \infty} (-1) = -1$$

Notice that we found this limit of -1 without specifying the sequence x_n and using only the fact that it is positive and converges to 0. So we conclude that the limit in (9.89) exists and its value is -1.

L'Hôpital's rule doesn't always work.

Let's consider the following limit:

$$\lim_{n \to \infty} \frac{\sqrt{n^2 + 1}}{n} \tag{9.91}$$

As n approaches ∞ so does $\sqrt{n^2 + 1}$ and we can use l'Hôpital's rule because the fraction is an ∞/∞ form.

If we apply l'Hôpital's rule then we get:

$$\lim_{n\to\infty} \frac{\sqrt{n^2+1}}{n} = \lim_{n\to\infty} \frac{(1/2)(n^2+1)^{-1/2}(2n)}{n} = \lim_{n\to\infty} \frac{n}{\sqrt{n^2+1}} \qquad (9.92)$$

The last fraction on the right is still an ∞/∞ form, so we may apply l'Hôpital's rule again to get:

$$\lim_{n\to\infty} \frac{n}{\sqrt{n^2+1}} = \lim_{n\to\infty} \frac{1}{(1/2)(n^2+1)^{-1/2}(2n)} = \lim_{n\to\infty} \frac{\sqrt{n^2+1}}{n} \qquad (9.93)$$

Back where we started! Clearly, even though we can keep applying l'Hôpital's rule, all we do is to go back and forth between the above two fractions.

Nevertheless, the limit in 9.91 *does exist* and we can find it by doing a little algebra. We put the n in the denominator up into the square root (remembering to square it inside) and simplify to obtain:

$$\lim_{n\to\infty} \frac{\sqrt{n^2+1}}{n} = \lim_{n\to\infty} \sqrt{\left(1 + \frac{1}{n^2}\right)} = \sqrt{1+0} = 1$$

Exercise 153 *Calculate each of the following limits (l'Hôpital's rule may or may not work)*

$$(a) \quad \lim_{n\to\infty} \frac{\sqrt{n^2+1}}{2n} \qquad (b) \quad \lim_{n\to\infty} \frac{\sqrt{n^2+1}}{n^2}$$

Finally, problems like the one above where l'Hôpital's rule applies but doesn't solve the problem must be distinguished from problems where the rule doesn't work because it is just not applicable to begin with. Recall that l'Hôpital's rule won't work unless it is applied to an indeterminate form.

For instance, if we try to calculate the limit

$$\lim_{x\to 0} \frac{x}{\cos x}$$

using l'Hôpital's rule then we get

$$\lim_{x\to 0} \frac{x}{\cos x} = \lim_{x\to 0} -\frac{1}{\sin x}$$

and the last fraction on the right diverges to $-\infty$ because $\sin x$ approaches 0 from the positive side as x does. However, this is not correct because the function $x/\cos x$ is actually continuous and its value at $x = 0$ is $0/1 = 0$. Therefore, the limit must also exist and be 0. In this case, the fraction $x/\cos x$ is not indeterminate so l'Hôpital's rule doesn't even apply.

9.8 Trigonometric series and Fourier expansions

We have seen that it is possible to express a function $f(x)$ as a power series by finding a Taylor series for $f(x)$. Recall that such a function $f(x)$ must have derivatives of *all* higher orders (that is, it must be infinitely differentiable) as required by the Taylor coefficients:

Series of Functions   361

$$c_n = \frac{f^{(n)}(a)}{n!}$$

But functions having a Taylor series are even more special: if $f(x)$ has a Taylor series then we completely know all of its values throughout the interval of convergence if we only have the list of numbers $f(a), f'(a), f''(a), \ldots, f^{(n)}(a), \ldots$ at a *single point* $x = a$!

Functions on intervals of real numbers are typically not *that* special; for example, a function like $|x|$ is continuous on the entire set $(-\infty, \infty)$ of real numbers but it does not have a derivative at $x = 0$ and there are no non-trivial Taylor series for it since $|x| = x$ if $x > 0$ and $|x| = -x$ if $x < 0$.

In applications of mathematics, we actually see even worse; electronic signals are often in the form of discontinuous, periodic functions like what we see in Figure 9.18.

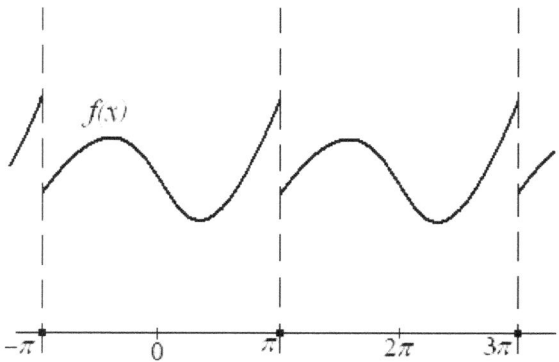

Figure 9.18: A discontinuous periodic function

These types of functions are not well represented by power series but because they are periodic we can represent them using infinite series of trigonometric functions, which are also periodic.

Trigonometric series.

A class of functions that has proven to be useful in applied mathematics, science and engineering is the class of *trigonometric series* where the functions involved are sine and cosine rather than power functions. Trigonometric series that are most commonly seen in practice are of the following variety:

$$\sum_{n=0}^{\infty} (a_n \cos(nx) + b_n \sin(nx)) = a_0 + a_1 \cos(x) + b_1 \sin(x) + \cdots \tag{9.94}$$

whose coefficients a_n and b_n are real numbers. Recall that $\cos 0 = 1$ and $\sin 0 = 0$ so when $n = 0$ we always get a_0 for the first term. For this reason, (9.94) is commonly written as:

$$a_0 + \sum_{n=1}^{\infty} (a_n \cos(nx) + b_n \sin(nx)) \tag{9.95}$$

The Weierstrass test for uniform convergence (WT) can be used to determine whether a trigonometric series such as (9.95) converges unifromly.

For instance, consider the trigonometric series

$$\sum_{n=0}^{\infty} \frac{1}{2^n} \cos(nx) = 1 + \frac{1}{2}\cos(x) + \frac{1}{4}\cos(2x) + \frac{1}{8}\cos(3x) + \cdots \qquad (9.96)$$

where $a_n = 1/2^n$ and $b_n = 0$ for $n = 0, 1, 2, \ldots$ Note that for every real number x in $(-\infty, \infty)$

$$\left|\frac{1}{2^n}\cos(nx)\right| = \frac{1}{2^n}|\cos(nx)| \leq \frac{1}{2^n}$$

and if we define the constants:

$$M_n = \frac{1}{2^n}$$

then the infinite series $\sum_{n=0}^{\infty} M_n$ converges since:

$$\sum_{n=0}^{\infty} \frac{1}{2^n} = \frac{1}{2^0} + \frac{1}{2^2} + \frac{1}{2^3} + \cdots$$

is a convergent geometric series. Therefore, the trigonometric series in (9.96) converges uniformly on the entire set of real numbers $(-\infty, \infty)$. Although we do not know what function $f(x)$ the series in (9.96) converges to, we know significant facts about it, like $f(x)$ is continuous and periodic with period 2π because $\cos(nx)$ has these properties for every n.

A similar argument works more generally: going back to (9.95) we notice that:

$$|a_n \cos(nx) + b_n \sin(nx)| \leq |a_n \cos(nx)| + |b_n \sin(nx)|$$

by the triangle inequality, and because $|\cos(nx)| \leq 1$ and $|\sin(nx)| \leq 1$ we may conclude that:

$$|a_n \cos(nx) + b_n \sin(nx)| \leq |a_r| + |b_n|$$

So with $M_n = |a_n| + |b_n|$ the (WT) gives the following result:

> If $\sum_{n=0}^{\infty} |a_n| + |b_n| < \infty$ then the trigonometric series (9.95) converges uniformly on $(-\infty, \infty)$ to a function $f(x)$ that is continuous and has period 2π.

Exercise 154 *Explain why the following trigonometric series converges uniformly:*

$$1 + \sum_{n=1}^{\infty} \left(\frac{1}{n^2}\cos(nx) - \frac{1}{2^n}\sin(nx)\right) = 1 + \cos(x) - \frac{1}{2}\sin(x) + \cdots$$

I mention in passing that trigonometric series are not limited to (9.95). There are other types of trigonometric series that converge; for example:

$$\sum_{n=0}^{\infty} \frac{1}{2^n} \cos^n x = 1 + \frac{1}{2}\cos x + \frac{1}{4}\cos^2 x + \frac{1}{8}\cos^3 x + \cdots$$

Series of Functions

is a geometric power series in disguise because we can write it as

$$\sum_{n=0}^{\infty} u^n, \quad u = \frac{1}{2^n}\cos^n x = \left(\frac{\cos x}{2}\right)^n$$

It converges uniformly on the entire set of real numbers $(-\infty, \infty)$ because $|u| = |(\cos x)/2| \leq 1/2$ for all real numbers x. We can even find its sum using the geometric series formula:

$$\sum_{n=0}^{\infty} \frac{1}{2^n}\cos^n x = \frac{1}{1 - \frac{\cos x}{2}} = \frac{2}{2 - \cos x}$$

This formula is valid for all x in $(-\infty, \infty)$ since $-1 \leq \cos x \leq 1$ for all real numbers x and therefore, the denominator on the right hand side is never 0.

Fourier series and coefficients.

We discovered earlier that it is possible to expand a given function as a Taylor series about a given center. Now, let's consider expanding a periodic function $f(x)$ in the form of a trigonometric series.

Periodic functions. We say that the function $f(x)$ has *period* p, where p is a positive real number if
$$f(x + p) = f(x)$$
for *all* real numbers x. Obvious examples are $\sin x$ and $\cos x$ each of which has period $p = 2\pi$.

The functions $\cos(nx)$ and $\sin(nx)$ in (9.95) also have period 2π for every value of the positive integer n but we can generalize (9.95) to a series with arbitrary period. A slight modification of (9.95) gives:

$$a_0 + \sum_{n=1}^{\infty} \left(a_n \cos \frac{n\pi x}{L} + b_n \sin \frac{n\pi x}{L}\right) \quad (9.97)$$

where L is a fixed positive real number. For each fixed value of n notice that

$$\cos \frac{n\pi(x + 2L)}{L} = \cos\left(\frac{n\pi x}{L} + \frac{2n\pi L}{L}\right) = \cos\left(\frac{n\pi x}{L} + 2n\pi\right) = \cos \frac{n\pi x}{L}$$

so $\cos(n\pi x/L)$ has period $2L$. Similarly, $\sin(n\pi x/L)$ has period $2L$ so the series in (9.97) has period $p = 2L$. In particular, if we set $L = \pi$ then this series reduces to (9.95).

Now suppose that a function $f(x)$ with period $p = 2L$ is given and we want to find the numbers a_n, b_n for which we can say that:

$$f(x) = a_0 + \sum_{n=1}^{\infty} \left(a_n \cos \frac{n\pi x}{L} + b_n \sin \frac{n\pi x}{L}\right), \quad -L \leq x \leq L \quad (9.98)$$

Notice that we need only consider the equality in (9.98) for x in the interval $[-L, L]$ since all other values of x are taken care of by periodicity. For instance, In Figure 9.18 we see that $f(2\pi) = f(0)$ where 0 in the interval $[-L, L]$ with $L = \pi$ even though 2π is not in $[-L, L]$.

To calculate the numbers a_n and b_n we take advantage of a property of the sine and cosine functions known as "orthogonality".[31] Let's multiply both sides of (9.98) by $\cos(m\pi x/L)$ where m is any unspecified positive integer to get

$$f(x)\cos\frac{m\pi x}{L} = a_0 \cos\frac{m\pi x}{L} + \sum_{n=1}^{\infty}\left(a_n \cos\frac{m\pi x}{L}\cos\frac{n\pi x}{L} + b_n \cos\frac{m\pi x}{L}\sin\frac{n\pi x}{L}\right)$$

Next, we integrate the above from $-L$ to L; the term by term integration of the series on the right hand side is possible if, for instance, the series in (9.98) converges uniformly to $f(x)$ on $[-L, L]$:

$$\int_{-L}^{L} f(x)\cos\frac{m\pi x}{L}dx = a_0 \int_{-L}^{L} \cos\frac{m\pi x}{L}dx + \qquad (9.99)$$

$$+ \sum_{n=1}^{\infty}\left(a_n \int_{-L}^{L}\cos\frac{m\pi x}{L}\cos\frac{n\pi x}{L}dx + b_n \int_{-L}^{L}\cos\frac{m\pi x}{L}\sin\frac{n\pi x}{L}dx\right)$$

The first integral on the right hand side is easy to evaluate by the substitution rule:

$$\int_{-L}^{L}\cos\frac{m\pi x}{L}dx = \frac{L}{m\pi}\sin\frac{m\pi x}{L}\Big|_{x=-L}^{x=L} = \frac{L}{m\pi}(\sin(m\pi)+\sin(m\pi)) = 0$$

The integrals inside the summation symbol can be evaluated using the *orthogonality formulas*:

$$\int_{-L}^{L}\cos\frac{m\pi x}{L}\cos\frac{n\pi x}{L}dx \int_{-L}^{L}\sin\frac{m\pi x}{L}\sin\frac{n\pi x}{L}dx = \begin{cases} L, & \text{if } m = n \\ 0, & \text{if } m \neq n \end{cases} \qquad (9.100)$$

and:

$$\int_{-L}^{L}\cos\frac{m\pi x}{L}\sin\frac{n\pi x}{L}dx = 0 \quad \text{for all } m, n \qquad (9.101)$$

Derivation of the orthogonality formulas. First consider (9.101). We use the trigonometric product formula (11.7) in Appendix 11.2:

$$\int_{-L}^{L}\cos\frac{m\pi x}{L}\sin\frac{n\pi x}{L}dx = \frac{1}{2}\int_{-L}^{L}\left[\sin\left(\frac{m\pi x}{L}+\frac{n\pi x}{L}\right)+\sin\left(\frac{m\pi x}{L}-\frac{n\pi x}{L}\right)\right]dx$$

$$= \frac{1}{2}\int_{-L}^{L}\sin\frac{(m+n)\pi x}{L}dx + \frac{1}{2}\int_{-L}^{L}\sin\frac{(m-n)\pi x}{L}dx$$

Next, we may calculate each of the integrals using the substitution method, or use the fact that $\sin(ax)$ is an odd function for every nonzero real number a so integrating it over a symmetric interval $[-L, L]$ gives a net value of zero:

$$\int_{-L}^{L}\sin\frac{(m\pm n)\pi x}{L}dx = 0$$

---

[31] Orthogonal is a technical term that is synonymous with perpendicular. Recall from calculus that two vectors are perpendicular if their "dot product" is zero. In metric spaces of functions, two functions are orthogonal if their "inner product" is zero. Integrals are used to define a type of inner product that is especially important in physics. The concept of inner product is discussed in most standard analysis textbooks that discuss metric spaces.

Series of Functions

This proves (9.101). The proof of (9.100) is similar but has one extra case. First, if $m \neq n$ then by (11.6)

$$\int_{-L}^{L} \cos\frac{m\pi x}{L} \cos\frac{n\pi x}{L} dx = \frac{1}{2}\int_{-L}^{L} \left[\cos\left(\frac{m\pi x}{L}+\frac{n\pi x}{L}\right) + \cos\left(\frac{m\pi x}{L}-\frac{n\pi x}{L}\right)\right] dx$$

$$= \frac{1}{2}\int_{-L}^{L} \cos\frac{(m+n)\pi x}{L} dx + \frac{1}{2}\int_{-L}^{L} \cos\frac{(m-n)\pi x}{L} dx$$

Here we use the substitution method: if we set $u = (m+n)\pi x/L$ then $du/dx = (m+n)\pi/L$ so $dx = [L/(m+n)\pi]du$ and further, $u = \pm(m+n)\pi$ if $x = \pm L$. So

$$\int_{-L}^{L} \cos\frac{(m+n)\pi x}{L} dx = \frac{L}{(m+n)\pi} \int_{-(m+n)\pi}^{(m+n)\pi} \cos u \, du$$

$$= \frac{L \sin u}{(m+n)\pi}\Big|_{u=-(m+n)\pi}^{u=(m+n)\pi} = 0$$

since $\sin(k\pi) = 0$ for every integer k. Finally, if $m = n \geq 1$ then

$$\int_{-L}^{L} \cos\frac{m\pi x}{L} \cos\frac{n\pi x}{L} dx = \int_{-L}^{L} \cos^2\frac{n\pi x}{L} dx$$

In this case, first we use a double-angle identity in Appendix 11.2 to get:

$$2\cos^2\frac{n\pi x}{L} - 1 = \cos\frac{2n\pi x}{L} \implies \cos^2\frac{n\pi x}{L} = \frac{1}{2}\left(1 + \cos\frac{2n\pi x}{L}\right)$$

Therefore,

$$\int_{-L}^{L} \cos^2\frac{n\pi x}{L} dx = \frac{1}{2}\int_{-L}^{L} \left(1 + \cos\frac{2n\pi x}{L}\right) du$$

Next, we substitute $u = 2n\pi x/L$ to get

$$\int_{-L}^{L} \cos^2\frac{n\pi x}{L} dx = \frac{L}{4n\pi} \int_{-2n\pi}^{2n\pi} [1 + \cos(u)] \, du$$

$$= \frac{L}{4n\pi} [u + \sin u]_{u=-2n\pi}^{u=2n\pi}$$

$$= \frac{L}{4n\pi} [2n\pi - (0) - (-2n\pi) - (0)]$$

$$= L$$

This completes the proof of the first equality in (9.100). The second is proved in a similar way.

Using these orthogonality results in (9.99) gives:

$$\int_{-L}^{L} f(x) \cos\frac{m\pi x}{L} dx = a_m L$$

Dividing by L and recalling that m was an arbitrary positive integer, this results in the following formula for any a_n when $n \neq 0$:

$$a_n = \frac{1}{L} \int_{-L}^{L} f(x) \cos \frac{n\pi x}{L} dx \qquad (9.102)$$

If $n = 0$ then $\cos(n\pi x/L) = 1$ and (9.99) with $m = 0$ we get:

$$\int_{-L}^{L} f(x) dx = a_0 \int_{-L}^{L} 1 dx = a_0(2L)$$

$$a_0 = \frac{1}{2L} \int_{-L}^{L} f(x) dx \qquad (9.103)$$

To find the formula for b_n we multiply by $\sin(m\pi x/L)$ and integrate; similar calculations to the above for the case $n \neq 0$ ultimately give the formula:

$$b_n = \frac{1}{L} \int_{-L}^{L} f(x) \sin \frac{n\pi x}{L} dx \qquad (9.104)$$

Formulas (9.102)-(9.104) were derived by Fourier (as well as by Euler earlier).

> **Fourier series expansion and coefficients.** The numbers a_n, b_n are determined by calculating the integrals in (9.102)-(9.104) and they are called the *Fourier coefficients*. With these numbers, the trigonometric series in (9.98) is called the *Fourier series expansion* of $f(x)$.

Let's find the Fourier expansion for the function $f(x) = |x|$ on the interval $[-1, 1]$ which with $L = 1$ looks like:

$$|x| = a_0 + \sum_{n=1}^{\infty} [a_n \cos(n\pi x) + b_n \sin(n\pi x)], \quad -1 \leq x \leq 1$$

First, using (9.103) we get the value of a_0:

$$a_0 = \frac{1}{2} \int_{-1}^{1} |x| dx = \frac{1}{2} \left[\int_{-1}^{0} (-x) dx + \int_{0}^{1} x dx \right] = \frac{1}{2} \left[-\frac{x^2}{2} \Big|_{-1}^{0} + \frac{x^2}{2} \Big|_{0}^{1} \right] = 1$$

Next, to calculate a_n where $n \neq 0$ we use (9.102):

$$a_n = \frac{1}{1} \int_{-1}^{1} |x| \cos(n\pi x) dx = \int_{-1}^{0} (-x) \cos(n\pi x) dx + \int_{0}^{1} x \cos(n\pi x) dx$$

Each of the integrals on the right hand side is found using integration by parts:

$$\int_{0}^{1} x \cos(n\pi x) dx = \frac{x}{n\pi} \sin(n\pi x) \Big|_{0}^{1} - \frac{1}{n\pi} \int_{0}^{1} \sin(n\pi x) dx$$

$$= 0 - \frac{1}{n\pi} \left[-\frac{1}{n\pi} \cos(n\pi x) \right]_{x=0}^{x=1}$$

$$= \left(\frac{1}{n\pi} \right)^2 [\cos(n\pi) - \cos 0]$$

Series of Functions

Recall that $\cos 0 = 1$ and $\cos \pi = -1$ so because $\cos x$ is periodic with period 2π we may conclude that:
$$\cos 2\pi = \cos(0 + 2\pi) = \cos 0 = 1, \quad \cos 3\pi = \cos(\pi + 2\pi) = -1$$
and so on, with signs alternating between positive when n is even and negative when n is odd. Therefore, $\cos(n\pi) = (-1)^n$ and we may write:
$$\int_0^1 x \cos(n\pi x) dx = \frac{(-1)^n - 1}{n^2 \pi^2}$$

A similar calculation gives the same value for the first integral $\int_{-1}^0 (-x) \cos(n\pi x) dx$ so ultimately,
$$a_n = \frac{2[(-1)^n - 1]}{n^2 \pi^2}$$

Next, using integration by parts again and proceeding as in the previous case, we find that:
$$b_n = \frac{1}{1} \int_{-1}^1 |x| \sin(n\pi x) dx = \int_{-1}^0 (-x) \sin(n\pi x) dx + \int_0^1 x \sin(n\pi x) dx = 0$$

Therefore, we have the Fourier series expansion for $-1 \leq x \leq 1$:
$$|x| = \frac{1}{2} + \sum_{n=1}^\infty \frac{2[(-1)^n - 1]}{n^2 \pi^2} \cos(n\pi x) = \frac{1}{2} - \frac{4}{\pi^2} \cos(\pi x) - \frac{4}{9\pi^2} \cos(3\pi x) - \cdots \quad (9.105)$$

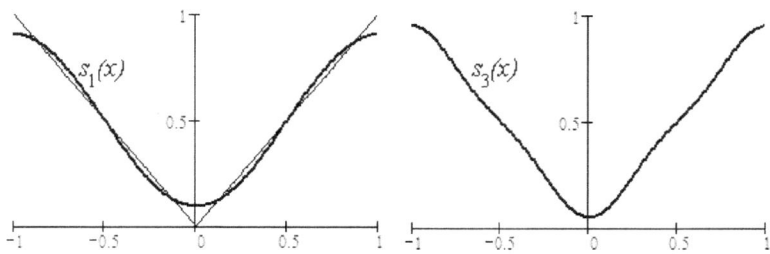

Figure 9.19: The first two partial sums of the Fourier series for $|x|$

Figure 9.19 illustrates the first two partial sums of the above series, $s_1(x)$ in the left hand side panel and and $s_3(x)$ in the other panel; we see that $s_3(x)$ looks more like $|x|$ than $s_1(x)$. Figure 9.20 shows the partial sum $s_{99}(x)$ which is visually indistinguishable from $|x|$ unless we zoom on some part of it, like near the origin as shown in the right hand side panel.

Exercise 155 *(a) Determine the Fourier series for $f(x) = |x|$ on the interval $[-2, 2]$ ($L = 2$).*
(b) Determine the Fourier series for $f(x) = x$ on the interval $[-1, 1]$ ($L = 1$).

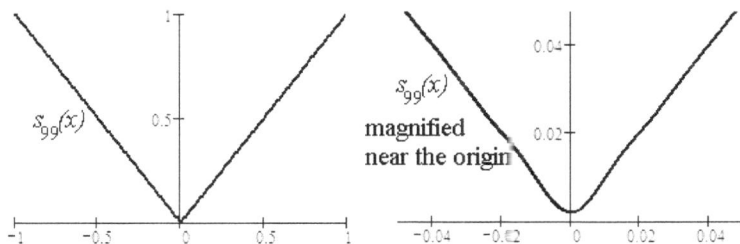

Figure 9.20: A higher order partial sum of the Fourier series for $|x|$

Deriving the series in (9.105) isn't just a good illustration for calculating Fourier series; it yields an unexpected bonus that I discuss next.

A nice little bonus: solving the Basel problem!

When discussing the infinite series of numbers, we discovered that the sum of the reciprocals of all positive integers, known as the harmonic series:

$$\sum_{n=1}^{\infty} \frac{1}{n} = 1 + \frac{1}{2} + \frac{1}{3} + \cdots$$

diverges. However, using the integral test we showed that the following series:

$$\sum_{n=1}^{\infty} \frac{1}{n^2} = \frac{1}{1^2} + \frac{1}{2^2} + \frac{1}{3^2} + \cdots \tag{9.106}$$

which is the sum of the reciprocals of the squares of all natural numbers, converges. But the integral test does not give the value of the number to which the series in (9.106) converges. Now we can use the Fourier series theory to find out what it converges to! Think of it as a nice, little application of our results. The Fourier theory is one of the most applicable branches of mathematics and if you have a science or engineering background, I probably don't need to convince you of that.

Finding the sum of the series in (9.106) was first posed as a problem by the Italian mathematician Pietro Mengoli in 1650. It was solved by Euler in 1734 who proved that the sum was $\pi^2/6$. The proof was ingenious although the rigorous proof of Euler's arguments was not completed until 100 years later when Weierstrass provided the last missing piece, namely, that Euler's representation of $\sin x$ as an infinite product was, in fact, valid (Euler's answer was known to be correct using partial sum approximations). In the mean time, Euler extended his ideas and the trend led later in 1859 to the famous zeta function of Riemann. Mengoli's problem was named after Basel, which was Euler's home town.

To find the value of the sum of the series in (9.106) we begin by setting $x = 1$ in (9.105) to get:

$$|1| = \frac{1}{2} - \frac{4}{\pi^2} \cos \pi - \frac{4}{9\pi^2} \cos(3\pi) - \frac{4}{25\pi^2} \cos(5\pi) - \cdots$$

Series of Functions

Since cosine of an odd multiple of π is -1 we end up with:

$$1 = \frac{1}{2} + \frac{4}{1^2\pi^2} + \frac{4}{3^2\pi^2} + \frac{4}{5^2\pi^2} + \cdots$$

Subtracting $1/2$ and multiplying by $\pi^2/4$ gives the equality:

$$\frac{\pi^2}{8} = \frac{1}{1^2} + \frac{1}{3^2} + \frac{1}{5^2} + \cdots \qquad (9.107)$$

This shows that *the sum of the reciprocals of the squares of all odd natural numbers is $\pi^2/8$*. This result is rather striking, no less so than the solution of the Basel problem itself.

To find the solution of the Basel problem using Fourier series we may look for a suitable function $f(x)$ to expand as a Fourier series and then use that series to get the answer in the same way that obtained (9.107).

Alternatively, we may use (9.107) and a little algebraic manipulation by way of a shortcut. Let:

$$S = 1 + \frac{1}{2^2} + \frac{1}{3^2} + \frac{1}{4^2} + \frac{1}{5^2} + \frac{1}{6^2} + \cdots$$

be the sum of the series in (9.106). Next, split the sum on the right hand side into two as follows:

$$S = \left(1 + \frac{1}{3^2} + \frac{1}{5^2} + \cdots\right) + \left(\frac{1}{2^2} + \frac{1}{4^2} + \frac{1}{6^2} + \cdots\right) \qquad (9.108)$$

This is legitimate because the series converges absolutely (recall Dirichlet's result that we discussed earlier). We found the value of the first series in (9.107); the second series can be written more suggestively as follows:

$$\frac{1}{2^2} + \frac{1}{4^2} + \frac{1}{6^2} + \cdots = \frac{1}{(2 \times 1)^2} + \frac{1}{(2 \times 2)^2} + \frac{1}{(2 \times 3)^2} + \cdots$$
$$= \frac{1}{2^2 \times 1} + \frac{1}{2^2 \times 2^2} + \frac{1}{2^2 \times 3^2} + \cdots$$
$$= \frac{1}{4}\left(1 + \frac{1}{2^2} + \frac{1}{3^2} + \cdots\right)$$

It follows that the value of the second series in (9.108) is $S/4$. Inserting what we have calculated so far in (9.108) gives:

$$S = \frac{\pi^2}{8} + \frac{S}{4}$$

We now find S from this equality by subtracting $S/4$ and simplifying to get:

$$\frac{3S}{4} = \frac{\pi^2}{8} \implies S = \frac{4}{3}\frac{\pi^2}{8} \implies S = \frac{\pi^2}{6}$$

This shows that

$$\frac{1}{1^2} + \frac{1}{2^2} + \frac{1}{3^2} + \cdots = \frac{\pi^2}{6}$$

and the Basel problem is solved!

Trigonometric polynomials and approximation of functions.

Calculating the coefficients of Fourier series *precisely* usually involves performing complicated integration. For example, it is quite simple to calculate the Taylor series for e^{-x^2} and we have discussed similar examples earlier, but not even a single nonzero coefficient of the Fourier series for e^{-x^2} can be precisely determined (see below). So what can we do in this case?

If you thought of *approximating* the coefficients then you are spot on! However, while we can approximate any finite number of the coefficients a_n and b_n we do have infinitely many of them in (9.98). The natural solution is to truncate the series and keep a finite number of terms, similarly to what we did for Taylor series.[32] This leaves us with a finite number of coefficients to estimate and leads to the following idea.

> **Trigonometric polynomials.** If we truncate the series in (9.98) to obtain a finite number of terms, say, m then the m-th partial sum $s_m(x)$ of the series is the Fourier *trigonometric polynomial of order m* for $f(x)$:
> $$f(x) \simeq s_m(x) = a_0 + a_1 \cos \frac{\pi x}{L} + b_1 \sin \frac{\pi x}{L} + \cdots - a_m \cos \frac{m\pi x}{L} + b_m \sin \frac{m\pi x}{L}$$

A "trigonometric polynomial" is a polynomial only by analogy. In practice we can usually only estimate the coefficients a_n, b_n so we just get *approximate* trigonometric polynomials.

To illustrate approximation by trigonometric polynomials, let's take a look at the (approximate) Fourier trigonometric polynomial of $f(x) = e^{-x^2}$ over the interval $[-2, 2]$. From (9.102) and (9.103) we get:

$$a_0 = \frac{1}{4} \int_{-2}^{2} e^{-x^2} dx, \quad a_n = \frac{1}{2} \int_{-2}^{2} e^{-x^2} \cos \frac{n\pi x}{2} dx$$

We may estimate the integrals using one of the methods discussed earlier, like the trapezoidal rule or the mid-point rule, or by integrating a Taylor polynomial for e^{-x^2}. The following table lists some of the coefficients a_n that I calculated electronically:

n	0	1	2	3	4
a_n	0.441	0.482	0.072	0.0055	-0.0014

We can use a similar procedure for the coefficients b_r using (9.104); however, looking closely at the integral in question, we deduce that it must be zero since the function inside it is an odd function.[33] Thus

$$b_n = \frac{1}{2} \int_{-2}^{2} e^{-x^2} \sin \frac{n\pi x}{2} dx = 0 \quad \text{for all } n \geq 1$$

Using the above estimates of coefficients, the fourth-order Fourier trigonometric polynomial $s_4(x)$ gives us the following approximation for e^{-x^2} over the interval $[-2, 2]$:

$$e^{-x^2} \simeq s_4(x) \simeq 0.441 + 0.482 \cos \frac{\pi x}{2} + 0.072 \cos \pi x + 0.0055 \cos \frac{3\pi x}{2} - 0.0014 \cos 2\pi x \quad (9.109)$$

---

[32] It is worth emphasizing that Fourier series involve *two different approximations*: one is the usual truncated infinite series and the other is the approximations of integrals that determine each coefficient.

[33] If $f(x)$ is an odd function then the integral over a symmetric interval $[-a, a]$ is zero because any part of $f(x)$ that is below (or above) the x-axis on the left half of $[-a, a]$ is matched by a part that is above (or below) the x-axis on the right half of $[-a, a]$. Thus "negative areas" over one half of $[-a, a]$ cancel areas over the other half.

Series of Functions

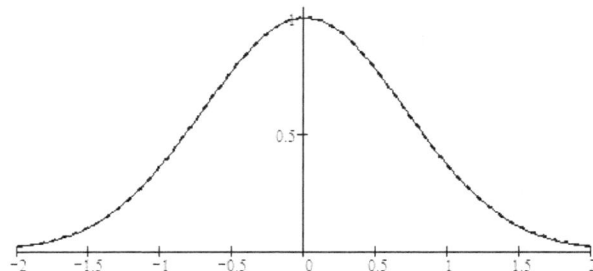

Figure 9.21: An exponential curve and its trigonometric series approximation

Figure 9.21 illustrates this approximation (solid curve) together with the graph of e^{-x^2} (dashed curve); we see that these curves are nearly identical and even zooming does not help tell them apart in this case. But there is a difference between the trigonometric polynomial and e^{-x^2} that we can exhibit clearly by graphing the difference between them as shown in Figure 9.22 that shows the difference $s_4(x) - e^{-x^2}$.

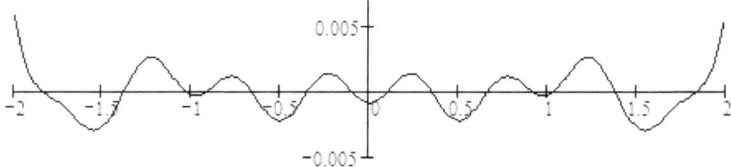

Figure 9.22: Caption for

Notice the wavy nature of (trigonometric) $s_4(x)$ that oscillates about the graph of e^{-x^2} in contrast to the partial sums of the (polynomial) Taylor series which do not wobble about the graph of e^{-x^2}.

An interesting difference between approximation using the Fourier trigonometric polynomials and approximation with the Taylor polynomials is noteworthy:

The non-centered convergence of trigonometric polynomials. Whereas a Taylor polynomial is exact at the center of its interval but tends to move away from the limit function, trigonometric polynomials have no center; they converge with the same general level of accuracy throughout the interval.

You may have noticed that the trigonometric series for $|x|$ and e^{-x^2} both involved only the cosine functions, with all the sine terms dropping out. There is a good reason for this! The cosine functions are all even since for all n

$$\cos(n\pi(-x)) = \cos(n\pi x)$$

However, sine functions are odd because

$$\sin(n\pi(-x)) = -\sin(n\pi x)$$

So if there is a nonzero sine term ($b_n \neq 0$ for some $n \geq 1$) then the function $f(x)$ that is represented by the series in (9.98) cannot be even. A similar observation verifies that an odd function cannot have a cosine term ($a_n \neq 0$ for some $n \geq 0$). These observations justify the following statement.

> **Fourier series for odd and even functions.** If $f(x)$ is an *even* function then its Fourier series (and all of its partial sums) contain only the cosine terms; such a series where $b_n = 0$ for all $n \geq 1$ is called a *Fourier cosine series*. If $f(x)$ is an *odd* function then its Fourier series (and all of its partial sums) contain only the sine terms; such a series where $a_n = 0$ for all $n \geq 0$ is called a *Fourier sine series*.

Not every function that is odd or even may look it. Consider $f(x) = \sin(x^2)$; the sine function may give the impression that this function is odd and periodic; but as we see in Figure 9.23 $f(x)$ is not periodic and it is actually even because:

$$f(-x) = \sin((-x)^2) = \sin(x^2) = f(x)$$

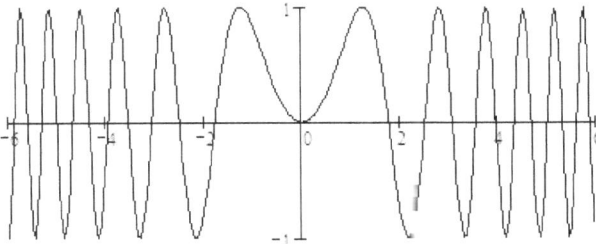

Figure 9.23: The non-periodic, even function $\sin(x^2)$

The squaring kills the negative sign and squeezes the oscillations of the basic sine function as the magnitude of x gets large. This function is continuous on $(-\infty, \infty)$ and its Fourier *cosine* series converges uniformly on every interval of type $[-L, L]$.

Discontinuous functions and the Gibbs phenomenon.

The examples of the previous section show that *finding the Fourier coefficients is done by integration rather than differentiation*. This means that unlike Taylor series, the functions that can be expanded as a Fourier series do not have to be differentiable–not even once!

In fact, *they don't even have to be continuous!*

To illustrate, consider a simple discontinuous function like the following step function:

$$f(x) = \begin{cases} 2, & \text{if } x \geq 0 \\ 0, & \text{if } x < 0 \end{cases} \qquad (9.110)$$

The Fourier coefficients in this case are easy to calculate; we need an interval of integration so

Series of Functions

consider the interval $[-1, 1]$. Then from (9.102)-(9.104) with $L = 1$ we obtain:

$$a_0 = \frac{1}{2}\int_{-1}^{1} f(x)dx = \frac{1}{2}\int_{0}^{1} 2dx = 1$$

$$a_n = \frac{1}{1}\int_{-1}^{1} f(x)\cos(n\pi x)dx = \int_{0}^{1} 2\cos(n\pi x)dx = \left.\frac{2}{n\pi}\sin(n\pi x)\right|_{x=0}^{x=1} = 0$$

$$b_n = \frac{1}{1}\int_{-1}^{1} f(x)\sin(n\pi x)dx = \int_{0}^{1} 2\sin(n\pi x)dx = \left.\frac{-2}{n\pi}\cos(n\pi x)\right|_{x=0}^{x=1} = \frac{2[1-(-1)^n]}{n\pi}$$

The integration bounds changed to 0 to 1 (from -1 to 1) because $f(x) = 0$ when $x < 0$. So we have the Fourier series expansion:[34]

$$f(x) = 1 + \sum_{n=1}^{\infty} \frac{2[1-(-1)^n]}{n\pi} \sin(n\pi x) = 1 + \frac{4}{\pi}\sin(\pi x) + \frac{4}{3\pi}\sin(3\pi x) + \cdots$$

for a function that is clearly discontinuous at $x = 0$. Figure 9.24 shows the partial sums $s_1(x)$ (dashed) and $s_{11}(x)$ (solid).

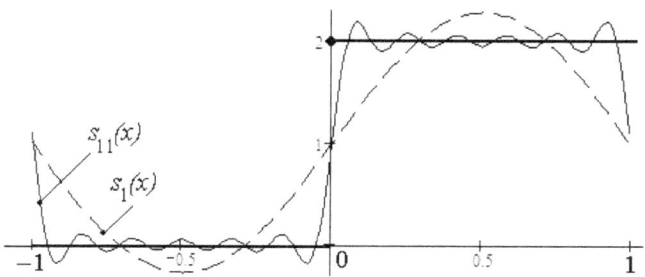

Figure 9.24: Two partial sums of the Fourier series for a function with jumps

Exercise 156 *Find the Fourier expansion of $f(x)$ in (9.110) over the interval $[-2, 2]$.*

In Figure 9.24 it is evident that the trigonometric polynomial $s_{11}(x)$ is a better estimate of $f(x)$ (displayed as thick lines). But even that is not a good estimate; we can improve upon our estimate by using larger partial sums. Figure 9.25 shows the graph of $s_{50}(x)$, which is really the same as $s_{49}(x)$; we see that the approximation has improved substantially.

But we also notice in Figure 9.25 that as we approach the point of discontinuity at $x = 0$ from either side, the graph of $s_{50}(x)$ deviates noticeably from the graph of $f(x)$ in the form of larger and more rapid oscillations. The same discrepancies exist at the two end-points. Using even larger partial sums does improve the approximation in the middle section but it does not remove the discrepancies near $x = 0, \pm 1$, which persist no matter how large a partial sum we use.

---

[34]Notice that this is *not* a sine series despite the absence of explicit cosine terms because the constant term 1 is a cosine term corresponding to $n = 0$. Also observe that $f(x)$ here is neither an odd function nor an even one.

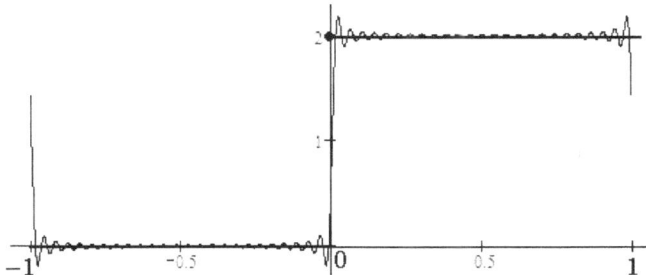

Figure 9.25: A higher order partial sum of the Fourier series for a function with jumps

The Gibbs phenomenon. *The persistent discrepancy is not limited to this particular function $f(x)$ and appears in the series for other discontinuous functions. It even has a name: the "Gibbs phenomenon", named after the American scientist Josiah Willard Gibbs who discovered it in 1899. The issue had been noted and explained in 1848 by the English mathematician Henry Wilbraham but Gibbs seems to have been unaware of Wilbraham's work.*

Evidently, something important is going on that shows up as discrepancies that we observe in Figure 9.25. If we attribute the discrepancy at $x = 0$ to the discontinuity somehow then why does the same thing seems to be happening at the end-points $x = \pm 1$ where there no discontinuities in $f(x)$?

A partial understanding comes from the nature of uniform convergence. If a trigonometric series converges uniformly then we know from earlier discussion that its limit function must be continuous. It follows that *if a trigonometric series converges to a (periodic) function that is not continuous then the convergence is not uniform.* This non-uniform convergence shows up as the Gibbs phenomenon, like the discrepancies that we see in Figure 9.25. We return to the issue of convergence after stating Fourier's convergence theorem below.

Trigonometric polynomials, or partial sums of the Fourier series for $f(x)$ above are all continuous but as more terms are added the sums approach a function that is not continuous. We have seen that this type of behavior may easily happen with pointwise-convergence: take a look at Figure 9.6 again.[35]

But what is happening at the end-points?

At these points we must consider the *periodicity*. Trigonometric polynomials (partial sums) are all periodic and even if they only converge pointwise, the limit function has to be periodic with the same period, say, $2L$. Here's why: suppose that every partial sum $s_n(x)$ has period $2L$ and that the partial sums converge pointwise to a function $\bar{f}(x)$. Then for each fixed, but arbitrarily chosen value of x,

$$\bar{f}(x + 2L) = \lim_{n \to \infty} s_n(x + 2L) = \lim_{n \to \infty} s_n(x) = \bar{f}(x)$$

---

[35]This raises some red flags; specifically, to find the coefficients of the Fourier series we had to integrate the infinite series term by term. This is justified if the convergence is uniform but not if it is pointwise. We will address this issue a little later.

Series of Functions   375

Since x was arbitrarily chosen, the above equality holds for every x and therefore, $\bar{f}(x)$ has period $2L$.

The function $\bar{f}(x)$ is not the same as $f(x)$ but they are equal on the interval $[-L, L]$. In fact:

> **The periodic extension.** $\bar{f}(x)$ is the Fourier series on the right hand side of (9.98) which is defined on $(-\infty, \infty)$. It is a periodic function that equals to the generally non-periodic function $f(x)$ only on $[-L, L]$, the interval for which the coefficients of the series were calculated. We call $\bar{f}(x)$ the *periodic extension* of $f(x)$ outside of the interval $[-L, L]$. By enlarging the interval $[-L, L]$ we can make the Fourier series agree with the given function $f(x)$ on a greater stretch of the x-axis.

In Figure 9.26 we see a part of the graph of $\bar{f}(x)$ if $f(x) = |x|$.

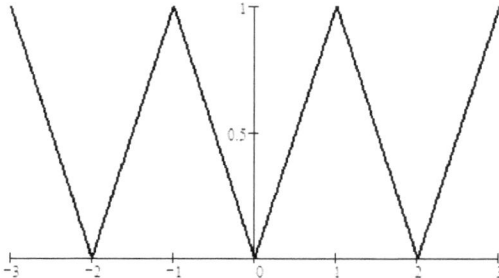

Figure 9.26: The Fourier series for $|x|$ converges to a periodic function

This figure is simply the graph of the function that is defined by the series in (9.105).[36] We may represent $|x|$ by a trigonometric series over a larger stretch of the x-axis by enlarging the interval of integration $[-L, L]$ (increasing L).

In Figure 9.27 we see the graph of the (approximate) trigonometric polynomial in (9.109) together with the graph of e^{-x^2} (thick curve) over a stretch of the x-axis.

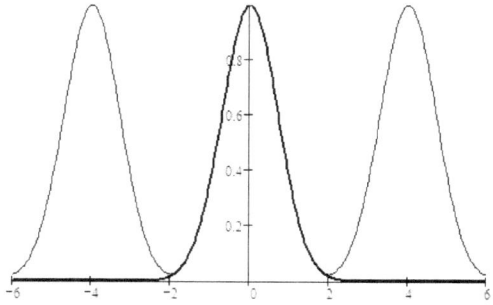

Figure 9.27: The Fourier series for e^{-x^2} converges to a periodic function

The essential thing to notice in Figures 9.26 and 9.27 is that $f(-L) = f(L)$. This means that $f(x)$ agrees with $\bar{f}(x)$ at the end-points $\pm L$.

---
[36]More accurately, it is the graph of a large partial sum of the series.

If $f(-L) \neq f(L)$ then $f(x)$ and $\bar{f}(x)$ disagree at the end points $x = \pm L$ because $\bar{f}(x)$ has period $2L$ so

$$\bar{f}(-L) = \bar{f}(-L + 2L) = \bar{f}(L) \tag{9.111}$$

Figure 9.28 shows what happens at the end-points: $f(x)$ disagrees with $\bar{f}(x)$ since $f(-1) = 0$ is not equal to $f(1) = 2$. Notice that the same type of behavior (Gibbs phenomenon) occurs at the function's point of discontinuity $x = 0$.

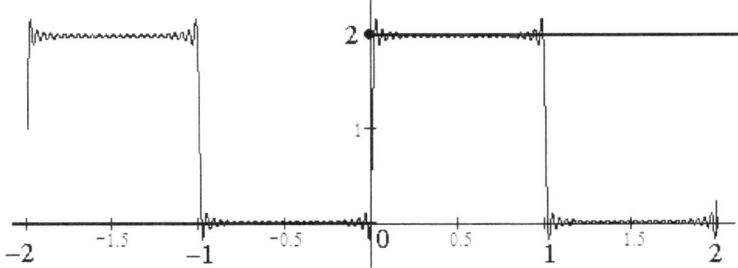

Figure 9.28: Convergence of a Fourier series in the presence of jumps

By extending the interval of integration $[-1, 1]$ as you might have done in Exercise 156 we obtain a series that agrees with $f(x)$ at ± 1 but not at whatever the new end-points are.

Now, we know that $\bar{f}(-1) = \bar{f}(1)$ in Figure 9.28 because $\bar{f}(x)$ that is shown is periodic with period 2. But *what is this common value?*

If we visually trace the continuous, oscillating curve as it falls from 2 to 0 when x passes 1 on the x-axis then we notice that its value at $x = 1$ appears to be 1. A similar conclusion holds at $x = -1$ and even at the point of discontinuity $x = 0$. So it seems that $\bar{f}(1) = 1$ and $\bar{f}(0) = 1$. So by periodicity, $\bar{f}(n) = 1$ for evey integer n.

It so happens that 1 is the *average* of the two function values 0 and 2 that are achieved at the end-points $x = -1$ and $x = 1$, respectively as well as at the point of discontinuity $x = 0$. This turns out to be no coincidence and is predicted by Fourier's theorem on the convergence of trigonometric series.

In order to explain this main result of the classical Fourier series, we need to define a few concepts.

> **Jump discontinuities.** A function $f(x)$ has a *jump discontinuity* at a point a if $f(x)$ is continuous on some interval to the left of a and on some interval to the right of a but there are sequences x_n^- and x_n^+ such that
>
> $$x_n^- < a \text{ for all } n \text{ and } \lim_{n \to \infty} x_n^- = a$$
> $$x_n^+ > a \text{ for all } n \text{ and } \lim_{n \to \infty} x_n^+ = a$$
>
> and both of the limits $\lim_{n \to \infty} f(x_n^-)$ and $\lim_{n \to \infty} f(x_n^+)$ exist (they are real numbers) but
>
> $$\lim_{n \to \infty} f(x_n^-) \neq \lim_{n \to \infty} f(x_n^+)$$

Series of Functions 377

We use a common shorthand notation for the left-hand limit and the right-hand limit:[37]

$$f(a^-) = \lim_{n\to\infty} f(x_n^-) \quad \text{and} \quad f(a^+) = \lim_{n\to\infty} f(x_n^+)$$

For instance, consider the function $f(x) = x$ for $-\pi \leq x \leq \pi$ whose periodic extension $\bar{f}(x)$ is shown in Figure 9.29.

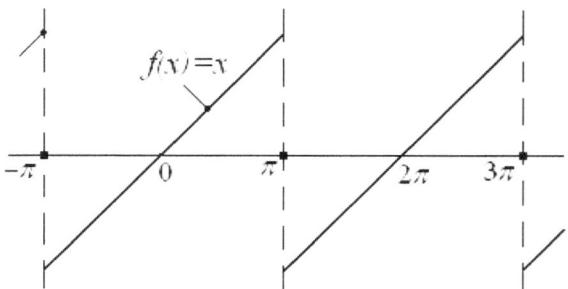

Figure 9.29: The periodic extension of $f(x) = x$

For this function we see that at $x = \pi$:

$$\bar{f}(\pi^-) = \pi, \quad \bar{f}(\pi^+) = -\pi$$

These numbers are fixed regardless of which sequences approach π from the left or the right. The same statement holds for all values of x that are odd multiples of π.

> **Piecewise smooth functions.** A function $f(x)$ is *piecewise smooth* on an interval $[a, b]$ if (a) $f(x)$ has at most a *finite number* (possibly zero) of points of discontinuity in $[a, b]$, (b) all discontinuities are jump discontinuities and (c) in between adjacent points of discontinuity (or a or b) the function $f(x)$ has a continuous derivative.

It is worth emphasizing that piecewise smooth functions are much better behaved than discontinuous functions generally; just remember Dirichlet's function that we discussed earlier.

Now we are ready to state the main result!

> **Fourier's convergence theorem.** If $f(x)$ is a piecewise smooth function on an interval $[-L, L]$ then the trigonometric series in (9.98) whose coefficients are given by (9.102)-(9.104) converges to the periodic extension $\bar{f}(x)$ wherever $\bar{f}(x)$ is continuous and to the average value $[\bar{f}(a^-) + \bar{f}(a^+)]/2$ at each point $x = a$ of (jump) discontinuity.

Note that the Fourier series for $f(x)$ converges to its periodic extension $\bar{f}(x)$ not just to $f(x)$ in the interval $[-L, L]$. If $f(x)$ is not periodic to begin with then its Fourier series, which is periodic, does not converge to it outside $[-L, L]$. We may represent $f(x)$ over arbitrary stretches of the x-axis by increasing the value of L (see Exercise 156) as long as the conditions of the convergence theorem are satisfied.

---
[37] As we discussed previously, the continuity of $f(x)$ before and after $x = a$ ensures that different choices of sequences x_n^- and x_n^+ do not alter the values of the limits of $f(x)$ from the left and from the right.

It is also worth stressing that if $f(-L) \neq f(L)$ then the periodic extension $\bar{f}(x)$ has points of discontinuity even when $f(x)$ is continuous in the interval $[-L, L]$; recall (9.111).

The above theorem is one form of the convergence theorem; there are others and they invariably involve technical conditions that ensure a particular type of convergence occurs. In the statement given above, the nature of convergence is left a little vague because of the various technical conditions.

Convergence how? *What do we mean by "convergence" when the periodic extension has discontinuities? If $f(x)$ is continuous on $[-L, L]$ but $f(-L) \neq f(L)$ or if $f(x)$ has any jump discontinuities inside $[-L, L]$, like the function in (9.110), then the periodic extension $\bar{f}(x)$ is discontinuous and cannot be a uniform limit of the continuous trigonometric partial sums.*

These observations suggest that the term by term integration of the trigonometric series has to be justified when $\bar{f}(x)$ has points of discontinuity; but this requires technical issues that we can't get into here. It is worth mentioning here that we cannot even assume that the convergence is pointwise everywhere; if $\bar{f}(x)$ has a jump discontinuity at $x = a$ then the trigonometric series converges to the average value $[\bar{f}(a^-) + \bar{f}(a^+)]/2$ which may be different from $\bar{f}(a)$. For instance, for the function in (9.110) note that:

$$\bar{f}(1) = f(1) = 2$$

but the trigonometric series converges to:

$$\frac{\bar{f}(1^-) + \bar{f}(1^+)}{2} = \frac{2 + 0}{2} = 1$$

Convergence is pointwise at points where there are no jumps in the value of $\bar{f}(x)$.

Historically, these technical problems caused some controversy regarding Fourier's results. A committee consisting of Lagrange, Laplace and others objected in a report on Fourier's memoir containing his original work that: "...the manner in which the author arrives at these equations is not exempt of difficulties and that his analysis to integrate them still leaves something to be desired on the score of generality and even rigor." It took a lot of significant work by mathematicians after Fourier, including his student Dirichlet, to clear up the various technical issues.

If there are no points of discontinuity for $\bar{f}(x)$, which means that $f(x)$ is continuous and satisfies $f(-L) = f(L)$ then we obtain the following much more clear (and restricted) version of the convergence theorem; this special case has already been proved by our earlier results on series of functions!

> **Convergence theorem: the continuous case.** *Let $f(x)$ be a piecewise smooth function on an interval $[-L, L]$. If $f(x)$ is continuous and $f(-L) = f(L)$ then the trigonometric series in (9.98) whose coefficients are given by (9.102)-(9.104) converges uniformly to the periodic extension $\bar{f}(x)$.*

The above continuous case is illustrated by Figures 9.26 and 9.27 and the functions pertaining to them; note that these functions were even functions. Notice further that the continuity of $f(x)$ in $[-L, L]$ does not imply that its derivative is continuous, or that it even exists throughout; a case in point is $|x|$ which is continuous on $[-1, 1]$ but its derivative does not exist at $x = 0$.

Also noteworthy is the fact that we used the Fourier series for $|x|$ to solve the Basel problem above; this function is continuous and its Fourier expansion is justified by the above *continuous* case.

9.9 Continuous yet nowhere differentiable: Koch's snowflake and Weierstrass's function

If you are asked to imagine a continuous curve you are likely to first think of something like a circle, or a wavy curve or perhaps a straight line. These curves are certainly continuous; but they are also much more: they are smooth curves meaning that they have no corners or sharp points where unique tangent lines (or derivatives) fail to exist.

It is not hard to imagine a continuous curve that fails to have derivative (or be non-differentiable) at some points; we may recall the graph of the absolute value function $|x|$ which is a v-shaped curve with its tip at the origin. By adjoining several such v-shaped curves we can create a sawtooth shaped curve that has a large number of points where derivative does not exist; there may be even an infinite number of such points at, say, every integer. We encountered functions whose graphs are sawtooth shaped in our study of Fourier series.

But can you imagine a continuous curve that fails to have a derivative at *any* point?

If you close your eyes and give this a try then you will realize that it is hard to do. In fact, it will be impossible to do because the only known examples of such curves are obtained as limits of an infinite sequence of simpler curves.

On the other hand, when we go back to our definition of continuity and derivatives in Chapter 7 then we notice that there is no reason why a continuous curve has to have derivatives at even a single point!

You may wonder why this fact is worthy of consideration; the reason has to do with the answer to the question that I asked above. We cannot imagine a curve that is continuous and yet fails to have a derivative at any point because such a curve can only exist at the end of an infinite process; the sort of thing that we like to talk about in this paper!

When analysis was not yet mature and understanding calculus relied on visual intuition it was hard to accept the existence of what are now called *nowhere differentiable functions*. For instance, in a 1839 calculus text by J.L. Raabe it was taken for granted that every continuous function can have at most a *finite* number of points in any bounded interval where it is *not* differentiable.

It is actually not hard to come up with a continuous function that has no derivatives at an infinite number of points in any bounded interval, no matter how tiny. For example, take the interval [0,1] and the sequence of numbers $1, 1/2, 1/3, \ldots$ For each positive integer n set up an isoceles triangle having the segment between $1/n$ and $1/(n+1)$ on the x-axis as its base and a height of 1 that is reached at the midpoint of the interval $[1/(n+1), 1/n]$. Now define a function $f(x)$ so that its value at any x is the vertical distance from the x-axis to the side of the triangle that sits above x (to ensure continuity, we define $f(0) = 0$). The graph of $f(x)$ is just the top sides of all the triangles; notice that this graph is a continuous (polygonal) curve that has an infinite number of sharp corners between 0 and 1. However, this curve also contains an infinite number of line segments, and each of these segments is smooth. In a sense, our function $f(x)$ is non-differentiable at infinitely many points of [0,1] but more importantly, it is smooth at most points of this interval.

It was not until 1872 that Weierstrass constructed a function that was continuous on all of $(-\infty, \infty)$ but didn't have a derivative at *any point*. This function was defined as the limit of a trigonometric series which by then were sufficiently well-understood to serve as tools in analysis.

Koch's snowflake.

We discuss Weierstrass's function shortly, but before that let's discuss a more intuitive construction, due to Helge von Koch (1870-1924). In 1904 von Koch constructed his curve, nowadays called *Koch's snowflake*. The title of his paper, "On a continuous curve without tangents constructible from elementary geometry" describes what was significant to him in this paper. Koch's construction might have been motivated in part by the fact that Weierstrass's function was difficult to describe without some analysis background.

The construction is quite simple. We start with an equilateral triangle. The first step is to remove the middle third of each side and insert two segments, each equal to 1/3 of the side of the original triangle, to form equilateral triangles on each of the three sides of the starting triangle; see Figure 9.30.

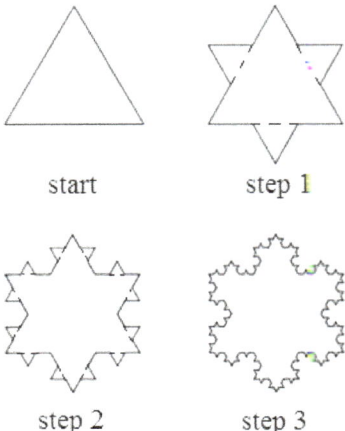

Figure 9.30: The first few steps in the construction of Koch's snowflake

In the second step of the construction, we repeat the process of removing the middle third of each of the new straight line segments and adding new segments of equal length to get a more spiky curve. The construction continues by repeating the above process at each stage, thus creating an increasingly spiky curve in each step.

The end result of this recursive process is impossible to visualize but it is easy to see why the limit of the recursion is a continuous curve: at no step in the recursion do we get a (spiky) curve that has a hole or a cut in it. Evidently, there can be no tangent lines to the limiting curve: a tangent line would have to coincide with a stretch in the form of a line segment, but there are no segments left at the end!

Koch's snowflake, that we have previously referred to in our discussion of the staircase curves, is considered a fractal shape. Let's establish two interesting facts about the snowflake. First, it has a finite area simply because the region inside the curve is contained within some disk of finite radius.[38] On the other hand, the perimeter of the snowflake is infinite. Let's calculate the length of the perimeter of the snowflake to see how it turns out to be infinite.[39]

---

[38]The exact value of the area is the limit of an infinite series with sum equal to 8/5 times the area of the starting triangle.

[39]This type of anomaly, an infinitely long curve enclosing a finite area, is not limited to fractals, as we saw earlier when discussing staircase curves. Also, you may find it fun to check that the graph of the function $f(x)$ that we

Series of Functions

For convenience suppose that each side of the starting triangle is 1 unit long. In Step 1, each of the two new segments that we added has length 1/3; this is easy to see in Figure 9.31 which shows the evolution of the top side of of the triangle.

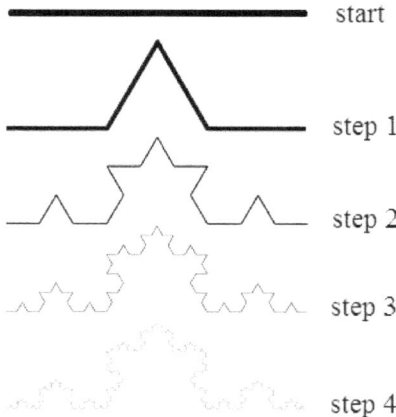

Figure 9.31: The evolution of one side of Koch's snowflake

Each side of the shape in Step 1 therefore has length

$$s_1 = \frac{1}{3} + 2\frac{1}{3} + \frac{1}{3} = \frac{4}{3}$$

In Step 2 we repeat the process so the length of a side increases to:

$$s_2 = \frac{4}{3}s_1 = \left(\frac{4}{3}\right)^2$$

See Figure 9.31. This pattern repeats so the total length of each side of the snowflake is ultimately given by the limit:

$$\lim_{n \to \infty} s_n = \lim_{n \to \infty} \left(\frac{4}{3}\right)^n = \infty$$

because $4/3 > 1$. Now, since the length of each side of the snowflake is infinitely large it follows that the perimeter, being 3 times as large, must also be infinite.

Weierstrass's nowhere differentiable function.

The snowflake is a popular and simple construction of a curve that is not differentiable at any point. Two points are worth emphasizing about this curve:

(a) The snowflake (the limiting curve) is continuous but it is not the graph of a continuous *function* on the x-axis.

---

defined earlier over the interval [0,1] has infinite length but the area between it and the x-axis is finite. Note that this graph is not a fractal.

(b) The continuous curves that are generated at each step of the process contain points (the corners) at which the derivative does not exist. In fact, it is these points that grow in number to yield a nowhere differentiable curve.

The above two features of the snowflake and its construction process are not problematic by any means, but it is an interesting fact that neither one has to be true!

A continuous, yet nowhere differentiable curve that is not only the graph of a function, but is also the uniform limit of a sequence of infinitely differentiable functions was constructed by Weierstrass using trigonometric series 32 years before Koch introduced his snowflake. As we discover shortly, the graph of this function looks rather like the snowflake in being self-similar and fractal-like.

Consider the trigonometric series:

$$\sum_{n=0}^{\infty} a^n \cos(k^n \pi x) \quad \text{where } 0 < a < 1 \text{ and } k \text{ is a positive integer} \qquad (9.112)$$

Because
$$|a^n \cos(k^n \pi x)| = a^n |\cos(k^n \pi x)| \leq a^n$$

and $\sum_{n=0}^{\infty} a^n = 1/(1-a)$ is a convergent geometric series, Weierstrass test for uniform convergence (WT) implies that the series in (9.112) converges uniformly on all of $(-\infty, \infty)$ to some function, say, $W(x)$ which is continuous on $(-\infty, \infty)$. But Weierstrass showed that if we choose k to be an odd integer large enough that:

$$k > \frac{3\pi + 2}{2a} \qquad (9.113)$$

then $W(x)$ fails to have a derivative at any real number x.[40] Notice that since $a < 1$ we have

$$\frac{3\pi + 2}{2a} > \frac{3\pi + 2}{2} \simeq 5.71$$

so the *least* odd integer for which Weierstrass's inequality in (9.113) is valid is $k = 7$. Weierstrass's inequality, which can also be written as:

$$ka > 1 + \frac{3\pi}{2}$$

is a sufficient condition for the non-existence of derivatives, but it is not necessary. English mathematician G.H. Hardy (1877-1947) improved on it in 1916 by showing that Weierstrass's series is nowhere differentiable as long as

$$ka \geq 1 \qquad (9.114)$$

where k doesn't have to be an integer but can be any (positive) *real* number. Aside from being more general and looking more natural than 9.113, Hardy's inequality 9.114 makes it much easier to conduct a graphical exploration.

Let's explore Weierstrass's series graphically for specific values of a and k; let $a = 0.6$ and $k = 2$ to satisfy Hardy's inequality (9.114). We then get the function:

$$W(x) = \sum_{n=0}^{\infty} 0.6^n \cos(2^n \pi x) \qquad (9.115)$$

Series of Functions 383

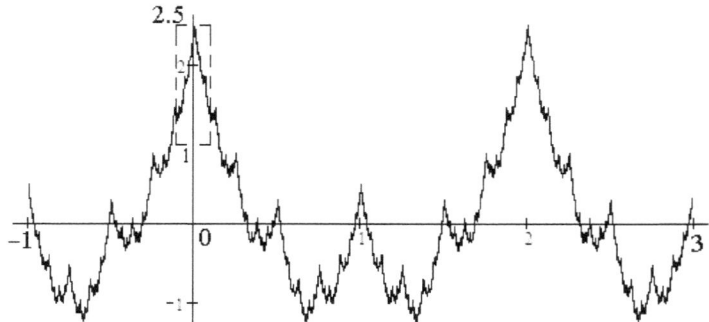

Figure 9.32: A partial sum of the Hardy-Weierstrass series for $W(x)$

Graphing this function is impossible so let's take a look at some of the partial sums of the series. Figure 9.32 shows the partial sum

$$s_{12}(x) = 0.6^0 \cos(2^0 \pi x) + 0.6^1 \cos(2^1 \pi x) + \cdots + 0.6^{12} \cos(2^{12} \pi x)$$
$$= \cos(\pi x) + 0.6 \cos(2\pi x) + 0.6^2 \cos(2^2 \pi x) + \cdots + 0.6^{12} \cos(2^{12} \pi x)$$

over the interval $[-1, 3]$. Even with these few terms of the infinite series in (9.115) we see a spiky, non-smooth graph. In Figure 9.33 we see three of the 13 individual terms of this partial sum.

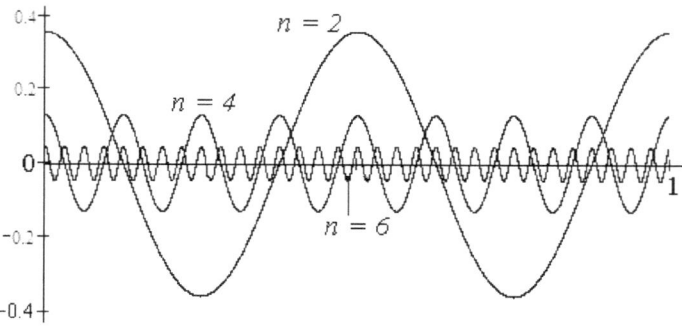

Figure 9.33: Three terms of the Hardy-Weierstrass series

To get a better sense of what is going on in Figure 9.33, we zoom in on a portion of it near $x = 0$; the result is shown in Figure 9.34, which represents a 20 fold magnification (plus a 10 fold increase in resolution to improve the image).

We still see a spiky curve at this magnification; if this was indeed the graph of $W(x)$ (the entire infinite series) then further magnifications would continue to show a spiky curve (like Koch's snowflake), essentially resembling the portion that we magnified. This self-similarity would be

---

[40] I will not discuss the proof of this statement; though not difficult, it requires using abstract analytical arguments that we need not get into here.

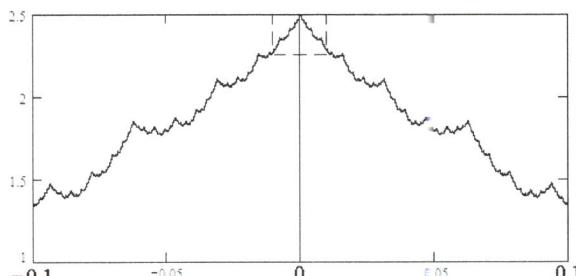

Figure 9.34: Zooming on the selected portion in Figure 9.32

preserved with further zooming in on the graph of $W(x)$. Figure 9.35 shows a 10 fold magnification of the portion near the tip in Figure 9.34.

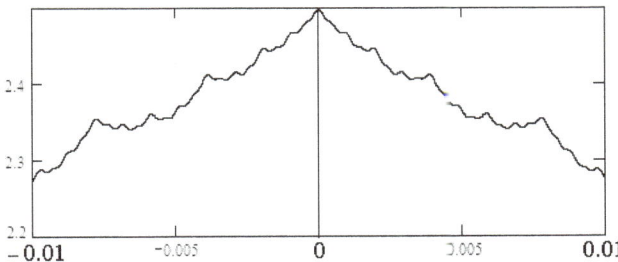

Figure 9.35: Zooming on the selected portion in Figure 9.34

We see the self-similarity feature at this magnification also, although you may notice that the curve in Figure 9.35 is not as spiky as in the previous figures. This is not surprising since we are not illustrating $W(x)$ in these figures but its partial sum $s_{12}(x)$ which is, of course, quite smooth (infinitely differentiable).

In fact, all finite partial sums of $W(x)$ are infinitely differentiable functions since each is a finite sum of cosine functions. The (uniform!) limit of this sequence of very smooth partial sums, however, is a function that is continuous and yet not differentiable anywhere!

The sudden change from infinitely differentiable everywhere to nowhere differentiable in this case shows that *uniform convergence can fail in the most spectacular way possible in preserving differentiability*, even though it does preserve continuity. We discuss the failure of uniform convergence with regard to the preservation of length in the context of a function space in Appendix 11.5.

Chapter 10

Infinity as the Link between Human Intuition and Reality

In science there are two pillars on which the human understanding of nature rests: theory and experiment. The basic idea is that once verified by experimentation, a theory that models and explains some aspect of nature constitutes our understanding of it. I would like to focus here on what is meant by *theory*.

10.1 Missing links in the human understanding of nature

I concetrate on physics where a theory is typically a mathematical model that falls within the domain of analysis, simply because physics is concerned with continuous space, time, motion as well as radiation and matter.[1] Consider the flow diagram below:

Nature $\to$ Theory (differential equations and analysis) $\to$ Numerical calculations

Numerical calculations are the end result of quantifying natural phenomena. Since they are typically performed with the help of digital devices, all numbers involved are rational (so that they can be stored in the device's finite memory). Further, there are finitely stages of calculation, such as iterations, and we stop at some point where a desired accuracy is achieved. Therefore, this final stage of our understanding has finite resolution.

On the other hand, infinity is a key component of analysis and its constructs: spaces of finite (and infinite) dimensions, differential equations and operators on spaces of functions, etc. These concepts decidedly have infinite resolution.

We might argue that the models come from the study of nature and the calculations are simply how we extract results from the models. But this argument is overly simplistic: infinity is neither a natural entity nor a practical idea. As far as we know infinity does not exist in nature (as infinite numbers of things or as infinitesimals) and we certainly do not use infinity in numerical calculations

---

[1] Other areas of mathematics, such as abstract algebra also play fundamental roles in physics but usually within the context of continuous space and time where analysis and differential equations are at play.

with digital devices. Whether there is such a thing as a continuous space (filled with "fields" of various kind) is a profound subject but we need not consider that here.

So there are missing links here: it seems that we should not need to use infinity either to study nature or carry out numerical calculations. The diagram in Figure 10.1 highlights the issue:

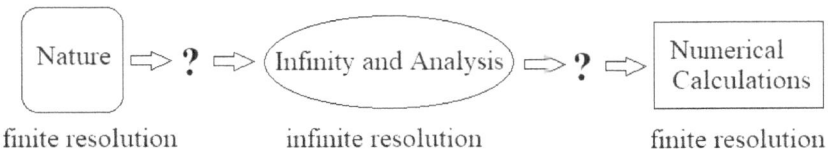

Figure 10.1: Analysis linking nature to human understanding

The question marks point to information that is missing in the figure.

Analysis requires continuous or smooth inputs (functions operations) to process and its output is typically also continuous or smooth. As we saw earlier in this book, analysis defines continuous concepts in terms of limit operations, hence infinity. But neither nature (outside of physics models) nor numerical calculations fundamentally involve continuity or infinity. We need to explain how analysis's output gets broken into discrete numerical calculations and how analysis and infinity get involved in the first place.

10.2 The second missing link and numerical algorithms

The second question mark in Figure 10.1 is where *discrete algorithms* come in.

These algorithms break down the continuous output of mathematical analysis into a finite (possibly very large) sequence of points that serve as the solution to some problem. For example, if the output is a differential equation then discrete algorithms transform such an equation into a recursion that can be solved by iteration. The resulting finite data (often sets of rational numbers) is either sufficient as is, or it verifies a presumed continuous or smooth solution.

In some cases, the analytical methods provide an "exact solution" to the problem (say, a differential equation). But even in such a case we do not use the full infinite resolution in practice. For example, if the solution is $\sin x$ then we only compute and retain approximations to the irrational values of this function for practical needs. This rounding off of irrational values is also a part of the numerical calculations.

10.3 The first missing link: human intuition

Now, let's consider the first question mark. There are two different issues involved here. First, I consider the technical one, which is more precisely understood.

Scientific observations and measurements usually involve equipment and experiments that generate discrete (finite) sets of data. In many cases, the discrete data are *interpolated* (extended) via a number of standard procedures that involve both statistical and analytical methods (like polynomial fits to data points). This interpolation is part of the first missing link in Figure 10.1 that gives infinite input into the core analytical part in the middle.

The interpolation step is not always necessary; it is required only when data has to be in the form of an infinite set or a continuous function. A more fundamental reason why infinity has to be brought into the picture has nothing to do with mathematics: the final missing link is *human intuition*.

Over the course of the last century we discovered that material objects are not really smooth (neither in shape nor in texture) because they are made of atoms and molecules. The sounds that our ears pick up are complex oscillation patterns in the surrounding air, which is made up of granular molecules. We do not see or feel the constituent atoms and molecules of matter because our eyes, skins, etc are not capable of resolving very small objects like atoms, molecules, etc.

Smooth shape or texture and continuous melody are the brain's interpretation of the jumble of signals from the environment that comes to it through the senses and the nerve system. The brain itself is a physical object, hence not infinite. It can process only a certain amount of information and must do it fast enough to be useful in preserving and protecting the body, including itself. In addition to all this, the brain routinely processes the body's own autonomic functions (regulating heart beat, breathing, etc).

If we have only a finite amount of memory and processing ability in our heads and must make decisions, big or small, in a limited amount of time then it makes sense for the brain to fudge things a bit. Why bother dealing with large, complex molecular structures or track, store and process an enormous number of oscillating air particles when there is little reaction time and insufficient memory available to store and process signals?

Evolution has taught the brain to fade out the unnecessary detail and come up with a useful illusion or cartoon of the surrounding reality. The outcome may not be scientifically accurate but it is suitably tailored to the limitations of our bodies (senses, muscles, etc) and of the usually limited reaction time available to the brain. This illusion is what I think of as *consciousness*.

When we watch a well-made movie with images, special effects and sound track, it may seem like real life but it is actually a large collection of digital images plus sound effects playing out in rapid succession in our eyes and ears. The brain's natural tendency to fade out details makes for an enjoyable period of make-believe indulgence in this case.

In analogous fashion, consciousness plays out a "movie" of our body and the surrounding environment that is different from reality, though usually not disconnected from it.[2]

Our *intuition* is a by-product of our consciousness. It is a way of compressing information (both storage and processing) to arrive at conclusions even more quickly. It develops and improves as we get older and better in performing tasks.

Intuition is extremely powerful in directing our thoughts and influencing our evaluation of the environment and the universe at large. In its absence we find ourselves at great disadvantage. For instance, if we try to imagine a three dimensional sphere (the surface of a four dimensional "ball") we run into an insurmountable obstacle without help from other sources at our disposal. In particular such an object is easy to describe in the abstract; here is the equation for it:

$$x^2 + y^2 + z^2 + w^2 = r^2$$

where r is the radius of the ball. We can also imagine the cross sections of this object with our three dimensional space (ordinary balls). But visualizing this object in its entirety is practically impossible because we have no brain-generated illusion (intuition) of a four dimensional space.

---

[2]There are times when the brain's movie is disconnected from reality; for instance, when we dream or hallucinate.

Calculus and much of analysis were developed before the 20th century when people relied largely on their intuition of nature to develop quantitative theories of material objects and natural phenomena.

They had to pay a price: base almost everything on infinity. Using infinity, analysis developed highly successful methods for quantifying our intuitive experiences using infinite series, derivatives and integrals that we discussed earlier.

10.4 Connecting the components

The take-away from the discussion in the last two sections is that *analysis quantifies our intuition of nature, not of nature itself*. Far from bypassing the intuition, we chose to stick to it by accepting infinity and all that comes with it; such is the power and influence of intuition! Even now, with the great experimental discoveries of the 20th century well behind us, it is quite hard for us to think of nature in terms that run counter to intuition.

The mathematical theories that we use to quantify motion and to model nature are also influenced by human intuition. They are overwhelmingly based on differential equations and smooth functions (having derivatives). That includes the theories of special and general relativity and the conceptually counter-intuitive quantum theory and its offsprings, the quantum field theories of the Standard Model of particle physics.

I now re-draw Figure 10.1 as you see in Figure 10.2.

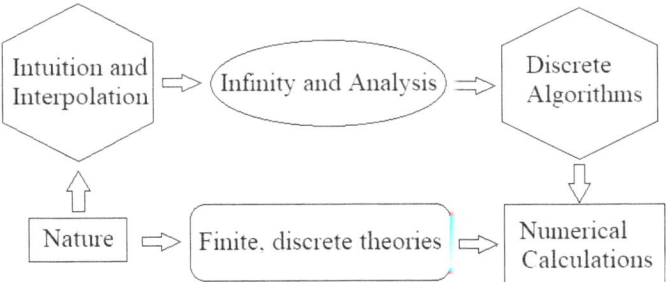

Figure 10.2: Quantitative understanding with and without infinity

At the bottom I have drawn a path from nature to numerical calculation that does not rely on intuition but instead uses discrete (as yet undeveloped) mathematical theories that do not involve infinity.

Is such a shortcut possible? I don't know, but if possible then it is undoubtedly difficult for us to imagine how to get there at this time.

Is it desirable?

I think so! After all, infinity itself is *not* a product of intuition; it is the cost of relying on it. We need it to quantify the intuitive world-view that includes continuity, smoothness, etc.

Maybe in the future our science and mathematics–and technology relating to digital computing and artificial intelligence–will have sufficiently advanced to allow us to bypass intuition as illustrated in Figure 10.2. This does not mean that we forget about human intuition in science; rather, we need to invent new mathematical methods to deal with the scientific challenges of today that seem to defy our intuition. Newton, Leibniz and others before and after them rose up to the challenges of their time by inventing calculus. The time has come once again to do something similar as outlined above.

Chapter 11

Appendices

11.1 Appendix: Archimedes's area argument and a modern derivation

The argument that Archimedes gave for deriving the formula for the area of a circle[1] was "the method of exhaustion" or "proof by exhaustion". This idea dates back to ancient times and was used through the middle ages.[2] Because he did not have at his disposal useful tools such as the Hindu-Arabic numerals that were developed over the course of centuries after him to simplify mathematical arguments and calculations, Archimedes's original writing is hard to understand in modern times. To make matters worse, like most Greeks of the era Archimedes made no mention of infinity and did not recognize a number zero. However, the gist of his argument is not hard to grasp when expressed in modern notation.

The method of exhaustion.

What Archimedes proves specifically is that the area of a circle is equal to its radius multiplied by half of its circumference; in modern notation, $(1/2)(2\pi R)R$ which simplifies to πR^2, the area of the circle. Archimedes did not use the number π in his argument in part because Greeks shied away from irrational numbers. Rather, he viewed π as the ratio of the circumference to the diameter of the circle.

Consider Figure 11.1 where the right triangle shown has one side equal to R and the other equal to the circumference of the circle $(2\pi R)$. Note that the area of this triangle is the height (radius R) multiplied by half of the length of the base (half of the circumference of the circle)–or πR^2 for short.

To prove that the circle has the same area as the triangle, Archimedes used a process of elimination. There are three possible cases:

(a) the area of the circle is smaller than that of the triangle, or:

(b) the area of the circle is larger than that of the triangle, or:

---

[1]Actually, the area of the region inside the circle, namely, the area of the disk. Circles being curves they have circumference but not area; the inaccurate terminology is common though so we stick with it.

[2]The method had been known before Archimedes's time. Eudoxus (408-355 BCE) has been credited with its use about a hundred years earlier. Logically, it is simply a trichotomy of possibilities; if two are ruled out then the third must be true. For instance, given two natural numbers m and n, exactly one of the three relations: $m < n$, $m = n$, $m > n$ must be true. The idea obviously extends to any finite number of possibilities.

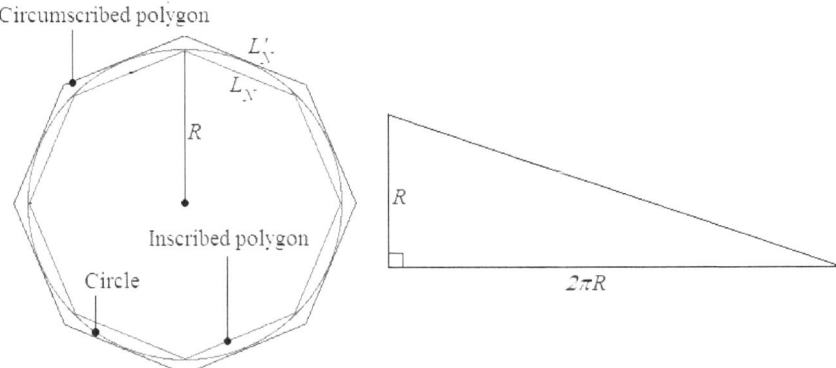

Figure 11.1: Illustrating Archimedes's argument

(c) the area of the circle is equal to that of the triangle.

He argued (using proof by contradiction) that (a) and (b) are not possible so that (c) must be true.

So, let's suppose that (a) is true and call the area of the circle A. Then the difference in areas is a positive number $\delta = \pi R^2 - A$. Recalling the N-gons from Chapter 1, notice that by choosing N large enough we can make the circumscribed N-gon to come as close to the circle as we wish. So we can choose N large enough that the difference between the area of the N-gon and the area of the circle is less than δ. Figure 11.2 shows one of the equal (or congruent) triangles that partition the N-gon, in both the circumscribed case (circle inside the N-gon) and the inscribed case (N-gon inside the circle).

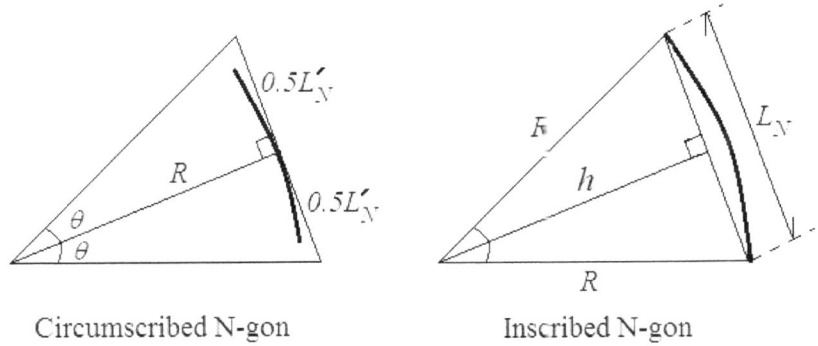

Figure 11.2: Partition triangles: circumscribed (left) and inscribed

In the circumscribed case, each side of the N-gon has length L'_N and the altitude of the triangle is the radius R of the circle, so the area of each constituent triangle in the N-gon is $L'_N R/2$.

Now, notice that for all large enough N:

$$N\left(\frac{1}{2}L'_N R\right) - A < \delta = \pi R^2 - A$$

$$\frac{1}{2}NL'_N R < \pi R^2$$

$$NL'_N < 2\pi R$$

The last inequality states that the perimeter of the circumscribed N-gon is less than the circumference of the circle; this contradicts the fact that the perimeter of a circumscribed N-gon is larger that the circumference of the circle, so (a) must be false.

Now suppose that (b) is true, in which case $A - \pi R^2 = \delta > 0$. We argue as in the previous case, but now use the inscribed N-gon. Choose N large enough that the difference between the area of the inscribed N-gon and the (presumed larger) area of the circle is less than δ. Figure 11.2 shows one of the equal (or congruent) triangles in Figure 1.5 with a perpendicular drawn to the side of polygon. The altitude or height is h which is (slightly) less than R, i.e. $h < R$. The area of each constituent triangle in the N-gon is

$$\frac{1}{2}L_N h < \frac{1}{2}L_N R$$

The area of the N-gon is N times the area of each triangle shown in Figure 1.5. So for all large enough N it is true that

$$A - N\left(\frac{1}{2}L_N R\right) < A - N\left(\frac{1}{2}L_N h\right) < \delta = A - \pi R^2$$

$$-\frac{1}{2}NL_N R < -\pi R^2$$

$$NL_N > 2\pi R$$

The last inequality states that the perimeter of the inscribed N-gon is greater than the circumference of the circle, which is a contradiction. Therefore, (b) must also be false. The only possibility left is (c) which must be true and the proof is complete!

Two points are worth stressing here. The first is historical: Archimedes started with a hexagon (6-gon) and doubled it to a 12-sided polygon and kept doubling until he reached a 96-sided polygon. He could then use the area of this polygon to estimate the area of a circle and use that answer to estimate the value of the "circumference over diameter", namely, π as

$$96L_{96} < \pi < 96L'_{96}$$

His estimate was accurate to over 99%; a remarkable feat, considering that he did not even have the advantage of using the convenient Hindu-Arabic numerals and had to work with the cumbersome Roman numerals instead. Centuries later in 1424, Kashi used the same idea to determine an estimate of π that was accurate to 16 figures. He calculated the perimeter of inscribed and circumscribed polygons with over 800 million sides (805,306,368 sides to be exact); a remarkable calculation for mid 15th century, even with the benefit of the Hindu-Arabic numerals!

The second point is that *Archimedes did not refer to infinity explicitly in his argument*; he does not say that the number N "approaches infinity". Ancient Greek mathematicians were uncomfortable with the concept of infinity and tried to avoid it in their arguments. However, the statement "by choosing N large enough we can make the circumscribed N-gon to come as close to the circle as we wish" and similar statements assume that a "sufficiently large" N exists and this is not guaranteed unless N can be made arbitrarily large; this is what we mean by $N \to \infty$. Possibly a deeper understanding of this process was had by the Chinese mathematician Liu Hui who in the mid-200's gave methods of finding both the circumference and the area of a circle by an argument similar to Archimedes's but displaying a greater recognition of the notion of limit as $N \to \infty$.[3] Just like Archimedes and Liu, we do not think of infinity as some vastly large quantity that N might be "getting close to".

Deriving formulas for area and circumference using limits.

Archimedes may have touched upon the concept of limit but the Greek hesitance in formalizing infinity prevented him from going deeper. For comparison, I discuss a straightforward calculation using limit theorems and basic trigonometry (also unknown to Archimedes) to simultaneously derive formulas for both the area and the perimeter of a circle of a given radius, without needing to have the perimeter as a given, as in Archimedes's argument.

Consider the graph of the circumscribed case in Figure 11.2 again (calculations for the inscribed case are only slightly longer). The line segment that splits the constituent triangle in half, also *bisects* the part of the central angle of the circle that is contained in the triangle. Since the N-gon is regular, the central angle of the entire circle is divided into N equal parts, each having a magnitude of $2\pi/N$ radians. Therefore,

$$\theta = \frac{1}{2}\left(\frac{2\pi}{N}\right) = \frac{\pi}{N}$$

Further, in either of the two right triangles, by the definition of the sine function,

$$\tan \theta = \frac{(1/2)L'_N}{R} = \frac{L'_N}{2R}$$

from which we obtain

$$L'_N = 2R \tan \frac{\pi}{N}$$

Since the perimeter and the area of the circumscribed N-gon are

$$L'_N N = 2RN \tan \frac{\pi}{N}, \qquad \frac{RL'_N}{2} N = R^2 N \tan \frac{\pi}{N}$$

it remains to calculate the limit

$$\lim_{N \to \infty} N \tan \frac{\pi}{N}$$

This involves the indeterminate form $\infty \times 0$ since π/N approaches 0 as $N \to \infty$ and $\tan 0 = 0$. We can calculate the limit using l'Hôpital's rule.[4] Although we did not discuss the $\infty \times 0$ indeterminate form in Chapter 9, the standard way of dealing with this form is by converting it to either $0/0$ or

---

[3] Liu also found the approximation $\pi \approx 3.1416$. He relied on the standard Chinese tools of his time, like "counting rods" which were considerably more cumbersome than modern tools.

[4] Alternatively, we can find this limit using the Maclaurin series for $\tan x$, the first few terms of which are

$$\tan x = x + \frac{x^3}{3} + \frac{2x^5}{15} + \cdots$$

Set $x = \pi/N$, multiply by N, simplify and take the limit as $N \to \infty$ to get π as the only term that remains.

Appendix

∞/∞ and then apply the procedure that we discussed in Chapter 9. Here we first convert to $0/0$ and then apply l'Hôpital's rule as before:

$$\lim_{N\to\infty} N\tan\frac{\pi}{N} = \lim_{N\to\infty} \frac{\tan(\pi/N)}{1/N} = \lim_{N\to\infty} \frac{d[\tan(\pi/N)]/dN}{d[1/N]/dN}$$

Now calculate the derivatives in the numerator and denominator

$$\frac{d}{dN}\left[\tan\frac{\pi}{N}\right] = \sec^2\frac{\pi}{N}\left[-\frac{\pi}{N^2}\right], \qquad \frac{d}{dN}\left[\frac{1}{N}\right] = -\frac{1}{N^2}$$

When substitute these into the above limit and simplify the expression we obtain:

$$\lim_{N\to\infty} N\tan\frac{\pi}{N} = \lim_{N\to\infty} \frac{\sec^2(\pi/N)[-\pi/N^2]}{-1/N^2} = \lim_{N\to\infty}\left(\sec^2\frac{\pi}{N}\right)[\pi] = \frac{\pi}{\cos^2 0} = \pi$$

This value now gives the circumference and area of the circle as

$$\lim_{N\to\infty} L_N N = \lim_{N\to\infty} 2R\left(N\tan\frac{\pi}{N}\right) = 2R\pi = 2\pi R$$

$$\lim_{N\to\infty} \frac{RL_N}{2} N = \lim_{N\to\infty} R^2\left(N\tan\frac{\pi}{N}\right) = R^2\pi = \pi R^2$$

11.2 Appendix: A Trigonometry refresher

Basic definitions and identities.

In a *right triangle* (with a 90 degree angle) we define the *trigonometric functions* of an acute angle (less than 90 degrees) as certain fractions, or ratios of its sides. Specifically, in the right triangle shown in Figure 11.3

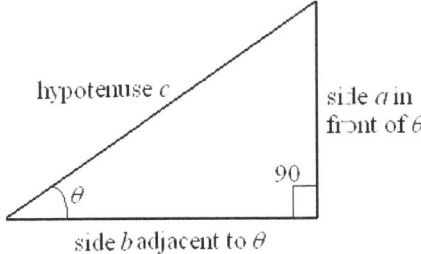

Figure 11.3: A right triangle with trigonometric labels

we define the *sine, cosine and tangent of the angle* θ as:

$$\sin\theta = \frac{a}{c}, \quad \cos\theta = \frac{b}{c}, \quad \tan\theta = \frac{a}{b}$$

The last function may be defined in terms of the first two:

$$\tan\theta = \frac{\sin\theta}{\cos\theta}$$

The reciprocals of the above three functions are also given their own names:

$$\frac{1}{\sin\theta} = \sec\theta \quad (\text{secant of } \theta)$$

$$\frac{1}{\cos\theta} = \csc\theta \quad (\text{cosecant of } \theta)$$

$$\frac{1}{\tan\theta} = \cot\theta \quad (\text{cotangent of } \theta)$$

From the Pythaogran identity for right triangles we deduce the following:

$$(\sin\theta)^2 + (\cos\theta)^2 = \left(\frac{a}{c}\right)^2 + \left(\frac{b}{c}\right)^2 = \frac{a^2+b^2}{c^2} = \frac{c^2}{c^2} = 1$$

This *trigonometric Pythagorean identity* is written as:

$$\sin^2\theta + \cos^2\theta = 1 \tag{11.1}$$

The square notation is a commonly used shorthand that is meant to distinguish between $(\sin\theta)^2$ and $\sin(\theta^2)$; the latter is often written as $\sin\theta^2$. The notation $\sin^2\theta$ shouldn't be interpreted as "the square of sin times θ" since "sin" and "cos" are just labels and have no meaning by themselves.

Appendix

Using (11.1) we can obtain other trigonometric Pythagorean identities; for instance,

$$\cos^2\theta = 1 - \sin^2\theta$$

or, if we divide the two sides of (11.1) by $\cos^2\theta$ the we get:

$$\frac{\sin^2\theta}{\cos^2\theta} + \frac{\cos^2\theta}{\cos^2\theta} = \frac{1}{\cos^2\theta}$$
$$\tan^2\theta + 1 = \sec^2\theta$$

Angle measures.

Angles are generally measured either in *degrees* (each degree being 1/360 of a full circle) or more commonly in calculus and analysis, in *radians*. In any circle, the radian measure of a central angle θ is the length of the arc of the circle that is in front of θ divided by the radius; see Figure 11.4.

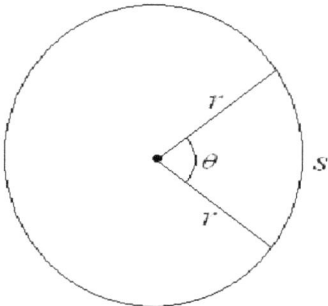

Figure 11.4: The radian measure

Using the notation in Figure 11.4 we may write:

$$\theta = \frac{s}{r} \quad \text{radians}$$

Since the circumference of a circle of radius r is $s = 2\pi r$ it follows that the radian measure of a full circle or 360 degrees (notation: 360°) is 2π. Therefore, half a circle or 180° is π radians and a quarter of a circle or 90° is $\pi/2$ radians. It is evident that one degree is just $\pi/180$ radians, which gives us our conversion factor. For instance,

$$45° = \frac{\pi}{180}(45) = \frac{\pi}{4} \quad \text{radians}$$
$$60° = \frac{\pi}{180}(60) = \frac{\pi}{3} \quad \text{radians}$$

Extending trigonometric functions using circles.

The right triangle definitions of trigonometric functions are valid for *acute* angles that are less than 90°. To define quantities like $\sin 90°$ or $\cos 120°$ we use right triangles within circles. In Figure 11.5 we have a circle of radius 1 (or a unit circle) where the angle θ is larger than 90°. We draw the

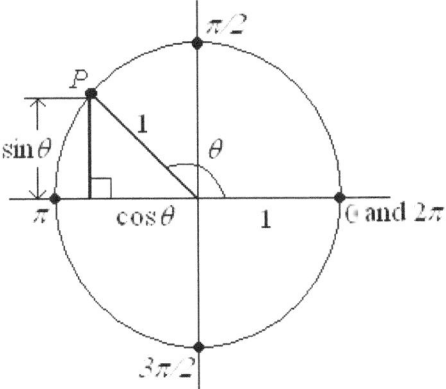

Figure 11.5: Trigonometric functions displayed in the unit circle

right triangle shown in the second quadrant, with the side adjacent to θ lying on the negative x-axis while the side in front of θ lies on the positive y-axis and therefore has a positive sign.

Since in this right triangle the hypotenuse is the radius and has length 1, it follows that $\cos x$ is *just the x-coordinate of the point P* shown on the circle and *$\sin x$ is the y-coordinate of P*. Using the unit circle in this way to define the sine and cosine functions, gives us functions that are defined for all real values of θ. After a full circle (going *counterclockwise*) everything repeats in periods of 2π radians; after one full rotation, θ goes from 0 to 2π, after two rotations to 4π radians and so on. *Clockwise* rotations define negative values for θ; after one full clockwise rotation θ reaches -2π, after two rotations -4π and so on.

For instance, if $\theta = \pi$ then the point P on the circle has coordinates $x = -1$ and $y = 0$ so $\cos \pi = -1$ and $\sin \pi = 0$. Similarly, $9\pi/2$ is greater than 2π specifically,

$$\frac{9\pi}{2} = 4\pi + \frac{\pi}{2}$$

and we are at the point $(0,1)$ on the circumference of the unit circle (after two full counterclockwise rotations). It follows that:

$$\cos \frac{9\pi}{2} = \cos \frac{\pi}{2} = 0 \quad \text{and} \quad \sin \frac{9\pi}{2} = \sin \frac{\pi}{2} = 1$$

For negative angles we simply go clockwise along the circumference of the unit circle. A right triangle is no longer required for defining the trigonometric functions![5]

Some special angles.

Certain angles are called *special angles* because we can evaluate their trigonometric functions precisely. A few of these are listed in the following table for easy reference:

---

[5]Trigonometric functions $\cos x$ and $\sin x$ are sometimes called "circular functions" because they are the x- and y-coordinates of a point on the circle $x^2 + y^2 = 1$. There are also the "hyperbolic functions" $\cosh x$ and $\sinh x$ that are given as the coordinates of points on the hyperbola $x^2 - y^2 = 1$. Such functions are used often in applications ranging from engineering calculations to the special relativity.

Appendix

θ	0	$\frac{\pi}{6} = 30°$	$\frac{\pi}{4} = 45°$	$\frac{\pi}{3} = 60°$	$\frac{\pi}{2} = 90°$	$\pi = 180°$	$\frac{3\pi}{2} = 270°$	$2\pi = 360$
$\cos\theta$	1	$\frac{\sqrt{3}}{2}$	$\frac{\sqrt{2}}{2}$	$\frac{1}{2}$	0	-1	0	1
$\sin\theta$	0	$\frac{1}{2}$	$\frac{\sqrt{2}}{2}$	$\frac{\sqrt{3}}{2}$	1	0	-1	0

The right half of this table can be read off the unit circle diagram in Figure 11.5. The left half is easy to calculate using an isosceles right triangle for the 45° case and (half of) an equilateral triangle for the 30° and 60° cases. The special angle values of the other four trigonometric functions $\tan\theta$, etc can be found using the above table and their definitions above. For instance,

$$\tan 60° = \frac{\sin 60°}{\cos 60°} = \frac{\sqrt{3}/2}{1/2} = \sqrt{3}$$

$$\sec 60° = \frac{1}{\cos 60°} = \frac{1}{1/2} = 2$$

Additional fundamental identities.

A number of identities involving trigonometric functions are used so frequently that it is useful to list them here. First, the definition of trigonometric functions via the unit circle makes *periodic functions* with period 2π; in particular, for every real number x:

$$\cos(x + 2\pi k) = \cos x \quad \text{and} \quad \sin(x + 2\pi k) = \sin x \quad \text{for all integers } k$$

Next, we note the *even and odd* identities:

$$\cos(-x) = \cos x \quad \text{and} \quad \sin(-x) = -\sin x$$

See Figure 11.6.

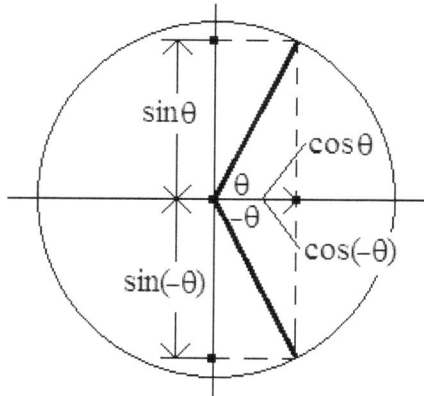

Figure 11.6: Odd and even trigonometric functions

Next, we have the angle sum identities:

$$\cos(x + y) = \cos x \cos y - \sin x \sin y \tag{11.2}$$
$$\sin(x + y) = \sin x \cos y + \cos x \sin y \tag{11.3}$$

that we derived in the main text; see (9.68) and (9.69). Using these and the even and odd identities we obtain the following *angle difference identities*:

$$\cos(x - y) = \cos x \cos y + \sin x \sin y \qquad (11.4)$$
$$\sin(x - y) = \sin x \cos y - \cos x \sin y \qquad (11.5)$$

An angle sum identity for the tangent function can be quickly derived from (11.2) and (11.3) as follows:

$$\tan(x + y) = \frac{\sin(x + y)}{\cos(x + y)} = \frac{\sin x \cos y + \cos x \sin y}{\cos x \cos y - \sin x \sin y}$$

To change the expression on the right to one that involves $\tan x$ and $\tan y$ let's divide the fraction by the product $\cos x \cos y$ and simplify to get:

$$\tan(x + y) = \frac{\frac{\sin x \cos y}{\cos x \cos y} + \frac{\cos x \sin y}{\cos x \cos y}}{1 - \frac{\sin x \sin y}{\cos x \cos y}} = \frac{\tan x + \tan y}{1 - \tan x \tan y}$$

Next, setting $y = x$ in (11.2) and (11.3) gives the *double-angle formulas*:

$$\cos(2x) = \cos^2 x - \sin^2 x, \quad \sin(2x) = 2 \sin x \cos x \quad \text{and} \quad \tan(2x) = \frac{2 \tan x}{1 - \tan^2 x}$$

Combining the formula on the left above with the trigonometric Pythagorean identity (11.1) gives the alternative forms:

$$\cos(2x) = 2 \cos^2 x - 1 = 1 - 2 \sin^2 x$$

Exercise 157 *(a) Notice that $3x = 2x + x$ and find formulas for $\sin(3x)$ and $\cos(3x)$.*
(b) Show that $\sin(x+\pi) = -\sin x$ and $\cos(x+\pi) = -\cos x$. What do $\sin(x+\pi/2)$ and $\cos(x+\pi/2)$ simplify to? How about $\tan(x + \pi)$ and $\tan(x + \pi/2)$?

The sum, difference and double-angle formulas together with the table of special angles above can be used to find exact values for many more angles. For example, to find $\sin 15°$ we use a double-angle formula as follows:

$$1 - 2 \sin^2 15° = \cos 30° = \frac{\sqrt{3}}{2}$$

so that

$$\sin^2 15° = \frac{1}{2}\left(1 - \frac{\sqrt{3}}{2}\right) = \frac{2 - \sqrt{3}}{4}$$

$$\sin 15° = \sqrt{\frac{2 - \sqrt{3}}{4}} = \frac{\sqrt{2 - \sqrt{3}}}{2}$$

Why not $\pm$ after taking the last square root? Because we know that the angle $15°$ is in the first quadrant so its sine must be positive. You may have also noticed that $15° = 45° - 30°$ so that we could use the angle difference formula (11.5). Doing so would give:

$$\sin 15° = \frac{\sqrt{6} - \sqrt{2}}{4}$$

Appendix

Both of these expressions give the same decimal result which you can check for yourself. This apparent difference between equal answers is a common problem when dealing with radicals. Here's another example:

$$\cos 75° = \cos(45° + 30°)$$
$$= \cos 45° \cos 30° - \sin 45° \sin 30°$$
$$= \frac{\sqrt{2}}{2}\frac{\sqrt{3}}{2} - \frac{\sqrt{2}}{2}\frac{1}{2}$$
$$= \frac{\sqrt{6} - \sqrt{2}}{4}$$

Is the fact that $\cos 75° = \sin 15°$ just a coincidence? No, and it is easy to see why:

$$\cos(90° - 15°) = \cos 90° \cos 15° + \sin 90° \sin 15° = 0 + (1)\sin 15° = \sin 15°$$

Exercise 158 *Find the exact values of each of the following quantities:*

$$\cos 105°, \quad \sin \frac{\pi}{12}, \quad \tan 120°, \quad \cos \frac{7\pi}{6}$$

Product formulas.

Identities (11.2)-(11.5) may be used to derive important identities for the *products* of trigonometric functions that are needed in the section on Fourier series.

We begin with the observation that by adding the two sides of identities (11.2) and (11.4) we get:

$$\cos(x+y) + \cos(x-y) = 2\cos x \cos y$$

So if we divide this by 2 we get the product formula:

$$\cos x \cos y = \frac{1}{2}[\cos(x+y) + \cos(x-y)] \tag{11.6}$$

Similarly, adding the two sides of identities (11.3) and (11.5) gives:

$$\sin(x+y) + \sin(x-y) = 2\sin x \cos y$$

so that:

$$\sin x \cos y = \frac{1}{2}[\sin(x+y) + \sin(x-y)] \tag{11.7}$$

Exercise 159 *Prove that*

$$\sin x \sin y = \frac{1}{2}[\cos(x-y) - \cos(x+y)] \tag{11.8}$$

Formulas (11.6)-(11.8) are used to derive the orthogonality formulas and Fourier's coefficients, among other things.

11.3 Appendix: The proof of Cantor's power-set theorem

I will now use Cantor's argument to show that $\#\mathcal{P}(S)$ is not equal to $\#S$ because there is no *onto* (or surjective) function $f : S \to \mathcal{P}(S)$ for every set S. So, then there are also no bijections between S and its power set $\mathcal{P}(S)$.

Cantor's Theorem: *There are no functions $f : S \to \mathcal{P}(S)$ that are onto $\mathcal{P}(S)$ (or surjective) for any set S.*

The proof is by contradiction. Suppose that there is a function $f : S \to \mathcal{P}(S)$ that is onto. Notice that the image of each element $a \in S$ is a set $f(a) \in \mathcal{P}(S)$ since $f(a)$ is a *subset* of S. Based on our understanding of *finite* sets we expect to find a lot more elements in $\mathcal{P}(S)$ than in S; yet we are assuming that f is onto, so *every subset* of S must be assigned some element of S. It seems reasonable to wonder about elements a in S that are in, or not in, their image sets $f(a)$. Cantor examined the following subset of S:

$$N = \{a \in S : a \notin f(a)\}$$

Recall that f is onto $\mathcal{P}(S)$ so there is an element $t \in S$ whose image is $N \in \mathcal{P}(S)$, so $f(t) = N$. It is natural to wonder now if $t \in N$ or $t \notin N$. If $t \in N$ then the definition of N implies that $t \notin f(t) = N$ which is a contradiction. If $t \notin N$ then the definition of N implies that $t \in f(t) = N$ and we reach a contradiction again. These contradictions are unavoidable if we assume that a function exists from S onto $\mathcal{P}(S)$. Therefore, we conclude that such a function does not exist and the theorem is proved.

11.4 Appendix: Cantor's construction of the real numbers

The material in this section is proof-oriented and more abstract than the coverage in the main text. This material is not essential reading for most readers. I include it for the benefit of readers with some math training so they can get their feet dirty; sort of an optional hiking tour excursion!

We start with a few preliminary results, beginning with the so-called *triangle inequality*[6] a simple, yet useful tool for proving statements about convergence.

Theorem 1. (the triangle inequality) *If a and b are rational numbers*[7] *then*

$$|a+b| \leq |a| + |b| \tag{11.9}$$

and

$$||a| - |b|| \leq |a - b| \tag{11.10}$$

Proof. Proving (11.9) is easy on a case by case basis: either a and b have the same signs, both positive or both negative, or one is positive and the other negative. If they are both non-negative then we can remove all absolute value signs and get the equality $a + b \leq a + b$ which is true.

If a, b are both negative then all absolute value signs can be replaced with negative signs which again gives a true statement

$$-(a+b) \leq -a - b$$

Now suppose that $a < 0 \leq b$. Then $|a+b|$ is a (positive) quantity that is less than b which is in turn less than $|a| + |b|$ because $|a| > 0$; symbolically,

$$|a+b| < b = |b| < |a| + |b|$$

which is consistent with (11.9). The last remaining case $b < 0 \leq a$ is argued similarly with the roles of a and b interchanged.

Now (11.10) is easy to prove. Let $c = a - b$ and use (11.9) to get

$$|a| = |b + c| \leq |b| + |c|$$

so that

$$|a| - |b| \leq |c| = |a - b| \tag{11.11}$$

Similarly, with $c = b - a$ and using (11.9) we get

$$|b| - |a| \leq |c| = |b - a| = |a - b| \tag{11.12}$$

Putting (11.11) and (11.12) together gives (11.10). ∎

The next result makes sense because as terms of a converging sequence approach a limit they cluster around it and therefore, must get close to each other. This is the idea behind the proof.

---

[6] This name comes from the observation that in every triangle the sum of the lengths of any two sides is greater than the length of the third side.

[7] The triangle inequality holds for all real numbers and is proved exactly as we see here. I assume that a, b are rational because we are still building the real numbers.

Theorem 2. *If a sequence q_n of rational numbers converges to a rational number q then q_n is a Cauchy sequence.*

Proof. First using the triangle inequality in Theorem 1 we have
$$|q_n - q_m| = |q_n - q + q - q_m| \leq |q_n - q| + |q - q_m|$$

Since $q_n \to q$ we can say that for every (rational) $\varepsilon > 0$ there is an integer index N large enough that both $|q_n - q| < \varepsilon/2$ and $|q - q_m| < \varepsilon/2$ if $m, n > N$. For such large index values then it is true that
$$|q_n - q_m| < \frac{\varepsilon}{2} + \frac{\varepsilon}{2} = \varepsilon$$

As ε can be arbitrarily small, we conclude that $|q_n - q_m| \to 0$ and q_n is Cauchy. ∎

Theorem 2 can be stated equivalently (in contrapositive) as:

If a sequence is not Cauchy then it does not converge.

This shows why we need to consider *Cauchy* sequences in building the real numbers. The next result tells us what happens if a Cauchy sequence of rational numbers does *not* converge to 0.

Theorem 3. *Let q_n be a sequence of rational numbers that does not converge to 0. Then the sequence is either eventually positive or eventually negative.*

Proof. Because q_n does not converge to zero there is $\varepsilon_0 > 0$ such that for every positive integer K there is $n \geq K$ such that $|q_n| \geq \varepsilon_0$. But q_n is also Cauchy so for each $\varepsilon > 0$ there is an index N such that $|q_n - q_m| < \varepsilon$ if $m, n > N$. Now suppose that q_r and q_m have opposite signs (they are on opposite sides of 0) for a pair of indices $m, n > N$. Then without loss of generality, assume that $q_m < 0 \leq q_n$. Then
$$|q_n - q_m| = q_n - q_m \geq q_n$$

By taking $K = N$ we may also assume that $q_n = |q_n| \geq \varepsilon_0$. Now if we choose the arbitrary ε to be less than ε_0, say, $\varepsilon = \varepsilon_0/2$ then we arrive at
$$\varepsilon_0 \leq q_n \leq |q_n - q_m| < \varepsilon = \frac{\varepsilon_0}{2}$$

which is impossible. Thus q_n and q_m must be on the same side of zero eventually, as claimed. ∎

The last of the aforementioned preliminary results states that every Cauchy sequence is bounded, which also makes sense; if terms of a sequence all get closer to each other then they tend to accumulate about a point and not march off to infinity. We use this insight to prove the next result.

Theorem 4. *Every Cauchy sequence is bounded.*

Proof. I start the proof by recalling that a sequence q_n is bounded if there is a rational number $M > 0$ such that $|q_n| \leq M$ for every index n. Now suppose that q_n is a Cauchy sequence. Then there is an index N large enough that $|q_n - q_m| < 1$ for all $m, n > N$ (pick $\varepsilon = 1$). In particular, for $m = N + 1$ we get
$$|q_n - q_{N+1}| < 1 \quad \text{if } n > N$$

Appendix

Since we can write, using the triangle inequality,

$$|q_n| = |q_n - q_{N+1} + q_{N+1}| \leq |q_n - q_{N+1}| + |q_{N+1}| < 1 + |q_{N+1}|$$

and this is true for every $n > N$ it follows that for all $n \geq 1$

$$|q_n| \leq \max\{|q_1|, |q_2|, \ldots, |q_N|, 1 + |q_{N+1}|\}$$

To complete the proof we need only define M to be the quantity to the right side of the above inequality. ∎

Both of the preceding theorems and their proofs are valid for all real numbers; but since the real numbers are not yet defined we assume that all numbers are rational.

[$\mathcal{C}$] is an algebraic field.

Consider the set $[\mathcal{C}]$ of all equivalence classes of rational Cauchy sequences as described in the text and define two operations: for each pair of equivalence classes $[q_n]$ and $[q'_n]$

$$[q_n] + [q'_n] = [q_n + q'_n] \quad \text{(addition)}$$
$$[q_n][q'_n] = [q_n q'_n] \quad \text{(multiplication)}$$

On the right hand sides of the above equalities we have the equivalence classes of the sum or product of two rational Cauchy sequences. Specifically, given any pair of sequences $q_n = q_1, q_2, q_3, \ldots$ and $q'_n = q'_1, q'_2, q'_3, \ldots$ in $[\mathcal{C}]$ we have

$$q_n + q'_n = (q_1, q_2, q_3, \ldots) + (q'_1, q'_2, q'_3, \ldots) = q_1 + q'_1, q_2 + q'_2, q_3 + q'_3, \ldots$$
$$q_n q'_n = (q_1, q_2, q_3, \ldots)(q'_1, q'_2, q'_3, \ldots) = q_1 q'_1, q_2 q'_2, q_3 q'_3, \ldots$$

Since we are dealing with equivalence classes rather than individual sequences, it is necessary to ensure that these operations are well-defined functions in the sense that *the result of addition or multiplication does not depend on which sequences in the equivalence classes we choose to add or multiply.*

Theorem 5. *If p_n and p'_n are Cauchy sequences of rationals such that $[q_n] = [p_n]$ and $[q'_n] = [p'_n]$ then $[q_n + q'_n] = [p_n + p'_n]$ and $[q_n q'_n] = [p_n p'_n]$. Therefore, addition and multiplication in $[C]$ are well-defined.*

Proof. Note that $[q_n] = [p_n]$ implies $|q_n - p_n| \to 0$ as $n \to \infty$; similarly, from $[q'_n] = [p'_n]$ it follows that $|q'_n - p'_n| \to 0$ as $n \to \infty$. Therefore, for every (rational) $\varepsilon > 0$ there is a large enough index N such that $|q_n - p_n| < \varepsilon/2$ and $|q'_n - p'_n| < \varepsilon/2$.

$$|(q_n + q'_n) - (p_n + p'_n)| = |(q_n - p_n) + (q'_n - p'_n)| \leq |q_n - p_n| + |q'_n - p'_n| < \frac{\varepsilon}{2} + \frac{\varepsilon}{2} = \varepsilon$$

where the first equality is implied by the triangle inequality. Since ε can be arbitrarily small, we have shown that

$$|(q_n + q'_n) - (p_n + p'_n)| \to 0 \quad \text{as } n \to \infty$$

This means that the sequences $q_n + q'_n$ and $p_n + p'_n$ are equivalent, which is saying the same thing as $[q_n + q'_n] = [p_n + p'_n]$. For products a little more work is needed; we start by writing

$$\begin{aligned}|q_n q'_n - p_n p'_n| &= |q_n q'_n - p_n q'_n + p_n q'_n - p_n p'_n| \\ &= |(q_n - p_n)q'_n + p_n(q'_n - p'_n)| \\ &\leq |(q_n - p_n)q'_n| + |p_n(q'_n - p'_n)| \qquad \text{(triangle inequality)} \\ &= |q_n - p_n||q'_n| + |p_n||q'_n - p'_n|\end{aligned}$$

Both of the quantities $|q'_n|$ and $|p_n|$ are bounded by Theorem 4 so there is (rational) $M > 0$ such that $|q'_n| \leq M$ and $|p_n| \leq M$. Further, we know that $|q_n - p_n| \to 0$ and $|q'_n - p'_n| \to 0$ so there is an index N large enough that for $n > N$

$$|q_n - p_n| < \frac{\varepsilon}{2M}, \qquad |q'_n - p'_n| < \frac{\varepsilon}{2M}$$

Therefore, for all $n > N$

$$|q_n q'_n - p_n p'_n| \leq \frac{\varepsilon}{2M}M + \frac{\varepsilon}{2M}M = \varepsilon$$

which, because ε can be arbitrarily small, is saying the same thing as $|q_n q'_n - p_n p'_n| \to 0$ as $n \to \infty$. Therefore, $[q_n q'_n] = [p_n p'_n]$. ∎

Okay, now we have two well-defined operations of addition and multiplication in $[\mathcal{C}]$. The next thing to do is verify that these are field operations; this means that it is necessary to verify that the nine field properties hold. This is a task that is more tedious than difficult.

The commutative property: Since addition and multiplication are commutative in the field $\mathbb{Q}$ we have

$$[q_n] + [q'_n] = [q_n + q'_n] = [q'_n + q_n] = [q'_n] + [q_n]$$
$$[q_n][q'_n] = [q_n q'_n] = [q'_n q_n] = [q'_n][q_n]$$

Hence, addition and multiplication are also commutative in $[\mathcal{C}]$.

The associative property: Since addition and multiplication are associative in the field $\mathbb{Q}$ we have

$$\begin{aligned}[q_n] + ([q'_n] + [q''_n]) &= [q_n] + [q'_n + q''_n] \\ &= [q_n + (q'_n - q''_n)] \\ &= [(q_n + q'_n) + q''_n] \\ &= ([q_n] + [q'_n]) + [q''_n]\end{aligned}$$

and similarly,

$$\begin{aligned}[q_n]([q'_n][q''_n]) &= [q_n][q'_n q''_n] \\ &= [q_n(q'_n q''_n)] \\ &= [(q_n q'_n)q''_n] \\ &= ([q_n][q'_n])[q''_n]\end{aligned}$$

Appendix

Hence, addition and multiplication are also associative in $[\mathcal{C}]$.

The distributive property: Since addition and multiplication have the distributive property in the field $\mathbb{Q}$ we have

$$[q_n]([q'_n] + [q''_n]) = [q_n][q'_n + q''_n]$$
$$= [q_n(q'_n + q''_n)]$$
$$= [q_n q'_n + q_n q''_n]$$
$$= [q_n q'_n] + [q_n q''_n]$$
$$= [q_n][q'_n] + [q_n][q''_n]$$

Hence, addition and multiplication also have the distributive property in $[\mathcal{C}]$.

The additive identity or zero element: Consider $[0]$, namely, the equivalence class of the constant sequence $0, 0, 0, \ldots$ Since for every $[q_n]$ in $[\mathcal{C}]$,

$$[q_n] + [0] = [q_n + 0] = [q_n]$$

it follows that $[0]$ is the zero element of $[\mathcal{C}]$. Note also that for every $[q_n]$

$$[q_n][0] = [q_n 0] = [0]$$

The multiplicative identity: Consider $[1]$, namely, the equivalence class of the constant sequence $1, 1, 1, \ldots$ Since for every $[q_n]$ in $[\mathcal{C}]$,

$$[q_n][1] = [q_n 1] = [q_n]$$

it follows that $[1]$ is the multiplicative identity in $[\mathcal{C}]$.

The additive inverses, or negatives: If to each sequence $q_1, q_2, q_3, \ldots$ in a class $[q_n]$ we add the sequence of negatives $-q_1, -q_2, -q_3, \ldots$ we obtain the zero sequence $0, 0, 0, \ldots$ Further, the sequence of negatives is again Cauchy since for all indices m, n

$$|(-q_m) - (-q_n)| = |-q_m + q_n| = |q_m - q_n|$$

So we simply define the additive inverse (the negative) of the class $[q_n]$ as

$$-[q_n] = [-q_n]$$

to obtain the desired result: $[q_n] + (-[q_n]) = [q_n + (-q_n)] = [0]$.

The multiplicative inverses, or reciprocals: If we multiply a sequence of *nonzero* rational numbers $q_1, q_2, q_3, \ldots$ by the sequence of reciprocals of its terms $1/q_1, 1/q_2, 1/q_3, \ldots$ we obtain the unit sequence $1, 1, 1, \ldots$ So it makes sense to define $[q_n]^{-1} = [1/q_n]$. But in every *class* $[q_n]$ there are sequences with lots of zero terms. In fact, if you take any sequence that is not equivalent to zero (say, $2, 2, 2, \ldots$) and add any number of zeros to the beginning of it then both the original sequence and the one starting with zeros (say, $0, 0, 0, 0, 2, 2, 2, \ldots$) are in the same equivalence class ($[2]$) because their tail ends certainly approach each other (actually meet). But of course, the rational number 0 has no reciprocal, so we need to resolve this issue of sequences containing zeros. It turns out that as long as $[q_n] \neq [0]$, every sequence in $[q_n]$ has at most a *finite* number of zeros, which makes the issue easy to resolve.

First, observe that if a Cauchy sequence $q_1, q_2, q_3, \ldots$ of rationals is in the class $[0]$, i.e. it is equivalent to the constant sequence $0, 0, 0 \ldots$ then q_n approaches 0. This means that for every (rational) $\varepsilon > 0$ there is an index N such that $|q_n| = |q_n - 0| < \varepsilon$ for $n > N$. From this we infer that if $q_1, q_2, q_3, \ldots$ is *not* equivalent to $0, 0, \ldots$ then there is a (rational) $\varepsilon_0 > 0$ such that $|q_n| = |q_n - 0| \geq \varepsilon_0$ infinitely often (i.e. for every positive integer N there is an index $n_0 \geq N$ such that $|q_{n_0}| \geq \varepsilon_0$). Since we are assuming that q_n is Cauchy, there is an index N such that $|q_m - q_n| < \varepsilon_0/2$ for $m, n > N$. Now, with $m = n_0$ and using the triangle inequality in the form (11.11), it follows that for all $n > N$

$$|q_n| \geq |q_{n_0}| - |q_{n_0} - q_n| > \varepsilon_0 - \frac{\varepsilon_0}{2} = \frac{\varepsilon_0}{2}$$

In summary:

If the sequence $q_1, q_2, q_3, \ldots$ is not in the class $[0]$ then there is a rational number $\varepsilon_0 > 0$ and an index N such that $|q_n| \geq \varepsilon_0$ for all $n > N$. In particular, $q_n \neq 0$ for $n > N$.

Now if $[q_n] \neq [0]$ then the above statement applies to *every* sequence in $[q_n]$, including the sequence $\tilde{q}_n$ such that

$$\tilde{q}_n = \begin{cases} q_n & \text{if } n > N \\ 1 & \text{if } n \leq N \end{cases}$$

Note that $[\tilde{q}_n] = [q_n]$ and further, *none* of the terms of $\tilde{q}_n$ are zeros. So we can define the multiplicative inverse of $[q_n]$ as (you may verify in Exercise 160 that the reciprocal sequence is Cauchy):

$$[q_n]^{-1} = [\tilde{q}_n]^{-1} = \left[\frac{1}{\tilde{q}_n}\right]$$

The above definition does not apply to one equivalence class in $[\mathcal{C}]$, namely, $[0]$. Although $[0]$ does contain sequences with no zero terms, like $1, 1/2, 1/3, \ldots$ it also contains sequences that are zeros infinitely often, like $1, 0, 1/2, 0, 1/3, 0, \ldots$ This means that

$$[0] = [1, 0, 1/2, 0, 1/3, 0, \ldots] = [1, 1/2, 1/3, \ldots]$$

But $1, 0, 1/2, 0, 1/3, 0, \ldots$ does not have a reciprocal sequence while $1, 1/2, 1/3, \ldots$ does, so it is not possible to define $[0]^{-1}$ unambiguously. It is also worth noticing that the reciprocal sequence of $1, 1/2, 1/3, \ldots$ namely, $1, 2, 3, \ldots$ is *not* Cauchy, and thus not a member of any equivalence class in $[\mathcal{C}]$ to begin with!

The preceding discussion verifies that $[\mathcal{C}]$ is indeed an algebraic field.

Exercise 160 *Suppose that $q_1, q_2, q_3, \ldots$ is Cauchy and not in $[0]$ i.e. not equivalent to 0. Assuming that $q_n \neq 0$ for all n, show that the reciprocal sequence $1/q_1, 1/q_2, 1/q_3, \ldots$ is Cauchy.*

$[\mathcal{C}]$ is a totally ordered field.

Recall that the set $\mathbb{Q}$ of rational numbers is totally ordered. If p and q are rational numbers then saying that $p < q$ is equivalent to saying that $q - p > 0$. This is the way we introduce the ordering of equivalence classes in $[\mathcal{C}]$ since we know that a difference of equivalence classes $[q_n] - [p_n]$ is just the equivalence class $[q_n - p_n]$.

With this in mind, we start with what is means to say something is "positive" in $[\mathcal{C}]$.

Appendix

The equivalence class $[q_n]$ of a rational Cauchy sequence $q_1, q_2, q_3, \ldots$ is *positive* if there is (rational) $\varepsilon > 0$ and an index N such that $q_n \geq \varepsilon$ for all $n > N$.

I had to use a positive lower bound ε here to ensure that q_n does not converge to 0. For example, $1/n > 0$ for all n but $q_n = 1/n$ is equivalent to the zero sequence. So its equivalence class is $[0]$ but it would be counterproductive to declare $[0]$ positive! Also starting from an index N is necessary, since as I mentioned above there are infinitely many sequences with a finite number of zeros in each member of $[q_n] \neq [0]$.

To turn the above definition into an ording $\prec$ in $[\mathcal{C}]$ we start by defining

$$[0] \prec [q_n] \text{ if } [q_n] \text{ is positive.}$$

Now if $q'_1, q'_2, q'_3, \ldots$ is *any* other sequence in $[q_n]$ then $[q'_n] = [q_n]$ so for the definition to make sense, there must be a (rational) $\varepsilon' > 0$ and an index N' such that $q'_n \geq \varepsilon'$ for $n > N'$. Since $q'_1, q'_2, q'_3, \ldots$ is equivalent to $q_1, q_2, q_3, \ldots$ there is an index N_0 such that $|q'_n - q_n| < \varepsilon/2$ for all $n > N_0$. This means that $-\varepsilon/2 < q'_n - q_n < \varepsilon/2$ for $n > N_0$ and since $q_n \geq \varepsilon$ for $n > N$, if I pick N' to be the larger of N_0 and N then for all $n > N'$ it is the case that

$$q'_n = q'_n - q_n + q_n > -\frac{\varepsilon}{2} + \varepsilon = \frac{\varepsilon}{2}$$

If I pick $\varepsilon' = \varepsilon/2$ then I have shown that $q'_n > \varepsilon'$ for $n > N'$ which was necessary to make $\prec$ well-defined. Now let us extend the relation $\prec$ to all pairs in $[\mathcal{C}]$ as follows:

$$[p_n] \prec [q_n] \text{ if } [0] \prec [q_n] - [p_n], \text{ (i.e. } [q_n - p_n] \text{ is positive).}$$

To prove that $\prec$ makes $[\mathcal{C}]$ a totally ordered field we need to establish the following: For all classes $[p_n]$ and $[q_n]$ in $[\mathcal{C}]$.

1. If $[p_n] \prec [q_n]$ then $[p_n] + [r_n] \prec [q_n] + [r_n]$ for every class $[r_n]$ in $[\mathcal{C}]$.

2. If $[p_n] \prec [q_n]$ then $[p_n][r_n] \prec [q_n][r_n]$ for every class $[r_n]$ in $[\mathcal{C}]$ such that $[0] \prec [r_n]$.

3. If $[p_n] \neq [q_n]$ then either $[p_n] \prec [q_n]$ or $[q_n] \prec [p_n]$ (not both).

4. The relation $\prec$ is transitive.

To prove Item 1 above it is necessary to show that

$$[0] \prec ([q_n] + [r_n]) - ([p_n] + [r_n])$$

Using the field properties of $[\mathcal{C}]$, the right hand side works out to $[q_n] - [p_n] = [p_n - q_n]$. This is assumed to be positive in Item 1 so we are done.

To prove Item 2 we must show that

$$[0] \prec [q_n][r_n] - [p_n][r_n]$$

The field properties and the definitions of addition and multiplication reduce the problem to proving:

$$0 \prec ([q_n] - [p_n])[r_n] = [q_n - p_n][r_n] = [(q_n - p_n)r_n]$$

So it remains to show that there is an ε_0 and an index N such that $(q_n - p_n)r_n \geq \varepsilon_0$ for all $n \geq N$. By the positivity assumptions, there are $\varepsilon_1, \varepsilon_2 > 0$ and indices N_1, N_2 such that $q_n - p_n \geq \varepsilon_1$ for $n \geq N_1$ and $r_n \geq \varepsilon_2$ for $n \geq N_2$. So now simply choose $\varepsilon_0 = \varepsilon_1 \varepsilon_2$ and N to be the greater of N_1 and N_2 to conclude that $[(q_n - p_n)r_n]$ is positive and finish the proof.

Next, I prove Item 3. Assume that $[p_n] \neq [q_n]$. This means that $q_n - p_n$ is a rational Cauchy sequence that does not converge to 0; so by Theorem 3 $q_n - p_n$ is eventually positive or eventually negative and stays some positive distance ε_0 above or below zero. If eventually positive then $[q_n - p_n]$ is positive and thus $[p_n] \prec [q_n]$. If eventually negative then then $[p_n - q_n]$ is positive and thus $[q_n] \prec [p_n]$.

Finally, I show that $\prec$ is transitive, Item 4. Suppose that there are classes $[p_n], [q_n]$ and $[r_n]$ in $[\mathcal{C}]$ such that $[p_n] \prec [q_n]$ and $[q_n] \prec [r_n]$. Then

$$[0] \prec [q_n] - [p_n] \quad \text{and} \quad [0] \prec [r_n] - [q_n]$$

Using the field properties of $[\mathcal{C}]$ and Item 1 above we have

$$[0] = [0] + [0] \prec [r_n] - [q_n] + [q_n] - [p_n] = [r_n] - [p_n]$$

It follows that $[p_n] \prec [r_n]$ and transitive property is proved.

The above discussion proves that $[\mathcal{C}]$ is a totally ordered field under the relation $\prec$ and thus also under its reflexive extension $\preceq$.

Rational numbers are dense in the set of real numbers.

We have shown so far that the set $[\mathcal{C}]$ of equivalence classes of rational Cauchy sequences is a totally ordered field that contains a copy of the set $\mathbb{Q}$ of all rational numbers in the sense that each rational number q is uniquely identified with the equivalence class $[q]$ of the constant sequence $q, q, q, \ldots$. Verifying this uniqueness is a good exercise; the preceding sections contain enough details to draw on here.

Exercise 161 *Show that if p, q are distinct rational numbers then $[p] \neq [q]$; in fact, show that the inequality $p < q$ in $\mathbb{Q}$ is equivalent to the inequality $[p] \prec [q]$ in $[\mathcal{C}]$.*

This copy of $\mathbb{Q}$ turns out to be a very important subset in calculus and beyond. To explore this further, it is helpful to blur the precise nature of elements in $[\mathcal{C}]$ now that we have some structure for $[\mathcal{C}]$ we can use a simpler (even if more vague) notation for more effective progress moving forward. We start by formally defining the set of real numbers:

> **The Real Numbers:** Each equivalence class of rational Cauchy sequences in $[\mathcal{C}]$ is a *real number*. We use the common notation $\mathbb{R}$ instead of $[\mathcal{C}]$ for the *set of all real numbers*.

We have thus shown that $\mathbb{R}$ is a totally ordered field. We now identify the rational numbers with the equivalence classes of constant rational sequences so as to simply say that $\mathbb{Q}$ is a subset of $\mathbb{R}$.

The concepts of positive and negative are easy to define. A real number $r = [q_n]$ is *negative* if $[q_n] \prec [0]$ and *positive* if $[0] \prec [q_n]$. Identifying $[0]$ with the rational number 0 and the relations $\prec$ and $\preceq$ with the usual orderings $<$ and $\leq$ of the rationals, we simply write $r < 0$ and $r > 0$ to indicate whether the real number r is negative or positive.

Appendix

The extension of the concept of absolute value to $\mathbb{R}$ is also straightforward:

$$|r| = \begin{cases} r & \text{if } r \geq 0 \\ -r & \text{if } r < 0 \end{cases}$$

From this definition you can see that $|r| \geq 0$ (i.e. $[0] \preceq |[q_n]|$) no matter what real number r we choose. It is well worth a mention that the triangle inequality readily extends to the real numbers now, since the proof of that theorem simply considered the signs of rational numbers and not their nature as rationals. Now we are ready to state an important result about the real numbers.

Theorem 6. (density of rational numbers). *For every pair of real numbers x and y such that $x < y$ there is a rational number q such that $x < q < y$.*

Proof. Suppose that $x = [q_n]$ and $y = [r_n]$ where q_n and r_n are Cauchy sequences of rationals. We are given that $[q_n] \prec [r_n]$ or equivalently, $[r_n - q_n] \succ [0]$. Therefore, there is a (rational) $\varepsilon_0 > 0$ and index N such that $r_n - q_n \geq \varepsilon_0$ for all $n \geq N$. On the other hand, q_n is a Cauchy sequence so there is an index N_0 such that $|q_n - q_m| < \varepsilon_0$ for all $m, n > N_0$. If I define K to be the larger of N and $N_0 + 1$ and the rational number $q = q_K$ then

$$q_n < q + \varepsilon_0 \leq r_n \quad \text{for all } n \geq K$$

In particular, $q_n < q < r_n$ for all $n \geq K$ which translates to $[q_n] < [q] < [r_n]$, or equivalently, $x < q < y$. ∎

This theorem shows that every real number can be approximated by rational numbers, and so it should come as no surprise; we in fact defined a real number essentially as something that is approximated by a sequence of rational numbers. The important difference is that now we can actually associate a sequence with a real number rather than the other way around. For instance, suppose that x is a real number and we want to find a sequence of rationals that approximates x. We with some rational number q_1 and then find a rational q_2 that is between q_1 and x. Clearly, q_2 is closer to x than q_1 was, in the sense that

$$|q_2 - x| < |q_1 - x|$$

We may go on to a rational q_3 that is between q_2 and x and so on, creating a sequence $q_1, q_2, q_3, \ldots$ in this way that converges to x. Such a sequence is Cauchy by Theorem 2 (this theorem, like all of the Theorems 1-4 readily extends to real numbers) so $x = [q_n]$.

The set of real numbers is complete.

We now come to the most important property of real numbers, and the reason for their invention: the set $\mathbb{R}$ of real numbers is complete, or put differently, $\mathbb{R}$ has no gaps or holes. First, it is convenient to have the extensions of Theorems 1-4 listed for easy reference. The proofs are essentially the same as before; I present the proof of one theorem in order to set the stage for the completeness theorem below. As in the previous section, we take advantage of all our work above and use the common notation for rational numbers instead of the more accurate but cumbersome equivalence class notation.

Theorem 7. (the triangle inequality) *If a and b are real numbers then*

$$|a + b| \leq |a| + |b| \tag{11.13}$$

and
$$||a| - |b|| \leq |a - b| \qquad (11.14)$$

Theorem 8. *If a sequence x_n of real numbers converges to a real number x then x_n is a Cauchy sequence.*

This theorem can be stated equivalently (in contrapositive form) as: *If a sequence is not Cauchy then it does not converge.*

Proof. It is given that for every (real, or rational) $\varepsilon > 0$ there is an index N such that if $n > N$ then $|x_n - x| < \varepsilon/2$. So for $m > N$

$$|x_n - x_m| = |x_n - x + x - x_m| \leq |x_n - x| + |x - x_m| < \frac{\varepsilon}{2} + \frac{\varepsilon}{2} = \varepsilon$$

It follows that x_n is a Cauchy sequence. ∎

Theorem 9. *Let x_n be a sequence of real numbers that does not converge to 0. Then the sequence is either eventually positive or eventually negative.*

Theorem 10. *Every Cauchy sequence of real numbers is bounded.*

It is worth mentioning the technical point that in proving statements about the real numbers, the numbers ε or ε_0 in all proofs can be chosen real or be kept rational since by Theorem 6 if ε is real then a rational ε' can be chosen that is arbitrarily close to it, and either smaller or larger, as needed.

A very useful result about sequences is the following that was stated as Theorem (BC) in the text.

Theorem 11. *Every non-decreasing sequence of real numbers that is bounded from above is Cauchy.*

Proof. Suppose that x_n is a non-decreasing sequence in $\mathbb{R}$ that is bounded from above, i.e. there is a fixed real number M such that

$$x_n \leq M \quad \text{for every } n = 1, 2, 3, \ldots \qquad (11.15)$$

I show that if x_n is *not* Cauchy then I get a contradiction. That will be grounds for declaring the sequence Cauchy. If x_n is not Cauchy then there must be some positive ε_0 and an increasing sequence of indices n_k

$$n_1 < n_2 < n_3 < \cdots$$

such that

$$x_{n_{j+1}} - x_{n_j} \geq \varepsilon_0 \quad \text{for every } j = 1, 2, 3, \ldots \qquad (11.16)$$

The absolute value was not required above since x_n is non-decreasing so that

$$x_{n_j} \leq x_{n_j+1} \leq x_{n_j+2} \leq \cdots \leq x_{n_{j+1}}$$

Appendix

Now using (11.16) I get

$$(x_{n_{j+1}} - x_{n_j}) + (x_{n_j} - x_{n_{j-1}}) + \cdots + (x_{n_2} - x_{n_1}) \geq \varepsilon_0 + \varepsilon_0 + \cdots + \varepsilon_0$$
$$x_{n_{j+1}} - x_{n_1} \geq j\varepsilon_0$$
$$x_{n_{j+1}} \geq x_{n_1} + j\varepsilon_0$$

There are infinitely many j so $j\varepsilon_0$ gets arbitrarily large as j does (this is a consequence of the Archimedean property that is discussed in the text). Therefore, there is J large enough that $J\varepsilon_0 > M - x_{n_1}$. This means that

$$x_{n_{J+1}} \geq x_{n_1} + J\varepsilon_0 > M$$

This contradicts the hypothesis (11.15); to avoid this x_n must be Cauchy. ■

The next result is the *converse* of Theorem 8. Up to now, when dealing with Cauchy sequences of rational numbers we could only assert that they accumulate around a certain value. If that value was not rational then we had reached a gap or hole in the set of rationals. The next theorem fills such gaps and says more than just Cauchy sequences of rational numbers converge to real numbers.

Theorem 12. (completeness) *Every Cauchy sequence of real numbers converges to a real number.*

Proof. Suppose that r_n is a Cauchy sequence of *real* numbers, i.e. $r_n \in \mathbb{R}$ for every index n. First we dispense with a trivial case: if there is an index N such that $r_n = r$ for all $n \geq N$ (i.e. if r_n is eventually constant) then clearly r_n converges to r and we are done.

So let us assume that r_n is not eventually constant. Then for each index n there another $n' > n$ such that $r_{n'} \neq r_n$. We may assume that n' is the smallest or least such index. Now by the density of rationals (Theorem 6) there is a rational number q_n between r_n and $r_{n'}$; in other words, q_n is closer to each of r_n and $r_{n'}$ than the two real numbers are to each other. In particular,

$$|r_n - q_n| < |r_n - r_{n'}|$$

Although I have blurred the notation in favor of better readability, it is worth remembering that each term of r_n is an equivalence class $[p_k]_n$ of rational Cauchy sequences and likewise, each term of the rational sequence q_n is the equivalence class of a constant sequence; i.e. q_1 is $[q_1, q_1, q_1, \ldots]$ and so on.

So now we know that there is a sequence (at least one) of *rational* numbers $q_1, q_2, q_3, \ldots$ whose terms are sandwiched in between the terms of the original sequence r_n. Let us verify that all such rational sequences are Cauchy; let $\varepsilon > 0$ and note that by the triangle inequality

$$|q_n - q_m| = |q_n - r_n + r_n - r_m + r_m - q_m|$$
$$\leq |q_n - r_n| + |r_n - r_m| + |r_m - q_m|$$
$$< |r_{n'} - r_n| + |r_n - r_m| + |r_m - r_{m'}|$$

Since r_n is Cauchy by assumption, there is an index N such that if $m, n > N$ then $|r_n - r_m| < \varepsilon/3$. This inequality is valid for $|r_{n'} - r_n|$ and $|r_m - r_{m'}|$ too because $m' > m$ and $n' > n$ by the construction above. Therefore, if $m, n > N$ then

$$|q_n - q_m| < \frac{\varepsilon}{3} + \frac{\varepsilon}{3} + \frac{\varepsilon}{3} = \varepsilon$$

and it follows that q_n is Cauchy. Therefore, its equivalence class $[q_n]$ is a real number; let us call it r. It remains to show that r_n converges to r. With ε and N defined above, we can state (with $m = N+1$) that $|r_n - r_{N+1}| < \varepsilon/3$ and $|q_{N+1} - r_{N+1}| < \varepsilon/3$. Further, since r is the equivalence class of q_n by taking N large enough we can ensure that $|q_{N+1} - r| < \varepsilon/3$ also. Therefore, using the triangle inequality again,

$$\begin{aligned}|r_n - r| &= |r_n - r_{N+1} + r_{N+1} - q_{N+1} + q_{N+1} - r| \\ &\leq |r_n - r_{N+1}| + |r_{N+1} - q_{N+1}| + |q_{N+1} - r| \\ &< \frac{\varepsilon}{3} + \frac{\varepsilon}{3} + \frac{\varepsilon}{3} = \varepsilon\end{aligned}$$

This verifies that r_n converges to r and concludes this proof. ∎

Appendix 415

11.5 Appendix: Discontinuity in a space of functions

The sudden jump of the staircase curves' lengths down to the length of the square's diagonal looks a lot like a jump discontinuity in the value of length: a sequence of functions, each with a well-defined length converges (uniformly) to a unique function, also having a well-defined length, but the length of the limit function is different from the limit of the lengths of the terms of the sequence.

If you go back to our definition of continuity in Chapter 7 then you may notice that this discrepancy bears a resemblance to what we get when we take the limit of a discontinuous function at a jump discontinuity. The "function" that is being discontinuous here is evidently the *length function*, say, L.

But the length function is not like the familiar functions whose domains are intervals or more generally, sets in the ordinary (three-dimensional) space. When we speak of length here we mean the length of *a curve*. If the curve in question is the graph of a function $f(x)$ then we write $L(f)$ as the length of $f(x)$.

This notation makes clear that L is a function that maps a function $f(x)$ to a number, namely, the length of the curve. In other words, *the domain of L is a set of ordinary functions* not a set of points on the x-axis or in ordinary space. For example, if $f(x) = \sqrt{1-x^2}$ whose graph is the upper half-circle of radius 1 for $-1 \leq x \leq 1$ then $L(f) = \pi$ is just half the circumference of the full circle of radius 1. Or if $f(x) = 2x$ whose graph is a straight line then the length of the stretch of this line from $x = 0$ to $x = 1$ is easily found using the Pythagorean theorem as $L(f) = \sqrt{5}$.

To understand the nature of the discontinuity in lengths and relate it to the earlier definition of continuity in Chapter 7, we need to understand spaces of functions. These are spaces in which individual points are real-valued functions. As we briefly noted earlier in Chapter 3, these are subsets of the infinite product of the set of real numbers with itself. These sets are technically complicated and not suitable for full discussion in this book. However, I give a brief, simplified discussion of one such space here to explain the discontinuity that we observe in the staircase curves' lengths.

Consider the set of all continuous functions (in the sense of Chapter 7) on the interval $[0,1]$ and let's denote this set of functions by the symbol $C[0,1]$. The polygonal staircase functions and the smooth sine functions that we encountered above are all continuous functions on $[0,1]$ and therefore, points or elements in $C[0,1]$. Since every function in $C[0,1]$ is also an ordinary real valued function on $[0,1]$ we see that each function in $C[0,1]$ is an element of the infinite product $\mathbb{R}^{[0,1]}$.[8] This is analogous to identifying every vector in the usual three-dimensional space $\mathbb{R}^3$ as a point of that space. The main difference is that the space of functions has infinite dimension.

Norm.

Just as vectors have lengths (their magnitudes) so can the functions in $C[0,1]$. But unlike $\mathbb{R}^3$, many essentially different length concepts are possible for $C[0,1]$ that turn out to be useful for different purposes. The one that we discuss here is one of the most important in analysis because it is associated with uniform convergence. Given any function f in $C[0,1]$ and any number x_0 in $[0,1]$, the absolute value $|f(x_0)|$ gives us the magnitude of $f(x_0)$ for the specific number x_0. This magnitude is different for different numbers in $[0,1]$. So let's consider $|f(x)|$ for *all* numbers x in $[0,1]$; the largest, or maximum of these numbers is a finite number that we denote by $\|f\|$. In symbols:

$$\|f\| = \max\{|f(x)| : x \in [0,1]\}$$

---

[8]$C[0,1]$ turns out to be a very small subset of $\mathbb{R}^{[0,1]}$; the latter has cardinal number $2^{2^{\aleph_0}}$ compared to the cardinal number of $C[0,1]$ which is a "mere" $2^{\aleph_0}$.

We call $\|f\|$ the *max norm*[9] of f which represents the size or magnitude of f relative to this norm. It so happens that $\|f\|$ has the same basic properties as the vector magnitude; in particular, it satisfies the important triangle inequality: for every pair of functions f, g in $C[0, 1]$

$$\|f + g\| \leq \|f\| + \|g\|$$

Recall that because of continuity, the Extreme Value Theorem implies that there is at least one number x^* in $[0,1]$ such that $\|f\| = |f(x^*)|$.

Calculating the norm of a function in $C[0, 1]$ is straightforward. For example, $\|x^2\|$ is the maximum value of x^2 over the interval $[0,1]$. This maximum occurs at $x = 1$ and $|1^2| = 1$ so $\|x^2\| = 1$. Can you figure out $\|2x\|$ the same way?

More generally, we simply find the absolute value and if necessary, use calculus to find the norm. For example, the maximum value of $\sin 2x$ is 1 which occurs at $x = \pi/4 \simeq 0.785$ where the derivative $2\cos 2x$ is 0. Therefore, $\|\sin 2x\| = 1$. Can you figure out $\|\sin x\|$? Keep in mind that $0 \leq x \leq 1$.

Distance and the uniform metric.

Just as for vectors $\vec{v}$ and $\vec{u}$ in $\mathbb{R}^3$ we define the distance between $\vec{v}$ and $\vec{u}$ as the magnitude of the difference $\vec{v} - \vec{u}$ we define *the distance between two functions* f, g in $C[0, 1]$ to be *the norm* of their difference, that is,

$$\text{distance between } f \text{ and } g \text{ is: } \|f - g\|$$

As we saw above, this distance can often be calculated using standard methods of calculus. For example, the distance between $f(x) = x$ and $g(x) = x^3$ is

$$\|x - x^3\| = \max\{|x - x^3| : x \in [0, 1]\}$$

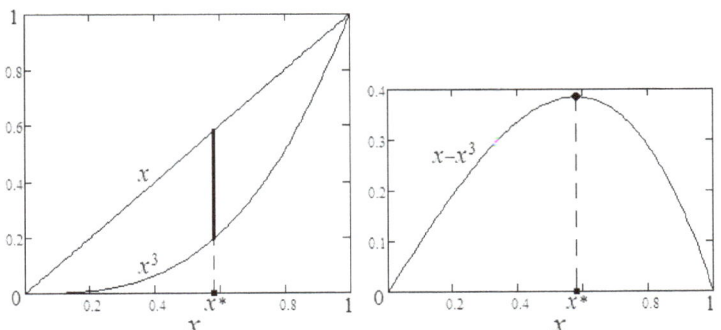

Figure 11.7: Uniform metric distance between two curves

Figure 11.7 shows that the difference $x - x^3$ is non-negative for all x in $[0,1]$; if we define $h(x) = x - x^3$ then the maximum value of $h(x)$ occurs where $h'(x) = 0$, that is, for the value of x that satisfies

$$h'(x) = 1 - 3x^2 = 0$$

---

[9] The standard term is the *sup norm*, where as we saw earlier in Chapter 8 "sup" is short for *supremum*, or the least upper bound. Recall that this term is synonymous with maximum when the largest value is attained at a specific number.

Solving the last equation gives a solution $x^* = 1/\sqrt{3} \simeq 0.58$ in the interval $[0,1]$; inserting this number in $x - x^3$ gives the distance between the two curves:

$$\|x - x^3\| = h(x^*) = \frac{1}{\sqrt{3}} - \left(\frac{1}{\sqrt{3}}\right)^3 = \frac{1}{\sqrt{3}} - \frac{1}{3\sqrt{3}} = \frac{2}{3\sqrt{3}} \simeq 0.38$$

The set $C[0,1]$ of continuous functions endowed with the distance function above is an example of a *metric space*. "Metric" is synonymous with distance. The max norm distance here is also called the *uniform metric* for the following reason. Recall that a sequence of functions f_n in $C[0,1]$ converges uniformly to a function f in $C[0,1]$ (or $f_n \xrightarrow{u} f$) if for every $\varepsilon > 0$ there is a positive integer N such that $|f_n(x) - f(x)| < \varepsilon$ for all $n > N$ *regardless of the choice of x* in $[0,1]$. If the absolute value inequality is true for every number x then the largest of the absolute value quantities must also be less than ε so that

$$\|f_n - f\| = \max\{|f_n(x) - f(x)| : x \in [0,1]\} < \varepsilon \quad \text{if } n > N \tag{11.17}$$

Therefore, we can conveniently define uniform convergence in terms of the uniform metric as:

$$f_n \xrightarrow{u} f \quad \text{if} \quad \|f_n - f\| \to 0 \quad \text{as } n \to \infty \tag{11.18}$$

Notice that $\|f_n - f\|$ is just a sequence of real numbers which is easier to deal with than a sequence of functions.

Functionals.

Before we can discuss continuity of functions on a metric space like $C[0,1]$ we need to define what we mean by functions on such a space. To keep technical details at a minimum, I consider only functions from $C[0,1]$ into the real numbers $\mathbb{R}$. These types of functions are called *functionals* because their range is the set of real numbers.[10] The norm or length of a function is a functional because for every f in $C[0,1]$ its norm $\|f\|$ is a real number.

We also worked with another important type of functional earlier: *the definite integral* $\int_a^b f(x)dx$. For instance, if f is in $C[0,1]$ then we may define the functional $I : C[0,1] \to \mathbb{R}$ as

$$I(f) = \int_0^1 f(x)dx$$

We may readily calculate the value of I for a variety of functions using the Fundamental Theorem of Calculus. For example,

$$I(x^n) = \int_0^1 x^n dx = \frac{x^{n+1}}{n+1}\bigg|_{x=0}^{x=1} = \frac{1}{n+1}$$

for all positive integer values of n. Here the functional I assigns to each function x^n the real number $1/(n+1)$; in particular, $I(x) = 1/2$, $I(x^2) = 1/3$ and so on. Similarly,

$$I(e^x) = \int_0^1 e^x dx = e^1 - e^0 = e - 1$$

---

[10]The range of a functional does not have to be $\mathbb{R}$; the set of complex numbers is also a useful range for various purposes.

A simple interpretation of the integral functional is to think of it as giving the *average value* of a function. For example, $I(x^2) = 1/3$ is the average value of x^2 on the interval [0,1] and $I(e^x) = e - 1$ is the average value of e^x.

The integral belongs to an important class of functionals known as *linear functionals*. A functional $F : C[0,1] \to \mathbb{R}$ is linear if for every pair of functions f, g in $C[0,1]$ and every pair of real numbers a, b it is true that

$$F(af + bg) = aF(f) + bF(g) \tag{11.19}$$

The integral I is a linear functional (recall the properties of integral that we discussed in Chapter 8 but the norm is not linear. To see why not, let $F(f) = \|f\|$ so that $F(x - x^3) = \|x - x^3\|$. We showed above that $F(x - x^3) = 2/3\sqrt{3}$. On the other hand,

$$F(x) - F(x^3) = \|x\| - \|x^3\| = 1 - 1 = 0 \neq \frac{2}{3\sqrt{3}}$$

which shows that (11.19) is not satisfied (with $a = 1$, $b = -1$, $f(x) = x$ and $g(x) = x^3$).

Continuous functionals.

A functional $F : C[0,1] \to \mathbb{R}$ is *continuous at a point f* in $C[0,1]$ relative to the max norm, if for every sequence f_n in $C[0,1]$ that converges uniformly to f we have $\lim_{n \to \infty} F(f_n) = F(f)$. Notice that this limit is just the limit of a sequence of real numbers If F is continuous at every point f in $C[0,1]$ then F is a *continuous functional*.

The integral I is a continuous functional because for every continuous function f and every sequence of continuous functions f_n that converges uniformly to f we have

$$\lim_{n \to \infty} I(f_n) = \lim_{n \to \infty} \int_0^1 f_n(x) dx = \int_0^1 \lim_{n \to \infty} f_n(x) dx = \int_0^1 f(x) dx = I(f)$$

With regard to the equality in the middle, recall that in the case of uniform convergence we can move the limit in or out of the integral. The norm itself is also a continuous functional; this is quickly proved using the triangle inequality. We are more interested here in the *length functional* which we consider next.

Lengths of curves are discontinuous functionals.

The graph of each continuous function on the interval [0,1] is a continuous curve. For each such curve f we define $L(f)$ to be the length of f. To ensure that L is a functional it is necessary that the graph of every function f have a finite, well-defined length that is a real number. For $C[0,1]$ this is technically complicated to deal with so we limit our attention to a subset of $C[0,1]$ for which it is easy to see that L is well-defined.

A function in $C[0,1]$ is *smooth* if it is *continuously differentiable* at every number in [0,1]; so f in $C[0,1]$ is smooth if its derivative $f'(x)$ exists *and* is continuous at every number x in [0,1]. We use the notation $C_1[0,1]$ for the set of all smooth functions in $C[0,1]$.

The set $C_1[0,1]$ is a proper subset of $C[0,1]$ because there are functions that are continuous but not smooth; a simple example is the absolute value function $|x|$ whose derivate at $x = 0$ cannot be defined continuously. An extreme example of a continuous function that is *not smooth at even a single point* is Weierstrass's function that we discuss later in this chapter.

On the other hand, $C_1[0,1]$ contains most functions that we find in calculus: all polynomials as well as the trigonometric functions $\sin x$ and $\cos x$ exponential functions are smooth; in fact, we

Appendix

discover later in this chapter that they are infinitely differentiable, a property that makes it possible to expand each of them as an infinite power series.

The set $C_1[0,1]$ endowed with the uniform metric is a metric space. It is called a *subspace* of $C[0,1]$ because the uniform metric on $C_1[0,1]$ is inherited from the larger space $C[0,1]$.

In calculus it is shown that if f is smooth then its length can be defined using an integral; specifically, if f is in $C_1[0,1]$ then

$$L(f) = \int_0^1 \sqrt{1 + [f'(x)]^2}\,dx$$

If $f'(x)$ is continuous then so is the square root expression inside the integral. It follows that the integral is well-defined. This formula shows that L is a well-defined functional on $C_1[0,1]$. *But is L continuous?*

Our discussion of the staircase curves shows that L is not continuous relative to the uniform metric. The functions

$$f_n(x) = \frac{\sin^2(n\pi x)}{2n}$$

are smooth functions on the interval [0,1] with continuous derivatives $(\pi/2)\sin(2n\pi x)$. Therefore, f_n is a point in $C_1[0,1]$ for every n. Using (11.17) or (11.18) we see that the sequence f_n converges uniformly to the zero function $f(x) = 0$ on [0,1] because the maximum value of $\sin^2(n\pi x)$ is simply 1 so that

$$\|f_n - 0\| = \left\|\frac{\sin^2(n\pi x)}{2n}\right\| = \frac{1}{2n} \to 0 \quad \text{as } n \to \infty$$

so $f_n \xrightarrow{u} 0$. The length of the zero function on the interval [0,1] is

$$L(0) = \int_0^1 \sqrt{1+0}\,dx = 1$$

But as we saw in our discussion of the staircase curves, for every positive integer n,

$$L(f_n) = \int_0^1 \sqrt{1 + \frac{\pi^2}{4}\sin^2(2n\pi x)}\,dx \geq \sqrt{2} > 1$$

It follows that the length functional L is not continuous on the space $C_1[0,1]$ relative to the uniform metric. Therefore, *the drop in the lengths of staircase curves in the limit that we observed earlier may be attributed to the discontinuity of the length functional!*

This is analogous to the jump discontinuity for ordinary functions that we discussed in Chapter 7 but the domain is now a metric space of functions so it is not possible to visualize the discontinuity using an ordinary graph.

Before closing, it is worth mentioning that other norms can be defined on $C_1[0,1]$ and some of those may be sensitive to variations in function values. One such norm is

$$\|f\|_1 = \max\{|f(x)| : x \in [0,1]\} + \max\{|f'(x)| : x \in [0,1]\}$$

This norm defines a metric or distance on $C_1[0,1]$ that takes into account the derivative, or the variation in the function as well as its raw values. It is therefore a refinement of the uniform norm. Specifically, if f and g are any two functions in $C_1[0,1]$ then the distance between them is:

$$\|f - g\|_1$$

Since derivatives are slopes of curves, the greater the oscillations, the larger the distance in the above norm is. Figures 9.9 and 11.8 illustrate this point.

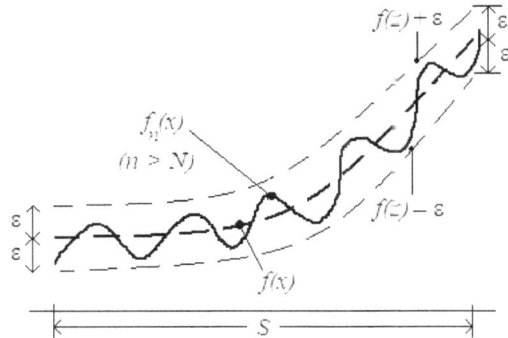

Figure 11.8: Curves that are near each other in value but not in variation

We see that the difference in values between $f_n(x)$ and $f(x)$ is smaller than ε throughout; however, the derivatives $f'_n(x)$ and $f'(x)$ can be very different from each other, especially at the points where $f_n(x)$ and $f(x)$ intersect since at these points the slopes of these curves are very different from each other. Near an intersection point x, the difference

$$|f'(x) - g'(x)|$$

is much larger than ε and this leads to a large value of $\|f - g\|_1$ overall.

Let's take a look at the norms of the earlier sine curves (with arbitrary c_n). Notice that

$$\|f_n\|_1 = \max\left\{\left|c_n \sin^2(n\pi x)\right| + |n\pi c_n \sin(2n\pi x)| : x \in [0,1]\right\} \geq n\pi c_n$$

since the maximum value of the sine function $\sin(2n\pi x)$ is 1. Therefore, the sequence f_n does not converge to the zero function (or any other function) in the new norm. In fact, we see that the new norm can be infinitely large if nc_n goes to infinity. The anomaly of staircase curves does not occur if we measure the distance between each of the staircase curves and the diagonal of the square using the new, refined norm!

11.6 Appendix: The derivative formula for $\sin x$

Our calculation of derivative formulas for the trigonometric functions in Chapter 7 was based on knowing the derivative of $\sin x$ at any number $x = a$. As you read through this section, it is important to keep some severe restrictions in mind: our knowledge of $\sin x$ (and $\cos x$, etc) is limited to what is stated in Appendix 11.2 and some numerically generated graphs. While these graphs suggest that $\sin x$ is smooth (and in particular continuous), strictly speaking we do not even know if $\sin x$ is actually continuous as defined in Chapter 7.

The derivative of $f(x) = \sin x$ is, by definition,

$$f'(a) = \lim_{n \to \infty} \frac{f(x_n) - f(a)}{x_n - a} = \lim_{n \to \infty} \frac{\sin x_n - \sin a}{x_n - a}$$

provided that the limit exists for every sequence x_n that approaches a (but $x_n \neq a$ for all n). A modification of the fraction in this limit makes it easier to calculate using trigonometric identities. Define $u_n = x_n - a$ for every n so that $u_n \to 0$ as $n \to \infty$ but $u_n \neq 0$ for all n. With this substitution, the derivative of $\sin x$ is given by the limit

$$f'(a) = \lim_{n \to \infty} \frac{\sin(a + u_n) - \sin a}{u_n}$$

Using the trigonometric identity $\sin(a + u_n) = \sin a \cos u_n + \cos a \sin u_n$ (see Appendix 11.2), a couple of limit theorems from Chapter 4 and doing a little algebra we get

$$\begin{aligned}
f'(a) &= \lim_{n \to \infty} \frac{\sin a \cos u_n + \cos a \sin u_n - \sin a}{u_n} \\
&= \lim_{n \to \infty} \frac{(\cos u_n - 1) \sin a}{u_n} + \lim_{n \to \infty} \frac{\sin u_n \cos a}{u_n} \\
&= \left(\lim_{n \to \infty} \frac{\cos u_n - 1}{u_n} \right) \sin a + \left(\lim_{n \to \infty} \frac{\sin u_n}{u_n} \right) \cos a
\end{aligned} \qquad (11.20)$$

We use the definitions of trigonometric functions to find the limits in (11.20). Consider Figure 11.9 which contains the geometric representation of the tangent function for an acute angle θ in addition to those of sine and cosine.[11]

In Figure 11.9 we see that in triangle OPQ the sine of the central angle θ is the length $|PQ|$ of the front side over hypotenuse (the radius of the circle) so

$$\sin \theta = \frac{|PQ|}{1} = |PQ|$$

The radian measure of θ is the length of the arc PQ' of the circle of unit radius. Evidently, the length of this arc is greater than the length of PQ so that

$$\sin \theta < \theta \qquad (11.21)$$

---

[11] For the sake of finding the limits, we are only interested in the values of θ that are near zero, so acute angles ($\theta < \pi/2$) are all that we need to consider. Although the figure shows a positive θ, this angle does not have to be positive; for negative values of θ we use the reflection of the triangles in Figure 11.9 across the horizontal axis and essentially the same arguments apply.

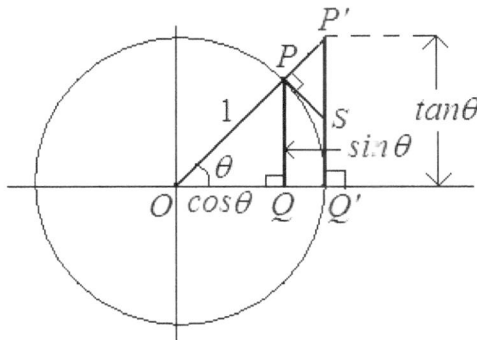

Figure 11.9: A geometrical representation of sine and tangent functions

Next, consider triangle $OP'Q'$ where the tangent of the central angle θ is the length of the front side $P'Q'$ over the adjacent side (the radius of the circle) so

$$\tan\theta = \frac{|P'Q'|}{1} = |P'Q'|$$

The length $|P'Q'|$ is greater than the length of the arc PQ' but this is not visually obvious. To see why, consider the line segment PS which is tangent to the circle at the point P and thus, it is also perpendicular to the radius OP. The small triangle SPP' is a right triangle so the length $|SP'|$ of its hypotenuse is greater than the length $|SP|$ of the side. It follows that the sum of the lengths of the two line segments $Q'S$ and SP is smaller that the length of the side $Q'P'$, that is,

$$|Q'S| + |SP| < |Q'P'|$$

The number on the left hand side of the above inequality is greater the length of the arc PQ', the radian measure of θ, so we conclude that

$$\theta < |Q'S| + |SP| < \tan\theta$$

When we put this inequality and (11.21) together we have

$$\sin\theta < \theta < \tan\theta$$

Now, divide by $\sin\theta$ and recall that $\tan\theta = \sin\theta/\cos\theta$ to obtain

$$1 < \frac{\theta}{\sin\theta} < \frac{1}{\cos\theta}$$

Flipping the fractions changes the directions of both inequalities and yields

$$\cos\theta < \frac{\sin\theta}{\theta} < 1$$

By repeating these arguments for the reflections of the triangles in Figure 11.9 we find that these inequalities are also true when $\theta < 0$; but this is actually rather plain with a little reflection, recalling that $\sin\theta$ is an odd function while $\cos\theta$ is even.

Now, with our sequence u_n above inserted in these inequalities we get

$$\cos u_n < \frac{\sin u_n}{u_n} < 1 \qquad (11.22)$$

The number 1 on the right hand side of (11.22) is fixed for all choices of u_n so if we show that the sequence $\cos u_n$ converges to 1 as $u_n \to 0$ then the Squeeze Theorem implies that

$$\lim_{n \to \infty} \frac{\sin u_n}{u_n} = 1 \qquad (11.23)$$

This must hold regardless of which sequence u_n is selected that converges to 0 and this is guaranteed if the function $\cos x$ is continuous at $x = 0$ as suggested by a numerically generated graph of $\cos x$ near the origin.

Proving that both $\sin x$ and $\cos x$ are continuous at $x = 0$ is quite simple. We go back to the inequality in (11.21) and use the fact that $\sin x$ is an odd function to conclude that

$$\sin(-\theta) = -\sin\theta > -\theta$$

We have shown that $\sin x > x$ for $x = -\theta < 0$ and $\sin x < x$ for $x = \theta > 0$; use the absolute value this fact can be stated succinctly as

$$|\sin x| < |x| \quad x \neq 0 \qquad (11.24)$$

So if θ_n is an arbitrary sequence that converges to 0 (but $\theta_n \neq 0$ for all n) then for any $\varepsilon > 0$ there is a positive integer N such that $|\theta_n| < \varepsilon$, so by (11.24) $|\sin \theta_n| < \varepsilon$ also. Hence,

$$\lim_{n \to \infty} \sin \theta_n = 0 = \sin 0 \quad \text{with} \quad \theta_n \to 0$$

So $\sin x$ is continuous at 0. From this fact and the identity $\cos x = \sin(x + \pi/2)$ (see Appendix 11.2) we can also see that $\cos x$ is continuous at $x = 0$ because

$$\lim_{n \to \infty} \cos \theta_n = \lim_{n \to \infty} \sin\left(\theta_n + \frac{\pi}{2}\right) = \sin\left(\lim_{n \to \infty} \theta_n + \frac{\pi}{2}\right) = \sin \frac{\pi}{2} = 1 = \cos 0$$

These conclusions did not depend on the choice of the sequence θ_n that converges to 0 so we have proven the continuity of both sine and cosine functions at $x = 0$. This completes our proof of (11.23).

To finish the derivation of the formula for the derivative of $\sin x$ we now calculate the other limit in (11.20) as follows: first, multiply and divide the fraction by $\cos u_n + 1$ and then use the trigonometric Pythagorean identity and a little algebra to get

$$\frac{\cos u_n - 1}{u_n} = \frac{\cos^2 u_n - 1}{u_n(\cos u_n + 1)} = \frac{-\sin^2 u_n}{u_n(\cos u_n + 1)} = -\frac{\sin u_n}{\cos u_n + 1} \frac{\sin u_n}{u_n}$$

We now take the limit:

$$\lim_{n\to\infty} \frac{\cos u_n - 1}{u_n} = \lim_{n\to\infty} \left(-\frac{\sin u_n}{\cos u_n + 1} \cdot \frac{\sin u_n}{u_n}\right)$$
$$= -\lim_{n\to\infty}\left(\frac{\sin u_n}{\cos u_n + 1}\right) \cdot \lim_{n\to\infty}\left(\frac{\sin u_n}{u_n}\right)$$
$$= -\frac{\lim_{n\to\infty} \sin u_n}{\lim_{n\to\infty} \cos u_n + 1} \quad (1)$$
$$= -\frac{0}{1+1} = 0$$

Inserting this number and using (11.23) in (11.20) gives the formula for the derivative of $\sin x$ at $x = a$

$$f'(a) = \cos a$$

just as we had asserted in Chapter 7!

11.7 Appendix: Proofs of some limit and derivative theorems in the text

In this section I prove the limit theorems for sequences in Chapter 4 and the differentiation theorems or derivative rules in Chapter 7. The discussion is straightforward though fairly technical. The limit theorems for sequences are naturally more abstract since the concept is a logically more primitive one. The proofs of the derivative rules are more computational because we can use the sequence limit theorems in our proofs.

Proving the limit theorems for sequences

We have two sequences s_n and s'_n that converge to real numbers s and s' respectively:

$$s_n \to s \quad \text{and} \quad s'_n \to s' \quad \text{as } n \to \infty$$

These statements mean that for every number $\varepsilon_1 > 0$ there are positive integers N and N' such that

$$|s_n - s| < \varepsilon_1 \quad \text{for all } n \geq N \tag{11.25}$$
$$|s'_n - s'| < \varepsilon_1 \quad \text{for all } n \geq N' \tag{11.26}$$

If N'' is the larger of N and N' then both of the above inequalities hold whenever the index n exceeds N''.

Sequence sum.

Let's consider the sum $s_n + s'_n$ of the two sequences; we show that

$$s_n + s'_n \to s + s' \quad \text{as } n \to \infty \tag{11.27}$$

Notice that

$$|s_n + s'_n - (s + s')| = |s_n - s + s'_n - s'| \leq |s_n - s| + |s'_n - s'|$$

The last inequality is just the triangle inequality. According to (11.25) and (11.26) each of the two quantities on the far right is less than ε_1 if $n \geq N''$ so if ε is any positive number and we set $\varepsilon_1 = \varepsilon/2$ then

$$|s_n + s'_n - (s + s')| < \frac{\varepsilon}{2} + \frac{\varepsilon}{2} = \varepsilon \quad \text{for all } n \geq N''$$

This proves (11.27), the limit theorem for sums.

Sequence product.

For the product we must show that

$$s_n s'_n \to ss' \quad \text{as } n \to \infty \tag{11.28}$$

To this end, let ε be an arbitrary (unspecified) positive number and notice that

$$|s_n s'_n - ss'| = |s_n s'_n - s_n s' + s_n s' - ss'| = |s_n(s'_n - s') + (s_n - s)s'|$$

The simultaneous addition and subtraction of the term $s_n s'$ inside the absolute value causes no change in value on the left hand side but it allows us to use our hypotheses in (11.25) and (11.26). Using the triangle inequality gives

$$|s_n s'_n - ss'| \leq |s_n||s'_n - s'| + |s_n - s||s'|$$

The sequence s_n is convergent and therefore, it is bounded (see Appendix 11.4); that is, there is a positive number M such that $|s_n| \leq M$ for all n. Now, let M' be the larger of M and $|s'|$ so that

$$|s_n s'_n - ss'| \leq M'(|s'_n - s'| + |s_n - s|)$$

The final step is to set $\varepsilon_1 = \varepsilon/(2M')$ so that $M' = \varepsilon/(2\varepsilon_1)$ and the last inequality above can be written as

$$|s_n s'_n - ss'| \leq \frac{\varepsilon}{2\varepsilon_1}(|s'_n - s'| + |s_n - s|) < \frac{\varepsilon}{2\varepsilon_1}(\varepsilon_1 + \varepsilon_1) = \varepsilon$$

This now proves (11.28), the limit theorem for products.

A special case of (11.28) is when one of the sequences is a constant sequence, say, $s'_n = c$ for all n where c is some fixed real number. For this case, (11.28) reads

$$cs_n \to cs \quad \text{as } n \to \infty$$

This *constant multiple rule* is useful result which says that if we multiply a convergent sequence by any number c then the new sequence converges to the old limit multiplied by c. As a quick application, we see that if s'_n converges to s' then

$$s_n - s'_n = s_n + (-1)s'_n \to s + (-s') = s - s'$$

Sequence quotient.
We prove that if $s'_n \neq 0$ for all n and also $s' \neq 0$ then

$$\frac{s_n}{s'_n} \to \frac{s}{s'} \quad \text{as } n \to \infty \tag{11.29}$$

We start with

$$\left|\frac{s_n}{s'_n} - \frac{s}{s'}\right| = \left|\frac{s_n s' - s'_n s}{s'_n s'}\right| = \left|\frac{s_n s' - ss' + ss' - s'_n s}{s'_n s'}\right| = \left|\frac{(s_n - s)s' + s(s' - s'_n)}{s'_n s'}\right|$$

The last fraction is split using the triangle inequality as follows

$$\left|\frac{s_n}{s'_n} - \frac{s}{s'}\right| \leq \left|\frac{s_n - s}{s'_n}\right| + \left|\frac{s(s' - s'_n)}{s'_n s'}\right| = \frac{1}{|s'_n|}|s_n - s| + \left|\frac{s/s'}{s'_n}\right||s'_n - s'|$$

Next, we need to examine the coefficient $|1/s'_n|$. Intuitively, since $s'_n \to s'$ and $s' \neq 0$ we see that s'_n cannot get arbitrarily close to 0 and from this observation conclude that $|1/s'_n|$ is bounded, that is, there is a positive number M such that $|1/s'_n| \leq M$ for all large values of n. To be precise, we know that as s'_n gets close to s' there is a positive integer, say, N_1 such that $|s'_n - s'| < |s'|/2$ for every $n \geq N_1$. Now the triangle inequality implies that (see Appendix 11.4)

$$\frac{|s'|}{2} > |s'_n - s'| = |s' - s'_n| > |s'| - |s'_n|$$

so we can write

$$|s'_n| > |s'| - \frac{|s'|}{2} = \frac{|s'|}{2} \quad \text{for all } n \geq N_1$$

Thus we see that
$$\left|\frac{1}{s'_n}\right| < \frac{2}{|s'|} \quad \text{for all } n \geq N_1$$

From this argument we see that the upper bound can be chosen as $M = 2/|s'|$ as long as $n \geq N_1$. The process now follows the same path as that of the products; we choose N_2 to be the larger of N_1 and N'' and given an arbitrary $\varepsilon > 0$ and notice that for all $n \geq N_2$

$$\left|\frac{s_n}{s'_n} - \frac{s}{s'}\right| < \frac{2}{|s'|}|s_n - s| + \left|\frac{2s}{s'^2}\right||s'_n - s'| < \frac{2}{|s'|}\varepsilon_1 + \left|\frac{2s}{s'^2}\right|\varepsilon_1 = A\varepsilon_1$$

where
$$A = \frac{2}{|s'|}\varepsilon_1 + \left|\frac{2s}{s'^2}\right|$$

So if we define $\varepsilon_1 = \varepsilon/A$ then we see that

$$\left|\frac{s_n}{s'_n} - \frac{s}{s'}\right| < \varepsilon \quad \text{for all } n \geq N_2$$

This proves (11.29), the limit theorem for quotients.

The Squeeze Theorem.

We are given three sequences of real numbers a_n, b_n and c_n such that $a_n \leq b_n \leq c_n$ for every index n.

We want to show that if the flanking sequences a_n and c_n both converge to the same limit L then b_n also converges to L.

To prove this, let's collect some facts. Since the sequences a_n and c_n both converge to the same limit L we conclude that for every $\varepsilon > 0$ there is a positive integer N such that for all $n > N$:

$$L - \varepsilon < a_n < L + \varepsilon \quad L - \varepsilon < c_n < L + \varepsilon$$

Therefore,
$$b_n \leq c_n < L + \varepsilon \quad b_n \geq a_n > L - \varepsilon$$

from which it follows that for all $n > N$:

$$L - \varepsilon < b_n < L + \varepsilon$$

Since the above argument holds for every $\varepsilon > 0$ it follows that b_n converges to L and the Squeeze Theorem is proved.

Proving the product and quotient rules for derivatives

In this part of the appendix we prove the derivative rules for products and quotients (the rule for sums was proved in the text). The proofs are mainly calculation of limits but in some steps we need to modify an expression so that it accommodates our hypotheses.

The product rule.

We first prove that the following identity, namely, the product rule, is valid:

$$(fg)'(a) = f'(a)g(a) + f(a)g'(a) \tag{11.30}$$

By assumption, $f'(a)$ and $g'(a)$ exist and are given as the limits

$$f'(a) = \lim_{n\to\infty} \frac{f(x_n) - f(a)}{x_n - a} \quad \text{and} \quad g'(a) = \lim_{n\to\infty} \frac{g(x_n) - g(a)}{x_n - a}$$

where as usual x_n is an arbitrary unspecified sequence that converges to a but $x_n \neq a$ for all n. The left hand side of (11.30) exists if the following limit exists:

$$\lim_{n\to\infty} \frac{f(x_n)g(x_n) - f(a)g(a)}{x_n - a}$$

We show that the above limit exists and actually equals the expression that is given by the left hand side of (11.30), thus proving the product rule.

Using a trick similar to what we used for the limit of a product earlier, lets subtract and add the term $f(x_n)g(a)$ and combine terms in the numerator to get

$$\lim_{n\to\infty} \frac{f(x_n)g(x_n) - f(a)g(a)}{x_n - a} = \lim_{n\to\infty} \frac{f(x_n)[g(x_n) - g(a)] + [f(x_n) - f(a)]g(a)}{x_n - a}$$

The expression on the right hand side can be split into two using the limit theorem for the sum of sequences:

$$\lim_{n\to\infty} \frac{f(x_n)g(x_n) - f(a)g(a)}{x_n - a} = \lim_{n\to\infty} \frac{f(x_n)[g(x_n) - g(a)]}{x_n - a} + \lim_{n\to\infty} \frac{[f(x_n) - f(a)]g(a)}{x_n - a}$$

Next, using the limit theorem for the product of sequences gives

$$\lim_{n\to\infty} \frac{f(x_n)g(x_n) - f(a)g(a)}{x_n - a} = \lim_{n\to\infty} f(x_n) \lim_{n\to\infty} \frac{g(x_n) - g(a)}{x_n - a} + g(a) \lim_{n\to\infty} \frac{f(x_n) - f(a)}{x_n - a}$$

We now recognize the limits on the right hand side as derivatives:

$$\lim_{n\to\infty} \frac{f(x_n)g(x_n) - f(a)g(a)}{x_n - a} = \lim_{n\to\infty} f(x_n)g'(a) + g(a)f'(a)$$

To finish the proof of (11.30) we need to make one more observation: by hypothesis, $f(x)$ is differentiable at $x = a$; therefore, $f(x)$ is continuous at $x = a$ and we conclude that

$$\lim_{n\to\infty} f(x_n) = f(\lim_{n\to\infty} x_n) = f(a)$$

This concludes our proof of the product rule.

The quotient rule.

The quotient rule is proved in a similar fashion. In addition to the existence of $f'(a)$ and $g'(a)$ we also assume that $g(a) \neq 0$. We must show that the following limit exist and its value is the one given by the quotient rule:

$$\lim_{n\to\infty} \frac{f(x_n)/g(x_n) - f(a)/g(a)}{x_n - a} \qquad (11.31)$$

The sequence x_n has its usual meaning here. If you are a careful reader then you may be wondering about the fraction $f(x_n)/g(x_n)$ whose denominator may well be zero for some values of

Appendix

x_n. This is an important issue but not hard to resolve. We are assuming that $g(x)$ is differentiable at $x = a$ so it is also continuous there. We are also assuming that $g(a)$ is not 0 so there is a small interval around a on the x-axis, say, $(a - \varepsilon, a + \varepsilon)$ where $g(x)$ is nonzero;[12] therefore, $g(x) \neq 0$ for all values of x in $(a - \varepsilon, a + \varepsilon)$. Since every sequence that converges to a has to enter this interval (no matter how small ε may be) we can discard the (finitely many) terms of x_n that are not in $(a - \varepsilon, a + \varepsilon)$ without affecting its limit.

Let's simplify the fraction in (11.31) to get

$$\lim_{n\to\infty} \frac{f(x_n)/g(x_n) - f(a)/g(a)}{x_n - a} = \lim_{n\to\infty} \frac{f(x_n)g(a) - g(x_n)f(a)}{(x_n - a)g(x_n)g(a)}$$

Next, we subtract and add $f(a)g(a)$ in the numerator of the last fraction and combine terms to get

$$\lim_{n\to\infty} \frac{f(x_n)/g(x_n) - f(a)/g(a)}{x_n - a} = \lim_{n\to\infty} \frac{[f(x_n) - f(a)]g(a) + [g(a) - g(x_n)]f(a)}{(x_n - a)g(x_n)g(a)}$$

Next, we split the last fraction to obtain

$$\lim_{n\to\infty} \frac{f(x_n)/g(x_n) - f(a)/g(a)}{x_n - a} = \lim_{n\to\infty} \frac{1}{g(x_n)g(a)} \left[\frac{f(x_n) - f(a)}{x_n - a} g(a) - \frac{g(x_n) - g(a)}{x_n - a} f(a) \right]$$

We may now apply the limit theorems for sequences to get

$$\lim_{n\to\infty} \frac{f(x_n)/g(x_n) - f(a)/g(a)}{x_n - a} = \frac{1}{\lim_{n\to\infty} g(x_n)g(a)} \left[\lim_{n\to\infty} \frac{f(x_n) - f(a)}{x_n - a} g(a) - \lim_{n\to\infty} \frac{g(x_n) - g(a)}{x_n - a} f(a) \right]$$
$$= \frac{1}{[g(a)]^2} [f'(a)g(a) - g'(a)f(a)]$$

In the last step we used the fact that $g(x)$ is continuous at $x = a$ to conclude that $\lim_{n\to\infty} g(x_n) = g(a)$. Since the last expression on the right hand side is a well-defined number we see that the limit on the left hand side exists. Therefore, the quotient $f(x)/g(x)$ is differentiable at $x = a$ and as we see from the above calculations, its value is the one given by the quotient rule.

---

[12] This is intuitively easy to see by drawing the graph of a continuous function whose value is not 0 at some $x = a$. A proof by contradiction goes like this: if $g(x)$ can be zero no matter how close the value of x is to a then we can find a sequence z_n that converges to 0 and $g(z_n) = 0$. Since $g(x)$ is continuous at a this would mean that $g(a) = \lim_{n\to\infty} g(z_n) = 0$ which contradicts our assumption that $g(a) \neq 0$.

References and Further Reading

Some of the publications in this section (identified with *) are also useful for further reading.

David Acheson, *The Calculus Story: A Mathematical Adventure*, Oxford University Press, Oxford, 2017

J.L. Berggren, *Episodes in the Mathematics of Medieval Islam*, Springer-Verlag, New York, 1986

*David Bressoud, *A Radical Approach to Real Analysis*, Mathematical Association of America, Washington DC, 1994

Eugenia Cheng, *Beyond Infinity: An Expedition to the Outer Limits of the Mathematical Universe*, Profile Books, London, 2017

T.G. Faticoni, *The Mathematics of Infinity: A Guide to Great Ideas*, Wiley, New York, 2006

*Patrick M. Fitzpatrick, *Advanced Calculus*, 2nd ed., Thomson, Brooks-Cole, Blemont, 2006

George Gamov, *1, 2, 3 ... Infinity: Facts and Speculations of Science*, Viking-Compass, New York, 1962

Luke Heaton, *A Brief History of Mathematical Thought*, Robinson-Little, Brown, London, 2015

*J.M. Henle and E.M. Kleinberg, *Infinitesimal Calculus*, MIT Press, Cambridge, 1979

*David Hilbert and S. Cohn-Vossen, *Geometry and the Imagination*, 2nd ed., Chelsea Publishing Company, New York, 1952

Morris Klein, *Calculus: An Intuitive and Physical Approach*, 2nd ed., Dover Publications, New York, 1998

*Erwin Kreyszig, *Introductory Functional Analysis with Applications*, Wiley, New York, 1978

*Serge Lang, *Analysis I*, Addison-Wesley, Reading, 1968

Eli Maor, *To Infinity and Beyond*, Princeton University Press, Princeton, 1991

Eli Maor, *e: The Story of a Number*, Princeton University Press, Princeton, 1994

*Abraham Robinson, *Non-standard Analysis*, Princeton University Press, Princeton, 1996

Rudy Rucker, *Infinity and the Mind: The Science and Philosophy of the Infinite*, Princeton University Press, Princeton, 2005

Charles Seife, *Zero: The Biography of a Dangerous Idea*, Penguin Books, New York, 2000

Ian Stewart, *Taming the Infinite*, Quercus Publishing, London, 2008

Ian Stewart, *Infinity: A Very Short Introduction*, Oxford University Press, Oxford, 2017

Ian Stewart, *Significant Figures: The Lives and Works of Great Mathematicians*, Basic Books, New York, 2017

James Stewart, *Calculus*, 6th ed., Thomson, Brooks-Cole, Blemont, 2008

*Manfred Stoll, *Introduction to Real Analysis*, Addison-Wesley, Reading, 1997

Steven Strogatz, *The Joy of X: A Guided Tour of Math, from One to Infinity*, Mariner Books, New York, 2012

Steven Strogatz, *Infinite Powers: How Calculus Reveals the Secrets of the Universe*, Houghton Mifflin Harcourt, New York, 2019

*Patrick Suppes, *Axiomatic Set Theory*, Dover Publications, New York, 1972

Patricia Barnes-Svarney and Thomas E. Svarney, *The Handy Math Answer Book: Your Smart Reference*, Visible Ink Press, Canton, 2006

*Terence Tao, *Analysis I*, 3rd ed., Hindustan Book Agency, New Delhi, 2014

*Terence Tao, *Analysis II*, 3rd ed., Hindustan Book Agency, New Delhi, 2014

Richard Webb, Editor, *How Numbers Work*, New Scientist Instant Expert, John Murray Learning, London and Nicholas Brealey Publishing, Boston, 2018

Index

Abel, N. H., 321
absolute convergence, 153, 160, 325, 328
absolute convergence test, 153
acceleration, 222
Al-Khwarizmi, M., 5, 25
aleph null, 34
algebra, of sequence limits, 89
algebraic number, 117
Alhazen, 25
alternating series test, 150, 290
alternating series, error estimate, 151
antiderivative, 243
antisymmetric relation, 41
approximation, 83, 255
    area, 232
    by sequences, 21, 173
    integral, 270
    numerical, 20
    of π, 10
    of $\sqrt{2}$, 96
    of $\sqrt{2}$ by the divide and average rule, 7, 98
    of e, 260
    rational, 98, 285
approximation error (area), 225
approximation error (Riemann sum), 231
Archimedean property, 109, 112, 125
Archimedes, 2, 11, 18
area function, 254
Aristotle, 11
average value (of a function), 256, 418
Avicenna, 25

Barrow, I., 16
Basel problem, 368, 369
Berkeley, G., 23
Bernoulli, J., 355
bijection, 53, 64, 73

binomial function, 344
binomial series, 345
Biruni, 25
bisection algorithm, 208
bounded function, 230
bounded set, 110, 111
Brahmagupta, 24

Cantor's construction (real numbers), 21, 97, 403
Cantor's diagonal argument, 114, 165
Cantor's power-set theorem, 67, 402
Cantor, G., 5, 34
Cardano, G., 27
cardinal number, 64
cardinality, 32, 64, 73, 166
Cauchy convergence criterion, 135, 153
Cauchy sequence, 99, 102, 112, 134, 153
    equivalent, 102
Cauchy, A-L., 99
Cavalieri, B., 16
centillion, 2
chain rule, 182
Champernowne number, 8
chaotic sequence, 82
closed interval, 38
comparison test, 138, 142, 156
complement, 35
completeness, 5, 203
composition (of functions), 46, 181
conditional convergence, 156, 157, 328
conjunction, 56
constant sequence, 78
constant-multiple rule (derivatives), 180
constant-multiple rule (integrals), 249
containment relation, 33
continuous function, 192, 217
continuous functional, 418

continuum, 11
continuum hypothesis, 68
contradiction, 60
contrapositive, 51, 59, 74, 307
convergence
    absolute, 153, 325, 328
    conditional, 156, 328
    pointwise, 300
    uniform, 14, 306
convergence (of sequences), 87
converse, 51, 59
Cotes, R., 343
countable, 94, 117
countable set, 69
countably infinite, 69
countably infinite union, 70
cusp, 189, 199

d'Alembert, J-B., 144
Darboux sums, 234
Darboux, J-G., 234
de Moivre's identity, 343
De Morgan laws, 36, 70
De Morgan, A., 36
decimal expansion, 162, 165
decreasing function, 52, 233
Dedekind cuts, 93
Dedekind, R., 93
degree measure (angles), 397
density of irrational numbers, 107
density of rational numbers, 107, 108
denumerable, 69
derivative, 90
    definition, 174
    function, 175
    notation, 175
    one sided, 174
derivative, function-limit definition, 217
Descartes, R., 5
diagonal argument (Cantor), 114
difference quotient, 197
differentiable, 175
differential notation, 175
Dirichlet's function, 196, 243, 377
Dirichlet, J.P.G.L., 196
discontinuous function, 192, 196

discrete algorithms, 386
disjoint sets, 37, 40
disjunction, 56
distance (between functions), 416
divergence test, 135, 150
divergence to infinity, 92
divergent sequence, 92
divide and average rule, 7, 100, 207
domain, 42, 48
double containment, 33, 36, 41

empty set, 67
epsilon, 86
epsilon and delta definition of limit, 216
epsilon-index, convergent sequences, 87
epsilon-index, function, 294
epsilon-index, general, 101
epsilon-strip, 307
equipollent, 66, 69
equivalence class, 40, 103
equivalence relation, 40, 102, 128
equivalent (logical), 56
Euclid, 11, 66
Euclid's theorem on primes, 66
Euler's identity, 343
Euler, L., 117
exponential function, general, 267
Extreme Value Theorem, 203, 232, 416

factorial, 144
Fermat, P., 5
Fibonacci (Leonardo of Pisa), 24
field, 95
finite resolution, 84
Fourier coefficients, 366
Fourier cosine series, 372
Fourier series, 366
Fourier sine series, 372
Fourier's convergence theorem, 377
    continuous case, 378
Fourier, J-B.J., 14
free fall motion, 221
function, 42
    affine, 53
    constant, 52
    decreasing, 52

Appendix

    domain, 42
    even, 44, 342, 372
    increasing, 52
    iteration, 48
    monotone, 52
    nonlinear, 203
    nowhere differentiable, 379
    odd, 44, 342, 372
    one to one, 50
    onto, 53
    periodic extension, 375
    piecewise smooth, 377
function limit, 217
functional, 417
    continuous, 418
    linear, 418
functional analysis, 75
Fundamental Theorem of Algebra, 117
Fundamental Theorem of Calculus, Part 1, 246
Fundamental Theorem of Calculus, Part 2, 255

Galileo, 172
Galileo's equations, 173, 224
Gamov, G., 64
general power rule (derivatives), 183
Generous (elf), 11
geometric power series, 286
geometric progression (finite), 131
geometric series, 130, 286
    common ratio, 130, 286
    formula, 132
Gibbs phenomenon, 374
Grateful (elf), 13
greatest lower bound, 111, 113, 238
Greedy (elf), 11, 305
Gregory, J., 334

Hardy's inequality, 382
Hardy, G.H., 382
harmonic series, 137
harmonic series (alternating), 149, 156, 332
Hermite, C., 117
Hilbert's hotel, 65
Hilbert, D., 64
horizontal line test, 51
hyperreal numbers, 28

iff, 57
image set, 48
implication, 50, 56
improper Riemann integral
    unbounded functions, 275
    unbounded intervals, 278
increasing function, 52, 233
indeterminate form, ∞/∞, 358
indeterminate form, $0/0$, 353, 356
indexed sets, 69
inference, rule, 60
infimum, 111, 238
infinite product (of sets), 72
infinite resolution lens, 84
infinite series, 9, 124
    convergence, 124
    divergence, 124
    equivalent, 127
    geometric, 119, 130
    harmonic, 120
    infinite values, 125
    n-th term, 124
    tail, 129
    telescoping, 127
    the integral test, 282
    the p-series test, 281
infinite zooming process, 84
infinitely differentiable, 360
infinitely differentiable function, 338
infinitesimal, 22
infinity, actual (sets), 63
infinity, potential, 11, 39
injective, 50
integers, 34
integral (Riemann), 239
integral test, 282
integration, 19
integration by parts, 251
integration constant, 244
Intermediate Value Theorem, 201, 208
intersection (of sets), 34
interval, 38, 45
    bounded, 39
    closed, 38
    end-points, 38
    half-closed, 38

half-open, 38
nested, 112
open, 38
unbounded, 39
intuition, 385, 387
inverse function, 55
inverse sine and its Maclaurin series, 348
irrational numbers, 93

jump discontinuity, 376

Kashi, J., 393
Kepler, J., 16
Khayyam, O., 25
Koch's snowflake, 315, 380
Koch, N.F.H. von, 380

l'Hôpital's rule, 356–359, 394
l'Hôpital, G., 355
law of excluded middle, 28, 58
least integer function, 87
least upper bound, 111, 113, 237
Least Upper Bound Property (sets), 203
Leibniz's notation (derivative), 22, 175
Leibniz, alternating series test, 150
Leibniz, G.W., 16
limit, 19, 216
algebraic properties, 89
epsilon and delta definition of, 216
notation, 217
of a function, 217
sequence, 85, 87
Lindemann, F., 117
linear functional, 418
linear property (integrals), 249
Liouville numbers, 166
Liouville's constant, 166
Liouville, J., 117, 166
Liu, Hui, 394
logarithm function, base 10, 269
logarithm, base e (natural), 259
logarithmic function (general), 268
logic, 55
logic (binary), 55
logical equivalence, 51
logistic equation, 82

lower bound, 110
lower integral, 238
lower sum 234

Maclaurin series, 340
$\sin x$, 340
$\sin^{-1} x$, 348
even function, 342
odd function, 342
Maclaurin, C., 340
Madhava of Sangamagrama, 334
Madhava-Gregory-Leibniz series, 10, 334
magnitude, 24
mapping, 43, 45
max norm, 416
Mean Value Theorem, 212
Mean Value Theorem (integrals), 255
metric space, 417
modus ponens, 60
monotone function, 52, 240
monotone sequence, 81
monotone sums property, 235

Napier, J., 17
natural exponential function, 264
natural logarithm, 259
natural logarithm, power series expansion, 331
natural numbers, 33
necessary and sufficient condition, 57
necessary condition, 198
negation, 59
neighborhood, 218
Nested Intervals Property, 85, 112, 203
Newton's gravitational force, 221
Newton's second law of motion, 221
Newton, I., 16
Newton-Raphson method, 7
Newton-Raphson recursion, 205, 207, 208
Newton-Raphson recursion function (or map), 209
non-standard analysis, 28
nonlinear dynamical systems, 17
nonlinear equation, 203
nowhere differentiable function, 379
numerical integration
the midpoint rule, 270
the trapezoidal rule, 272

open interval, 38
order relation, 41
ordered field, 95
ordered set, 42
ordering, 41
    partial, 41
    reverse, 41
    strict, 41
    total, 41
orthogonality formulas, 364
oscillating sequence, 82

p-series, 141
p-series (alternating), 151
p-series test, 141, 152, 154, 281
parabola, 43
paradox
    elves' gold, 305
    Gabriel's Horn, 16, 274
    painting a column, 127
    Tower of Boxes, 146, 274
    Tower of Boxes, finite height, 16, 147
    Zeno's Achilles and the tortoise, 8
partial ordering, 41
partial sum, 124, 291
partition
    mesh, 229
    of a closed interval, 228
    refinement, 229
    regular, 228, 229, 231
    subinterval, 229
    union, 230
Peano, G., 34
periodic sequence, 79, 82, 92, 100
point function, 239
pointwise convergence and limit, 300
polynomial, 116, 338
    degree, 116
    roots, 116
position, 169
power function, 178
power rule (derivatives), 178
power rule (integrals), 247
power series, 324
    center, 324
    centered at zero, 324
    coefficients, 324
    geometric, 286
    interval of convergence, 325
    radius of convergence, 325
power set, 67
power set theorem (Cantor), 67
prime (numbers), 66
product rule (derivatives), 179, 251
product, direct or Cartesian (of sets), 37
proof
    countability argument, 116
    indirect (or by contradiction), 60
proof by contradiction, 28, 55, 60, 66, 67, 81, 97,
    114, 116, 159, 219, 392, 402
Pythagorean identity, 396
Pythagorean theorem, 14, 313

quantifiers (logical), 58
quotient rule (derivatives), 179

radian measure, 397
range, 48
ratio test, 144, 155, 290
ratio test (limit), 147, 325
rational numbers, 94
real number, 104, 162
    base b expansion, 162
    binary expansion, 161
    decimal expansion, 162
real numbers, 6, 38
    Cantor's construction, 97
real numbers, field, 105
rearrangement theorem (series), 10, 157, 160
recursion, 97
regular polygon, 18
relation, 39
    antisymmetric, 41
    equivalence, 40
    ordering, 41
    reflexive, 40
    symmetric, 40
    transitive, 40
relative complement, 35
Riemann sum, 230
    lower (underestimates), 232
    upper (overestimates), 233

Robinson, A., 28

secant line, 176, 211
sequence, 77, 94, 173, 174
    bounded, 92
    Cauchy, 99, 112, 134
    chaotic, 82
    constant, 79
    convergent, 134
    divergent, 92
    eventually monotone, 82
    eventually periodic, 79
    infimum, 111
    limit, 85
    n-th term, 78
    oscillating, 82
    partial sums, 124, 291
    plot, 79
    range, 77
    supremum, 111
    terms, 78
set, 31
    bounded, 110
    co-finite, 68
    complement, 34
    countable, 69
    element, 31
    empty, 32
    infinite, 34, 65
    infinite Cartesian product, 72
    intersection, 34
    notation, 31
    operations, 34
    ordered, 42
    product, 34
    totally ordered, 42
    union, 34
    universal, 34
set inclusion, 33, 41
shrinking tail, 129
sigma notation, 124
singularity, infinite, 194, 196, 204, 209
singularity, non-removable, 194, 198, 245, 291
singularity, removable, 194, 197
slope, 176
smooth function, 418

smooth infinitesimal analysis, 28
special angles, 398
speed, 169
Squeeze Theorem, 91, 112
staircase curves, 312, 316
subinterval, 229
subset, 33
    improper, 33
    proper, 33, 65
subspace, 419
sufficient condition, 198
sum rule (derivatives), 179
sum rule (integrals), 249
supremum, 111, 237
surjective function, 53
syllogism (hypothetical), 61
symmetry
    relative to the origin, 44
    relative to the y-axis, 44

tangent line, 176, 212
tangent line (equation), 177, 204
tangent line (slope), 177
tautology, 58
Taylor coefficients, 360
Taylor polynomial, 349
Taylor series, 340
Taylor, B., 340
telescoping series, 126, 127, 129, 139
Thompson's lamp, 17
Torricelli, E., 16
total ordering, 41, 95
transcendental number, 117, 166
transient part, 80
transitive
    implication, 61
    relation, 40
triangle inequality, 153, 160, 362, 416
triangular numbers, 127
trichotomy, 42
trigonometric functions (derivatives), 184
trigonometric polynomial, 370
trigonometric series, 361
truth table, 57
Tusi, 25

uncountable set, 69, 116
undecidable, 68
uniform convergence, 306
uniform metric, 417
union (of sets), 34
upper bound, 110
upper integral, 238
upper sum, 234

variables (logical), 58
variation, unbounded, 195
vector, 42, 73
velocity, average, 169, 211
velocity, instantaneous, 172, 211
Venn diagram, 34

Wallis, J., 22, 204
Weierstrass test, uniform convergence, 361, 382
Weierstrass's nowhere differentiable function, 382
Weierstrass, K., 379

Zeno, 8, 127
zillion, 2
zoom factor, 84
zooming sequence (of intervals), 85

www.ingramcontent.com/pod-product-compliance
Lightning Source LLC
Chambersburg PA
CBHW080450220526
45465CB00006B/2215